Elementary Reaction Steps in Heterogeneous Catalysis

NATO ASI Series

Advanced Science Institutes Series

A Series presenting the results of activities sponsored by the NATO Science Committee, which aims at the dissemination of advanced scientific and technological knowledge, with a view to strengthening links between scientific communities.

The Series is published by an international board of publishers in conjunction with the NATO Scientific Affairs Division

A Life Sciences
B Physics

Plenum Publishing Corporation
London and New York

C Mathematical and Physical Sciences
D Behavioural and Social Sciences
E Applied Sciences

Kluwer Academic Publishers
Dordrecht, Boston and London

F Computer and Systems Sciences
G Ecological Sciences
H Cell Biology
I Global Environmental Change

Springer-Verlag
Berlin, Heidelberg, New York, London, Paris and Tokyo

NATO-PCO-DATA BASE

The electronic index to the NATO ASI Series provides full bibliographical references (with keywords and/or abstracts) to more than 30000 contributions from international scientists published in all sections of the NATO ASI Series.
Access to the NATO-PCO-DATA BASE is possible in two ways:

– via online FILE 128 (NATO-PCO-DATA BASE) hosted by ESRIN, Via Galileo Galilei, I-00044 Frascati, Italy.

– via CD-ROM "NATO-PCO-DATA BASE" with user-friendly retrieval software in English, French and German (© WTV GmbH and DATAWARE Technologies Inc. 1989).

The CD-ROM can be ordered through any member of the Board of Publishers or through NATO-PCO, Overijse, Belgium.

Series C: Mathematical and Physical Sciences - Vol. 398

Elementary Reaction Steps in Heterogeneous Catalysis

edited by

R. W. Joyner

Leverhulme Centre for Innovative Catalysis,
Department of Chemistry,
University of Liverpool,
Liverpool, United Kingdom

and

R. A. van Santen

Schuit Institute of Catalysis,
Eindhoven Technical University,
Eindhoven, The Netherlands

Springer Science+Business Media, B.V.

Proceedings of the NATO Advanced Research Workshop on
Elementary Reaction Steps in Heterogeneous Catalysis
Bédoin, Vaucluse, France
November 1-7, 1992

Library of Congress Cataloging-in-Publication Data

NATO Advanced Research Workshop on Elementary Reaction Steps in Heterogeneous Catalysis (1992 : Bédoin, France)
Elementary reaction steps in heterogeneous catalysis : proceedings of the NATO Advanced Research Workshop on Elementary Reaction Steps in Heterogeneous Catalysis, Bédoin, Vaucluse, France, November 1-7, 1992 / edited by R.W. Joyner and R.A. van Santen.
p. cm. -- (NATO ASI series. Series C, Mathematical and physical sciences ; no. 398)
Includes index.
ISBN 978-94-010-4741-8 ISBN 978-94-011-1693-0 (eBook)
DOI 10.1007/978-94-011-1693-0
1. Heterogeneous catalysis--Congresses. 2. Chemical reaction, Conditions and laws of--Congresses. I. Joyner, R. W. II. Santen, R. A. van (Rutger A.) III. Title. IV. Series.
QD505.N3644 1992
541.3'95--dc20 93-17085

ISBN 978-94-010-4741-8

Printed on acid-free paper

CONTENTS

Preface

This book comprises the proceedings of a NATO sponsored Advanced Research Workshop held from 1st November to 6th November 1992 in the delightful Chateau de Florans, Bédoin, Vaucluse, France and entitled 'Elementary Reaction Steps in Heterogeneous Catalysis.' The organisers are grateful to the Science Committee of NATO for their support of this meeting. This is believed to be the first wide ranging NATO ARW in the field of heterogeneous catalysis for 20 years, following a previous venture organised in Sardinia by Basolo and Burwell, of Northwestern University, Illinois, USA [1]. This volume collects the lecture presentations and reports on the lively Panel discussions.

The idea for the meeting evolved from a series of International Symposia on Quantum Chemistry and Mechanism in Heterogeneous Catalysis. The first of these was held in Lyon, France in 1986, the second in Krakow, Poland in 1988 and the third in Berkeley, California in 1990. The organising committee of the present meeting was Bernard Bigot, France, Tony Farragher, Netherlands, Richard Joyner, UK, Mme. Danielle Olivier, France, and Rutger van Santen, Netherlands, (Chairman). We wish to thank all members of the committee but in particular Bernard Bigot, who undertook the very extensive work involved in the local organisation with consummate skill and made our stay in Provence a great pleasure. Bernard Bigot's secretary, Mme. Marie-Noëlle Coscat and Richard Joyner's secretary, Mrs. Pat Gibbs, also deserve our considerable thanks.

There were fifty-four participants from eleven countries. About two thirds of these came from academic laboratories and the rest from the European and American chemical and oil industry. The backgrounds of the participants varied from chemical engineering to theoretical physics and surface physics to organic chemistry. We would like to thank all those who took part.

Richard Joyner
Liverpool

Rutger van Santen
Eindhoven

1) Basolo, F. and Burwell, R.L Jr., Eds., (1973) 'Catalysis, Progress in Research,' Plenum New York.

Elementary Reaction Steps in Heterogeneous Catalysis

Richard W. Joyner,
Leverhulme Centre for Innovative Catalysis,
Department of Chemistry,
University of Liverpool, PO Box 147,
Liverpool L69 3BX, UK

Rutger A. van Santen,
Schuit Institute of Catalysis,
Eindhoven Technical University,
P.O. Box 513, 5600 MB Eindhoven, Netherlands.

Heterogeneous catalysis is rapidly developing from a black box technology employed by the chemical engineer and becoming a frontier area of modern physical science. Progress is encouraged by parallel developments in surface science, in *in situ* methods suitable for studies of the working catalyst, in our understanding of reactivity, kinetics and mechanism, and of relevant and cognate theoretical studies. This improved insight into catalysis at the molecular level has increased the rate of catalyst development and is changing catalytic practise. This is therefore an appropriate time to review the field and to set the agenda for future progress.

Understanding and applying catalysis is increasingly an interdisciplinary activity, requiring the expertise of the chemical engineer as well as the surface physicist and from the theoretician as well as the practical organic chemist. In this workshop relevant expertise is available in abundance.

The presentations and discussions are structured around five major themes:

1) Catalytic reactivity. Reactivity represents the *zeitgeist* of catalysis and is therefore an appropriate starting point. Reactivity and mechanism can be viewed from many different standpoints and a synthesis of the different approaches will clearly be valuable but may require considerable dialogue.

2) Surface science studies. Application of the methods of surface science provides the molecular level information on elementary reaction steps at surfaces which our understanding of catalysis requires. Our picture may be considered complete when the appropriate n-dimensional potential energy surfaces of the reacting adsorbed molecules can be constructed.

3) In situ studies of catalysis. Spectroscopic and structural studies can now be carried out under in situ conditions and provide direct and detailed information on the catalyst surface under reaction conditions.

4) Theoretical studies. Computational chemistry has reached a point where appropriate systems and models can be studied with useful accuracy. Appropriate levels of approximation, from molecular simulation up to full ab initio calculation, are available.

5) Chemical kinetics and chemical engineering. If the catalytic reaction mechanism is known and the reaction can be broken down into its elementary steps, a kinetic model

R. W. Joyner and R. A. van Santen (eds.), Elementary Reaction Steps in Heterogeneous Catalysis, 1–2.

which describes the catalytic reaction may be constructed based on microkinetic information. The model is of use to the chemical engineer who is responsible for incorporating the new catalyst into a useful process.

A central purpose of the Workshop is to identify the keys to progress in the next five to ten years. It is important to remember that all descriptions of the catalytic events are approximations. For some situations the molecular picture is more useful, while others may be better described by an extended solid picture. Much discussion of the validity and usefulness of these contrasting approaches is expected. The completeness with which the *catalytic active site* can be described is crucial, and developments of better descriptions of active sites must be a key aim. In the future we must focus on ways to *modify* active sites to develop new catalysts or to improve the properties of present catalysts, for example activity, selectivity or lifetime.

The techniques available to study catalysis have broadened and improved markedly in the recent past. Because of the rapidity of technique development, another focus of discussion is expected to be the advances in experimental methods that are still required to describe catalysis, so that it can be properly treated, as a *dynamic phenomenon.*

The speakers were asked expected to pinpoint the most important questions in catalysis. Each was also invited to indicate the best ways these crucial questions can be addressed now and during the next five to ten years. All were conscious that catalysis is an exciting academic discipline and at the same time of enormous practical benefit to humanity.

SURFACE SCIENCE AS A BASIS FOR UNDERSTANDING HETEROGENEOUS CATALYSIS

G.A. SOMORJAI
Department of Chemistry, and
Materials Sciences Division, Lawrence Berkeley Laboratory
University of California
Berkeley, California 94720

ABSTRACT. Surface Science provides new techniques of catalyst characterization on the molecular level and uncovers new phenomena that correlates with catalyst behavior. These include trends of reactivity and bonding across the periodic table; the observation that rough surfaces and defect sites have high reaction rates and more strongly bind the adsorbates; the reaction and chemisorption induced restructuring of metal surfaces; and the presence of a strongly chemisorbed overlayer during catalysis along with co-adsorbates that alter the rates of reactions. There is evidence from surface science studies of the cluster-like bonding of molecular adsorbates of thermal activation and the unique reactivity of oxide-metal interfaces. Correlations of catalytic reaction and surface science studies are providing new concepts and models of heterogeneous catalysis.

1. Introduction.

Surface Science, during the past two decades, provided Catalysis Science with new techniques of characterization and new concepts based on new phenomena that were uncovered by molecular level studies of surfaces. The surface characterization techniques employed most frequently are listed in Table 1. Most of these can be utilized to determine the composition of the surface, often the oxidation state or charge distribution of surface atoms, and the structure of the surface and adsorbed molecules, on the molecular scale. The techniques are continually being developed to detect smaller concentrations of surface species, and to obtain ever improved spatial, energy, and time resolution. Nevertheless, they have many shortcomings. Often they are capable of detecting the properties only on one side of the surface chemical bond; on the adsorbed molecule side (like infrared spectroscopy) or on the solid surface side (like EXAFS). Most frequently the catalyst surface can only be characterized before or after the catalytic process because the conditions of surface analysis require low pressures or some special experimental configuration that is not compatible with the experimental conditions required to carry out the catalytic reaction. Many techniques can only interrogate the properties of external surfaces and cannot obtain information about the internal surfaces of microporous, high surface area catalysts.

To overcome the difficulties in matching the experimental conditions needed for catalytic reaction studies and surface characterization, surface scientists invented model catalysts for carrying out these studies in combination [1]. These are usually single crystals of small surface area (~1cm^2), or thin films and polycrystalline foils. The external surface of these model catalysts carries out the reaction. Reaction chambers have been developed

R. W. Joyner and R. A. van Santen (eds.), Elementary Reaction Steps in Heterogeneous Catalysis, 3–38.

that permit both surface characterization and preparation of the model catalysts at low pressures or in vacuum, and the performance of reaction studies at high pressures and temperatures [2]. Under these conditions, the cleanliness of the surface, and its atomic structure, can be well controlled. Catalyst additives can be deposited from the vapor phase and complex multicomponent catalytic systems can be built this way [3]. Some of the catalyst systems studied in this model catalyst configuration, and the reactions investigated in my laboratory, are listed in Table 2.

It is necessary to compare chemical behavior for any model system with the real system, to assure its credibility. This has been carried out successfully for model metal catalysts ranging from cyclopropane ring opening, hydrogenation reactions of olefins, N_2, CO, and for more complex isomerization and cyclization reactions [4].

New features of surface structure, composition, and bonding have been discovered as a result of these studies. These new features either correlate with catalytic reactivity (reaction rate, selectivity, and other rate parameters) or are present on the working catalyst. This paper will present these as phenomena or new concepts that can be added to the classical chemical concepts of heterogeneous catalysis developed before the widespread use of modern surface science techniques. The roles these phenomena play during catalysis can often be explained or rationalized based on our knowledge of atomic surface structure, the dynamics of surface atoms, and of adsorbed molecules and the nature of the surface chemical bonds. Other phenomena, while they exhibit excellent correlation between certain surface properties and catalytic reactivity, await explanation through studies (experimental and theoretical) in the future.

2. Trends of Transition Metal Reactivity and Bonding Across the Periodic Table.

Transition metals are uniquely important in carrying out catalytic reactions. There are orders of magnitude changes in the reaction rates from metal to metal that are well demonstrated for several reactions [5] shown in Figure 1. Here the reaction rates are plotted for a series of transition metals (Re, Os, Ir, Pt and Au) for the hydrogenolysis of the C-C bond in ethane, the C-N bond in methylamine and the C-Cl bond in methyl chloride. The influence of the electronic structure of surface atoms shows up in producing volcano-shaped curves. If the surface bonds are too strong, the reaction intermediates block the adsorption of new reactant molecules because of their long surface residence times and the reaction stops. For too weak adsorbate-surface bonds, the necessary bond-scission processes may be absent. Hence the catalytic reaction will not occur. A good catalyst is thought to able to form chemical bonds of intermediate strength. These bonds should be strong enough to

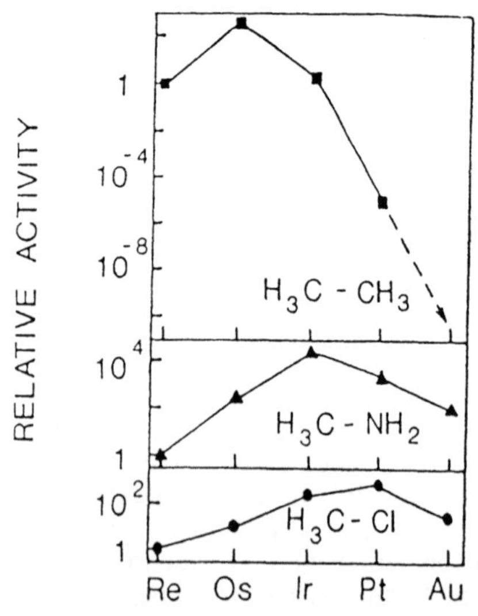

Figure 1. Catalytic activities of transition metals across the Periodic Table for a)the hydrogenolysis of the C-C bond in ethane; b)the C-N bond in methylamine; and the c)C-Cl bond in methylchloride. [Sinfelt, J.H. (1991) *Cat. Ltrs.*, 9, 159.]

For Table 2, see pp 38

induce bond dissociation in the reactant molecules. However, the bond should not be too strong to assure only short residence times for the surface intermediates and rapid desorption of the product molecules so that the reaction can proceed with a large turnover number. The volcano-shaped reactivity curves clearly reflect this compromise in requirements for intermediate bond strength.

Most available data on heat of adsorption on transition metal surfaces concern atoms (for example, H, O, N, K), and small molecules (for example, N_2, O_2, CO, and CO_2) [6]. Some of the data has been displayed in several review papers. The heat of adsorption generally increases from right to left in the periodic table, across the series of transition metals. This trend has been explained by the chemisorption model developed by Norskov [7] using the effective-medium theory, whereby the interaction of the adsorbate with the metal is primarily determined by the so-called one-electron energy term. This term is defined as due to changes in the charge distribution of the atom or molecule when it is taken from a homogeneous electron gas (uniform charge) onto the surface. The main difference in a transition-metal surface is due to the d-electrons that lie in a band around the Fermi level. These d-states can interact with the adsorbate states, and thus hybridization can occur, giving rise to bonding and anti-bonding shifts. The d-electron contribution to the bonding scales with $(1-f_d)$, where f_d is the degree of filling of the d-band. The d-electron contribution to the surface chemical bond depends on the degree of filling of the anti-bonding states (since the bonding states are already filled). Thus, early transition metals with fewer d-electrons form stronger chemical bonds. This effect has been described in several review papers for hydrogen and oxygen adsorbed on the 3d transition metals. The effective-medium theory provides good agreement between experimental data and calculated energies. Similar trends and agreement are found for the heats of adsorption of CO and N_2 as well.

One striking success of this theory to predict reaction rates across the periodic table was achieved for ammonia synthesis [8]. There are two factors that are all important in determining the ammonia synthesis rate. One is the N_2 dissociative sticking probability. N_2 dissociation turns out to be rate limiting and at low conversions the total rate of the reaction equals the dissociation rate of N_2. The other factor is the nitrogen atom chemisorption energy. Chemisorbed atomic nitrogen is by far the most stable reaction intermediate; therefore, the surface is mainly covered by nitrogen atoms up to 90% of a monolayer and the number of free sites on the surface where the nitrogen can adsorb is proportional to $(1-\theta_N)$ where θ_N is the nitrogen coverage.

Using a kinetic model that is reported by A. Nielsen et al, the ammonia formation rate can be calculated as a function of the number of d-electrons in the transition metals. The results are shown in Figure 2. It produces a volcano curve similar to that observed experimentally by Ozaki and Aika who have plotted the variation of the activity of various transition metals for the ammonia synthesis reaction as a function of the degree of filling of the d-band. The calculated results, and those found by experiments, overlap very well indeed. On the right side of the maximum in the volcano curve, the ammonia production decreases because the rate of N_2 dissociation drops as a consequence of the increase in the activation energy for dissociation. To the left of the tip of the volcano, the dissociation rate increases but since the nitrogen chemisorption bond also increases in strength, the number of empty surface sites where the nitrogen molecule can dissociate decreases so fast that the overall rate decreases.

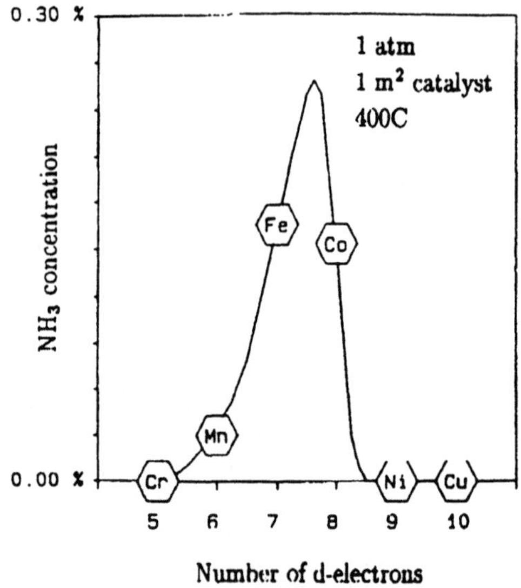

Figure 2a. The calculated ammonia concentration for a fixed set of reaction conditions as a function of the number of d-electrons. [Norskov, J.K.; Stoltze, P. (1987) *Surf. Sci.*, 189/190, 91.]

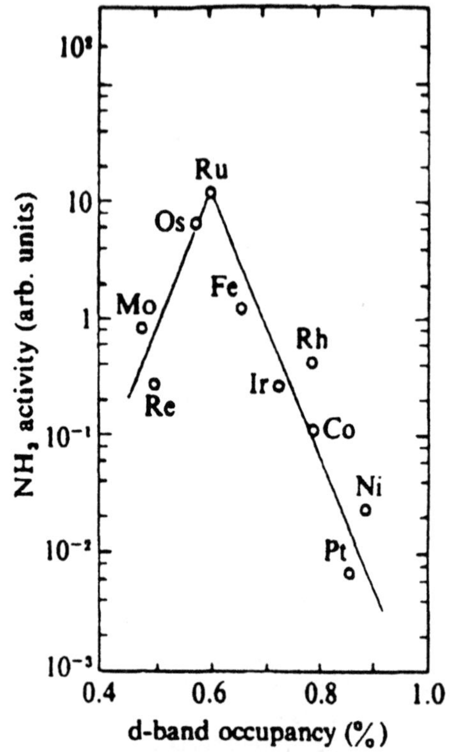

Figure 2b. The activity of various transition metals for ammonia synthesis as a function of the degree of filling of the d-bond. [Ozaki, A.; Aika, K. (1981) *Catalysis* Vol. 1., J. Anderson and M. Boudart (eds.), Springer-Verlag, Berlin, 87]

3. High Reaction Rates and Strong Bonding at Defects and on Rough Surfaces.

The structure sensitivity of catalytic reactions have clearly been demonstrated by studies of ammonia synthesis from nitrogen and hydrogen [9]. This reaction has been studied over various single crystal surfaces of iron. This is a particularly surface structure sensitive reaction; the (111) and (211) surface orientations are about an order of magnitude more active than the (100) and (210) faces and 2 orders or magnitude more active than the close-packed (110) face, this surface being the least active of all those studied.

A somewhat more complicated example of surface structure sensitivity and insensitivity is the hydrodesulfurization reaction [10], a very important process used to remove sulfur from an oil feed. This reaction may be modeled by the hydrodesulfurization of thiophene to butane, butenes, and butadiene. When this reaction is carried out on molybdenum and rhenium single-crystal surfaces, it exhibits structure insensitivity over molybdenum but significant structure sensitivity over rhenium. For both of these reaction studies, more open crystal faces that have lower atomic density and therefore greater surface roughness are more active to carry out the catalytic reaction. (Figure 3)

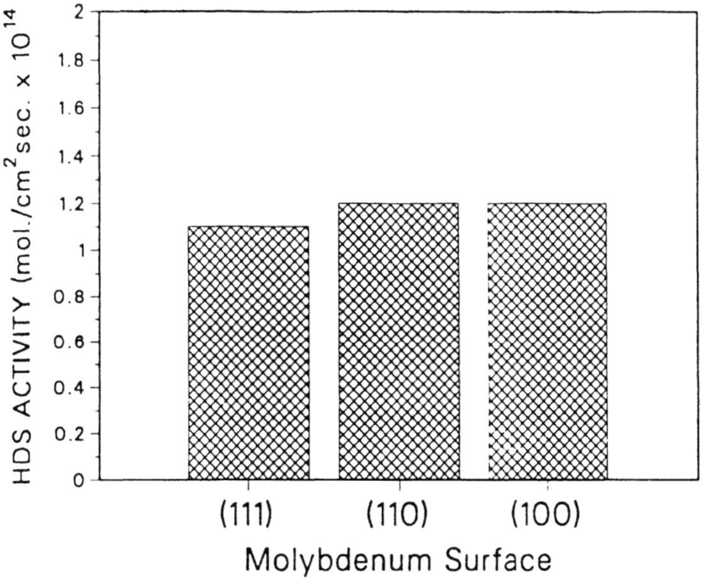

Figure 3a. The structure insensitivity of the hydrodesulfurization of thiophene over molybdenum crystal surfaces.

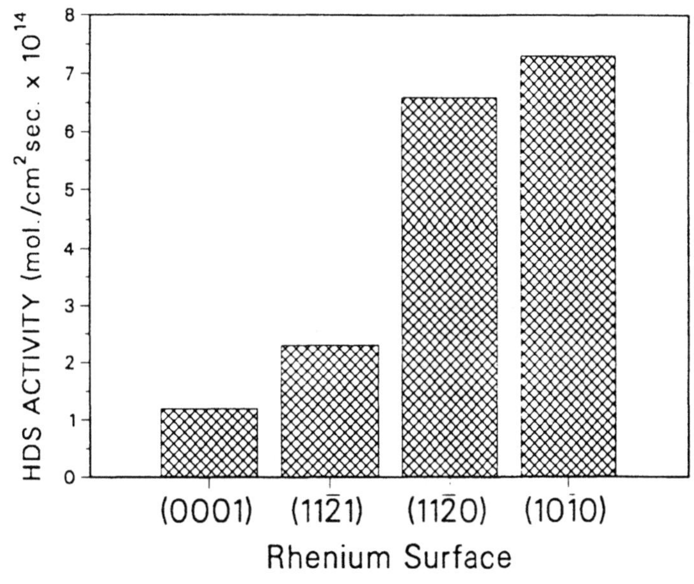

Figure 3b. The structure sensitivity of thiophene hydrodesulfurization to butene on rhenium single-crystal surfaces.

Rough surfaces are not only catalytically more active, they are also more active in breaking chemical bonds. For example, a stepped surface of Ni decomposes ethylene to smaller fragments at much lower temperatures (below 150 K) while the decomposition of ethylene on the Ni(111) face occurs about 250 K [11]. During the temperature programmed desorption (TPD) of H_2 from flat, stepped, and kinked surfaces of Pt. H_2 desorbs at maximum rates at the highest temperature from kink sites, then at somewhat lower temperatures from step sites, and at even lower temperatures from the flat (111) terraces [12]. This indicates higher heats of adsorption of the H atom at these defect sites. Thus, the thermodynamic driving force for dissociation is certainly greater at these sites, which can explain their enhanced bond-breaking activity.

It is difficult to understand, however, that these same strongly adsorbing sites are also very active sites for catalysis. The reaction probability of H_2/D_2 exchange on stepped surfaces is near unity at low pressures [13] on a single scattering while the reaction probability is below the detection limit ($<10^{-3}$) on the flat (111) crystal face as shown by molecular beam-surface scattering studies. How is it possible that the strongly adsorbing step sites, where H has a long residence time because of its high binding energy, are also the sites of rapid reaction turnover?

One possible explanation is that the strongly adsorbed hydrogen restructures the surface near the step, thereby creating the active site for the catalytic exchange process. At the low hydrogen pressures of these molecular beam scattering experiments, the low coverage keeps the structure of the flat part of the surface unaltered.

4. Metal Restructuring Controlled Reaction Rates. Adsorbate Induced Restructuring of Metals.

Chemisorption induced restructuring can be very well seen using a small metal tip and field ion microscopy [14]. In Figure 4 the field ion microscope picture of a rhodium tip is shown when clean and after exposure to carbon monoxide at 420 K at low pressures ($\approx 10^{-4}$ Pa). The metal tip has been completely reshaped as a result of CO chemisorption. The tip becomes faceted and rougher, the step density is reduced and extended low Miller index terraces are formed.

Scanning tunneling microscopy (STM) that can be operated at high pressure (~2 atm) and temperatures (≤240 C) can also be used to detect surface reaction induced restructuring of metal surfaces [15]. Figure 5 shows the structure of a Pt(110) crystal face when heated in hydrogen then heated in oxygen around one atmosphere. The platinum surface exhibits atomic details of its surface structure and ordered domains when exposed to H_2. In the presence of oxygen, the platinum surface becomes faceted and exhibits large areas composed of different crystal faces. These changes of surface structure are completely reversible, as shown when platinum is heated alternately in H_2 or in O_2.

Adsorption-induced restructuring can occur on the chemisorption time scale ($\approx 10^{-15}$ s for charge transfer or $\approx 10^{-12}$ s for vibrational times). There is evidence, however, that adsorbate-induced restructuring can occur on the time scale of catalytic reactions (seconds). CO oxidation to CO_2 or ammonia reacting with NO to produce N_2 and H_2O show oscillatory behavior under certain circumstances of temperature and reactant partial pressures [16,17]. The reaction rate alternates periodically between two values. One reason for the oscillation is the periodic restructuring of the surface. In this circumstance the sticking probability of one of the reactants is greater on one type of surface structure, while sticking probability of the other reactant is greater on the surface structure of the other type. Thus, the reaction rate alternates between the two branches of

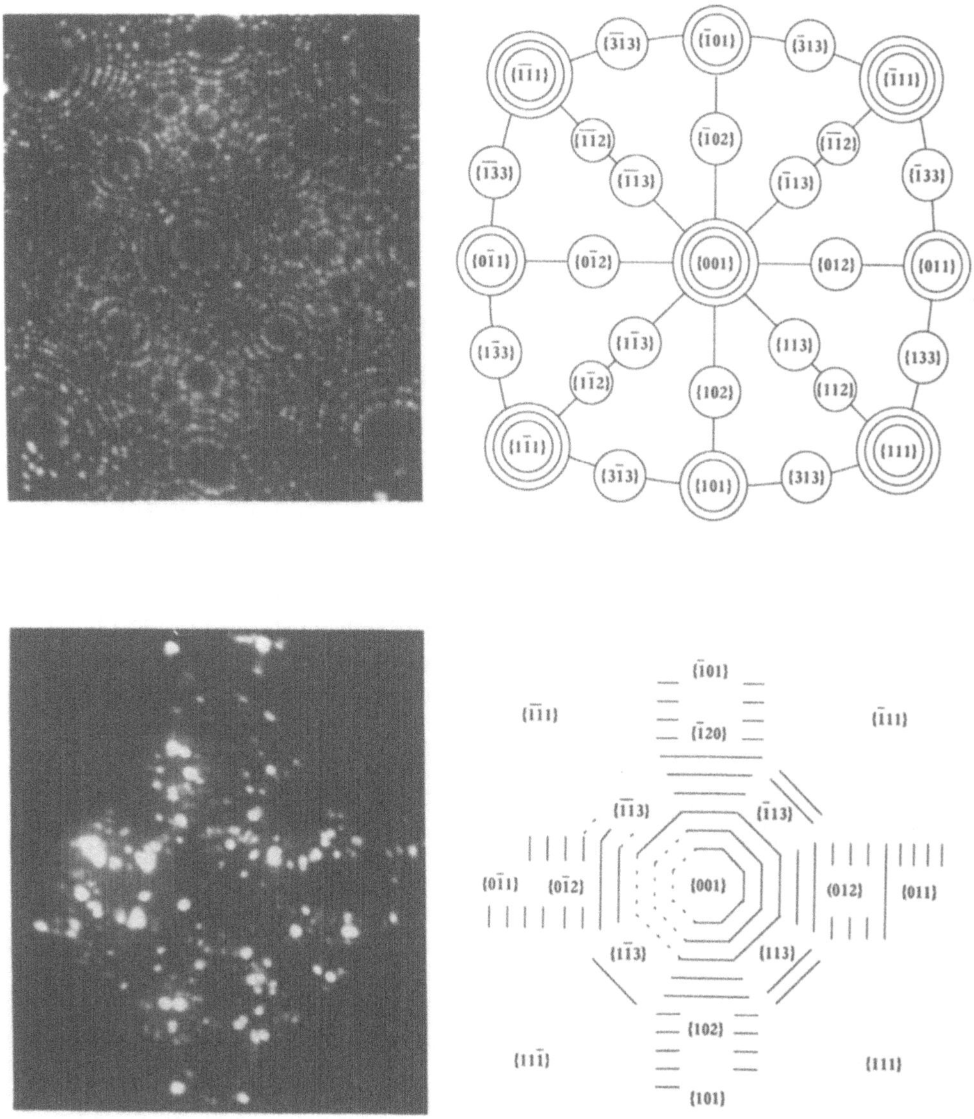

Figure 4. Field ion micrographs (image gas: Ne; , T=85 K) of a (001) oriented Rh tip before (top left) and after reaction with 10^{-4} Pa CO during 30 minutes at 420 K (bottom left). [Gaussmann, A.; Kruse, N. (1991) *Cat. Ltrs.*, 10, 305.]

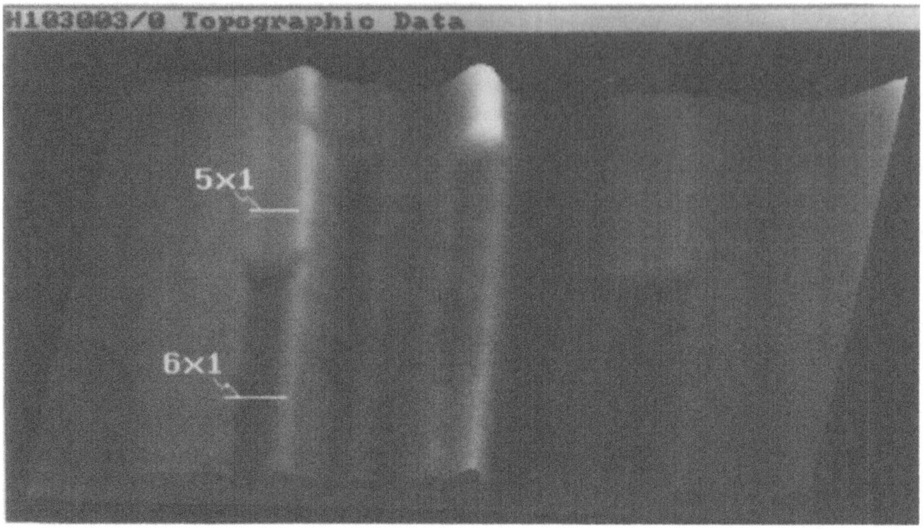

Figure 5a. STM topographic image of a 47nm x 67nm region of Pt(110) in 1.6 atm of hydrogen, after heating to 425 K for 5 hours, showing n x 1 reconstructions (n=3,4,5,6) and defects. dz=1.5nm

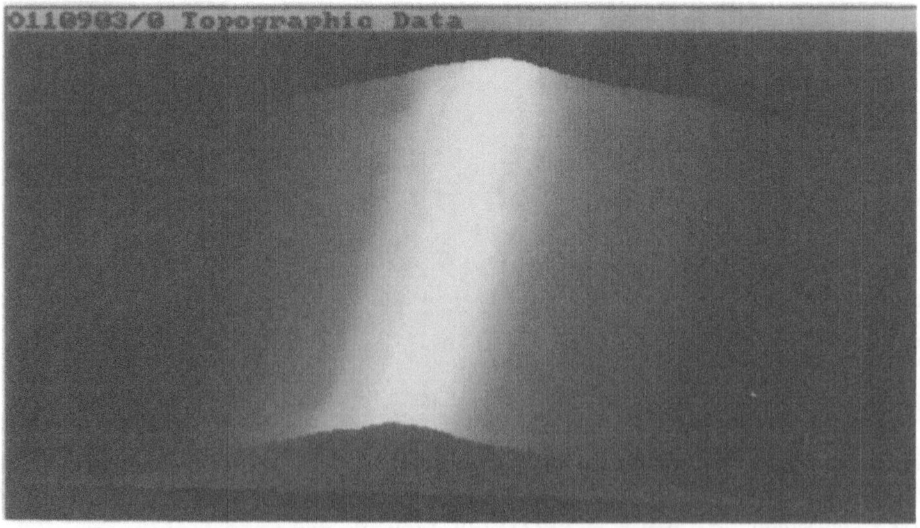

Figure 5b. Topographic image of a 51nm x 66nm region of Pt(110) in 1 atm of oxygen, after heating to 425 K for 5 hours.

the reaction, one taking place on the CO- or NO-covered metal surface, and the other taking place on the oxygen- or ammonia-covered metal surface.

Adsorbate-induced restructuring can occur on even longer time scales (hours), involving massive restructuring of the surface by atom transport. For example, sulfur restructures the (111) crystal face of nickel until the metal surface assumes the (100) orientation. Alumina restructures iron through the formation of an iron-aluminate phase to produce (111) crystal faces during ammonia synthesis, regardless of the original crystallite orientation [18, 19]. In this circumstance the chemisorption-induced restructuring can be viewed as the initial phase of a solid-state reaction whose kinetics are controlled by atom transport (by diffusion). When the adsorbate induced restructuring creates crystal faces that are beneficial to catalytic activity these adsorbates are called "structure modifiers" or structural promoters. If the reaction rate is decreasing, due to adsorbate induced restructuring, the adsorbate is called a "catalyst poison."

Restructuring occurs in order to maximize the bonding and stability of the adsorbate-substrate complex. Thus it is driven by thermodynamic forces and is most likely to occur when the stronger adsorbate-substrate bonds that form compensate for the weakening of bonds between the substrate atoms, that is the consequence of the chemisorption-induced restructuring process.

Substrate restructuring occurs during the chemisorption of molecules as well. The metal surface atoms that are "relaxed" by moving inward when the surface is clean, move outward during the formation of the chemisorption bond. When ethylidyne forms on the Platinum(111) surface, in addition to outward movement of the surface atoms [20], the nearest-neighbor metal atoms move closer to the adsorption site, and the next-nearest-neighbor Pt atom moves inward, causing a slight corrugation of the surface, while the Pt atom underneath the adsorption site moves down, away from the carbon atom. (Figure 6)

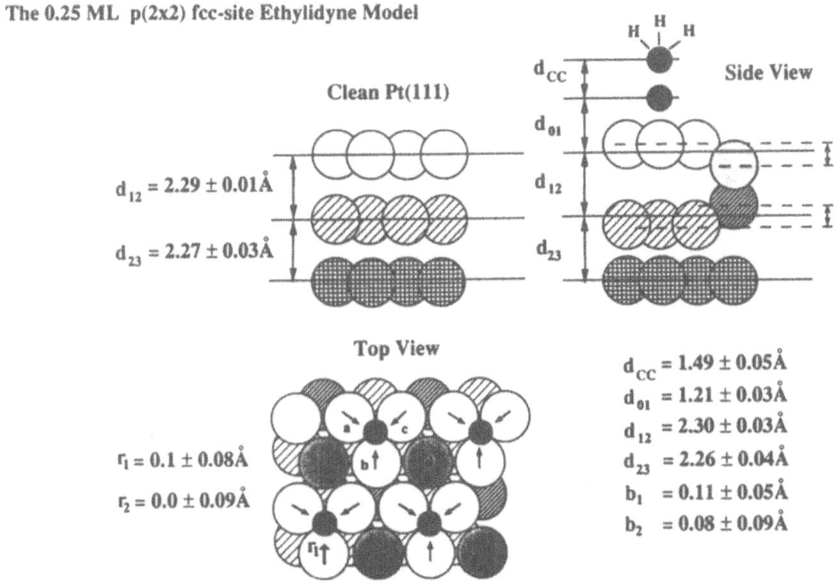

Figure 6. The surface structure of ethylidyne on the Pt(111) crystal face. The adsorption site, as well as the restructuring of the metal during chemisorption, is displayed.

Structure modifiers are often introduced as important additives when formulating the complex catalyst systems. Structural promoters can change the surface structure that is often the key to catalyst selectivity. Alloy components may not participate in the reaction chemistry but modify structure and site distribution on the catalyst surface. Site blocking could improve selectivity as has been proven for many working catalyst systems. Sulfur and silicon or other strongly adsorbed atoms that seek out certain active sites can block undesirable side reactions.

5. The Catalytically Active Metal Surface is Covered with a Strongly Chemisorbed Overlayer. Coverage Dependence of Bonding. Co-adsorption.

During most reactions the surface of the metal catalyst is covered with a strongly chemisorbed overlayer that remains tenaciously bound to the surface for 10^2-10^6 turnovers [21]. During hydrocarbon reactions, this is a carbonaceous overlayer with a composition of about (H/C) ≈ 1, during ammonia synthesis it is chemisorbed nitrogen, and during the hydrodesulfurization of thiophene to butenes it is a mixture of sulfur and carbon. Isotope labeling of the adsorbed species (using ^{14}C, ^{35}S) has been used to determine their long residence times [22].

During hydrocarbon conversion reactions over platinum, over 80% of the metal surface is covered with the carbonaceous deposit. While there are suggestions that the remaining uncovered metal sites are the only active sites for catalysis, there is increasing evidence that the carbonaceous overlayer is an active part of the working catalyst. Only when it is totally dehydrogenated will it deactivate the metal by forming a cross-linked graphite coating. There are suggestions that structure insensitive reactions (hydrogenation of ethylene, for example) take place on top of the strongly bound deposit, and the metal only participates in the reaction indirectly by aiding the dissociation of molecular hydrogen.

There is recent evidence, from scanning tunneling microscopy studies in our laboratory, that the strongly bound adsorbed atoms and molecules are mobile on metal surfaces. It has been known that the activation energy for surface diffusion of chemisorbed species is low as compared to their heats of desorption and therefore they migrate along the surface visiting various adsorption sites during their surface residence time (two dimensional phase approximation). The STM studies provide the first direct evidence of the rearrangement of the chemisorbed layer. In Figure 7a we show an STM picture of an ordered sulfur overlayer on rhenium(0001) single-crystal surface. In Figure 7b the same surface is shown in the presence of CO. As the molecule co-adsorbs on the metal surface it does so by compressing the sulfur adsorbate into a higher coverage structure around the adsorption site. Thus CO molecules adsorb by laterally displacing the sulfur atoms.

It is likely that the adsorption of reactants on a strongly chemisorbed overlayer covered surface occurs by laterally displacing the overlayer thereby creating a reaction site. As long as the activation energy for the surface diffusion (or lateral displacement) of the deposit is small, this process can always create new sites for adsorption and reaction. Only when the activation energy for surface diffusion of the overlayer becomes large will the catalyst deactivate. This happens when the overlayer polymerizes or forms a graphitic network, that occurs during hydrocarbon reactions at high temperatures.

The heat of chemisorption per atom or per molecule declines with increasing coverage for most chemisorption systems. This has been shown for potassium on a rhodium(111) crystal face [23] and for CO on a palladium(100) face [24], respectively. At low coverages, potassium is strongly bound to the transition metal as it transfers electrons to it to become positively charged. With increasing coverage, adsorbate-adsorbate

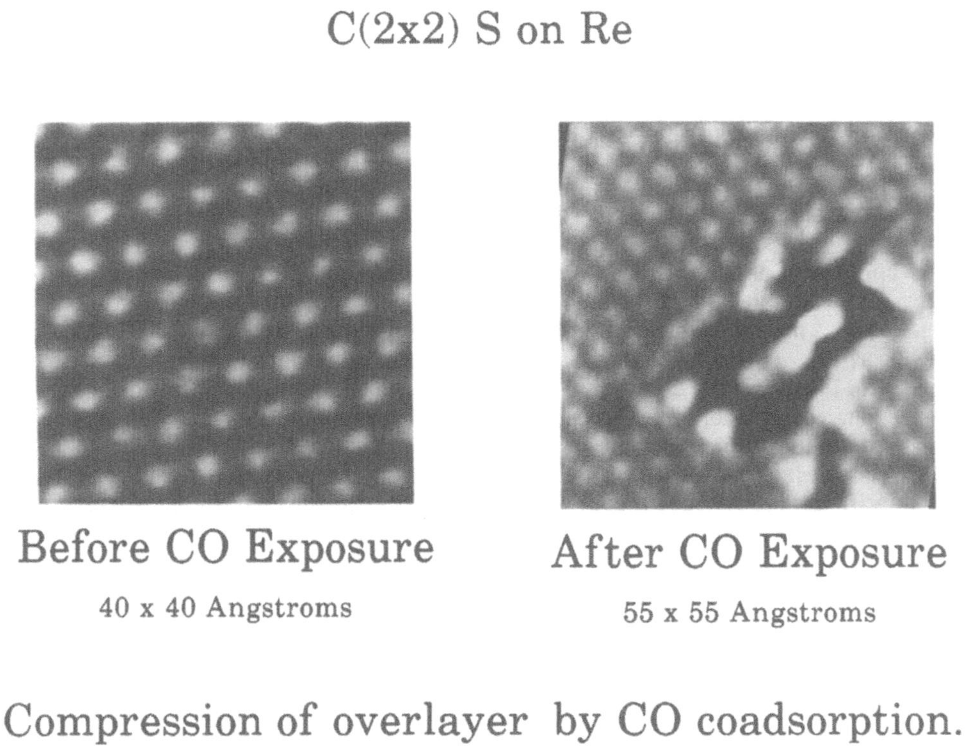

Figure 7a) Scanning tunneling microscopy picture of the high-coverage sulfur structure on the rhenium(0001) crystal face. **b)** The same structure after co-adsorption of carbon monoxide.

interaction causes repulsion among the charged species, leading to depolarization and much weakened adsorption bonds until the heat of adsorption becomes equal to the heat of sublimation of metallic potassium. Carbon monoxide chemisorbs with its C-O bond perpendicular to the metal surface, occupying on-top and bridge sites until about one-half of a monolayer coverage is reached. The heat of adsorption stays relatively constant with coverage in this coverage range, indicating that very little adsorbate-adsorbate interaction is influencing the bonding of the molecule to the metal. At higher coverages, however, the molecules strongly repel each other, forcing the on-top site CO molecules to relocate to maximize adsorbate-adsorbate distances, and ΔH_{ads} declines rapidly until it reaches about one-third of its value at low CO coverages.

Thus, increasing coverage of chemisorbed species not only leads to sequential filling of binding sites (the stronger binding sites filling first), as shown in the literature, but can also weaken the adsorbate-substrate bonds markedly. This effect of coverage influences the surface residence times of adsorbates and subsequently their behavior during chemisorption and surface chemical reactions.

Most catalytic reactions are carried out at high pressures under conditions of high surface coverages of adsorbed reactants and reaction intermediates. In this circumstance the heat of desorption is low as compared to the situation at low coverages. In most surface science studies adsorbates are scrutinized at low coverages because their stability makes their study easier, and because of the low pressure conditions necessary to utilize many techniques of surface science. Thus, correlations in chemical behavior are hindered because of the differences of experimental conditions that are applied.

6. Coadsorption.

The co-adsorption of two different species can lead to either attractive or repulsive adsorbate-adsorbate interaction. The coadsorption of ethylene and carbon monoxide demonstrates the attractive interaction that can occur in the adsorbed layer. CO and C_2H_4 chemisorbed together [25] on the Rh(111) crystal face form the structure shown in Figure 8. There are two different molecules per unit cell, indicating attraction among the molecular species of different type. Ethylene adsorption decreases the work function of rhodium while CO increases the work function of rhodium upon chemisorption. Thus C_2H_4 is an electron-donor while CO is an electron-acceptor on the transition metal, resulting in an attractive donor-acceptor interaction among the two types of adsorbates.

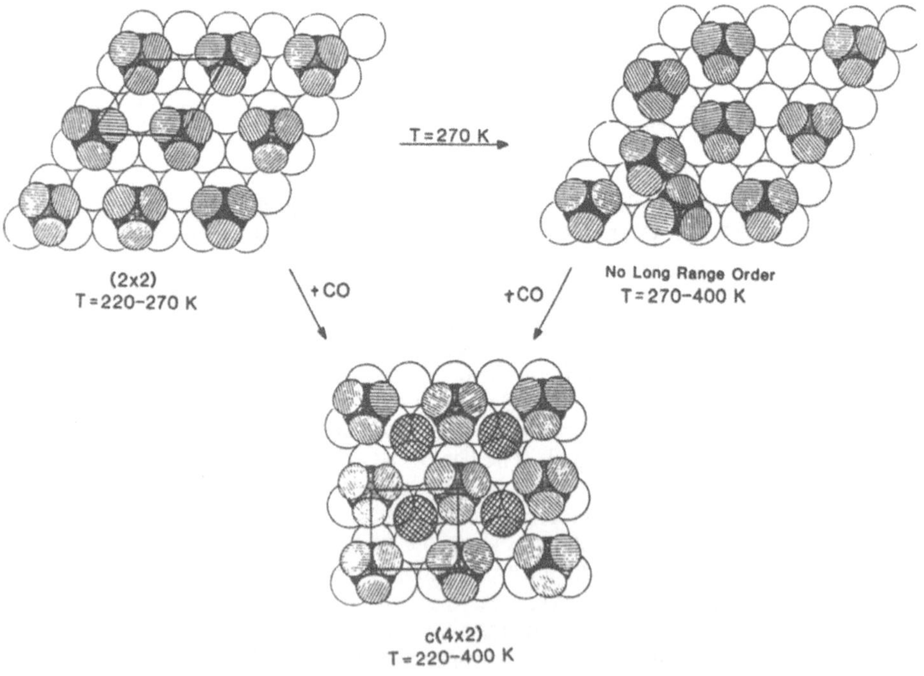

Figure 8. The ordering of ethylidyne, with and without carbon monoxide co-adsorption, on the rhodium(111) crystal face.

The ordering of one adsorbate by the co-adsorption of another through donor-acceptor interaction is commonly observed, as shown for several co-adsorbed systems. For Rh(111) the magnitude of the adsorbate-adsorbate attractive interaction is about an order of magnitude weaker [about 4 - 6 kcal/mol (17 - 25 kJ/mol)] than most adsorbate-substrate chemisorption bonds [about 30 - 60 kcal/mol (125 - 250 kJ/mol)]. Repulsive interaction between two donor or two acceptor coadsorbed molecules leads to separation of the adsorbates by island formation or disorder in the adsorbed layer.

Strong attractive interaction among adsorbates can lead to dissociation of the molecular species. This is observed during the co-adsorption of potassium (donor) and CO (acceptor) on several transition-metal surfaces [26]. Thermal-desorption data indicate CO desorbing at much higher temperatures than normal in the presence of the adsorbed alkali metal, often showing a 17 kcal/mol (71 kJ/mol) increase in its heat of adsorption. The CO stretching frequency decreases with increasing dipole moment of coadsorbed donors. Isotope-labeling studies (using $^{12}C^{18}O$ and $^{13}C^{16}O$) indicate scrambling of the two isotopic species in the presence of potassium, signaling molecular dissociation, while no dissociation is apparent in the absence of potassium on rhodium at low pressures. Up to three CO molecules dissociate per potassium atom at an alkali metal coverage of 20% of a monolayer [27].

Repulsive interaction is also observed with the coadsorption of potassium and ammonia [28]. Both species are electron donors to transition metals. On iron, a 4 kcal/mol (17 kJ/mol) decrease in the heat of chemisorption of NH_3 is observed due to co-adsorbed potassium.

Alkali metals are often used as additives during catalytic reactions. They are "bonding modifiers"; that is, they influence the bonding and thus the reactivity of the co-adsorbed molecules. Potassium is an important promoter in CO hydrogenation reactions where CO dissociation is desired and is one of the elementary reaction steps. The alkali metal also reduces the hydrogen chemisorption capacity of the transition metal. Potassium is an important promoter in ammonia synthesis for the opposite reason, because it weakens the NH_3 product molecule bonding to the metal, thereby reducing its surface concentration, which would block important reaction sites. It also aids in the dissociation of dinitrogen.

Halogen species can also be important bonding modifiers, as they are powerful electron acceptors. Indeed, they are used as promoters in several catalytic processes (for example, ethylene oxidation to ethylene oxide over silver, or during partial oxidation of methane). Nevertheless, their molecular and atomic chemisorption behavior has been studied less and therefore is not as well-understood as the role of co-adsorbed alkali metal ions.

We would like to present three more new phenomena of catalysis and surface science that are more difficult to correlate. Nevertheless, the structure of surfaces and the dynamics of surface restructuring are the likely reasons for their existence.

7. High Reaction Rates at Oxide-Metal Interfaces.

Oxide metal interfaces can exhibit very high activity as compared to the metal alone whether the metal is placed on the oxide or the oxide on the metal as shown by Schwab et al. in the 1930's [29]. More recently Touster has reinvestigated this phenomenon using titanium oxide surfaces and found unique high catalytic activity due to oxide metal interfaces and this phenomenon is called strong metal support interaction [30]. The typical oxide induced changes in the metal activity can be readily shown for nickel if one compares the reaction rates for forming methane on this metal, from carbon monoxide and hydrogen, on various nickel catalysts that are deposited on different oxide supports.

There are orders of magnitude changes in the reaction rate when the nickel particles are deposited on silica as compared to nickel on titanium oxide [31]. This is shown in Figure 9. The typical activation treatment of catalysts of this type involve reduction in hydrogen during which the oxide encapsulates the metal resulting in much reduced carbon monoxide or hydrogen chemisorption activity followed by oxygen heat treatment that activates the catalysts during which the oxide forms small islands in such a way as to maximize the oxide-metal periphery area.

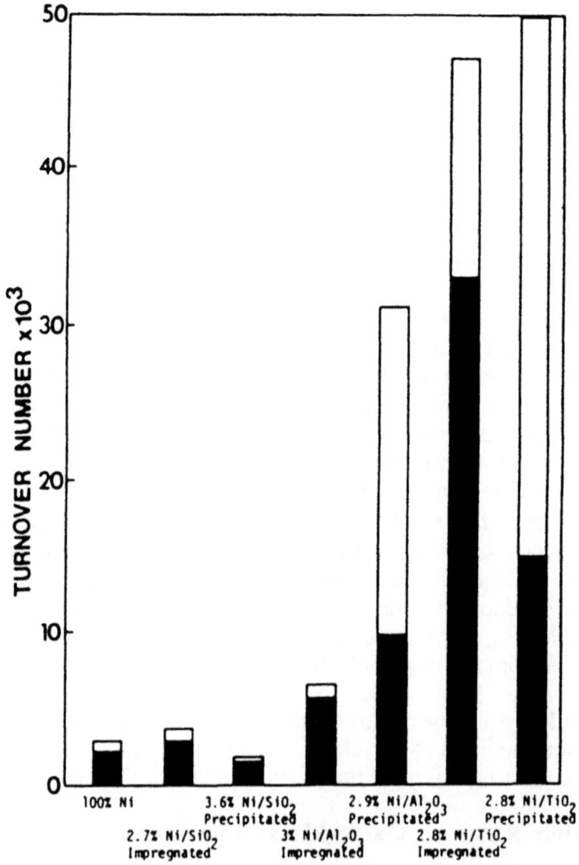

Effects of support and preparation on methane turnover number at 525 K for Ni: ■ CH_4 turnover No.; □ C_{2+} hydrocarbon turnover No.; total bar length is CO turnover No.

Figure 9. Effect of support on CO hydrogenation over nickel catalysts. [Bartholemew, C.H.; Penell, R.B.; Butler, J.L. (1980) *J. Catalysis.*, 65, 335.]

A typical reaction pattern reflecting the behavior of an active oxide metal interface system has been reported for the CO_2/H_2 reaction on Rh on to which a TiO_x layer is deposited, and the catalytic activity is measured as a function of the oxide coverage [32]. The oxide is deposited by condensing the metal from the vapor phase and then oxidized to form TiO_x. the reaction rate is at a maximum at ~0.5 monolayer oxide coverage. The reaction rate increases by 14-fold with respect to the clean metal surface. If we were to give the specific turnover rates by dividing it with the surface area we should use the oxide-metal interface area instead of the total area which is the metal plus the oxide surface area since the oxide is inactive. Since the oxide metal periphery area is about 10% of the total area, at 50% oxide coverage the specific turnover rate is two orders of magnitude greater as compared to the rate on the clean Rh metal. (Figure 10)

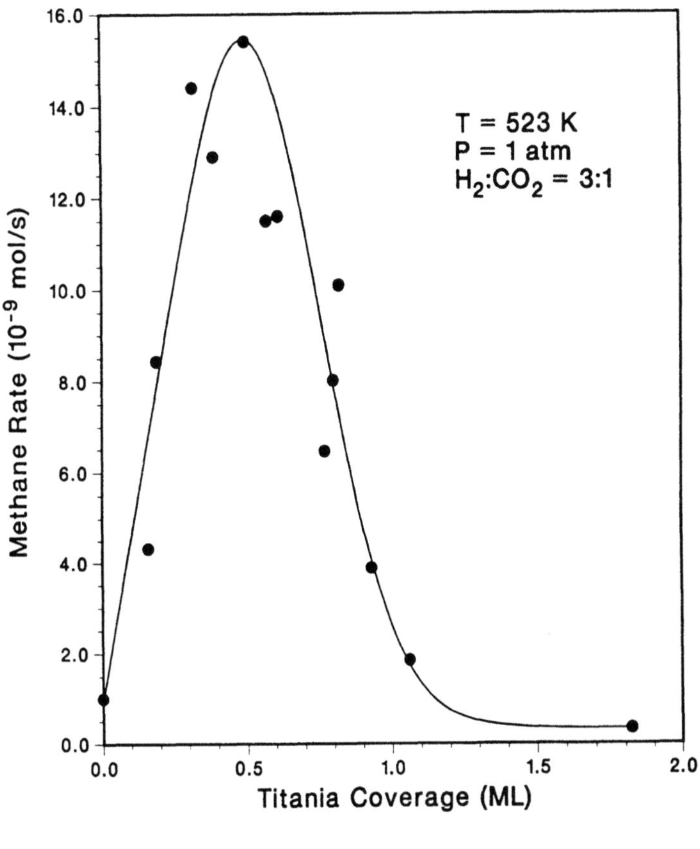

Figure 10a. CO_2 hydrogenation rate over the rhodium-titanium oxide catalysts as a function of oxide coverage of the metal.

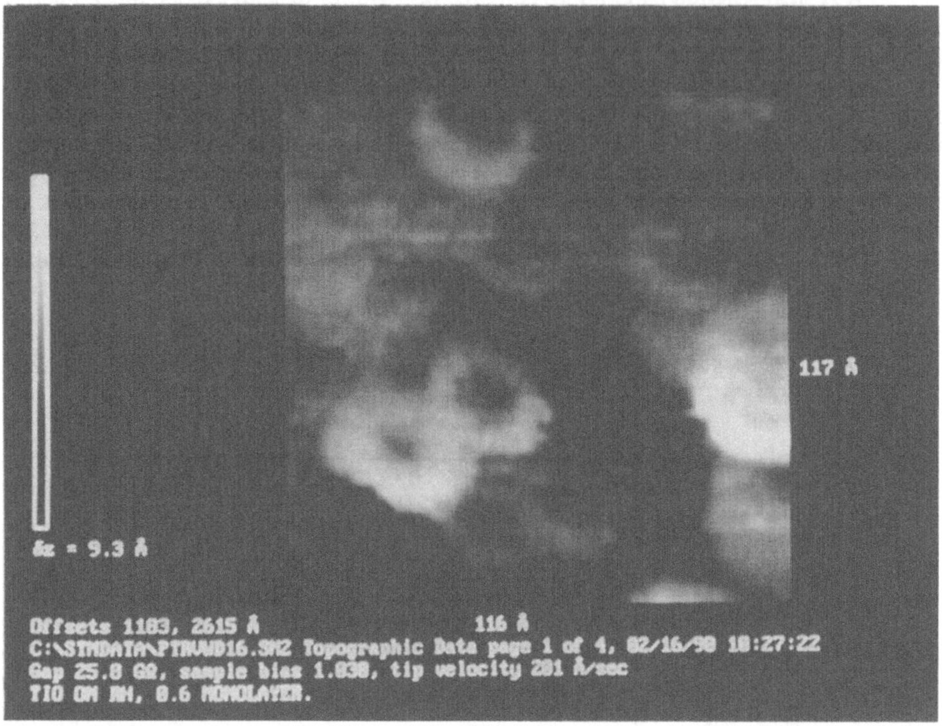

Figure 10b. Scanning tunneling microscopy picture of the titanium oxide islands on rhodium. The donut-like shape is due to the increased charge density at the periphery of the oxide islands, because of the concentration of Ti^{3+} ions.

Similar behavior is found for CO/H_2 and acetone/H_2 while the hydrogenation of ethylene and the hydrogenolysis of ethane remained unaffected by the presence of TiO_x indicating that the carbon-oxygen bond is being activated at the oxide-metal interface [33]. Thus any molecules that contain CO bonds are benefiting from the presence of the oxide-metal interface catalytic sites. Other oxides deposited on Rh such as vanadium oxide show similar effects, however, the increase in reaction rate is not as large in the presence of vanadium oxide as it is with the TiO_x/Rh interface.

I propose that the increased rate of chemisorption induced restructuring of the metal at the oxide interface is responsible for the increase in reaction rates. The lower coordination metal sites at the periphery of the oxide islands can restructure with higher frequency thereby greatly enhancing the turnover rate. We carried out scanning tunneling microscopy studies to image the TiO_x islands and the oxide-metal interface. Figure 10 shows the STM picture. The bright rings that appear at the oxide island periphery indicating large charge density associated with the presence of Ti^{3+} ions in the partially reduced oxide at the metal interface.

8. Cluster-like Bonding of Molecular Adsorbates.

When ethylene chemisorbs at ≈300K on the (111) crystal faces of various transition metals (Pt, Rh, Pd), it chemically rearranges to form the molecule-surface compound which we call ethylidyne. Its structure is determined by LEED-surface crystallography and is very similar to those of the multinuclear organometallic complexes [34]. The rearranged ethylene, which has also lost a hydrogen, is called ethylidyne and belongs to the alkylidyne group, $\equiv C_nH_{2n-1}$, a common substituent in surface and in organometallic chemistry. The vibrational spectrum of chemisorbed ethylidyne is nearly identical to that in the organometallic cluster which contains three metal atoms. The C-C bond distance is slightly less than the single carbon-carbon bond length of 0.154 nm (1.54 Å), as in the cluster compounds. Thus the surface chemical bond of chemisorbed ethylene can, as a first approximation, be viewed as a cluster-like bond that contains at least three metal atoms. The C-C bond order present in gaseous ethylene is reduced from two to nearly one upon chemisorption. This reduction in bond orders of alkenes and alkynes upon chemisorption on metal surfaces is commonly observed, indicating charge transfer from the molecules into the metal. In fact, the metal work function usually decreases when organic molecules are adsorbed, further proving the direction and magnitude of the charge transfer as the chemisorption bonds form.

There are many chemisorbed organic groups whose surface bonding can be viewed as identical to that of organometallic clusters. There are equivalent bonding arrangements of methylidyne (-CH) and acetylide ($-C_2H$) groups, respectively, on surfaces and in organometallic clusters.

Benzene usually chemisorbs on metals with its ring parallel to the surface (although it may adsorb in a different configuration when it loses hydrogen). Because of charge transfer to the metal, C-C bond elongations occur with respect to the gas-phase configuration, with periodic distortions of the C-C distance that reflect the symmetry of the adsorption site [35]. The ring may even bend, with two of the opposing carbon atoms closer to the metal surface than the other four carbon atoms [36]. Distortions and elongations of C-C bonds are also found when benzene is bound to clusters of metal atoms in organometallic complexes. Thus the cluster-like bonding model appears to be valid for chemisorbed benzene as well.

The bonding picture of adsorbed molecules becomes more complicated if there are more bonding sites available on the same molecule. For example, pyridine (C_5H_5N) may bind through the lone electron pairs of its nitrogen or through the p-electrons of its carbon ring. Thus, depending on the metal, the binding geometry of the substrate, the temperature, or the adsorbate coverage, the molecule may be tilted with respect to the substrate surface, its ring may be parallel with it, or it may be upright with bonding solely through nitrogen.

It appears that the restructuring of the metal substrate in the presence of the chemisorbed molecule in the proximity of the adsorption sites provides the necessary driving force for the formation of cluster-like bonding. By optimizing the strength of the adsorbate-substrate bond the metal-metal bonds can be weakened enough to permit the reallocation of the metal atoms in the cluster-like configuration of the adsorbate-substrate surface compound.

9. Thermal Activation.

When molecules adsorb on a solid surface of low enough temperature (say 20-25 K), they maintain their gas-phase-like structure and remain chemically intact even on the most reactive metal surfaces. As the temperature is increased, either chemical rearrangement of the adsorbed molecule or bond breaking occurs at a certain temperature or narrow temperature range. Each adsorbate-substrate system has a characteristic temperature of bond activation. As the temperature is increased further, another bond-breaking or molecular rearrangement occurs; and sequential bond scission continues at characteristic temperatures until the molecule breaks up into its atomic constituents, which then desorb or diffuse into the bulk. An example, ethylene chemisorption on the platinum(111) surface [37]. The thermal-desorption spectrum indicates sequential hydrogen evolution, while the vibrational spectra taken in the different temperature ranges indicate that molecular rearrangements and chemical bond breaking occur simultaneously, as follows:

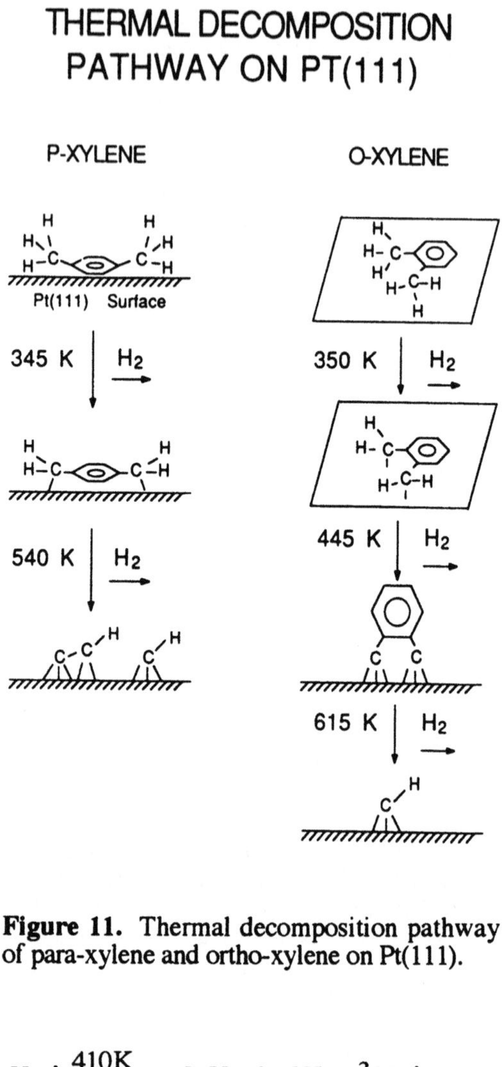

Figure 11. Thermal decomposition pathway of para-xylene and ortho-xylene on Pt(111).

$$2C_2H_{4(ads)} \xrightarrow{300K} 2-C_2H_3 + H_2\uparrow \xrightarrow{410K} -C_2H + 2\text{-}CH + \tfrac{3}{2}H_2\uparrow \rightarrow$$

$$\xrightarrow{>900K} 4C + \tfrac{3}{2}H_2\uparrow$$

In Figure 11 the bond-breaking sequences for ortho-xylene and para-xylene chemisorbed on the Pt(111) surface are shown [38]. At low temperatures the structures of the chemisorbed molecules are different. As the temperature is increased, hydrogen desorption occurs at well defined temperatures. At high temperatures, the molecular fragments produced from the two molecules are the same. The molecular fragmentation sequence for other hydrocarbons, propylene and propyne (methyl-acetylene), for example, are also very similar.

From the examples above, it is clear that molecular rearrangement or bond breaking on the surface has to be "activated" by increasing the temperature. Perhaps the first experimental observation of this phenomenon was the activated dissociation of dinitrogen (N_2) on iron surfaces, a phenomenon that gave rise to the suggestion of "physisorption" to "chemisorption" transition.

Recently a molecular mechanism has been proposed that considers surface restructuring as the driving force for dissociative chemisorption. This model also predicts a coverage dependence for the phenomenon as the chemisorption of more molecules should lower the activation energy for the surface restructuring.

10. The Missing Links.

Contradictory behavior patterns emerge from combined catalysis and surface science studies. The high activity and strong binding at defect sites, rough surfaces, and the presence of strongly chemisorbed stagnant overlayers on the active catalyst surfaces are two examples of this pattern. In addition, other important surface phenomena; the high reactivities at certain oxide-metal interfaces and thermal activation awaits explanation on the molecular level.

Several new directions of research can be suggested that could provide the missing links necessary to explain many of the new surface phenomena. Studies of <u>atomic surface structure and bonding in the adsorbed monolayer at high coverages</u> should be carried out. The scanning tunneling microscope and structural studies using synchrotron radiation (EXAFS for example) could provide this information. <u>Time resolved investigation of short lived adsorbates</u> could help us to determine the nature of reaction intermediates and the concentration of active sites. Photon spectroscopy techniques have the time resolution and can be operated at high pressures to help us carry out studies of this type (infrared spectroscopy, sum frequency generation, for example). <u>Time resolved studies of the dynamics of substrate structure and adsorbate structure</u> could yield data on the importance of surface restructuring in heterogeneous catalysis. These types of measurement await innovation in the techniques of surface characterization.

Acknowledgments.

This work was supported by the Director, Office of Energy Research, Office of Basic Energy Science, Materials Science Division, of the U.S. Department of Energy under Contract No. DE-AC03-76SF00098.

References

[1] Somorjai, G.A. (1990) *J. Phys. Chem.*, 94, 1013.
[2] Blakely, D.W., Kozak, E.I., Sexton, B.A., and Somorjai, G.A. (1976) *J. Vac. Sci. Technol.*, 13, 1091. Goodman, D.W. (1984) *Acc. Chem. Res.*, 17, 194. Campbell, C.T., and Paffett, M.T. (1984) *Surf. Sci.*, 193, 396. Rucker, T.G., Frank, K., Columb, D., Logan, M.A., and Somorjai, G.A. (1987) *Rev. Sci. Instrum.*, 58, 2292.
[3] Somorjai, G.A., Kim, C.M., and Knight, C. (1992) 'Building of Complex Catalysts on Single-Crystal Surfaces', ACS volume *Surface Science of Catalysis: In Situ Probes and Reaction Kinetics.* D.J. Dwyer and F. Hoffmann (eds.), ACS Books, Washington, DC.
[4] Somorjai, G.A. (1981)*Chemistry in Two Dimensions: Surfaces.* Cornell University Press, Ithaca, New York.
[5] Sinfelt, J.H. (1991) *Cat. Ltrs.*, 9, 159.
[6] Toyoshima, I. and Somorjai, G.A. (1970) *Catal. Rev. Sci. Eng.*, 19, 105.
[7] Norskov, J.K., Stoltze, P., and Nielsen, U. (1991) *Cat. Ltrs.*, 9, 173.
[8] Norskov, J.K. and Stoltze, P. (1987) *Surf. Sci.*, 189/190, 91.
[9] Strongin, D.R., Carrazza, J., Bare, S.R., and Somorjai, G.A. (1987) *J. Catal.*, 103, 213.
[10] Bussell, M.E., Gellmann, A.J., and Somorjai, G.A. (1988) *J. Catal.*, 110, 423.
[11] Somorjai, G.A., Van Hove, M.A., and Bent, B. (1988) *J. Phys. Chem.*, 92, 973.
[12] Somorjai, G.A. (1992)in *Perspectives in Catalysis*, J.M. Thomas and K.I. Zawarsev (eds.), Blackwell Sci. Pub., London.
[13] Somorjai, G.A., Salmeron, M., and Gale, R.J. (1979) *J. Chem. Phys.*, 70, 2807.
[14] Gaussmann, A. and Kruse, N. (1991) *Cat. Ltrs.*, 10, 305.
[15] McIntyre, B.J., Salmeron, M.B., and Somorjai, G.A. (1992) *Cat. Ltrs.*, 14, 263.
[16] Ertl, G. (1991) *Cat. Ltrs.*, 9, 219.
[17] Katona, T. and Somorjai, G.A. (1992) *J. Phys. Chem.*, 96, 5465.
[18] Strongin, D.R. and Somorjai, G.A. (1989)*J. of Catal.*, 118, 99.
[19] Bare, S.R., Strongin, D.R., and Somorjai, G.A. (1986) *J. Phys Chem.*, 90, 4726.
[20] Materer, N., Starke, U. Van Hove, M.A., Barbieri, A., and Somorjai, G.A. (1992) *Surface Science.* (to be published)
[21] Davis, S.H., Zaers, F., and Somorjai, G.A. (1982) *J. Catal.*, 77, 439.
[22] Bussell, M.E. and Somorjai, G.A. (1987) *J. of Catal.*, 106, 93.
[23] Somorjai, G.A. and Garfunkel, E.L. (1989) in *Alkali Adsorption on Metals and Semiconductors.* H.P. Bonzel, A.M. Bradshaw and G. Ertl (eds.), Elsevier, New York, 319.
[24] Nix, R.M. and Somorjai, G.A. (1989) *Perspectives in Quantum Chemistry*, J. Jortner and B. Pullman (eds.), Kluwer Academic Publishers, 97.
[25] Somorjai, G.A., Slavin, A.J., Bent, B.E., and Kao, C.-T. (1988) *Surf. Sci.*, 202, 388.
[26] Somorjai, G.A., Blackman, G.S., Kao, C.-T., Bent, B.E., Mate, C.M., and Van Hove, M.A. (1988) *Surf. Sci.*, 207, 66. Somorjai, G.A., Mate, C.M., and Kao, C.-T. (1988) *Surf. Sci.*, 206, 145.
[27] Somorjai, G.A. and Crowell, J.E. (1984) *Appl. Surf. Sci.*, 19, 73.
[28] Strongin, D.R. and Somorjai, G.A. (1991) *Catalytic Ammonia Synthesis: Fundamentals and Practice*, J.R. Jennings (ed.) Plenum Publishing Company.
[29] Schwab, G.-M. (1981)*Catalysis, Science and Technology.* J.R. Anderson and M. Boudart (eds.), Springer Verlag, 1.

References: cont.

[30] Tauster, S.J. and Fung, S.C. (1978) *J. Catal.*, 55, 29.
[31] Batholomew, C.H., Powell, R.B., and Butler, J.L. (1977) *J. Catal.*, 65, 461.
[32] Somorjai, G.A., Williams, K.J., Boffa, A.B., Levin, M.E., Salmeron, M., and Bell, A.T. (1990) *Cat. Ltrs.*, 5, 385.
[33] Kelly, D.G., Gellman, A.J., Salmeron, M., Maurice, V., Huber, M., Oudar, J., and Somorjai, G.A. (1988) *Surf. Sci.*, 204, 1.
[34] Somorjai, G.A. and Bent, B. (1989) *J. Colloid Interface Sci.*, 29, 223.
[35] Ohtani, H., Van Hove, M.A., and Somorjai, G.A. (1988) *Applied Surface Science*, 33/34, 254.
[36] Wander, A., Held, G., Hwang, R.Q., Blackman, G.S., Xu, M.L., de Andres, P., Van Hove, M.A., and Somorjai, G.A. (1991) *Surf. Sci.*, 249, 21.
[37] Somorjai, G.A. and Salmeron, M. (1982) *J. Phys. Chem.*, 86, 341.
[38] Wilk, D.E., Stanners, C.D., Shen, Y.R., and Somorjai, G.A. (1992) accepted *Surf. Sci.*

TABLE 1: Surface Science Techniques that are Used Frequently

NAME	ACRONYM	DESCRIPTION	PRIMARY SURFACE INFORMATION
Adsorption or Selective Chemisorption	Adsorption	Atoms or molecules are physisorbed, and their concentration measures total surface area. Chemisorption of atoms or molecules on sites yields surface concentration of selected elements or atomic sites.	surface area site concentration composition
Atom or Helium Diffraction	AD	Monoenergetic beams of thermal energy neutral atoms are elastically scattered off ordered surfaces and detected as a function of scattering angle. This gives structural information on the outermost layer of the surface. Atom diffraction is extremely sensitive to surface ordering and defects.	atomic structure
Auger Electron Appearance Potential Spectroscopy	AEPS	The AEAPS cross-section is monitored by Auger electron intensity. Also known as APAES.	electronic structure
Auger Electron Spectroscopy	AES	Core-hole excitations are created, usually by 1-10 KeV incident electrons, and Auger electrons of characteristic energies are emitted through a two-electron process as excited atoms decay to their ground state. AES gives information on the near-surface chemical composition.	composition

TABLE 1: cont.

NAME	ACRONYM	DESCRIPTION	PRIMARY SURFACE INFORMATION
Atomic Force Microscopy	AFM	Similar to STM. An extremely delicate mechanical probe is used to scan the topography of a surface by measuring forces exerted by surface atoms. Light interference is used to measure the deflection of the mechanical surface probe. This is designed to provide STM-type images of insulation surfaces or to detect mechanical properties at the molecular level.	atomic structure
Appearance Potential Auger Electron Spectroscopy	APAES	see AEAPS	
Appearance Potential X-ray Photoemission Spectroscopy	APXPS	The EAPFS excitation cross-section is monitored by fluorescence from core-hole decay (also known as SXAPS).	electronic structure
Angle-Resolved Auger Electron Spectroscopy	ARAES	Auger electrons are detected as a function of angle to provide information on the spatial distribution or environment of the excited atoms (see AES).	composition
Angle-Resolved Photo-Emission Fine Structure	ARPEFS	Electrons are detected at given angles after being photoemitted by polarized synchrotron radiation. The interference in the detected photoemission intensity as a function of electron energy ~100-500 eV above the excitation threshold gives structural information.	electronic structure
Angle-Resolved Photo-Emission Spectroscopy	ARPES	A general term for structure-sensitive photoemission techniques including ARPEFS, ARXPS, ARUPS, and ARXPD.	electronic structure

TABLE 1: cont.

NAME	ACRONYM	DESCRIPTION	PRIMARY SURFACE INFORMATION
Angle-Resolved Ultraviolet Photoemission Spectroscopy	ARUPS	Electrons photoemitted from the valence and conduction bands of a surface are detected as a function of angle. This gives information on the dispersion of these bands (which is related to surface structure), and also structural information from the diffraction of the emitted electrons.	valence band structure
Angle-Resolved X-ray Photoemission Diffraction	ARXPD	Similar to ARXPS and ARPEFS. The angular variation in the photoemission intensity is measured at a fixed energy above the excitation threshold to provide structural information.	electronic structure
Angle-Resolved X-ray Photoemission Spectroscopy	ARXPS	The diffraction of electrons photoemitted from core levels gives structural information on the surface.	electronic structure
Conversion-Electron Mössbauer Spectroscopy	CEMS	A surface-sensitive version of Mössbauer spectroscopy. Like Mössbauer spectroscopy, this technique is limited to some isotopes of certain metals. After a nucleus is excited by γ-ray adsorption, it can undergo inverse β-decay, creating a core hole. The decay of core holes by Auger processes within an electron mean free path of the surface produces a signal. Detecting emitted electrons as a function of energy gives some depth-profile information, because of the changing electron mean-free path.	oxidation state

TABLE 1: cont.

NAME	ACRONYM	DESCRIPTION	PRIMARY SURFACE INFORMATION
Dis-Appearance Potential Spectroscopy	DAPS	The EAPFS cross section is monitored by variations in the intensity of electrons elastically back-scattered from the surface.	electronic structure
Electron Appearance Potential Fine-Structure	EAPFS	A fine-structure technique (see EXAFS). Core holes are excited by monoenergetic electrons at ~1 eV. The modulation in the excitation cross section may be monitored through adsorption fluorescence, or Auger emission.	atomic structure
Electron-Energy-Loss Near-Edge Structure	ELNES	similar to NEXAFS, except monoenergetic high-energy electrons ~60-300 KeV excite core holes.	atomic structure
Electron-Energy-Loss Spectroscopy	ELS	Monoenergetic electrons ~5-50 eV are scattered off a surface and the energy losses are measured. This gives information on the electronic excitations of the surface and adsorbed molecules (see HREELS). Sometimes called EELS.	electronic structure
Electron Spectroscopy for Chemical Analysis	ESCA	Now generally called XPS.	composition, co-oxidation state
Electron-Stimulated Ion Angular Distribution	ESDIAD	Electrons break chemical bonds in adsorbed atoms or molecules, causing ionized atoms or radicals to be ejected from the surface along the axis of the broken bond by Coulomb repulsion. The angular distribution of these ions gives information on the bonding geometry of adsorbed molecules.	molecular orientation

TABLE 1: cont.

NAME	ACRONYM	DESCRIPTION	PRIMARY SURFACE INFORMATION
Extended X-ray Adsorption Fine Structure	EXAFS	Monoenergetic photons excite a core hole. The modulation of the adsorption cross-section with energy. 100-500 eV above the excitation threshold yields information on the radial distances to neighboring atoms. The cross section can be monitored by fluorescence as core holes decay or by the attenuation of the transmitted photon beam. EXAFS is one of many "fine-structure" techniques.	local atomic structure coordination no.
Extended X-ray Energy Loss Fine Structure	EXELFS	A fine-structure technique similar to EXAFS, except that 60-300 eV electrons rather than photons excite core holes.	atomic structure
Field-Ionization Microscopy	FIM	A strong electric field-volts/Angstrom is created at the tip of a sharp, single-crystal wire. Gas atoms, usually He, are polarized and attracted to the tip by the strong electrostatic field, and then ionized by electrons tunneling from the gas atoms into the tip. These ions, accelerated along radial trajectories by Coulomb repulsion, map out the variations in the electric-field strength across the surface with atomic resolution, showing the surface topography.	atomic structure and surface diffusion
Fourier-Transform Infra-Red Spectroscopy	FTIR	Broad-band IRAS experiments are performed, and the IR adsorption spectrum is deconvoluted by using a Doppler-shifted source and Fourier analysis of the data. This technique is not restricted to surfaces.	molecular structure

TABLE 1: cont.

NAME	ACRONYM	DESCRIPTION	PRIMARY SURFACE INFORMATION
High-Energy Ion Scattering	HEIS	High-energy ions, above ~500 KeV, are scattered off of a single-crystal surface. The "channeling" and "blocking" of scattered ions within the crystal can be used to triangulate deviations from the bulk structure. HEIS has been especially used to study surface reconstructions and the thermal vibrations of surface atoms (see also MEIS, ISS).	atomic structure
High-Resolution Electron-Energy-Loss Spectroscopy	HREELS	A monoenergetic electron beam, usually ~2-10 eV, is scattered off a surface, and energy losses below ~0.5 eV to bulk and surface phonons and vibrational excitations or adsorbates are measured a s function of angle and energy (also called EELS).	molecular structure
Ion-Neutralization Spectroscopy	INS	Slow ionized atoms, typically He^+, are incident on a surface where they are neutralized in a two-electron process that can eject a surface electron, a process similar to Auger emission from the valence band. The ejected electrons are detected as a function of energy, and the surface density of states can be determined from the energy distribution. The interpretation of the data is more complicated than for SPI or UPS.	valence band

TABLE 1: cont.

NAME	ACRONYM	DESCRIPTION	PRIMARY SURFACE INFORMATION
Infrared Reflection Adsorption Spectroscopy	IRAS	Monoenergetic IR photons are reflected off a surface, and the attenuation of the IR intensity is measured as a function of frequency. This yields a spectrum of the vibrational excitations of adsorbed molecules. Recent improvements in the sensitivity of this technique allow IRAS measurements to be made on single-crystal surfaces.	molecular structure
Infra-Red Emission Spectroscopy	IRES	The vibrational modes of adsorbed molecules on a surface are studied by detection the spontaneous emission of infrared radiation from thermally excited vibrational modes as a function of energy.	molecular structure
Ion-Scattering Spectroscopy	ISS	Ions are inelastically scattered from a surface, and the chemical composition of the surface is determined from the momentum transfer to surface atoms. The energy range is ~1 KeV to 10 MeV, and the lower energies are more surface sensitive. At higher energies this technique is also known as Rutherford Back-Scattering (RBS).	composition
Low-Energy Electron Diffraction	LEED	Monoenergetic electrons below ~500 eV are elastically back-scattered from a surface and detected as a function of energy and angle. This gives information on the structure of the near surface region.	atomic structure and molecular structure

TABLE 1: cont.

NAME	ACRONYM	DESCRIPTION	PRIMARY SURFACE INFORMATION
Low-Energy Ion Scattering	LEIS	Low-energy ions, below ~5 eV, are scattered from a surface, and the ion "shadowing" gives information on surface structure. At these low energies the surface-atom ion-scattering cross section is very large, resulting in large surface sensitivity. Accuracy is limited because the low-energy ion-scattering cross sections are not well known.	atomic structure
Low-Energy Positron Diffraction	LEPD	Similar to LEED with positrons as the incident particle. The interaction potential for positrons is somewhat different than for electrons, so the form of the structural information is modified.	atomic structure
Medium-Energy Electron Diffraction	MEED	Similar to LEED, except the energy range is higher, ~300-1000 eV.	atomic structure
Medium-Energy Ion Scattering	MEIS	Similar to HEIS, except that incident ion energies are ~50-500 KeV.	atomic structure
Neutron Diffraction		Neutron diffraction is not an explicitly surface-sensitive technique, but neutron diffraction experiments on large surface-area samples have provided important structural information on adsorbed molecules, and also on surface phase transitions.	molecular structure

TABLE 1: cont.

NAME	ACRONYM	DESCRIPTION	PRIMARY SURFACE INFORMATION
Near-Edge X-ray Adsorption Fine Structure	NEXAFS	A core-hole is excited as in fine structure techniques (see EXAFS), except the fine structure within ~30 eV of the excitation threshold is measured. Multiple scattering is much stronger at low electron energies, so this technique is sensitive to the local 3-dimensional geometry, not just the radial separation between the source atom and its neighbors. The excitation cross section may be monitored by detecting the photoemitted electrons or the Auger electrons emitted during the core-hole decay.	atomic structure
Nuclear Magnetic Resonance	NMR	NMR is not an explicitly surface-sensitive technique, but NMR data on large surface-area samples ($\geq 1m^2$) have provided useful data on molecular adsorption geometries. The nucleus magnetic moment interacts with an externally applied magnetic field and provides spectra highly dependent on the nuclear environment of the sample. The signal intensity is directly proportional to the concentration of the active species. This method is limited to the analysis of magnetically active nuclei.	molecular structure
Normal Photoelectron Diffraction	NPD	Similar to ARPEFS with a somewhat lower energy range.	atomic structure

TABLE 1: cont.

NAME	ACRONYM	DESCRIPTION	PRIMARY SURFACE INFORMATION
Rutherford Back-Scattering	RBS	Similar to ISS, except the main focus is on depth-profiling and composition. The momentum transfer in back-scattering collisions between nuclei is used to identify the nuclear masses in the sample, and the smaller, gradual momentum-loss of the incident nucleus through electron-nucleus interactions provides depth-profile information.	composition
Reflection High-Energy Electron Diffraction	RHEED	Monoenergetic electrons of ~1-20 KeV are elastically scattered from a surface at glancing incidence, and detected as a function of angle and energy for small forward-scattering angles. Back-scattering is less important at high energies, and glancing incidence is used to enhance surface sensitivity.	atomic structure
Surface Electron-Energy-Loss Fine Structure	SELFS	A fine-structure technique similar to EXELFS, except the incident electron is more surface sensitive because of the lower excitation energy.	atomic structure
Surface-Enhanced Raman Spectroscopy	SERS	Some surface geometries (rough surfaces) concentrate the electric fields of Raman scattering cross section so that it is surface sensitive. This gives information on surface vibrational modes, and some information on geometry via selection rules.	molecular structure

TABLE 1: cont.

NAME	ACRONYM	DESCRIPTION	PRIMARY SURFACE INFORMATION
Surface-Extended X-ray Adsorption Fine-Structure	SEXAFS	A more surface-sensitive version of EXAFS where the excitation cross-section fine structure is monitored by detecting the photoemitted electrons (PE-SEXAFS), Auger electrons emitted during core-hole decay (Auger SEXAFS), or ions excited by photoelectrons and desorbed from the surface (PSD-SEXAFS).	atomic structure
Sum Frequency Generation	SFG	Similar to SHG. One of the lasers has a tunable frequency that permits variation of the second harmonic signal. In this way the vibrational excitation of adsorbed molecules is achieved.	molecular structure
Second-Harmonic Generation	SHG	A surface is illuminated with a high-intensity laser, and photons are generated at the second-harmonic frequency through nonlinear optical process. For many materials only the surface region has the appropriate symmetry to produce an SHG signal. The nonlinear polarizability tensor depends on the nature and geometry of adsorbed atoms and molecules.	electronic structure, molecular structure
Secondary-Ion Mass Spectroscopy	SIMS	Ions and ionized clusters ejected from a surface during ion bombardment are detected with a mass spectrometer. Surface chemical composition and some information on bonding can be extracted from SIMS ion fragment distributions.	composition

TABLE 1: cont.

NAME	ACRONYM	DESCRIPTION	PRIMARY SURFACE INFORMATION
Surface Penning Ionization	SPI	Neutral atoms, usually He, in excited states are incident on a surface at thermal energies. A surface electron may tunnel into the unoccupied electronic level, causing the incident atom to become ionized and eject an electron, which is then detected. This technique measured the density of states near the Fermi-level, and is highly surface sensitive.	electronic structure
Spin-Polarized Low-Energy Electron Diffraction	SPLEED	Similar to LEED, except the incident electron beam is spin-polarized. This is particularly useful for the study of surface magnetism and magnetic ordering.	magnetic structure
Scanning Tunneling Microscopy	STM	The topography of a surface is measured by mechanically scanning a probe over a surface with Angstrom resolution. The distance from the probe to the surface is measured by the probe-surface tunneling current. Also known as Scanning Electron Tunneling Microscopy (SETM).	atomic structure
Soft X-ray Appearance Potential Spectroscopy	SXAPS	Another name for APXPS.	

TABLE 1: cont.

NAME	ACRONYM	DESCRIPTION	PRIMARY SURFACE INFORMATION
Transmission Electron Microscopy	TEM	TEM can provide surface information for carefully prepared and oriented bulk samples. Real images have been formed of the edges of crystals where surface planes and surface diffusions have been observed. Diffraction patterns of reconstructed surfaces, superimposed on the bulk diffraction pattern, have also provided surface structural information.	atomic structure
Thermal-Desorption Spectroscopy	TDS	An adsorbate-covered surface is heated, usually at a linear rate, and the desorbing atoms or molecules are detected with a mass spectrometer. This gives information on the nature of adsorbate species and some information on adsorption energies and the surface structure.	composition, heat of adsorption, surface structure
Temperature-Programmed Desorption	TPD	Similar to TDS, except the surface may be heated at a non-uniform rate to obtain more selective information on adsorption energies	composition, heat of adsorption
Ultraviolet Photoemission Spectroscopy	UPS	Electrons photoemitted from the valence and conduction bands are detected as a function of energy to measure the electronic density of states near the surface. This gives information on the bonding of adsorbates to the surface (see ARUPS).	valence band structure

TABLE 1: cont.

NAME	ACRONYM	DESCRIPTION	PRIMARY SURFACE INFORMATION
Work-Function Measurements	WF	Changes in the work function during the adsorption of atoms and molecules provide information on charge-transfer and chemical bonding.	electronic structure
X-ray Adsorption Near-Edge Structure	XANES	Another name for NEXAFS	
X-ray Photoemission Spectroscopy	XPS	Electrons photoemitted form atomic core levels are detected as a function of energy. The shifts of core-level energies give information on the chemical environment of the atoms (see ARXPS, ARXPD).	composition, oxidation state
X-Ray Diffraction	XRD	X-ray diffraction has been carried out at extreme glancing angles of incidence where total reflection assures surface sensitivity. This provides structural information that can be interpreted by well-known methods. An extremely high x-ray flux is required to obtain useful data from single-crystal surfaces. Bulk x-ray diffraction is used to determine the structure of organo metallic clusters, which provide comparisons to molecules adsorbed on surfaces. X-ray diffraction has also given structural information on large surface-area samples.	atomic structure

TABLE 2: Model Catalyst Systems

1) Hydrocarbon conversion over platinum catalyst systems. (n-heptane, n-hexane, methyl-cyclopentane, cyclohexane, ethylene, benzene)

SUBSTRATE	PROMOTERS
Platinum crystal faces	Rhenium Rhenium, Sulfur Potassium Gold Copper
Platinum foil	Alumina Alumina, Chlorine
Rhenium crystal face	Platinum Platinum, Sulfur
Alumina film deposited on gold foil	Platinum Platinum, Chlorine

2) Ammonia synthesis over iron catalyst systems.

SUBSTRATE	PROMOTERS
Iron crystal faces	Alumina Potassium Alumina, Potassium

3) Hydrodesulfurization of thiophene over molybdenum catalyst systems.

SUBSTRATE	PROMOTERS
Molybdenum crystal faces	Sulfur Cobalt Cobalt, Sulfur Carbon Cobalt, Carbon

4) CO, CO_2, Acetone hydrogenation over rhodium catalyst systems.

SUBSTRATE	PROMOTERS
Rhodium foil	Titanium Oxide Vanadium Oxide

ORGANOMETALLIC CHEMISTRY AS A BASIS FOR UNDERSTANDING HETEROGENEOUS CATALYSIS

J. M. BASSET, S. L. SCOTT, A. CHOPLIN, M. LECONTE, F. QUIGNARD, C. SANTINI and A. THEOLIER
Ecole Supérieure de Chimie Industrielle de Lyon
43 boulevard du 11 Novembre 1918
69100 Villeurbanne Cedex
France

At the present time, our knowledge of the mechanism of action of most heterogeneous catalysts is limited. At the most primitive level, only the overall product distribution is known, while the nature of the catalyst-reactant interactions remains obscure. In a few cases, more detailed mechanistic proposals have been advanced, but they remain modest in scope compared to the detailed mechanisms that have been developed in molecular organic chemistry, and more recently, in organometallic chemistry. Indeed, even the apparently well-established catalytic mechanisms (for example, hydrogenation of olefins) are sometimes contested, indicating the fragility of the evidence on which these proposals are based. It must be recognized, however, that by comparison to strictly molecular systems, heterogeneous catalysts are intrinsically much more complicated. In addition, the application of physical methods to structural and mechanistic problems of heterogeneous systems, as well as interpretation of the results, can be difficult.

A common problem encountered in heterogeneous catalysis is the complexity of the surfaces of solid catalysts. The so-called "active sites", a concept proposed by Sir H. S. Taylor [1], may be very small in number relative to the overall surface and, consequently, their structure is almost unknown at an atomic level. It is probably this complexity which has inspired the development of surface science applied to catalysis. New concepts have emerged slowly from this approach: surface reconstruction in the presence of adsorbed molecules, surface mobility and structural reorganization of molecular-like species. It has been demonstrated that the binding of chemisorbed molecules resembles organometallic ligation (*e.g.*, η^1, η^2 and η^3-adsorbed CO, π-bound ethylene and η^1, η^2 and η^3-ethylidene species) [2]. Naturally, the concepts derived from surface science have been, for the most part, applied to the chemisorption process and not to the reactivity pattern, although the latter is, in fact, the key to catalytic mechanisms.

At a fundamental level, we are interested in the elementary steps of heterogeneous catalytic mechanisms, that is, the succession of chemical events in which bonds of a substrate molecule are broken or created in the proximity of the surface. The number of examples in which this succession of events has been demonstrated on a well-defined surface is still limited. However, a growing awareness of this deficiency has spurred the recent development of surface organometallic chemistry [3]. In this relatively new field, the organometallic character of the "active site" is exploited in order to model reaction mechanisms.

During a catalytic cycle, substrates which interact with the surface form one or several chemical bonds with one or several surface atoms. Hence the "active site" is a

R. W. Joyner and R. A. van Santen (eds.), Elementary Reaction Steps in Heterogeneous Catalysis, 39–49.

supramolecule which consists of both the substrate and one or a few atoms from the surface. If the substrate is an organic molecule, as is very often the case, the "active site" has surface organometallic character and, in principle, the rules of organometallic chemistry are applicable.

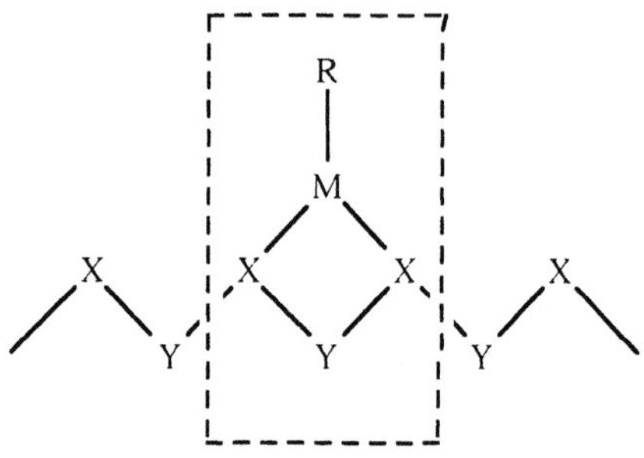

Our goal is to construct and characterise well-defined supramolecules which resemble intermediates in surface reactions. Studies of this kind have the potential to advance understanding of heterogeneous catalysts in the same way that organometallic chemistry has been at the origin of the formulation of mechanisms in homogeneous catalysis. This motivation, among others (*e.g.*, synthesis of tailor-made catalysts), is the driving force of surface organometallic chemistry (SOMC). The list of possible organometallic complexes includes compounds of main group elements, transition metals, lanthanides and actinides. The available surfaces encompass oxides and zero-valent metals, either of which may be highly divided materials, polycrystalline or well-defined single crystals.

1. Reactions of Organometallic Complexes with Surfaces

In the last decade, we and others have begun to establish the basic rules which govern the reactivity of various ligands with the functional groups which are present at the surfaces of oxides [4]. The concepts invoked are simple ones : oxidative addition, nucleophilic or electrophilic attack at the ligand or the metal center, ligand dissociation or association, electrophilic cleavage of metal carbon bonds, Brønsted and/or Lewis acid-base interactions, and redox reactions (including disproportionation), Table 1. The determination of these rules is a necessary prerequisite to the goal of constructing a desired coordination sphere for a given reaction.

Conventional surface chemistry has revealed the diversity of surface functionalities: redox centers, acido-basic centers, nucleophilic hydroxyl groups and electrophilic protons. Surface organometallic chemistry is a new approach to probe such functionalities. In particular, the observed product distribution frequently depends on the method of pretreatment of the surface, and apparently conflicting results from different laboratories can often be attributed to variability in the support. For example, the reactions of either $Fe_3(CO)_{12}$ or $Fe(CO)_5$ with magnesia produce different products depending on the degree of hydration of the oxide. The chemistry of hydrated magnesia is dominated by the reactivity of strongly (Brønsted) basic surface hydroxyl groups. For both organometallic complexes, nucleophilic attack by a surface hydroxyl on coordinated CO, followed by proton transfer, leads to the formation of $[HFe_3(CO)_{11}]^-$ and adsorbed CO_2 [5]. The

Table 1. Reactions of organometallic complexes with oxide surfaces, classified by mechanism

Mechanism	Substrate	Surface	SOM Product	Ref
Nucleophilic attack on coordinated CO	$M_3(CO)_{12}$ M = Fe, Ru, Os	hydroxylated MgO	$[HM_3(CO)_{11}]^-$	[6]
	$Fe_3(CO)_{12}$ $Fe(CO)_5$	dehydrated MgO	$[Fe(CO)_4(CO_2)]^{2-}$	[7, 8]
Electrophilic cleavage of M-C bonds	$Rh(\eta^3\text{-}C_3H_5)_3$	SiO_2 Al_2O_3 TiO_2	$(\blacksquare SiO)Rh(\eta^3\text{-}C_3H_5)_2$	[9]
Oxidative addition	$M_3(CO)_{12}$ M = Ru, Os	SiO_2 Al_2O_3	$M_3(\mu\text{-}H)(\mu\text{-}OSi\blacksquare)(CO)_{10}$	[10]
Brønsted acid-base	$H_2Os(CO)_4$	MgO	$[HOs(CO)_4]^-$	[11]
Lewis acid-base	$[HFe_3(CO)_{11}]^-$	MgO	$HFe_3(CO)_{10}(\mu\text{-}CO)\rightarrow[Mg^{2+}]$	[5]
Redox	$Rh_6(CO)_{16}$ $Rh_4(CO)_{12}$	Al_2O_3	$(\blacksquare AlO)Rh^I(CO)_2(OHAl\blacksquare)_2$	[12]
Disproportionation	$Co_2(CO)_8$	MgO	$Co^{2+}[Co(CO)_4]_2$	[13]
Promoted insertion	$CH_3Mn(CO)_5/CO$	Al_2O_3	$(OC)_5Mn(C(CH_3){=}O\rightarrow Al\blacksquare)$	[14]

$[HFe_3(CO)_{11}]^-$ ion is bound to the surface by a Lewis acid-base interaction between the bridging CO ligand and a surface Mg^{2+} ion, reflected by the decrease in frequency of the bridging ν(CO) mode. In contrast, magnesia dehydrated at 800 °C contains few surface hydroxyls. Reaction of a Lewis basic surface oxide ion with either $Fe_3(CO)_{12}$ or $Fe(CO)_5$ gives surface-bound $Fe(CO)_4(CO_2)^{2-}$ [7, 8] (Figure 1).

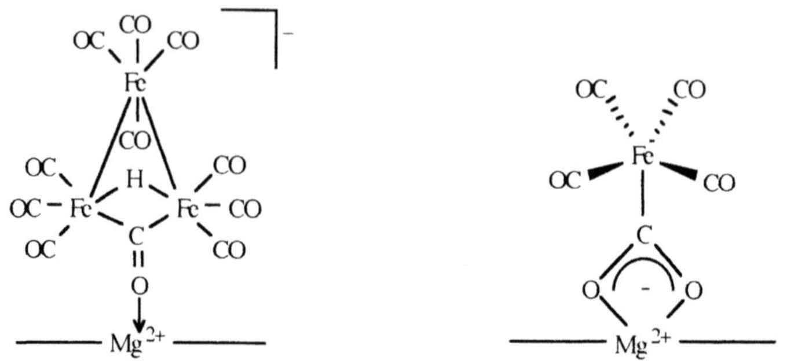

Figure 1. SOM products from the interaction of iron carbonyl clusters with magnesia.

2. Structure and Electronic Configuration of Surface Organometallic Fragments

Structural characterisation of surface organometallic fragments is necessary for tuning the catalytic activity of modern "tailor-made" catalysts. Due to the supramolecular character of the SOM fragment, it is necessary to use both "molecular" approaches (*e.g.*, FTIR, solid-state NMR) and atomic determination of the structure (*e.g.*, XPS, EXAFS). Knowledge of the precise coordination sphere of the metal will facilitate the *prediction* of catalytic reactions, for example, the presence of a hydride ligand for electrophilic alkane activation, a carbene for olefin metathesis, an alkyl ligand for olefin polymerization, or an oxo group for olefin oxidation. If the surface organometallic fragment has a unique structure (unfortunately not always the case), it is straightforward to deduce the electronic configuration of the metal. Based on the (6, 8, 10, 12, 14, 16, 18, 20) electron rule, one can predict the electrophilic or nucleophilic character of the metal. The following examples demonstrate how characterisation of surface organometallic species has enabled us to rationalize their structure and electronic configuration using concepts from organometallic chemistry.

The supported cluster $(\mu\text{-O-Si}\equiv)(\mu\text{-H})Os_3(CO)_{10}$, **1** [10], is an 18-electron cluster in which a surface oxygen atom of silica behaves as a 3-electron ligand (in the MLH Green formalism) to the cluster. The magnesia-supported ruthenium cluster $[HRu_3(CO)_{11}]^-$, **2**, represents a case in which the surface does not participate in formation of a covalent bond, but merely acts as a counter-cation which neutralizes the charge of the supported cluster [6]. In contrast to the analogous supported iron cluster [5], the stretching frequency of the bridging CO ligand is nearly unperturbed by interaction with the alumina surface, implying little direct bonding between the carbonyl oxygen and surface Al^{3+} cations.

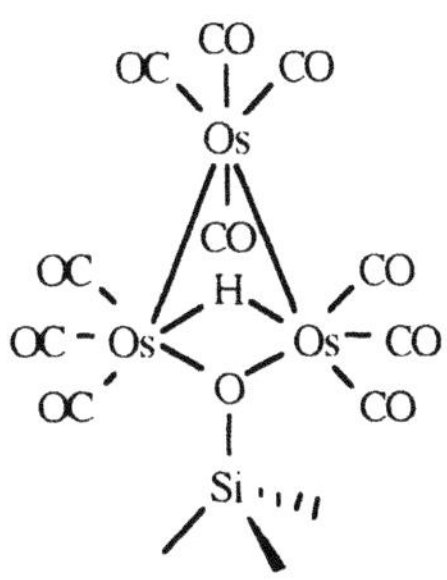

1

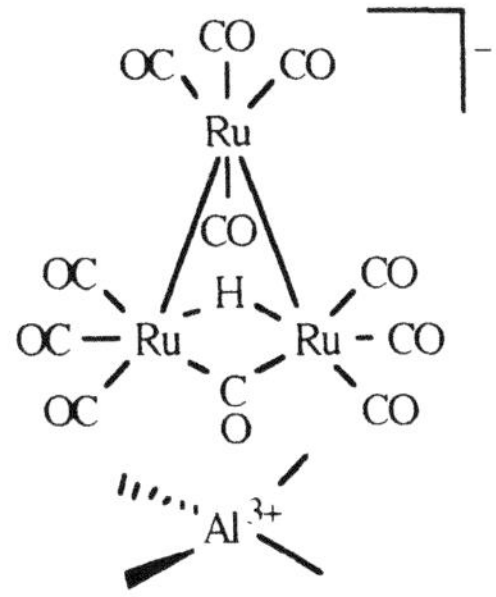

2

Tris(neopentyl)zirconium supported on silica, **3**, is a highly electron-deficient surface species in which a surface oxygen atom behaves as a one-electron ligand [15]. In the presence of H_2, a silica-supported zirconium hydride, **4**, is formed [16]. The latter, highly electrophilic, is formally an 8-electron species which has no equivalent in molecular chemistry.

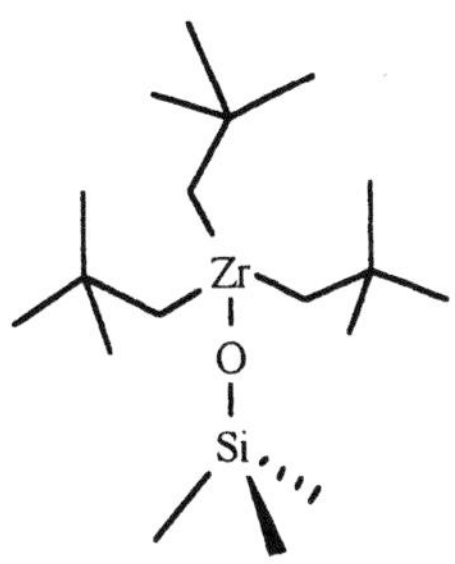

3

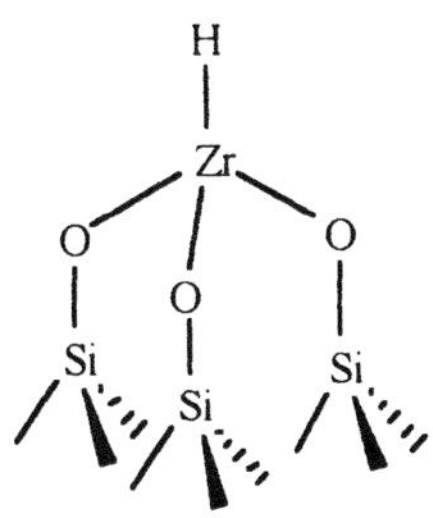

4

Bis(allyl)rhodium supported on silica, **5**, has one covalent bond to a surface siloxyl. The rhodium attains an 18-electron configuration by coordination to a surface hydroxyl group [9]. On alumina and titania, which have fewer surface hydroxyls, electron donation from a surface bridging oxygen serves the same purpose, **6**.

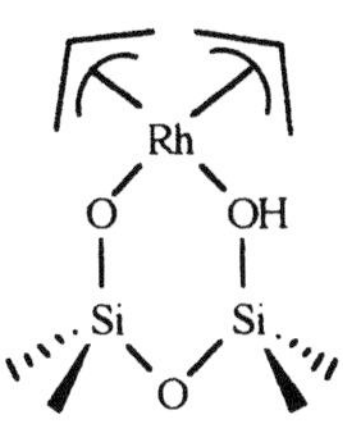

5

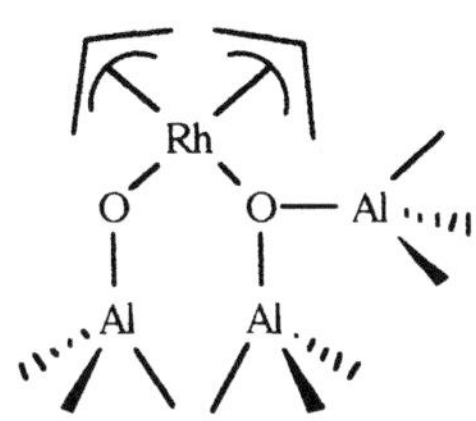

6

It is also possible to graft main group elements onto oxides by simple covalent bonds. A typical example is the formation of surface-bound Sn(n-$C_4H_9)_3$ by the reaction of Sn(n-$C_4H_9)_4$ with silica, **7** [17]. In contrast, the same substrate reacts with the surface of metallic rhodium supported on silica to give a grafted organometallic fragment with the empirical formula $Sn(n\text{-}Bu)_2$. The latter has the structure shown in **8** [18].

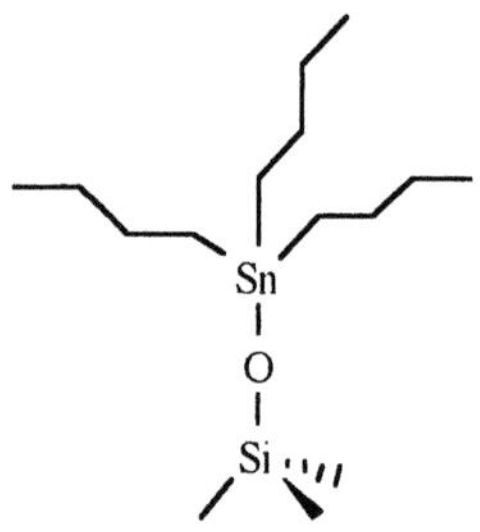

7

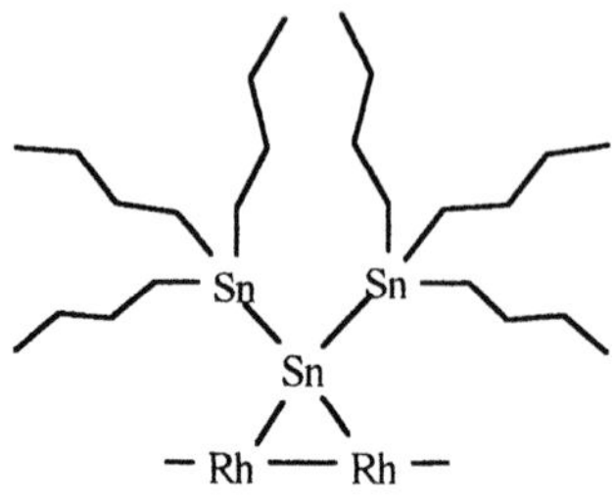

8

3. Stoichiometric Reactivity of Surface Organometallic Fragments

Elementary steps of homogeneous catalysis are usually deduced from known elementary steps of stoichiometric organometallic reactions: oxidative addition, insertion, reductive coupling, α, β and γ-hydrogen abstraction, metallocycle formation, *etc.* In heterogeneous catalysis, the elementary steps are usually unknown, and even well-identified chemisorbed molecules or molecular fragments are not necessarily reaction intermediates because of the complexity of the reaction mechanism (*vide infra* § 4). A well-chosen surface organometallic fragment may constitute an intermediate for one of these mechanisms, if the step-by-step transformation into products can be demonstrated. An illustration of this approach is the selective hydrogenolysis of alkanes by zirconium hydride supported on silica [16]. The σ-bond metathesis of C-H and Zr-H bonds, which takes place at room temperature, occurs by a 4-centre electrophilic mechanism [19].

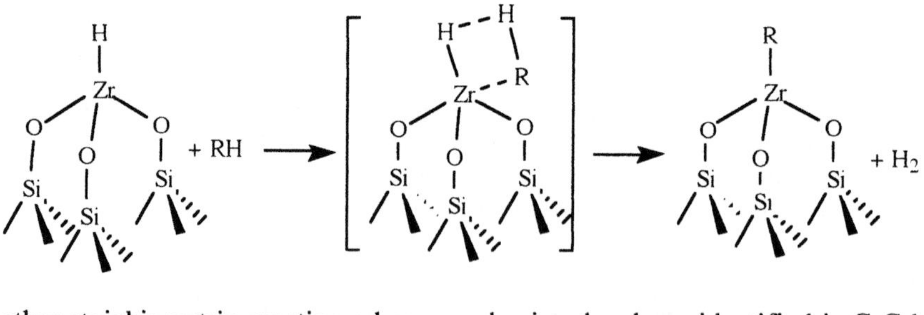

Another stoichiometric reaction whose mechanism has been identified is C-C bond formation by reductive coupling of two η^3-allyl ligands coordinated to surface-bound Rh(III) [20]. The reaction is initiated by addition of CO to the SOM fragment, causing the η^3-allyl ligands to become η^1-bound. The product, 1,5-hexadiene, is formed by reductive elimination.

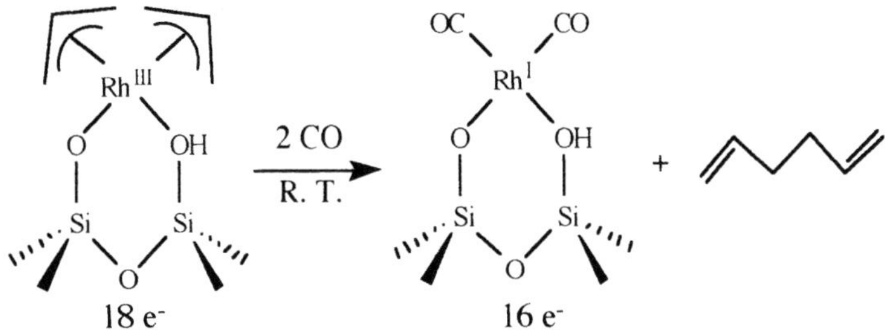

A parallel reaction in this system is CO insertion into the η^1-allyl ligand, followed by reductive elimination of allyl and acyl ligands. The presence of 1,6-heptadien-4-one was detected among the products.

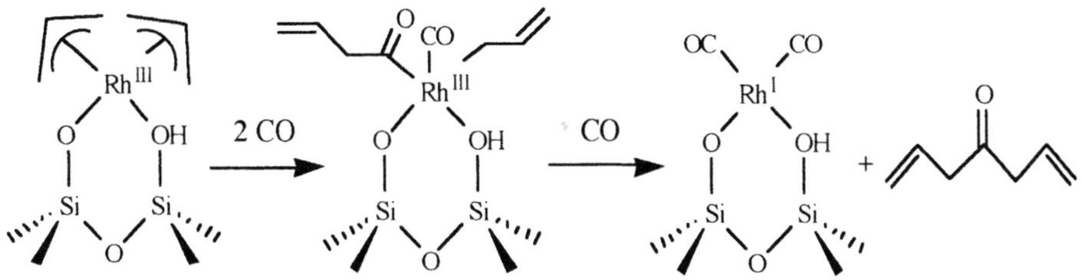

Ethylene coordinates reversibly to the supported cluster $Os_3(CO)_{10}(\mu\text{-H})(\mu\text{-OSi}\equiv)$. At the same time, the silica ligand alternates between 3-electron and 1-electron donation as needed to maintain the optimum electron count of the cluster [10].

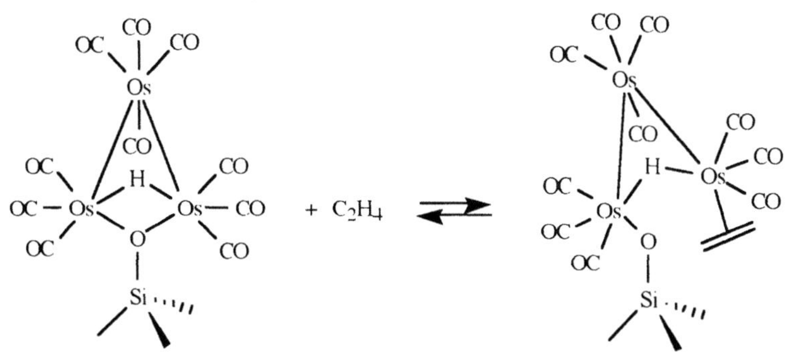

4. The Molecular Approach to Mechanisms in Heterogeneous Catalysis

The determination of reaction mechanisms in heterogeneous catalysis is an enormous task which will not be solved by surface organometallic chemistry alone. However, the general strategy should be based on the following considerations. In both homogeneous and heterogeneous catalysis, it is essential to extract an atomic-level understanding of catalytic processes. The surface of the heterogeneous catalyst interacting with the substrate must be viewed as a supramolecule (surface organometallic fragment). Preliminary knowledge about reactants and products must be refined with experimental evidence about intermediates and elementary steps.

To achieve this supramolecular level of understanding, and in order to identify and characterize various intermediates formed in the conversion from reactants to products, we must develop models which reflect both structural and electronic characteristics of the catalyst. Given these structures, the rates of transformation between various intermediates must be determined, in order to reduce complicated reactions to fundamental or elementary steps that can be used in constructing reactivity models. As in homogeneous organometallic chemistry, elementary steps established for one reaction provide useful concepts for analyzing many others. The ultimate goal in these studies is, of course, to construct a mechanism which allows us to predict how to modify the rates and/or products and to design new catalysts and new processes.

Consider a generalized catalytic mechanism such as the one shown in Figure 2. The sequence of steps $A + R \rightarrow B \rightarrow C \rightarrow A + P$, including the detailed structures (geometric and electronic) of intermediates A, B, and C and the rates of transformation between them, constitutes the mechanism.

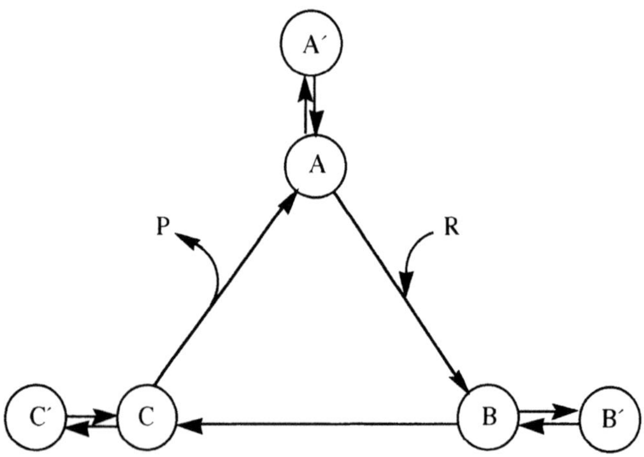

Figure 2. A typical mechanism in catalysis.

However, the actual starting point may be A´, which must be transformed into the active catalyst A. Even though B´ and C´ are not involved directly in the catalytic process, they may be the dominant species observed in spectroscopic and structural experiments. It is also possible that the presence of any of the various species A, A´, B, B´, C or C´ affects the structures and rates of transformation of the other species, so that the visualization of a reaction in terms of independent species is simplified at best. Even so, such models are crucial to extracting the concepts used to design experiments and new catalysts.

Mechanisms can be established by either (1) an experimental approach, using the techniques of homogeneous mechanistic chemistry applied to surfaces, or (2) a theoretical approach, which involves testing a mechanistic hypothesis by simulation. In order to illustrate the former, we consider the evolution of the role played by carbenes in both homogeneous and surface mechanisms. In Fischer-Tropsch catalysis, the original mechanism, published in 1926, invoked surface carbene complexes which polymerise to give higher hydrocarbons [21]. A more complicated mechanism involving hydroxy-carbene intermediates was subsequently proposed [22]. Later, a CO insertion mechanism gained wide acceptance [23]. However, studies of soluble models and more precise surface studies did not support the new proposals. During this period, organometallic chemists showed that carbene complexes were much more stable and prevalent than originally thought. The role of such species in homogeneous metathesis was established. Pettit and Brady carried out the experiments with solid catalysts that established carbene complexes as the accepted intermediates in Fisher-Tropsch catalysts [24, 25].

The CH_2 fragment now appears to be a key intermediate in a variety of other reactions of hydrocarbons on surfaces, including hydrogenolysis and homologation of olefins and hydrocarbons (Figure 3) [26, 27, 28].

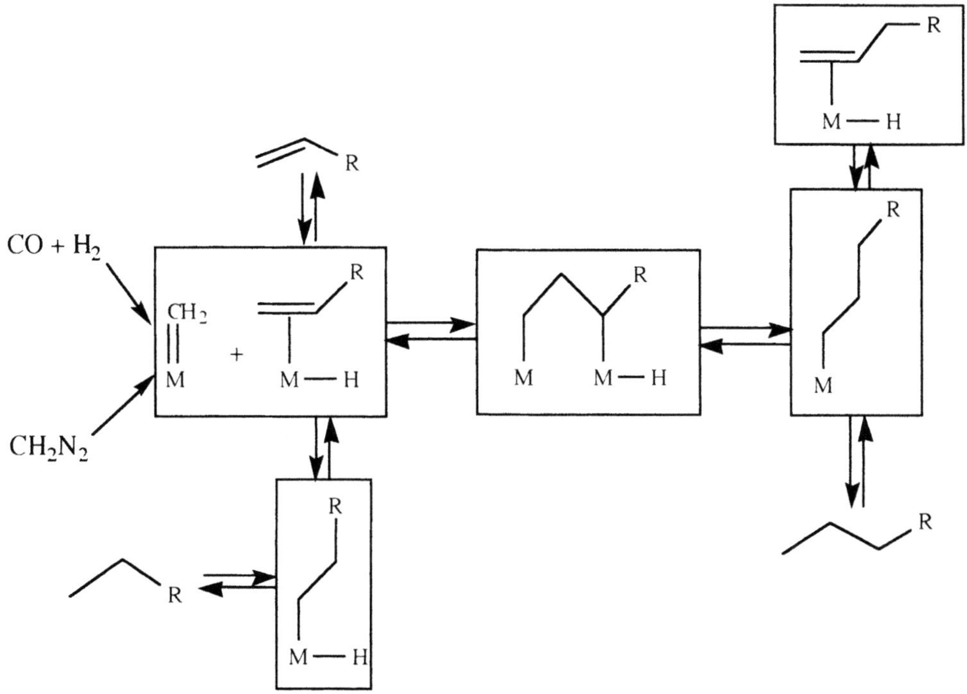

Figure 3. General mechanism for C-C bond formation and cleavage on metal surfaces.

Theory has potentially three important roles to play in elucidating the atomic-level mechanisms. First, and most important, is the establishment of general qualitative concepts concerning various elementary steps. The second role is calculation of quantitative data characterizing structures and barriers for intermediates too short-lived for detailed experimental study. Because of the complexity of surface processes and the limitations of experimental techniques, such quantitative information must, for the time being, be derived from theory. The third role of theory is the extraction of force fields from detailed molecular cluster calculations, for use in molecular dynamics and statistical simulations of complex models including all known intermediates and mechanistic steps. Such simulations, involving the dynamics of thousands of atoms, provide a strenuous test for the completeness of our molecular understanding of heterogeneous processes. Ultimately, the test of the usefulness and correctness of a theoretical approach is whether the theory suggests new approaches to improving a particular catalyst, or better, entirely new strategies for developing catalysts.

5. Summary and Future Prospects

In the course of this review, we have considered the impact of molecular chemistry, organometallic chemistry and surface organometallic chemistry on the understanding of the mechanisms of action of heterogeneous catalysts. If one defines a heterogeneous mechanism as a succession of elementary steps leading from reactants to products, it is clear that a molecular approach, or more precisely an organometallic approach, will be particularly influential in elucidating those elementary steps. The primary reason is the "surface organometallic character" of the active sites of heterogeneous catalysts.

The classical approach to heterogeneous catalysis involves the study of surface structure, surface reconstruction, diffusion and chemisorption processes, but not true elementary steps. Thus, information is obtained about the "static" catalyst, and may not be relevant to the molecular transformations which occur in the catalytic cycle. In addition, classically-prepared heterogeneous catalysts generally have a very small number of active sites and, consequently, the structures being studied may not be those which are relevant to the reactivity of the catalyst. In contrast, if one grafts a carefully chosen organometallic fragment onto a surface, then the number of active sites can be quite large and the stoichiometric transformation into products is easily followed. The tools of surface science are readily applied, and those which give information at a molecular level (*e.g.* solid state NMR) about the structure and reactivity of supramolecular fragments are especially useful.

It appears that, in several cases, the rules of molecular organometallic chemistry can be applied to surface organometallic fragments and thus the elementary steps of heterogeneous catalysis can be described using concepts from homogeneous mechanisms. Obviously, surface studies require supplementary parameters which do not exist in molecular chemistry but which are already integrated into surface organometallic chemistry, *e.g.*, ligand rigidity, limited translational mobility and bulk reactivity (electron reservoir properties of metallic particles, for example).

Knowledge of the rules which govern the reactivity of organometallic compounds with surfaces must precede the preparation of tailor-made catalysts. We are now in a position to design a given coordination sphere for a given catalytic reaction. The results can be quite spectacular: for example, catalytic cleavage of the C-C bonds of alkanes has been achieved at room temperature under hydrogen with a supported zirconium hydride complex. The activity and selectivity of a catalytic reaction can be deliberately adjusted (as in molecular chemistry), in contrast to highly empirical approach to catalyst modification of classical heterogeneous catalysis. While the future of surface organometallic chemistry seems assured by the growing interest in tailor-made catalysts, further advances in the field will require a thorough fundamental characterisation of surface intermediates and elementary steps.

References

1. H. S. Taylor *J. Am. Chem. Soc.* **53**, 578 (1931).
2. N. D. S. Canning and R. J. Madix *J. Phys. Chem.* **88**, 2437-2446 (1984).
3. J. M. Basset, J. P. Candy, A. Choplin, B. Didillon, F. Quignard and A. Théolier, in *Perspectives in Catalysis* J. A. Thomas, K. I. Zamaraev, Eds. (Blackwell Scientific Publications, Oxford, 1992).
4. J. M. Basset, J. P. Candy, P. Dufour, C. Santini and A. Choplin *Catalysis Today* **6**, 1 (1989).
5. F. Hugues, J. M. Basset, Y. Ben Taarit, A. Choplin, M. Primet, D. Rojas and A. K. Smith *J. Am. Chem. Soc.* **104**, 7020-4 (1982).

6. L. D'Ornelas, A. Théolier, A. Choplin and J. M. Basset *Inorg. Chem.* **27**, 1261-5 (1988).
7. E. Guglielminotti and A. Zecchina *J. Mol. Catal.* **24**, 331-44 (1984).
8. J. M. Basset, B. Besson, A. Choplin, F. Hugues, M. Leconte, D.Rojas, A. K. Smith, A. Théolier, Y. Chauvin, D. Commeruc, R. Psaro, R. Ugo and G. M. Zanderighi, in *Fundamental Research in Homogeneous Catalysis* M. Graziani, M. Giongo, Eds. (Plenum, New York, 1984), vol. 4, pp. 19-54.
9. P. Dufour, C. Houtman, C. C. Santini, C. Nédez, J. M. Basset, L. Y. Hsu and S. G. Shore *J. Am. Chem. Soc.* **114**, 4248-4257 (1992).
10. A. Choplin, B. Besson, L. D'Ornelas, R. Sanchez-Delgado and J. M. Basset, *J. Am. Chem. Soc.* **110** 2783-7 (1988).
11. H. H. Lamb and B. C. Gates, *J. Am. Chem. Soc.* **108**, 81-9 (1986).
12. J. M. Basset, A. Théolier, D. Commeruc and Y. Chauvin *J. Organomet. Chem.* **279**, 147 (1985).
13. N. Homs, A. Choplin, P. Ramirez de la Piscina, L. Huang, E. Garbowski, R. Sanchez-Delgado, A. Théolier and J. M. Basset *Inorg. Chem.* **27**, 4030-3 (1988).
14. S. B. Butts, S. H. Strauss, E. M. Holt, R. E. Stimson, N. W. Alcock and D. F. Shriver *J. Am. Chem. Soc.* **102**, 5093-5100 (1980).
15. F. Quignard, C. Lecuyer, C. Bougault, F. Lefebvre, A. Choplin, D. Olivier and J. M. Basset *Inorg. Chem.* **31**, 928-30 (1992).
16. C. Lecuyer, F. Quignard, A. Choplin, D. Olivier and J. M. Basset *Angew. Chem. Int. Ed. Engl.* **30**, 1660-1 (1991).
17. C. Nédez, A. Choplin, F. Lefebvre, B. Benazzi and J. M. Basset *J. Am. Chem. Soc.*, accepted for publication.
18. B. Didillon, O. Clause, F. Lefebvre, H. Lamb, C. Houtman, T. Shay, J. P. Candy and J. M. Basset, submitted for publication.
19. F. Quignard, A. Choplin and J. M. Basset *J. Chem. Soc., Chem. Comm.*, 1589 (1991).
20. P. Dufour, S. L. Scott, C. Santini, F. Lefebvre and J. M. Basset, submitted for publication.
21. F. Fischer and H. Tropsch *Brennst.-Chem.* **7**, 97-116 (1926); Chem. Ber. **59**, 830-6 (1926).
22. H. H. Storch, N. Golumbic and A. Anderson *The Fischer-Tropsch and Related Syntheses* (Wiley, New York, 1951).
23. H. Pichler and H. Schultz *Chem. Ing. Tech.* **12**, 1160-74 (1970). C. Masters *Adv. Organomet. Chem.* **17**, 61-103 (1979). G. Henrici-Olivé and S. Olivé *Angew. Chem., Int. Ed. Engl.* **15**, 136-141 (1976) and references therein.
24. R. C. Brady III and R. Pettit *J. Am. Chem. Soc.* **102**, 6182 (1980).
25. W. T. Osterloh, M. E. Cornell and R. Pettit *J. Am. Chem. Soc.* **104**, 3759 (1982).
26. M. Leconte, A. Théolier, D. Rojas and J. M. Basset *J. Am. Chem. Soc.* **106**, 1141 (1984).
27. E. Rodriguez, M. Leconte, J. M. Basset, K. Tanaka and K. I. Tanaka *J. Am. Chem. Soc.* **110**, 275 (1988).
28. E. Rodriguez, M. Leconte and J. M. Basset *J. Catal.* **132**, 472 (1991).

THE ORGANIC CHEMICAL APPROACH TO HETEROGENEOUS CATALYSIS

JOHN J. ROONEY
School of Chemistry
The Queen's University
David Keir Building
Stranmillis Road
BELFAST BT9 5AG
N. Ireland

ABSTRACT. The organic chemical probe approach to heterogeneous catalysis by metals is illustrated here by many examples taken from research extending over the past half century. A convincing case is provided that the site densities are low and very often mononuclear rather than ensembles of two or more contiguous atoms. The mechanisms of reaction are then due to the interconversion of various transient intermediates as ligands of these mononuclear centres, which by implication are coordinatively highly unsaturated and thus at corners and edges, etc., i.e. they are "defect" atoms. The bonding of the transients to the surface is weaker and much more reactive than is generally considered from a surface science viewpoint, which is more relevant to drastic bond-fission reactions; real turnover numbers are much higher than usually estimated. All of this is part of a general thesis where the catalytic behaviour of metals not only relates directly to that of metalla-enzymes and synthetic organometallic complexes but also shows many fundamental similarities to catalysis by refractory oxides.

Introduction

In the earlier part of my research career, from 1957 onwards, I was greatly influenced in my approach to catalysis by two factors. The first was the project given to me, a study of catalytic cracking on silica-alumina, and the second was the arrival in Belfast of the late Francois Gault who wished to learn deuterium exchange techniques then operating in C. Kemball's laboratories. Francois had with him a series of novel polymethylcycloalkanes, synthesized in J.E. Germain's laboratories, and it was his ambition to study the exchange reactions of these model substrates.

My instructions were to confine my attention to the catalytic cracking of cyclopentene and cyclopentane on silica-alumina since it was conjectured that larger C_6-ring compounds would rapidly form benzene(s) and then coke.

R. W. Joyner and R. A. van Santen (eds.), Elementary Reaction Steps in Heterogeneous Catalysis, 51–72.

Paradoxically I soon learned to my cost that cyclopentene was about the worst possible choice from this point of view as it exhibited a remarkable propensity to form practically every aromatic hydrocarbon and extensive graphitic coke. The reason soon became clear since cyclopentene oligomerizes first on the Brønsted sites and the oligomers then in part crack back to smaller units, but much of them, e.g. the dimer, rapidly converted to napthalene etc. according to the following carbonium ion reactions.

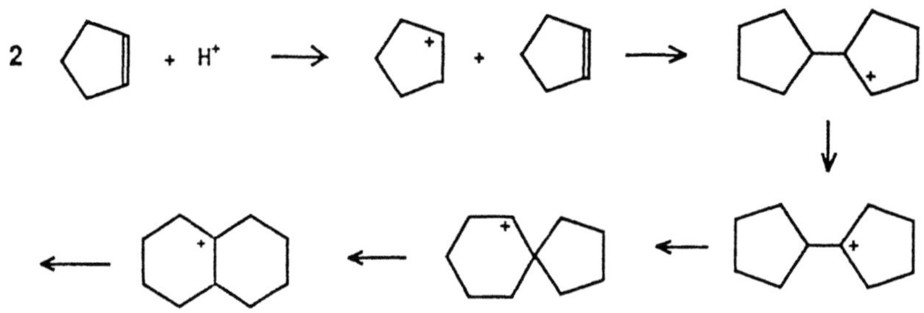

However, I became greatly interested as were others at that time in the nature and numbers of the active sites involved for this and other reactions on such high surface area refractory oxides including γ-Al_2O_3. The most striking general conclusion later extended to zeolites was that the active sites for a variety of types of chemisorption and different catalytic reactions only constitute <1% of the total area with numbers often less than 10^{12} cm^{-2}. Model chemical reactions were found to be the best probes of the active sites at these low levels, the numbers then being ascertained by titration using selective poisons. Paradoxically the organic substrate usually has to be rather large and complex in order to contain sufficient information within its structure and functionality such that a meaningful simple question may be answered concerning the surface sites. A major benefit from this approach in the acidic area was that the relationships between heterogeneous and homogeneous catalysis soon emerged. The following few examples illustrate these points and at the same time highlight the further problem that the species which are the most abundant in solutions or surfaces may be the least significant for catalysis.

Acid catalysed reactions of aryl-olefins and aryl-alcohols

When 1,1-diphenylethylene ($\phi_2C = CH_2$) is added at room temperature in dry benzene to activated silica-alumina a bright yellow colour first develops but the surface rapidly changes to green. U.v. reflectance spectroscopy shows two bands with λ_{max} at 423 and 605 mμ respectively (Fig. 1) [1].

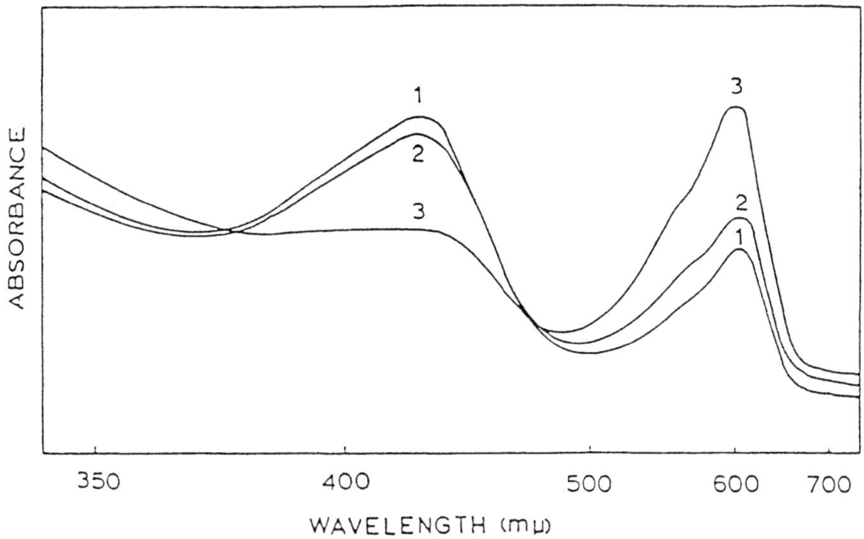

Fig. 1. Electronic spectra of 1,1-diphenylethylene adsorbed on silica-alumina; (1) initial spectrum; (2) change in (1) after period of 25 min; (3) immediate effect of added water on (2).

On addition of small amounts of water or alcohol the yellow band disappears leaving a bright blue colour. The same behaviour is also noted when conc. sulphuric acid is added slowly to a glacial acetic acid solution of this olefin. The yellow colour is due to $\phi_2\overset{+}{C}{-}CH_3$ formed through the action of protons with an acid strength equivalent to that of 70% conc. H_2SO_4 in water. This initial carbonium ion participates in a very rapid catalysis leading to formation of a stable dimer (IV) as shown in Scheme 1.

However, none of this including an alternative potential pathway which may ensue from the initial coupling was able to account for the more abundant and more stable blue species. Now silica-alumina possesses oxidizing properties as shown by the surface reaction of perylene and other π-bases [2].

$$\text{Perylene} + \text{Surface} \longrightarrow [\text{Perylene}]^{+\cdot} + [\text{Surface}]^{-\cdot}$$

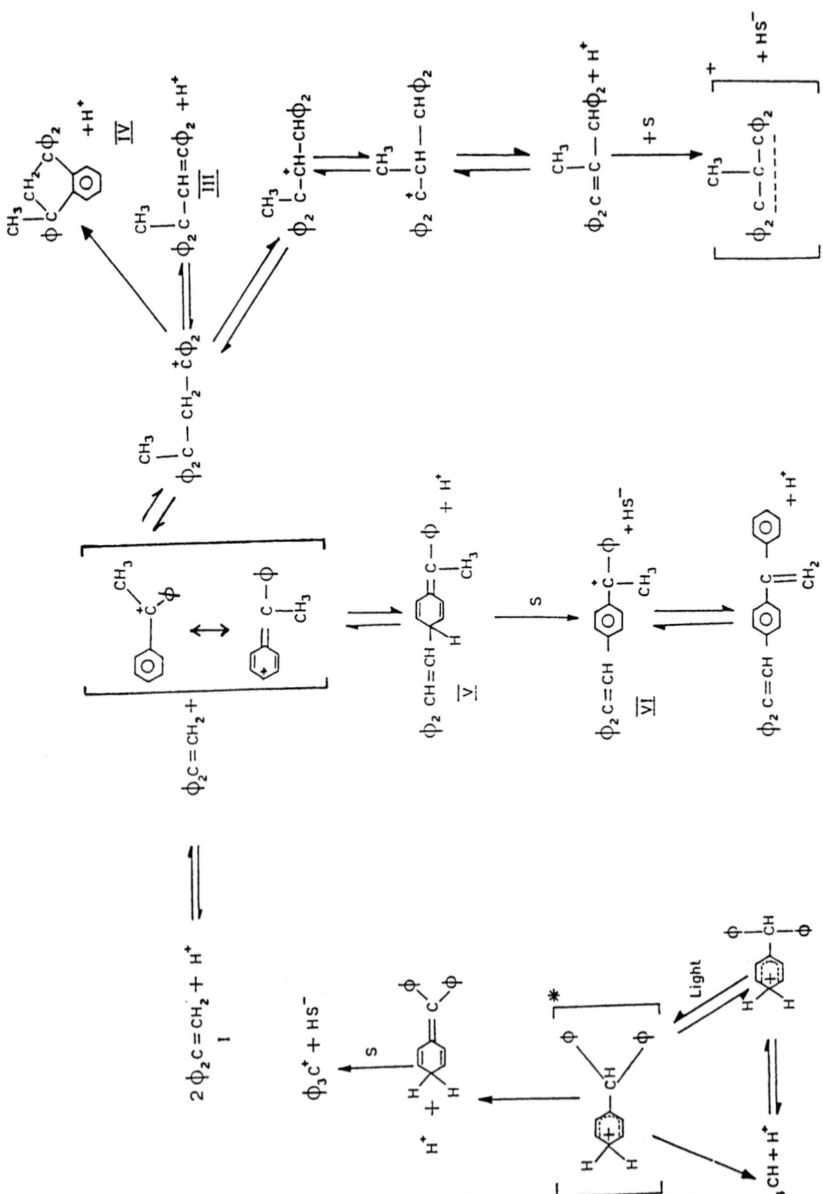

SCHEME 1

An esr examination of the $\phi_2C = CH_2/SiO_2\text{-}Al_2O_3$ system showed a small concentration of cation-radicals, $10^{11}\text{-}10^{12}$ cm^{-2}. This confused the issue for several years but eventually the blue species was shown to be a novel dimeric cation formed through the less stable canonical form II of $\phi_2\overset{+}{C}\text{–}CH_3$, e.g.,

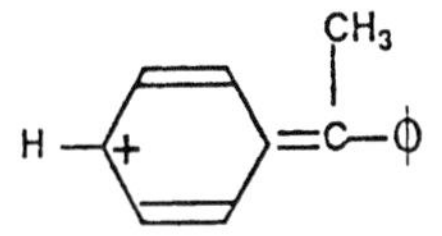

II

The latter reacted with $\phi_2C{=}CH_2$ to give eventually dimer V which rapidly loses H^- to give carbonium ion VI, now identified as the blue species. We synthesized the dimer VII as the corresponding alcohol VIII, and showed that immediately on addition to acids, including adsorption of VIII on anhydrous $Mg[ClO_4]_2$, the same blue complex is formed [3].

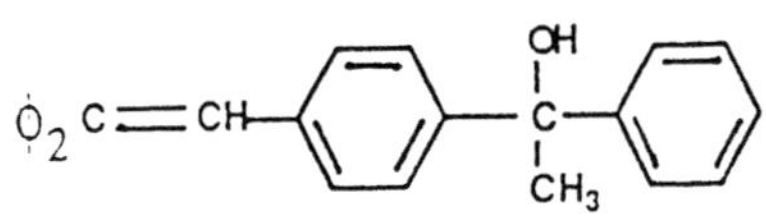

VIII

Incidentally the classical Gomberg free radical $\phi_3\dot{C}$ does not dimerize to hexaphenylethane, as thought for many years, but also reacts via the less stable canonical form.

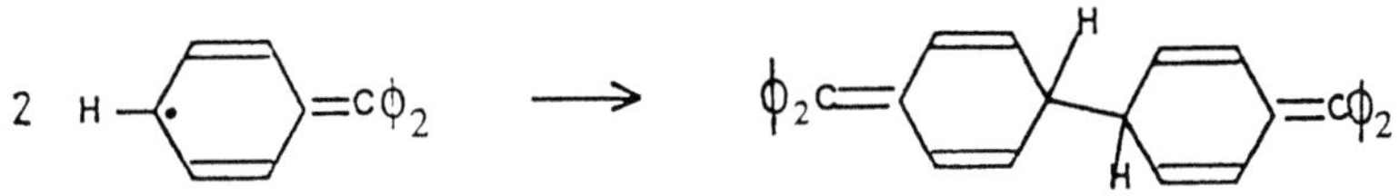

The blue species on silica-alumina is a beautiful example of an important chemisorptive complex, yet catalytically insignificant since the amount of dimerization via the less stable canonical form is tiny.

Even the final loss of H^- from $\underline{V}$ on Lewis acid sites may not be as simple as first thought since the following is the mechanism of conversion of ϕ_3CH into ϕ_3C^+ ions on silica-alumina, a reaction catalysed by light and O_2 [4]. The first step (Scheme 1) is again protonation, followed by excitation of the carbonium ion by light, and then elimination of the tertiary proton. The overall process results in the production of an endothermic isomer of ϕ_3CH which rapidly peroxidises in the presence of O_2 and is then converted into ϕ_3C^+ and O_2H^- by either Lewis or Brønsted acidity.

These sort of observations can be put to good use in syntheses, e.g. **tetraphenylallene,** a compound of considerable experimental and theoretical interest. In Brønsted acids the reaction, ϕ_2CHOH with protons occurs at lower acid strength than the corresponding reaction of $\phi_2C{=}CH_2$, even though $\phi_2\overset{+}{C}{-}CH_3$ is intrinsically more stable than ϕ_2CH^+.

$$\Phi_2CHOH + H^+ \longrightarrow \Phi_2\overset{+}{C}H + H_2O$$

$$\Phi_2C{=}CH_2 + H^+ \longrightarrow \Phi_2\overset{+}{C} - CH_3$$

Now if one uses a 50/50 mixture of $\phi_2C{=}CH_2$ and ϕ_2CHOH and reacts either on silica-alumina or in liquid Brønsted acids the following reactions occurs quantitatively giving 1,1,3,3-tetraphenylpropylene, which is then converted conveniently into tetraphenylallene by bromination and dehydrobromination.

$$\Phi_2\overset{+}{C}H + \Phi_2C{=}CH_2 \longrightarrow \Phi_2CHCH_2-\overset{+}{C}\Phi_2 \longrightarrow \Phi_2CHCH{=}C\Phi_2 + H^+$$

By way of contrast γ-Al_2O_3 has only weak Bronsted acidity which does not colour aryl-alcohol or aryl-olefin indicators. However, it possesses redox sites which can operate synergistically in the simultaneous conversion of π-bases and π-acids into cation and anion-radicals respectively, e.g.

$$\text{Perylene} + \text{Tetracyanoethylene} \longrightarrow [\text{Perylene}]^{+\cdot} + [\text{Tetracyanoethylene}]^{-\cdot}$$

These remarkable dual sites are capable of activating paraffins even at 298 K, as shown by deuterium exchange reactions. Very elegant work on these and analogous reactions of alkenes using selective reversible poisons such as CO_2 and H_2S has been done, especially by W.K. Hall and coworkers [5]. Again the site densities are very low, 10^{11}-10^{12} cm^{-2}.

Transition metal catalysis

The mechanistic problems here are much more formidable than in the refractory oxide area, but the answers to key general questions are turning out to be much the same. The active sites are only a small fraction of the total metal atoms exposed and are often mononuclear, i.e. ensemble sites are not significant, with much weaker and less extensive bonding of the intermediates to the surface than most people think. The close relationship to homogeneous catalysis by organometallic complexes is therefore quite striking as I do not know of any significant catalytic reaction in the homogeneous area which is even binuclear. The *cis*-ligand insertion reaction in mononuclear complexes is undoubtedly the cornerstone in both fields. Of course, relatively stable chemisorbed complexes occur as well during catalysis by metals but they tend to be catalytically inert as evidenced by the well-documented ethylidyne species, with tripod bonding to three contiguous metal atoms, which is formed during reactions of ethane, ethylene or acetylene on several surfaces [6]. The success of surface science during the past two decades in identifying such stable species has been quite remarkable but has unfortunately distorted the perspective of heterogeneous catalysis in terms of mechanism and innovation. The following examples will illustrate these points.

C-H bond cleavage and formation

The most fundamental surface reaction here is the interconversion of σ-bonded alkyl and 1,2-diadsorbed alkene which has had a long history as the classical Horiuti-Polanyi mechanism for olefin hydrogenation, and as the $\alpha\beta$ multiple exchange process of paraffins with D_2.

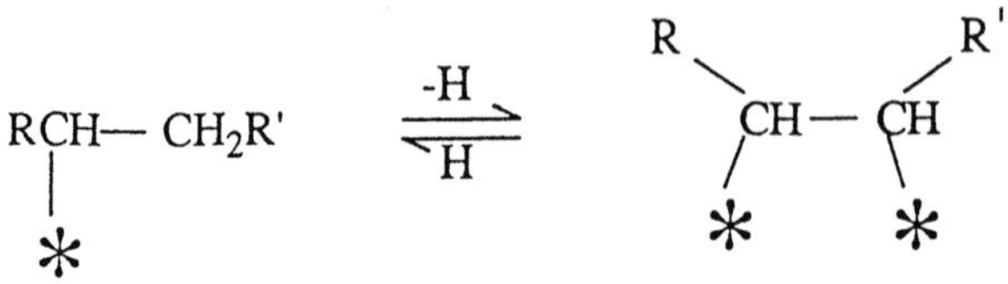

This mechanism has been very extensively probed using many model systems so what can we now say about it after so much labour over more than half a century?

The initial distributions for exchange of paraffins with D_2 are characteristic of the identify of the metal **and** not of its shape or form. Thus a single crystal of Pt (111 face) at 550 K seems to exhibit exactly the same behaviour in the ethane/D_2 reaction as Pt films and blacks at 450 K [7]. The initial distribution is U-shaped with maxima in C_2H_5D and in C_2D_6 (Table 1), which shows that two distinct sites are operating, one with an αβ process of low chance of alkyl reversal and the other where this process has a high chance of alkyl reversal.

Table 1. Initial product distributions in ethane exchange with D_2

	D_1	D_2	D_3	D_4	D_5	D_6
Pt film	20	17	12	10	15	26
Pt (111)	30	9	1	5	11	44
Pt black	42	18	12	7	10	12

As Boudart [8] has pointed out for Fe catalysts and ammonia synthesis, the single crystal is a true microcosmic model, both mechanistically and kinetically, for conventionally prepared catalysts for the exchange reaction.

The exchange of cyclopentane with D_2 on Pt/TiO_2 in the SMSI state gives much the same multiple initial distributions of deuteroisomers as Pt/SiO_2 even though Pt/TiO_2 (SMSI) chemisorbs negligible amounts of H_2 or D_2 [9]. This is a really remarkable result because it means that the site densities for paraffin activation are very low and that a large surface excess of rapidly scrambling H and D atoms is not essential. Since the characteristic multiplicity and kinetics require that the availability of the excess D in the whole system is never limiting we now have to think of sites which can activate paraffins, with the resulting surface complexes reacting directly with D_2 molecules. It would appear therefore that only corner and/or edge metal atoms, i.e. those which are coordinatively highly unsaturated can be really active in order to accommodate this new information. Furthermore, the exchange behaviour of numerous model compounds strongly indicates that the best description now of the Horiuti-Polanyi mechanism is the interconversion of σ-bonded alkyl and π-bonded alkene as ligands of the same metal atom. Thus the exchange of cyclopentane with D_2 on Pt and on several other metal gives initial distributions with maxima in the D_1, D_5 and D_{10}-isomers, and sometimes a small maximum in the D_8 isomer as well (Fig. 2) [10]. Rollover of the 1,2-diadsorbed species is required to give the D_8-isomer initially and multiple rollover to provide the D_{10}-isomer. The most convenient way of understanding rollover is to envisage π-bonded cyclopentene with cleavage of the π-bond and consequent mobility of the detached cycloalkene (X = H or D).

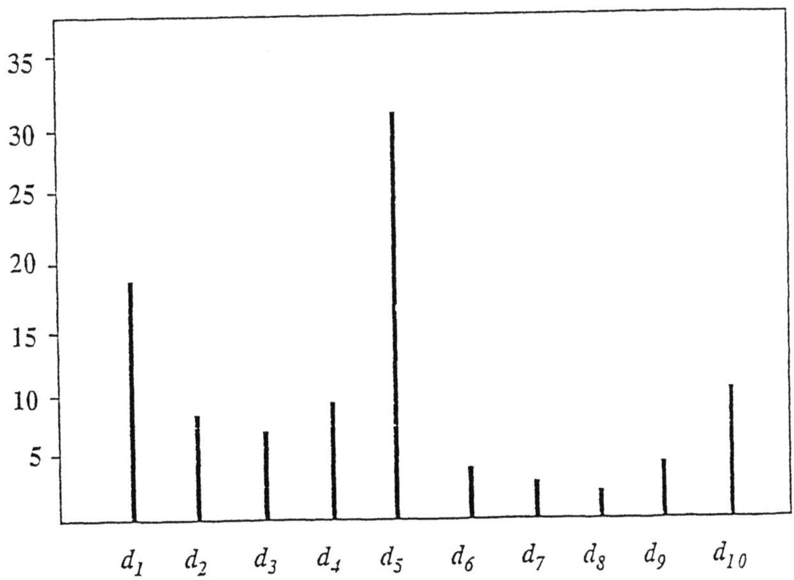

Fig. 2 Initial Distribution for Cyclopentane/D_2 exchange on 1% Pt/SiO_2 at 273 K.

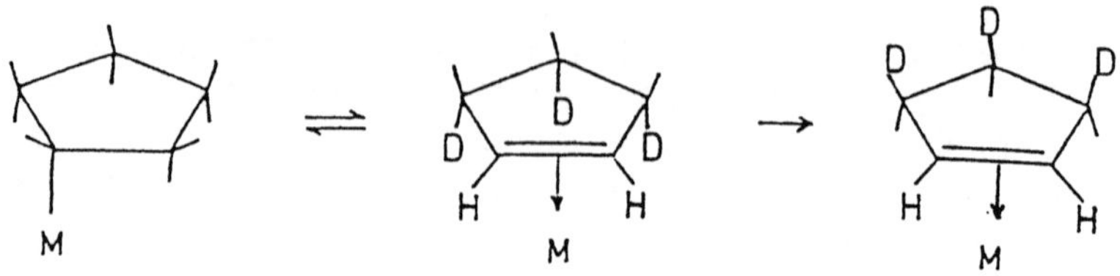

A wide variety of model cyclic compounds beginning with those brought to Belfast by F.G. Gault have been used to test this fundamental idea. Thus in order to have multiple initial exchange on both sides of the rings in the following substrates, as observed, it is necessary to have transient formation and rollover of the corresponding symmetrical olefins shown [10].

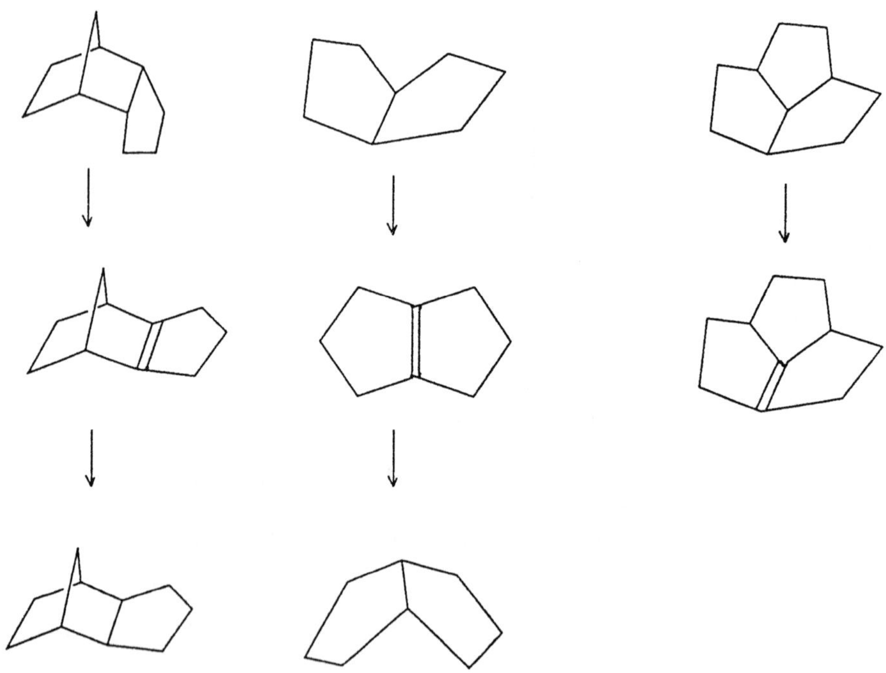

Isomerisation of *endo*- to *exo*-tetrahydrodicyclopentadiene occurs during the exchange reaction as expected if this mechanism is correct. On the other hand the dish-shaped tricyclic isomer shown only undergoes initial exchange on the convex side because the symmetrical olefin required for possible rollover is too strained and cannot invert. The requirement of π-bonding and not simply of eclipsing of vicinal H atoms was further emphasized by the following comparison. Norbornane exchanges 2H atoms initially on Pd but the heptacyclic dimer of norbornadiene only exchanges one H atom. Norbornane can readily form π-bonded norbornene but the heptacyclic substrate cannot form the corresponding olefin even though it has four pairs of eclipsed H atoms [10].

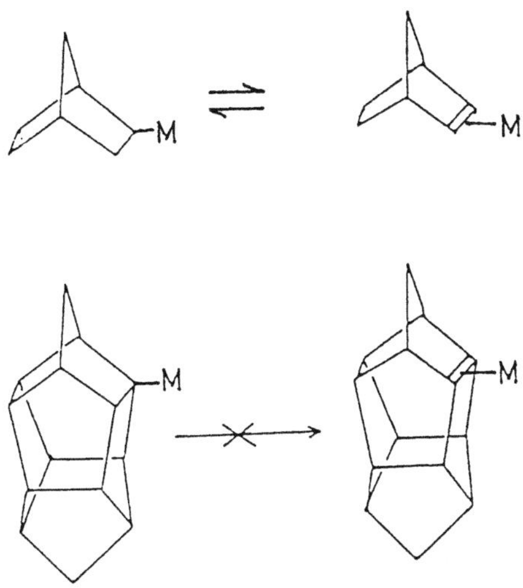

All of this shows that the surface mechanism for alkene hydrogenation, etc., by metals is closely akin to that for homogeneous hydrogenation by metal complexes. Furthermore the Pt/TiO_2 results [9] indicate that the addition and removal of H or D atoms may occur as follows involving only the same mononuclear centre.

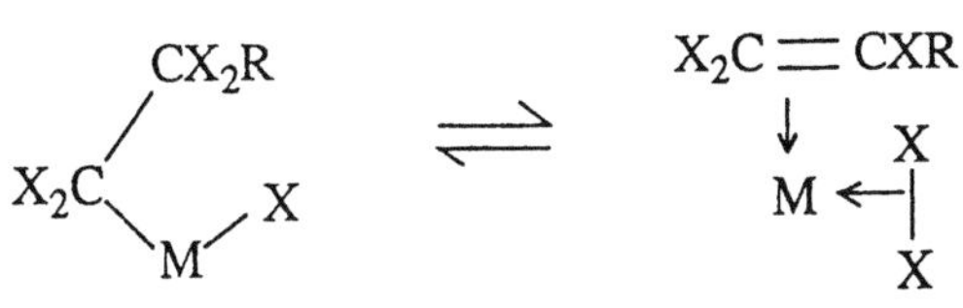

Analogous reactions of alkenes and ketones, etc. with a weakly held molecular H_2 ligand have recently been observed in homogeneous hydrogenation systems [11].

Electronic Effects

Since the metal surface seems to subtend relatively few mononuclear active complexes associated with defect atoms, e.g. corners and edges, nearest neighbour atoms can be regarded as permanent ligands of the central active atom in the same way as e.g. stannous and phosphine ligands in homogeneous catalysis. The question then arises can these "ligand" atoms be electronically modified and the effect relayed via the central atom to influence the catalysis? The characteristic $\alpha\beta$ process in cyclopentane/D_2 exchange is remarkably insensitive to metal dispersion and form either as films or on supports such as SiO_2, γ-Al_2O_3 or TiO_2 [9]. However, very recently we have noted that when Pt, Pd or Rh are dispersed on MgO the initial distributions are significantly shifted towards increased multiple exchange as can be seen by the relative amounts of the D_5 and D_{10} isomers (Fig. 3) [12].

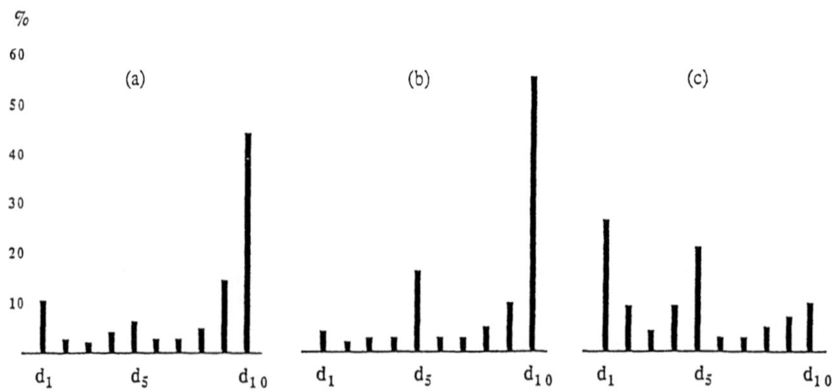

Fig. 3 Initial distributions for Cyclopentane/D_2 exchange at 372 K:
(a) 2%Pt/Al^{3+} - MgO (k_o = 2380); (b) 2%Pt/MgO(I) (k_o = 1478);
(c) 0.05%Pt/Al_2O_3 (k_o = 15443). k_o/% $min^{-1}g^{-1}$Pt.

Both the chances of alkyl reversal and of rollover of intermediate cyclopentenes are obviously greatly increased. This result has a direct bearing on an understanding of the observation by Derouane *et al.* [13] that Pt or Pd dispersed on MgO are highly selective catalysts for aromatisation of hexane. A key mechanism in this reaction is cyclisation of unsaturated chains so MgO is beneficial since as shown by the exchange results it induces a greater propensity in the metal for intermediate dehydrogenation. The abundance of the D_{10} isomer indicates a relative shift towards π-bonding verses σ-bonding and this could arise if there is electron donation from the surface of MgO to dπ-orbitals in the metal. Charge transfer complexes of this type involving π-acids and π-bases and the surfaces of refractory oxides have already been mentioned. Thus pure SiO_2 is inert, γ-Al_2O_3 and TiO_2 have redox properties but MgO is simply a donor, so this may be the only one of these supports capable of a net transfer of charge to the metal. This is a novel concept for support interactions but one that has far reaching implications both for catalyst fabrication and theory.

C-C Bond Reactions

These occur at higher temperatures when the C-H reactions are usually too fast for exchange with D_2 to be a useful probe. Francois Gault turned to ^{13}C labelling and showed that a multiplicity of reactions, 1,2-bond shift, cyclization, hydrogenolysis, aromatization, etc., may occur in affording initial products [14]. Many transient species which are frequently mobile, spilling over from one type of site to another on the metal are involved so mechanistic studies are difficult. It is worth stressing that ^{13}C labelling tells us essentially nothing about the mechanism of any individual reaction where a knowledge of the C_xH_y formula of any proposed surface intermediate complex is essential.

1,2-Bond Shift

The rearrangement of neopentane to isopentane on Pt is now a classical reaction in catalysis. The mechanism is undoubtedly the following where the half-reaction-state in the rearrangement of the initial free neopentyl radical is stabilized by π-bonding. In other words this is a metal assisted isomerization of a free radical.

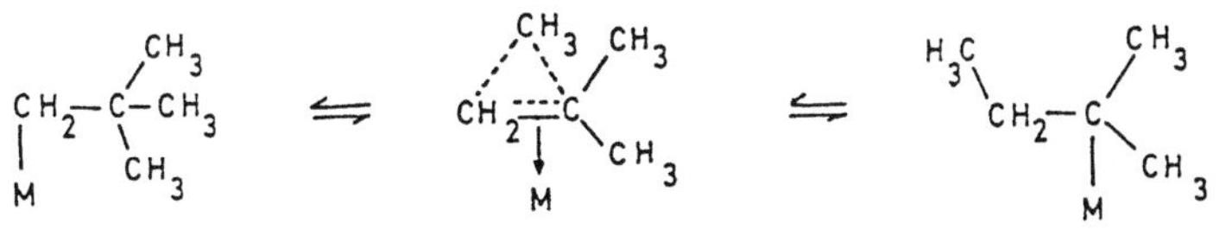

All of the following model reactions prove this mechanism [10].

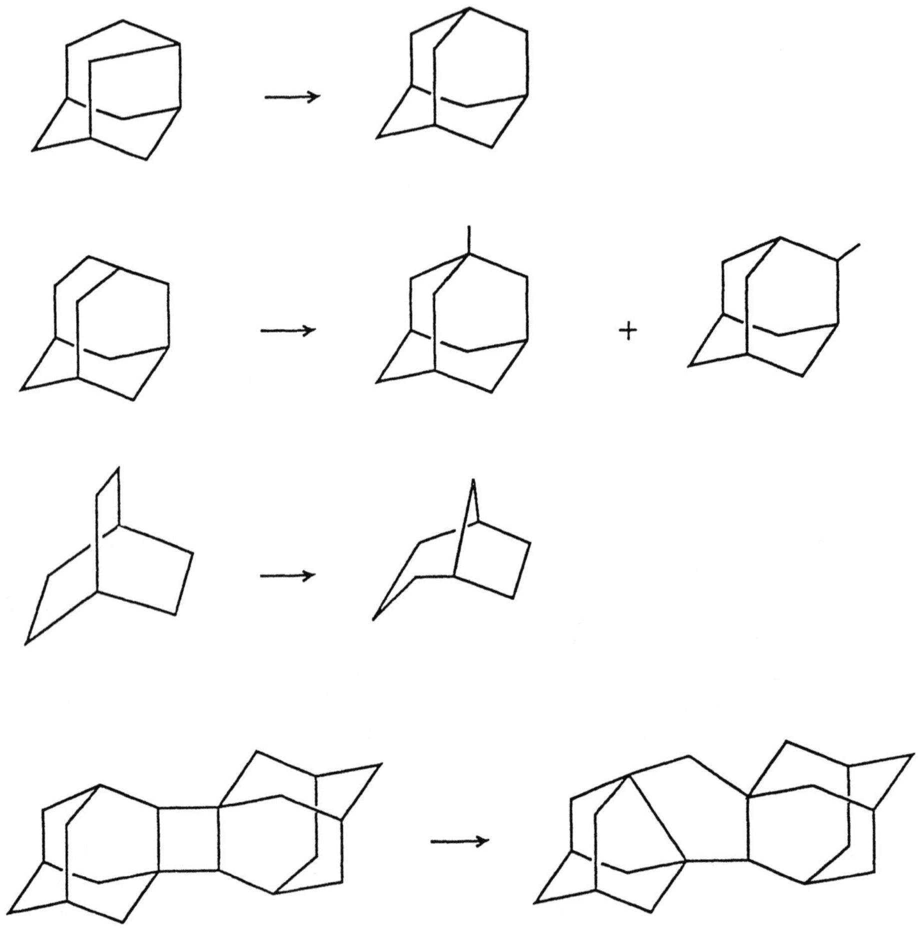

The alternative idea of a metathesis-type mechanism involving reversible cleavage of metallacyclobutane intermediates and metallacarbene complexes is simply not possible here. Furthermore, in one case we were able to show using D_2 that removal of only one H is necessary for 1,2-bond shift in an alkane and removal of only 2H for 1,5-cyclization on Pt, Pd, Ir and Rh [15]. This rules out carbenes and *a fortiori* carbynes and ensembles of metal atoms as essential mechanistic requirements. There is no possibility of the αβ process for the model substrate shown and this simplifies the use of D_2.

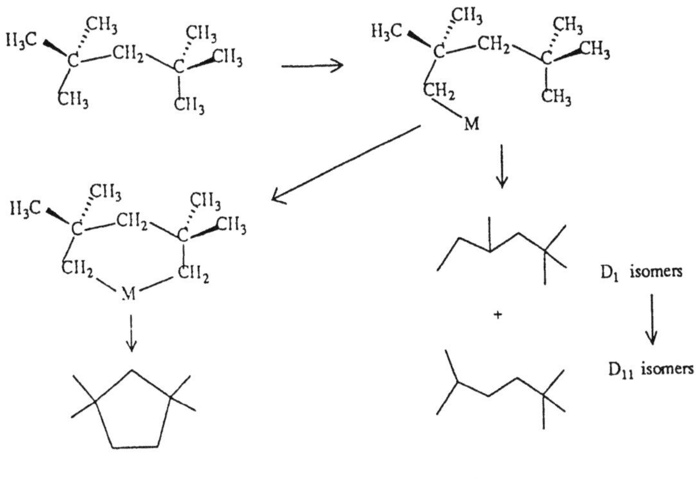

D_0 - D_2 isomers

All of this type of information is being put to good use in the synthesis of various cage hydrocarbons [16], as shown below.

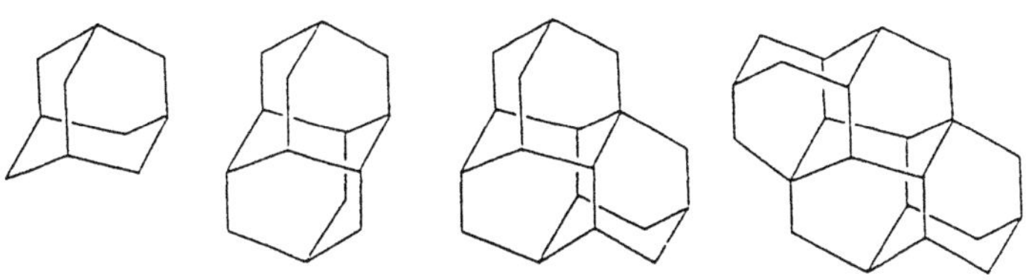

$C_{10}H_{16}$ $C_{14}H_{20}$ $C_{18}H_{24}$ $C_{22}H_{28}$

The first two are readily available by isomerization of cheap starting materials using acid catalysis. The last two were made by attaching one set of 4 C atoms and two sets of 4 C atoms respectively to the diamantane cage using standard synthetic procedures and then passing in excess H_2 over Pt catalysts at elevated temperatures. Hydrogenation and the crucial 1,6-cyclization plus bond shift occurred as expected giving the desired products, (Scheme 2).

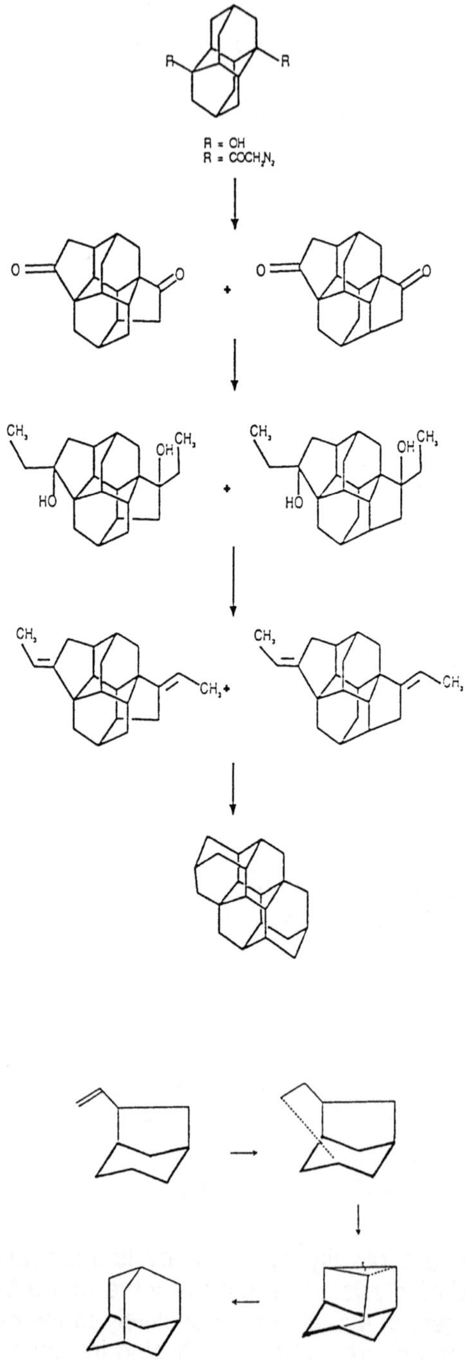

SCHEME 2

For many years attempts have been made to synthesize dodecahedrane $C_{20}H_{20}$ and this has recently been achieved in a most elegant fashion using metal catalysts [17].

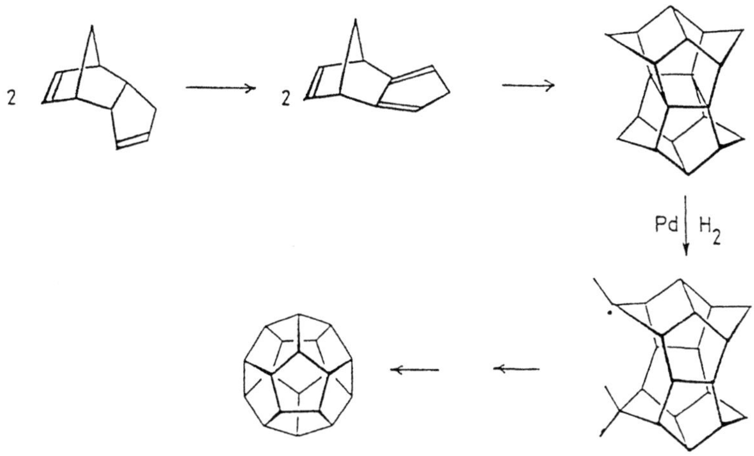

Mobility of transients

Formation and movement of intermediate alkenes including both rollover and reversible spillover apparently occur during exchange of alkanes with D_2 even at ambient temperatures. Free or weakly held alkenes therefore move over the metal surface. At elevated temperatures homolysis of metal-alkyl σ-bonds must also be quite rapid with formation and movement of free radical intermediates as well as alkenes. Thus the 1,5-cyclization required in the final step of the synthesis of dodecahedrane is most likely a free radical reaction, while 1,2-bond-shift rearrangements are apparently surface assisted free radical processes. This idea has also been applied to vitamin B_{12} catalysis and indeed Salem *et al.* [18] postulated the following Co π-complex for the 1,2-diol rearrangement, without being aware that their mechanism is exactly analogous to our earlier suggestion that neopentyl radicals rearrange via half-reaction-state π-complexes on Pt.

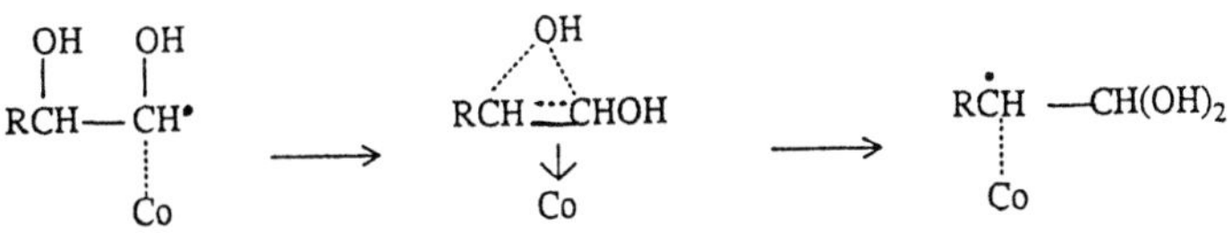

The general idea of mobility of transients in surface catalysis especially at elevated temperatures has hitherto not been given sufficient attention but has far reaching consequences for both kinetic and mechanistic understanding in the field. The concept of spillover of H and O atoms in certain systems can be extended to rollover and spillover of organic transients as described here, and even to surface initiated homogeneous free radical processes.

Polarity of intermediates

In some of our studies of bond-shift reactions we noticed [19] that a Pd catalyst which was not capable of activating paraffinic C-H bonds (no activity for adamantane/D_2 exchange) hydrogenolysed the following hydroxymethyl-protoadamantane under the same conditions to give 1-methyladamantane.

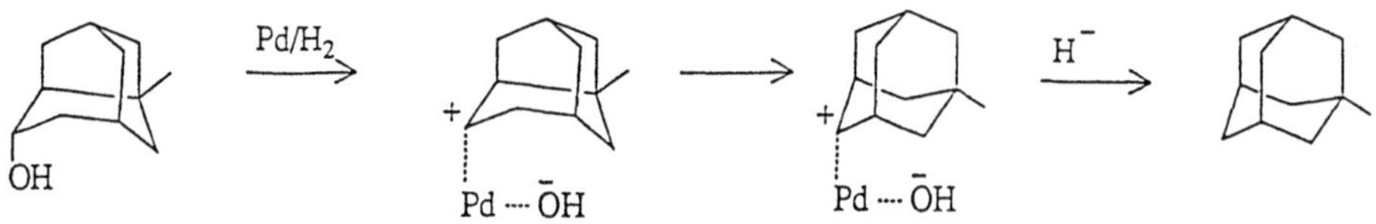

On the other hand a fresh Pd catalyst, highly active for adamantane/D_2 exchange, hydrogenolysed the same alcohol to give the much less stable isomeric 2-methyladamantane.

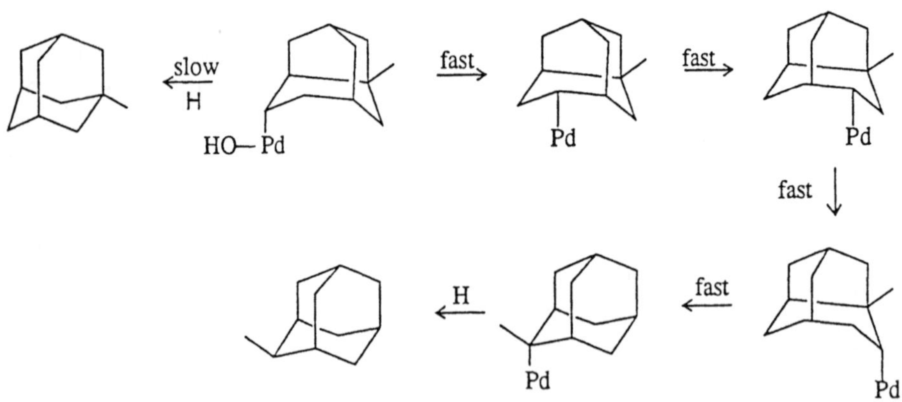

Another interesting example of this type of distinction is the observation [20] that protoadamantanone undergoes hydrogenolysis on a highly active ICI Cu/ZnO methanol synthesis catalyst to give mainly protoadamantane but the major products from reaction on a poor Cu/SiO_2 catalyst are adamantane, 1- and 2-adamantanols, and 2-adamantanone.

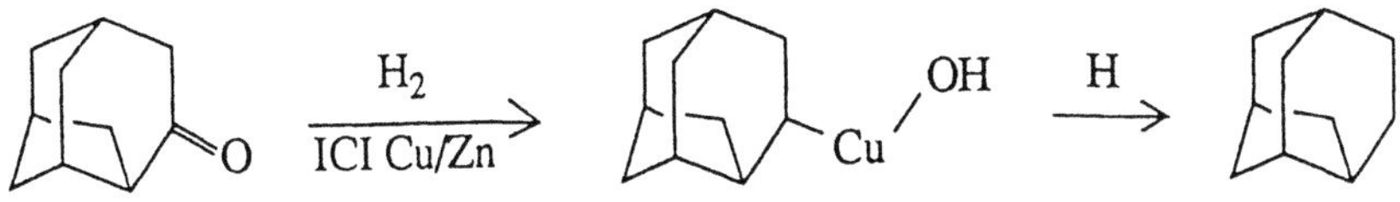

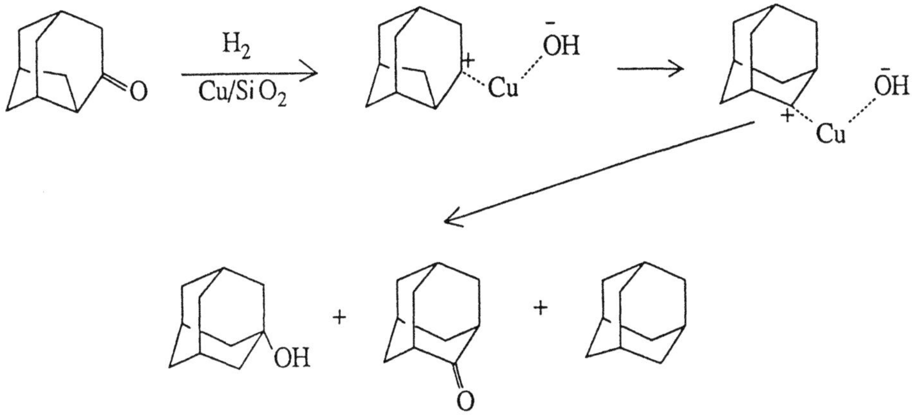

In both cases the major general distinction is the way in which alcohols react with the surface, e.g.

$$ROH + M \longrightarrow R{-}M{-}OH$$

$$ROH + M \longrightarrow \overset{+}{R} \cdots M \cdots \overset{-}{OH}$$

Cationic metal could well be involved in the second reaction which is a heterolytic bond fission, whereas the first reaction is homolytic at zero oxidation state metal atoms.

This type of fine distinction also extends to alkene hydrogenation. Thus norbornadiene hydrogenates on several methanol synthesis catalysts [20] including Cu and Au on ZnO and MgO, and Ag on ZrO_2, but not Ag/MgO which is inert, to give the following products where the surprising one is nortricyclene.

The poorer the Ag/ZrO_2 catalyst as governed by the pretreatment the greater is the relative yield of norticyclene, whereas when very active catalysts, e.g. Pd or Pt on SiO_2, are used this product is never observed. The formation of norticyclene in the homogeneous hydrogenation or norbornadiene has also been noted using $Mo(CO)_6$ in H_2 under the action of light [21]. A dipolar complex where norbornadiene as a ligand reacts with weakly held molecular hydrogen is believed to be responsible.

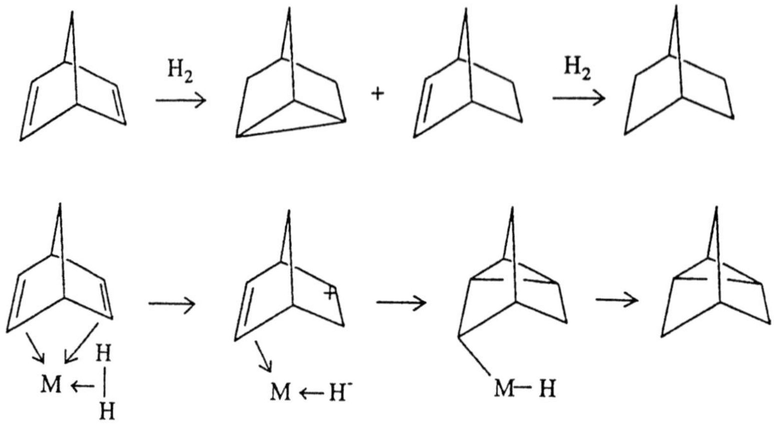

Again there is a distinction with the more polar character manifesting itself with the poorer catalysts.

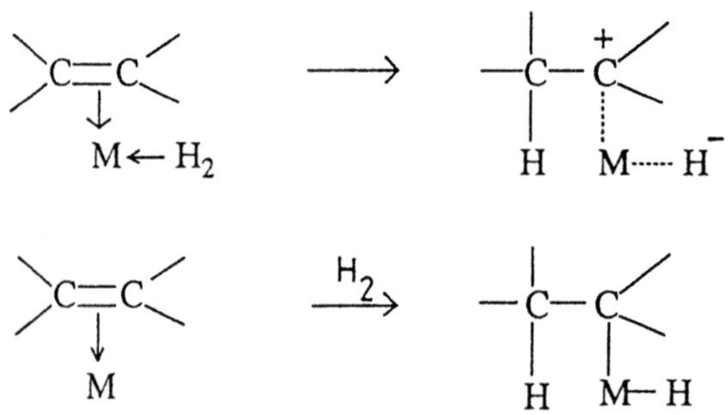

The norbornadiene hydrogenation is therefore proving to be an excellent probe of methanol synthesis catalysts, where Ag/ZrO_2 is also claimed to be active [22] in contrast to Ag/ZnO, so the hydrogenation of CO to methanol probably also involves H_2 and again a dipolar mechanism.

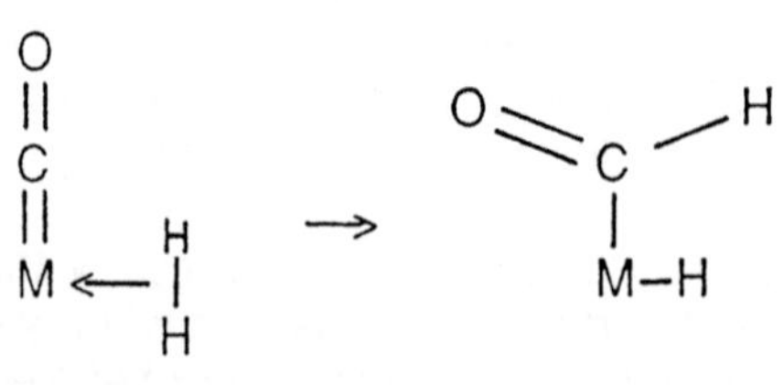

According to this mechanism the metal ions and atoms are very sensitive to electronic effects from the oxide matrix in which they are embedded, e..g. Cu in ZnO, or Ag in ZrO_2. This coordinative complex approach with ligand field effects, etc., is a more chemical way of appreciating junction effects between metal and oxide [23].

Conclusion

If heterogeneous catalysis in the direction of fine chemicals is to be developed less attention and resources should be given to surface physics and more to an understanding and exploitation of the factors involved in both organic and inorganic syntheses. The active sites on catalytic surfaces are much fewer and the energetics governing activities and selectivities much more subtle and delicate than surface science would indicate which is at its best for catalysis of rather brutal bond-fission reactions, e.g. methanation, where even the solid itself is in a quasi-steady-state of reconstitution. The picture developed here is rather that heterogeneous catalysis has much in common with enzymatic catalysis in site densities and real turnover numbers, etc. Attention has not been given here to dynamic and static shape selectivity, both chiral and achiral, which is already an important aspect of all areas of catalysis. It will undoubtedly become increasingly researched in the future with solid and semi-solid catalysts based on combined organic and inorganic polymers and composites.

References

1. Rooney, J.J. and Hathaway, B.J., *J. Catal.*, 3 (1964) 447.

2. Rooney, J.J. and Pink, R.C., *Trans. Faraday Soc.*, 58 (1962) 1632.

3. Garrett, B.R.T. and Rooney, J.J., *unpublished results* (1969).

4. Garrett, B.R.T. and Rooney, J.J., *J. Chem. Soc. Perkin II*, (1974) 960.

5. Hightower, J.W. and Hall, W.K., *Trans. Faraday Soc.*, 66 (1970) 477. For a review see, John, C.S. and Scurrell, M.S., *Specialist Periodicals Reports, Catalysis, Chem. Soc.*, 1 (1977) 136.

6. Somorjai, G.A., *Chem. Soc. Rev.*, 13 (1984) 375.

7. Hegarty, B.F. and Rooney, J.J., *J. Chem. Soc., Faraday Trans. I*, 85 (1989) 1861.

8. Boudart, M. and Loffler, D.G., *J. Phys. Chem.*, 88 (1984) 5763.

9. Clarke, J.K.A., Hegarty, B.F. and Rooney, J.J., *J. Mol. Catal.*, 62 (1990) L39.

10. Clarke, J.K.A. and Rooney, J.J., *Adv. Catal.*, 25 (1976) 125.

11. Amir-Ebrahimi, V. and Rooney, J.J., *J. Mol. Catal.*, 67 (1991) 339.

12. Baird, T., Kelly, E.J., Patterson, W.R. and Rooney, J.J., *J. Chem. Soc., Chem. Commun.*, (1992) 1431.

13. Davis, R.J. and Derouane, E.G., *J. Catal.*, 132 (1991) 269.

14. Gault, F.G., *Adv. Catal.*, 30 (1981) 1.

15. Finlayson, O.E., Clarke, J.J. and Rooney, J.J., *J. Chem. Soc., Faraday Trans. I*, 80 (1984) 191.

16. McKervey, M.A. and Rooney, J.J., Cage Hydrocarbons, Ed., Olah, G.A., John Wiley and Sons (1990) 39.

17. Fessner, W.-D. and Prinzbach, H., Cage Hydrocarbons, Ed., Olah, G.A., John Wiley and Sons (1990) 353.

18. Salem, L., Eisenstein, O., Anh, N.T., Burgi, H.B., Devaguet, A., Segal, G. and Veillard, A., *Nouv. J. Chem.*, 1 (1977) 335.

19. Dale, W.R. and Rooney, J.J., *unpublished results* (1976).

20. Amir-Ebrahimi, V. and Rooney, J.J., *unpublished results* (1992).

21. Hodges, P.M., Jackson, S.A., Jacke, J., Poliakoff, M., Turner, J.J. and Grevels, F.-W., *J. Amer. Chem. Soc.*, 112 (1990) 1234.

22. Baiker, A., Kilo, M., Maciejewski, M. and Wokaun, A., *Paper 086, 10th Int. Congr. Catal.*, Budapest (1992).

23. Frost, J.C., *Nature*, 334 (1988) 577.

ELECTROCHEMICAL ACTIVATION OF CATALYSED REACTIONS

C.G. VAYENAS
Institute of Chemical Engineering &
High Temperature Chemical Processes
University of Patras, Patras GR-26110
Greece

ABSTRACT. The catalytic activity and selectivity of metals can be altered dramatically and reversibly by interfacing the metal with a solid electrolyte such as Y_2O_3-doped-ZrO_2, an O^{2-} conductor, or β''-Al_2O_3, a Na^+ conductor, and by supplying or removing ions to or from the metal catalyst surface via application of an external voltage. The induced change in catalytic rate can be up to $3x10^5$ times higher than the rate of ion supply and up to 70 times higher than the catalytic rate when no voltage is applied. This new effect of Non-Faradaic Electrochemical Modification of Catalytic Activity (NEMCA) appears not to be limited to any particular type of metal or solid electrolyte and has revealed a surprisingly simple exponential relationship between catalytic rates and catalyst work function. The latter can be varied in situ by varying the catalyst potential. This interfacing of catalysis and electrochemistry appears to offer some exciting theoretical and technological possibilities.

1. INTRODUCTION

The use of Electrochemistry to activate Catalytic processes is a new idea which has emerged due to the existence of solid electrolytes. These materials are specific anionic or cationic conductors and, depending on their composition, have useful conductivities at temperatures 200° to 800°C. It thus becomes possible to use electrochemical techniques to affect heterogeneous catalysis on metals interfaced with solid electrolyte components. This has led to the discovery of the effect of Non-Faradaic Electrochemical Modification of Catalytic Activity (NEMCA) (1-22), for which the term "Electrochemical Promotion" has also been proposed (23).

In brief it has been found that the catalytic activity and selectivity of metal films deposited on solid electrolytes can be altered in a dramatic, reversible, and, to some extent, predictable manner by carrying out the catalytic reaction in solid electrolyte cells of the type:

gaseous reactants, metal catalyst |solid electrolyte| metal, O_2

and by applying currents or potentials to the cell, with a concomitant supply or removal of ions, e.g. O^{2-} or Na^+, to or from the catalyst surface.

The experimental setup is shown schematically on Fig. 1. The metal catalyst, usually in the form of a porous metal film 3-20 μm in thickness, is deposited on the surface of a ceramic solid electrolyte (e.g. Y_2O_3-stabilized-ZrO_2, (YSZ) an O^{2-} conductor, or

R. W. Joyner and R. A. van Santen (eds.), Elementary Reaction Steps in Heterogeneous Catalysis, 73–92.

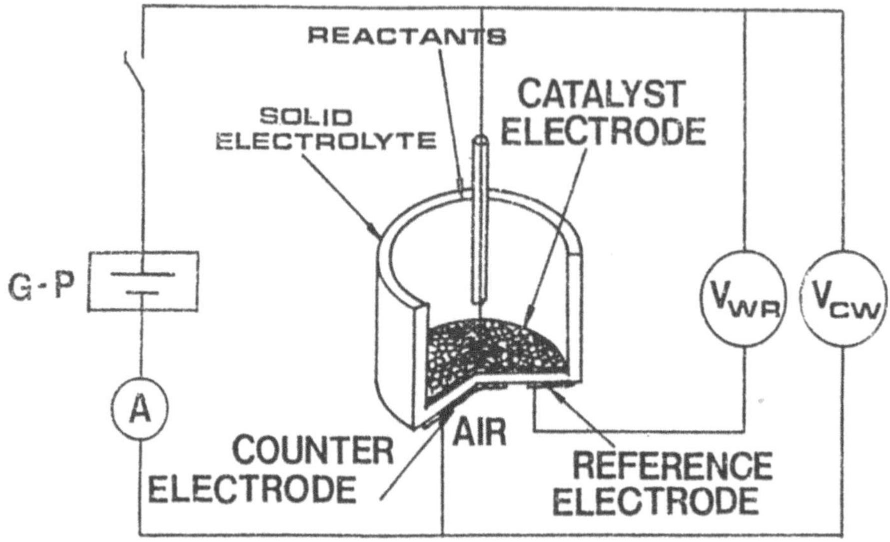

Figure 1. Catalytic reactor configuration for NEMCA studies. G-P : Galvanostat-Potentiostat

β''-Al_2O_3, a Na^+ conductor). Figure 2 shows SEMs of a Ag catalyst (a) and of a section perpendicular to the Ag catalyst-stabilized zirconia interface (b). Catalyst preparation and characterization details have been presented in detail elsewhere (2,5,6).

When a current or voltage is applied between the catalyst and the counter electrode, ions (O^{2-} in the case of doped ZrO_2, Na^+ in the case of β''-Al_2O_3) are supplied from (or to) the solid electrolyte to (or from) the catalyst surface. There is compelling evidence that these ions (together with their compensating charge in the metal thus forming surface dipoles) migrate (spillover) onto the catalyst surface (2,18,19). Thus the solid electrolyte acts as an <u>active catalyst support</u>.

Up to 70-fold steady-state reversible enhancement in catalytic rate has been observed (2,5,24). The steady-state increase in the rate of catalytic reactions can be up to $3x10^5$ times higher than the steady state rate of supply or removal of ions (1,2,5). More than 25 catalytic reactions have been shown already to exhibit the NEMCA effect (1-22). In this paper the main common features of previous NEMCA studies are summarized and the origin of the effect is discussed.

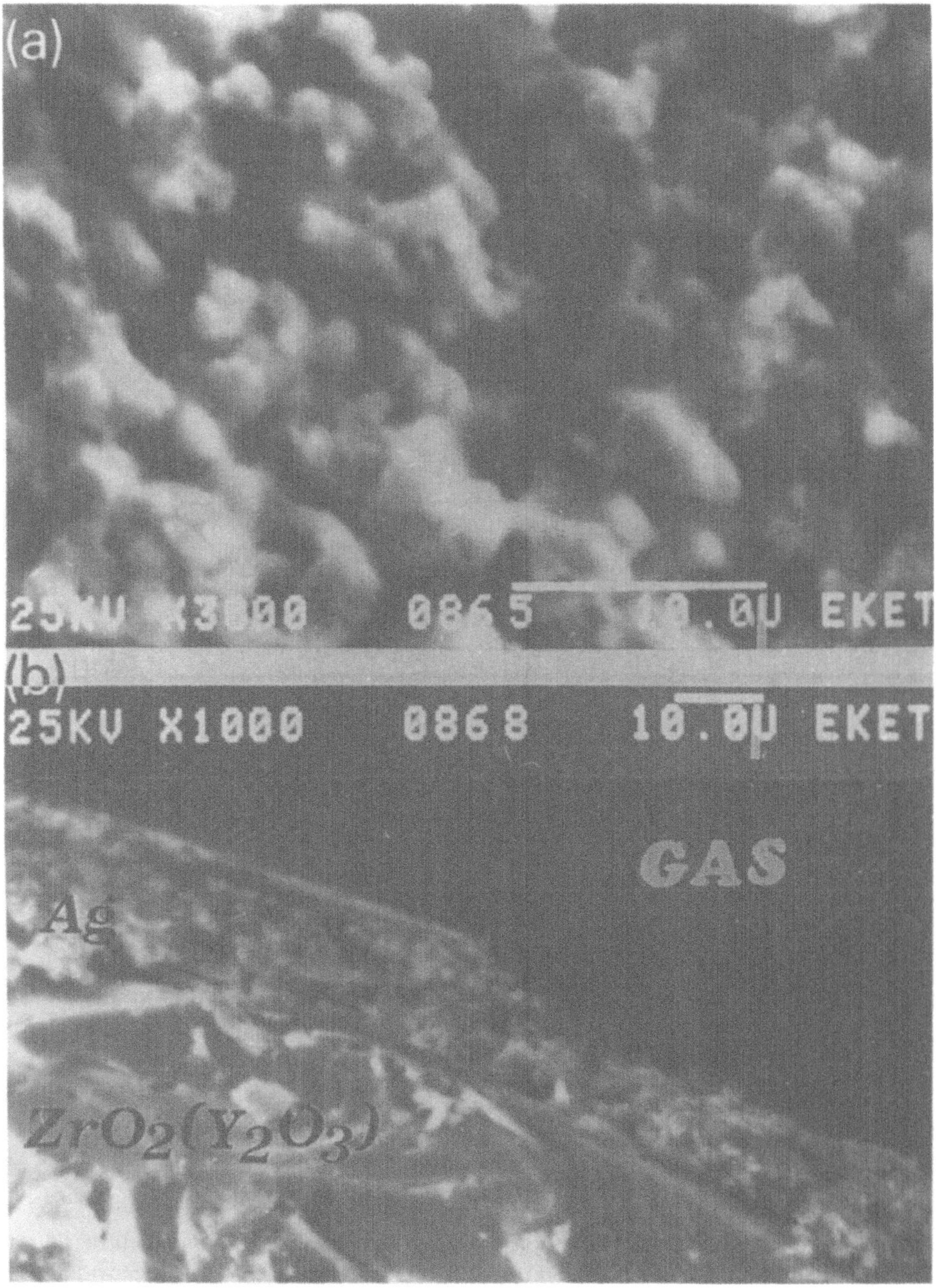

Figure 2. Scanning electron micrographs of a porous Ag catalyst film (a) and of a section perpendicular to the Ag catalyst-stabilized zirconia interface (b).

2. RESULTS AND DISCUSSION

2.1 Transient effect of applied constant current on catalytic rate

Figure 3 shows a typical NEMCA experiment carried out in the setup depicted on Figure 1. The catalytic reaction under study is the complete oxidation of C_2H_4 on Pt (5):

$$CH_2 = CH_2 + 3O_2 \rightarrow 2CO_2 + 2H_2O \tag{1}$$

The figure shows a typical galvanostatic transient, i.e., it depicts the transient effect of a constant applied current on the rate of C_2H_4 oxidation (expressed in g-atom O/s).

The Pt catalyst film with a surface area corresponding to $N=4.2 \cdot 10^{-9}$ g-atom Pt, as measured by surface titration techniques (2,3,5,6), is deposited on Y_2O_3-doped-ZrO_2 and

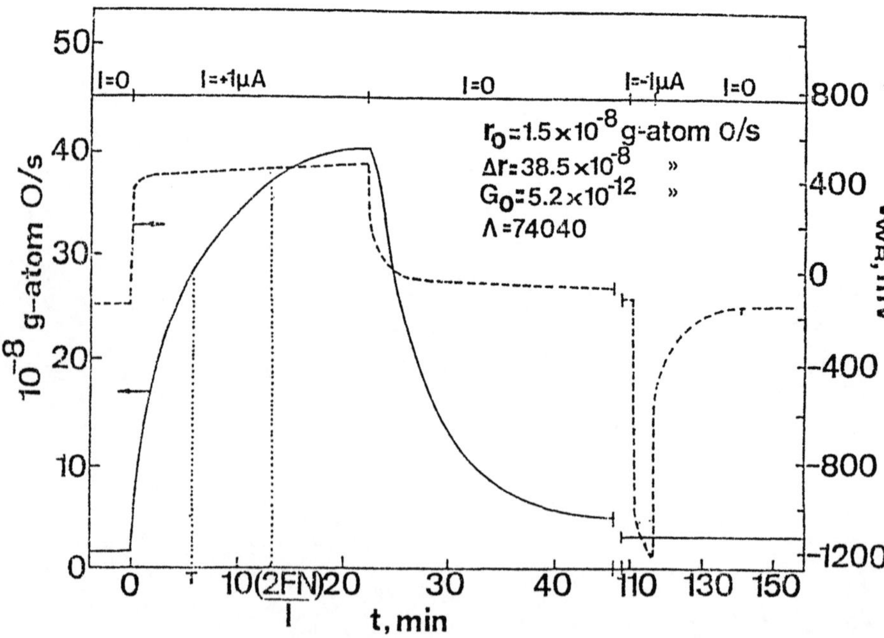

Figure 3. Rate and catalyst potential response to step changes in applied current during C_2H_4 oxidation on Pt ; T=370°C, P_{O_2}=4.6 kPa, $P_{C_2H_4}$=0.36 kPa. The steady-state rate increase Δr is 74,000 times higher than the steady-state rate of supply of O^{2-} to the catalyst (Λ=74,000). The catalytic rate relaxation time constant τ is of the order of 2FN/I.

is exposed to P_{O_2}=4.6 kPa, $P_{C_2H_4}$=0.36 kPa in the CSTR-type flow reactor depicted schematically on Fig. 1. Initially (t<0) the circuit is open (I=0) and the open-circuit catalytic rate r_o is $1.5 \cdot 10^{-8}$ g-atom O/s. The corresponding turnover frequency (TOF), i.e., oxygen atoms reacting per site per s is 3.57 s^{-1}.

Then at t=0 a galvanostat is used to apply a constant current of +1μA between the catalyst and the counter electrode (Fig. 1). Now oxygen ions O^{2-} are supplied to the catalyst-gas-solid electrolyte three-phase-boundaries (tpb) at a rate G_O=I/2F=$5.2 \cdot 10^{-12}$ g-atom O/s. (Current is defined positive when anions are supplied to or cations removed from the catalyst). The catalytic rate starts increasing (Fig. 3) and within 25 min gradually reaches a value r=$40. \cdot 10^{-8}$ g-atom O/s, which is 26 times larger than r_o. The new TOF is 95.2 s^{-1}. The increase in catalytic rate $\Delta r = r - r_o = 38.5 \cdot 10^{-8}$ g-atom O/s is 74,000 times larger than I/2F. This means that each O^{2-} supplied to the Pt catalyst causes at steady-state 74,000 additional chemisorbed oxygen atoms to react with C_2H_4 to form CO_2 and H_2O. This is why this novel effect has been termed Non-Faradaic Electrochemical Modification of Catalytic Activity (NEMCA).

There is an important observation to be made regarding the time required for the rate to approach its steady-state value. Since catalytic rate transients obtained during galvanostatic (i.e. constant current) operation are found in NEMCA studies to be usually, but not always (2), of the type:

$$\Delta r = \Delta r_{max} [1 - \exp(-t/\tau)] \qquad (2)$$

i.e., similar to the response of a first order system with a characteristic time constant τ, one can define the NEMCA time constant τ as the time required for Δr to reach 63% of its maximum, i.e., steady-state value. As shown on Fig. 3, τ is of the order of 2FN/I and this turns out to be a general observation in NEMCA studies utilizing doped ZrO_2, i.e.:

$$\tau \approx 2FN/I \qquad (3)$$

What this observation shows is that NEMCA is a catalytic effect, i.e., it takes place over the entire gas-exposed catalyst surface and is not an electrocatalytic effect localized at the three-phase-boundaries (tpb) metal-solid electrolyte-gas. This is because 2FN/I is the time required to form a monolayer of an oxygen species on a surface with N sites when it is supplied at a rate I/2F. The fact that τ is found to be shorter than I/2F, but of the same order of magnitude, shows that only a fraction of the surface is occupied by oxygen spillover species, as discussed in detail elsewhere (2).

It is worth noting that if NEMCA were restricted to the tpb, i.e., if the observed rate increase were due to an electrocatalytic reaction, then τ would be practically zero during galvanostatic transients.

As shown on Figure 3 NEMCA is reversible, i.e., upon current interruption the catalytic rate returns to its initial value within roughly 100 min. Negative current application has practically no effect on the rate of this particular reaction.

2.2 Effect of gaseous composition on regular (open-circuit) and NEMCA-induced reaction rate

Figure 4 shows the effect of O_2 to C_2H_4 ratio on the regular (open-circuit) steady-state rate of C_2H_4 oxidation on Pt (5) and on the NEMCA-induced rate when the same catalyst film is maintained at a potential of +1V (V_{WR}=+1V) with respect to the reference Pt/air electrode (Fig. 1). It can be seen that the effect is much more pronounced for high $P_{O_2}/P_{C_2H_4}$ ratios, where the NEMCA-induced reaction rate or TOF values is a factor of sixty higher than the corresponding open-circuit value.

A quantitative description and explanation of the NEMCA behaviour of C_2H_4 oxidation on Pt can be found elsewhere (2,5).

2.3 Definitions and the role of the exchange current I_0

The NEMCA effect has been already studied for more than twenty catalytic reactions on Pt, Rh, Pd, Ag, Ni and Au surfaces and using O^{2-}, Na^+ (1,2) and, very recently, F^--conducting solid electrolytes (25). Most of these reactions are listed in Table 1.

In order to compare different catalytic reactions, it is useful to define two dimensionless parameters, i.e., the enhancement factor Λ and the rate enhancement ratio ρ (2,5). The former is defined from :

$$\Lambda = \Delta r/(I/nF) \tag{4}$$

where the change in catalytic rate Δr is expressed in terms of g-atom or g-equivalent of oxygen consumed or produced and n is the ion charge, i.e. 2 for the case of doped ZrO_2 and 1 for β''-Al_2O_3. The current I is taken to be positive when the catalyst is positive with respect to the counter electrode, i.e., when anions are supplied to the catalyst or cations removed from it.

A catalytic reaction is said to exhibit NEMCA when $|\Lambda|>1$. When $\Lambda>1$ as, e.g. in the case of C_2H_4 oxidation on Pt, the reaction is said to exhibit positive or electrophobic NEMCA behaviour. When $\Lambda<-1$ then the reaction is said to exhibit electrophilic behaviour. The terms electrophobic and electrophilic are essentially synonymous with the terms "donor reaction" and "acceptor reaction" introduced by Wolkenstein in his pioneering work (26).

The rate enhancement ratio ρ is defined from :

$$\rho = r/r_o \tag{5}$$

In the C_2H_4 oxidation example presented on Figure 2 and discussed above the Λ and ρ values at steady-state are Λ=74000 and ρ=26. Table 1 shows the range of Λ and ρ values measured in previous NEMCA studies. Figure 5 shows the effect of applied current on the increase Δr in catalytic rate of C_2H_4 oxidation and on Λ which, for this reaction, is up to 3×10^5. Measured $|\Lambda|$ values are typically one to10^5 (Table 1 and Figure 6).

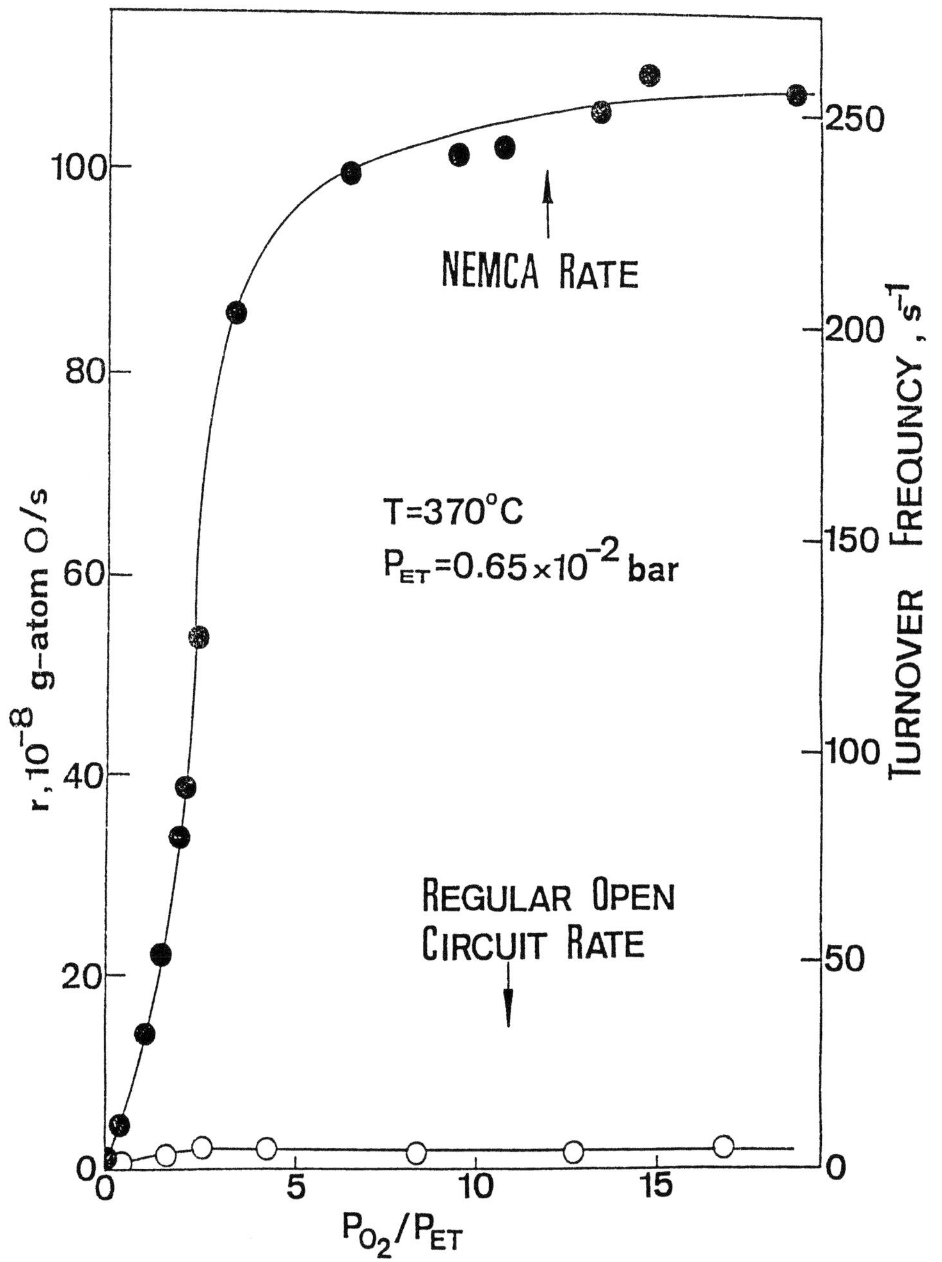

Figure 4. Effect of gaseous composition on the regular (open-circuit) steady-state rate of C_2H_4 oxidation on Pt and on the NEMCA-induced catalytic rate when the same catalyst film is maintained at V_{WR}=1 V (ref. 5).

TABLE 1. Catalytic reactions found to exhibit the NEMCA effect

I. Positive (Electrophobic) NEMCA effect ($\Delta r>0$ with $I>0$*, $\Delta e\Phi>0$)

Reactants	Products	Catalyst	Electrolyte	T [°C]	Λ	$\rho=r/r_0$
C_2H_4, O_2	C_2H_4O, CO_2	Ag	ZrO_2-Y_2O_3	320-470	[0,300]	<3**
C_3H_6, O_2	C_3H_6O, CO_2	Ag	ZrO_2-Y_2O_3	320-420	[0,300]	<2**
C_2H_4, O_2	CO_2	Pt	ZrO_2-Y_2O_3	260-450	$[0,3.10^5]$	<55
C_2H_4, O_2	CO_2	Pt	β''-Al_2O_3	180-300	$[0,5.10^4]$	<4
CO, O_2	CO_2	Pt	ZrO_2-Y_2O_3	300-550	$[0,2.10^3]$	<3
CO, O_2	CO_2	Pt	β''-Al_2O_3	300-400	$[0,10^4]$	12
CO, O_2	CO_2	Pd	ZrO_2-Y_2O_3	400-550	$[0,10^3]$	<1.5
CH_3OH,O_2	H_2CO, CO_2	Pt	ZrO_2-Y_2O_3	300-500	$[0,10^4]$	<4**
CH_4,O_2	CO_2,C_2H_4,C_2H_6	Ag	ZrO_2-Y_2O_3	650-850	[0,5]	<30**
CH_4,O_2	CO_2	Pt	ZrO_2-Y_2O_3	600-750	[0,5]	<70**
CO, O_2	CO_2	Ag	ZrO_2-Y_2O_3	350-450	[0,20]	<5
C_2H_4, O_2	C_2H_4O, CO_2	Ag	β''-Al_2O_3	350-410	[0,3.103]	<4**
CO_2, H_2	CH_4,CO	Rh	ZrO_2-Y_2O_3	390-450	[0,40]	<2**
CH_4, H_2O	CO, CO_2, H_2	Ni	ZrO_2-Y_2O_3	600-900	[0,3]	<2**

II. Negative (Electrophilic) NEMCA effect ($\Delta r>0$ with $I<0$*, $\Delta e\Phi<0$)

Reactants	Products	Catalyst	Electrolyte	T [°C]	Λ	$\rho=r/r_0$
CO, O_2	CO_2	Pt	ZrO_2-Y_2O_3	300-550	[0,-500]	<6
CO, O_2	CO_2	Pt	β''-Al_2O_3	300-400	$[0,-10^4]$	3
CH_3OH, O_2	H_2CO, CO_2	Pt	ZrO_2-Y_2O_3	300-550	$[0,-10^4]$	<15**
CH_3OH	H_2CO, CO, CH_4	Ag	ZrO_2-Y_2O_3	550-750	[0,-25]	<6**
CH_3OH	H_2CO, CO, CH_4	Pt	ZrO_2-Y_2O_3	400-500	[0,-10]	<3**
CH_4, O_2	CO_2	Pt	ZrO_2-Y_2O_3	600-750	[0,-5]	<30
CO, O_2	CO_2	Ag	ZrO_2-Y_2O_3	350-450	[0,-800]	<15

* Current is defined positive when O^{2-} are supplied to or Na^+ removed from the catalyst surface

** Change in product selectivity observed

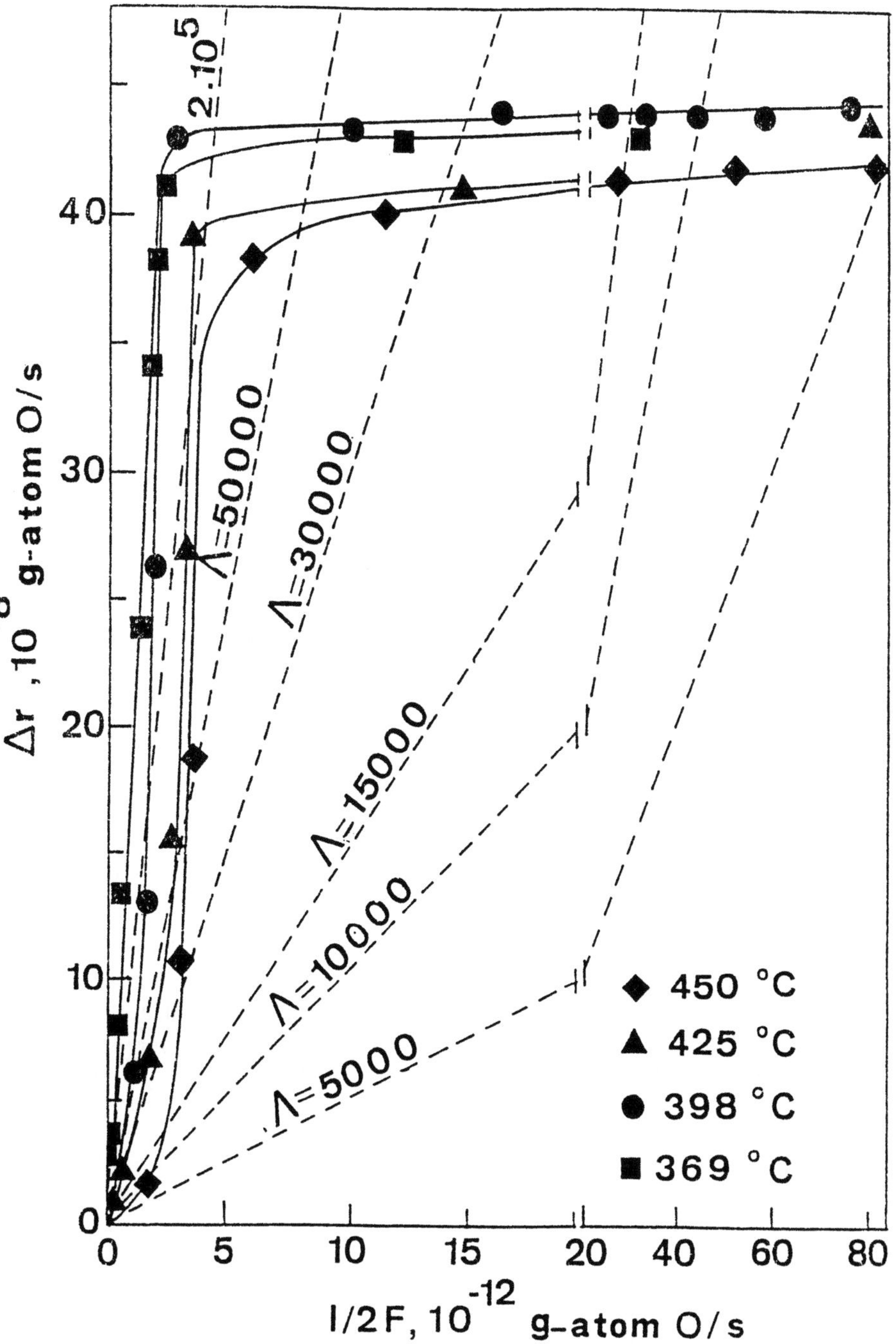

Figure 5. Steady-state effect of applied positive current on the increase in the rate of C_2H_4 oxidation ($P_{C_2H_4}$=0.4 kPa, P_{O_2}=6.4 kPa, ref. 5).

As it turns out experimentally (Figure 6) and can be explained theoretically (2,5) one can estimate the order of magnitude of the absolute value $|\Lambda|$ of the enhancement factor Λ for any given reaction, catalyst and catalyst-solid electrolyte interface from :

$$|\Lambda| \approx 2Fr_o/I_0 \qquad (6)$$

where I_0 is the exchange current of the metal-solid electrolyte interface.

The parameter I_0 can be easily determined from standard ln I vs η (Tafel) plots (2,5). The overpotential η is defined from :

$$\eta = V_{WR} - V^o_{WR} \qquad (7)$$

where V_{WR} is the catalyst (working electrode, W) potential with respect to a reference (R) electrode. The overpotential η is related to current I via the classical Butler-Volmer equation (2,5):

$$(I/I_0) = \exp(\alpha_a F\eta/RT) - \exp(-\alpha_c F\eta/RT) \qquad (8)$$

where α_a and α_c are the anodic and cathodic transfer coefficients, respectively. Thus by measuring η as a function of I one can extract I_0, α_a and α_c. Physically I_0 expresses the (equal) rates of the electronation and deelectronation reaction at the tpb, e.g.:

$$O^{2-} \leftrightharpoons O(a) + 2e^- \qquad (9)$$

where O(a) stands for oxygen adsorbed on the metal catalyst in the vicinity of the tpb. Thus, the exchange current I_0 is a measure of the non-polarizability of the metal-solid electrolyte interface (2).

As shown on Figure 6, equation (6) is in very good agreement with experiment for all catalytic reactions studied so far. The agreement extends for more than five orders of magnitude. Thus, contrary to fuel cell applications where non-polarizable, i.e., high I_0 electrode-electrolyte interfaces are desirable to minimize activation overpotential losses, exactly the opposite is true for NEMCA applications, i.e., I_0 must be low in order to obtain high Λ values, i.e., a strong Non-Faradaic rate enhancement.

Although Λ is an important parameter for determining whether a reaction exhibits NEMCA, it is not a fundamental one. The reason is that for the same catalytic reaction on the same catalyst material one can obtain significantly different $|\Lambda|$ values by varying I_0 (Equation 6). The parameter I_0 is proportional to the tpb length (2,27) and can be controlled during catalyst film preparation by varying the sintering temperature and thus metal crystallite size and tpb length (2,27).

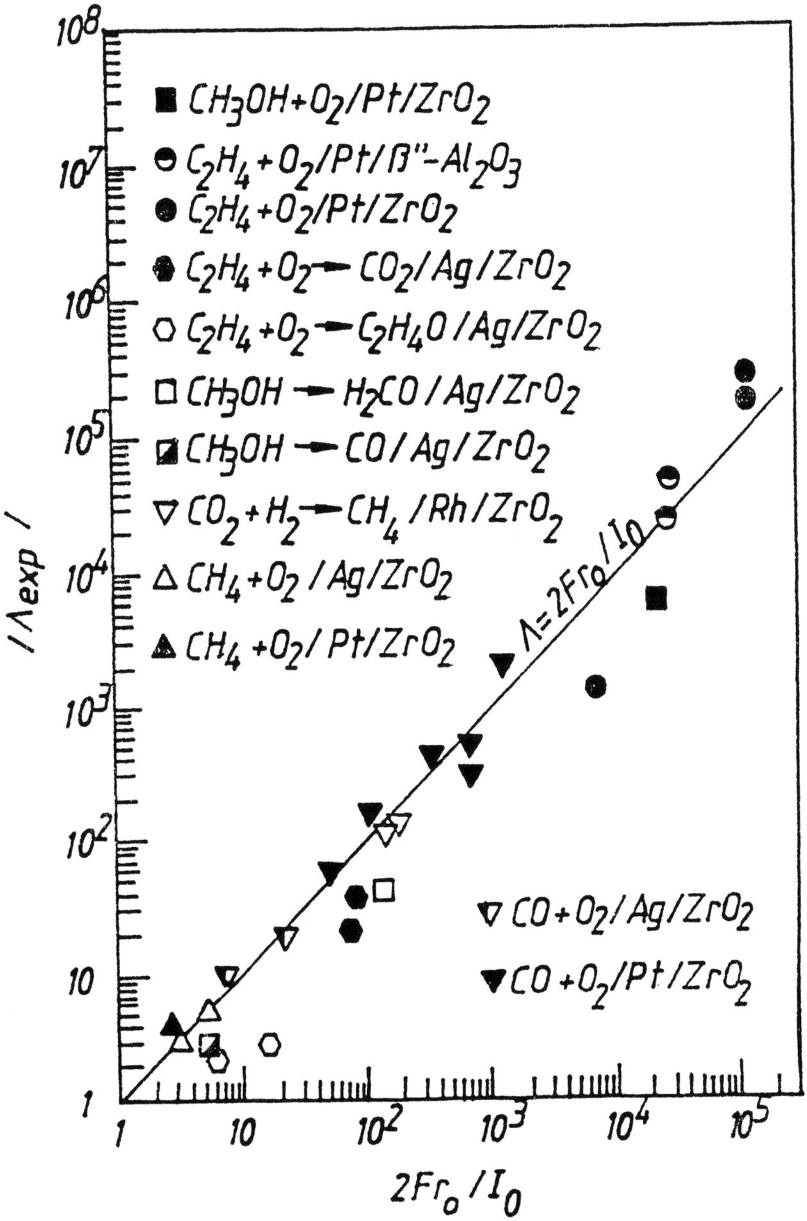

Figure 6. Comparison of predicted and measured enhancement factor Λ values for the catalytic reactions found to exhibit the NEMCA effect.

2.4. The work function of catalyst films interfaced with solid electrolytes

One of the key steps in understanding the origin of NEMCA was the realization that solid electrolyte cells can be used both to monitor and to control the work function of the gas-exposed surfaces of their electrodes (1,11). It was shown both theoretically (2,5) and experimentally (1,11) that:

$$eV^{o}_{WR} = e\Phi_W - e\Phi_R \quad (10)$$

and

$$e\Delta V_{WR} = \Delta\, e\Phi_W \quad (11)$$

where $e\Phi_W$ is the catalyst surface work function and $e\Phi_R$ is the work function of the reference electrode surface. The reference electrode must be of the same material as the catalyst for Equation (10) to hold, but Equation (11) is not subject to this restriction.

The validity of Equations (10) and (11) has been demonstrated by using a Kelvin probe to measure in situ $e\Phi$ on catalyst surfaces subject to electrochemical promotion (Figure 7).

Therefore by applying currents or potentials in NEMCA experiments and by thus varying V_{WR}, one is also varying the average catalyst surface work function $e\Phi$ (Eq. 11). Positive currents increase $e\Phi$ and negative currents decrease it. Physically the variation in $e\Phi$ is due to spillover of ions to or from the catalyst surface.

In view of Equation (11) it follows that NEMCA experiments permit to directly examine the effect of catalyst work function $e\Phi$ on catalytic rates.

2.5. Dependence of Catalytic rates and activation energies on catalyst work function $e\Phi$

From a fundamental viewpoint the most interesting finding of all previous NEMCA studies is that over wide ranges of catalyst work function $e\Phi$ catalytic rates depend exponentially on $e\Phi$ and catalytic activation energies vary linearly with $e\Phi$ (1,2).

A typical example is shown on Fig. 8 for the catalytic oxidation of C_2H_4 and of CH_4 on Pt. Both reactions exhibit electrophobic behaviour which is due to the weakening of the Pt=0 chemisorptive bond with increasing $e\Phi$ (2,5). Chemisorbed atomic oxygen is an electron acceptor, thus increasing $e\Phi$ causes a weakening in the Pt=0 bond, cleavage of which is involved in the rate-limiting-step of the catalytic oxidation, and thus a linear decrease in activation energy and an exponential increase in catalytic rate.

2.6. Selectivity modification

One of the most promising applications of NEMCA is in product selectivity modification. An example is shown on Fig. 9 for the case of CH_3OH oxidation on Pt. The selectivity to H_2CO can be varied between 35% and 60% by varying V_{WR}, and thus $e\Phi$. The results of Fig. 7 are plotted vs the dimensionless catalyst potential Π, defined from :

$$\Pi = FV_{WR}/RT \quad (12)$$

which, in many cases (2,6) is found to unify results obtained at different temperatures.

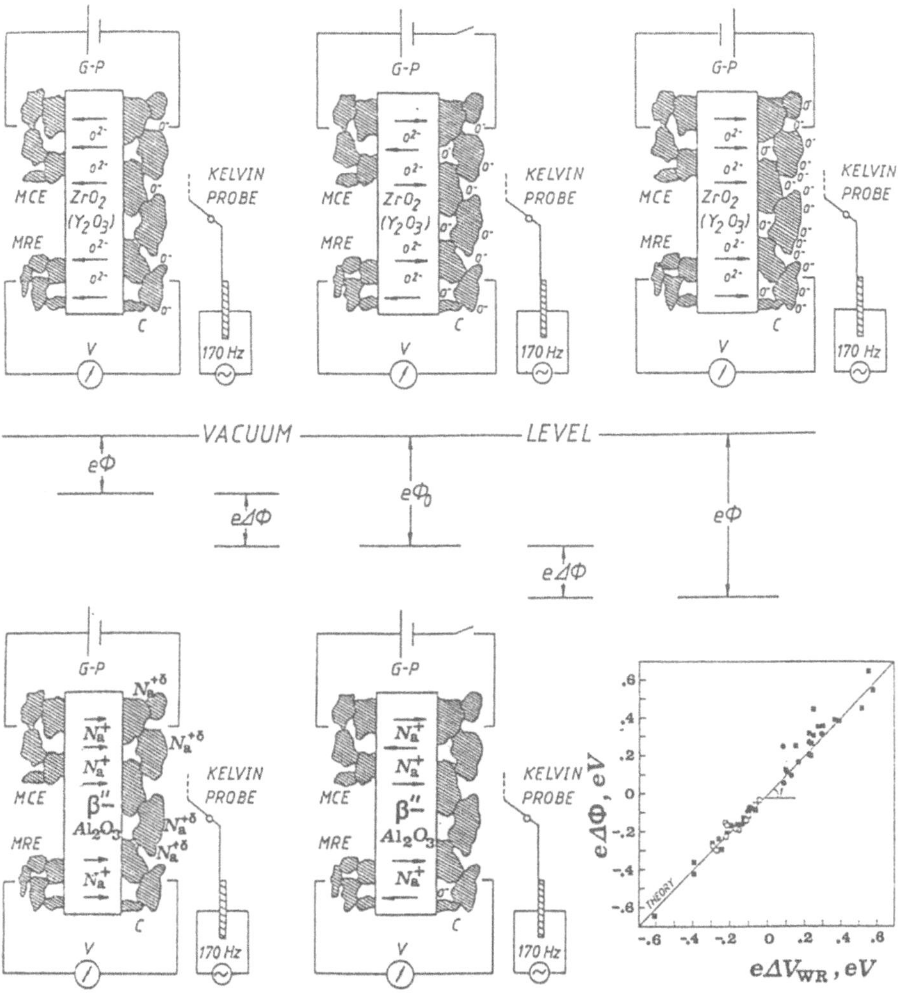

Figure 7. Principle of NEMCA : A metal counter electrode (MCE) in conjunction with a galvanostat (G) is used to supply or remove ions (O^{2-} for the doped ZrO_2 (a), Na^+ for β''-Al_2O_3 (b)) to or from the polarizable solid-electrolyte/catalyst (or working electrode, W) interface. The electrochemically induced spillover of ions (O^- in (a), $Na^{+\delta}$ in (b)) produced or consumed at the boundaries between the three phases solid electrolyte/catalyst/gas causes an increase (right) or decrease (left) in the catalyst work function $e\Phi$. In all cases $\Delta e\Phi = e\Delta V_{WR}$ (c), where ΔV_{WR} is the overpotential measured between the catalyst and the metal reference electrode (MRE). The insert Figure shows the comparison of $e\Delta V_{WR}$ and $\Delta e\Phi$ measured via a Kelvin probe; V_{WR} and $e\Phi$ were varied by varying the gaseous composition (open symbols) or the polarizing current (closed symbols). Pt catalyst; squares : YSZ support, circles : β''-Al_2O_3 support.

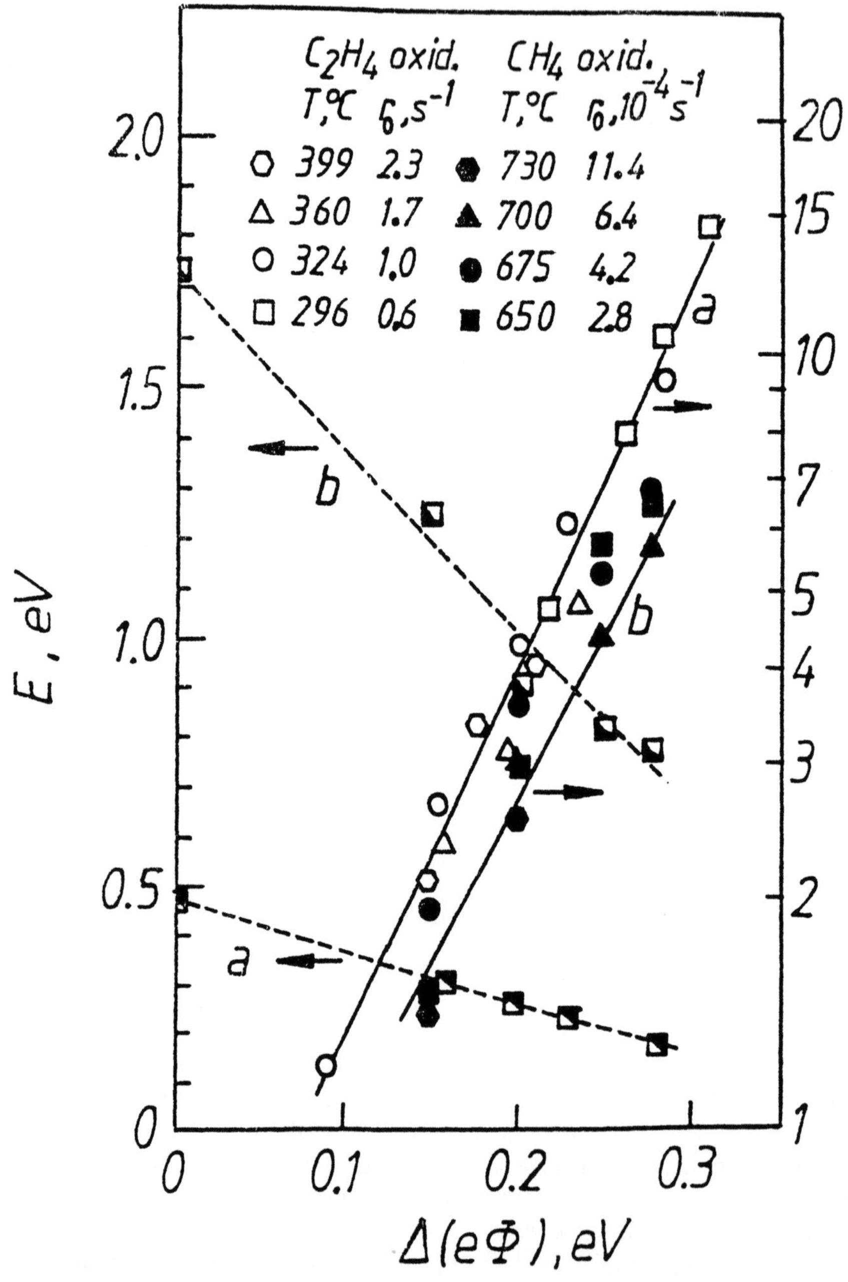

Figure 8. Effect of catalyst work function eΦ on the activation energy E and catalytic rate enhancement ratio r/r_o for C_2H_4 oxidation on Pt (a) and CH_4 oxidation on Pt(b).

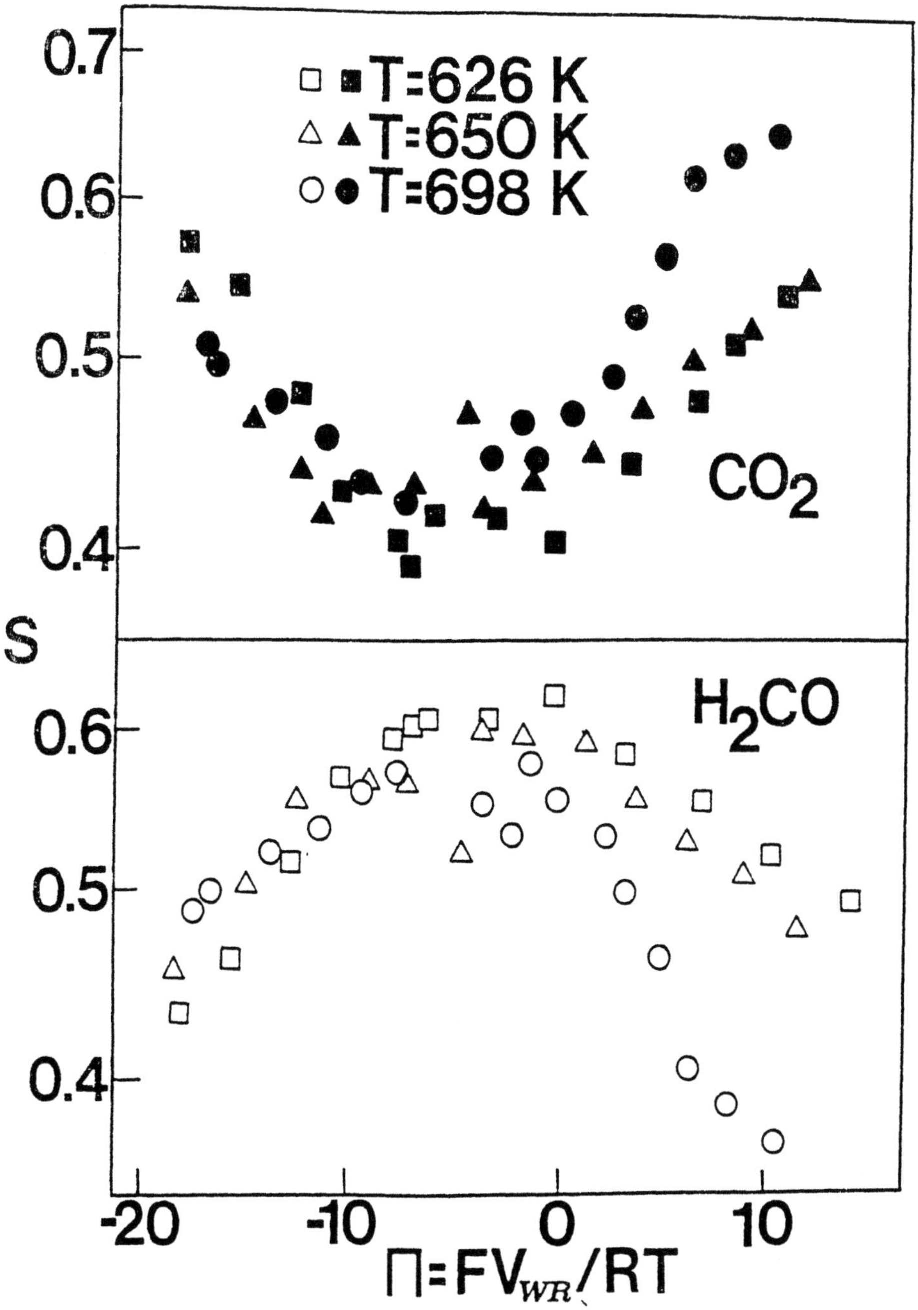

Figure 9. Effect of dimensionless catalyst potential $\Pi = FV_{WR}/RT$ on the selectivity to CO_2 and H_2CO during CH_3OH oxidation on Pt (ref. 6). Inlet conditions $P_{CH_3OH}=0.9$ kPa, $P_{O_2}=19$ kPa.

2.7. Induction and modification of oscillatory states

NEMCA can be used to induce, stop or modify and control the frequency of catalytic rate oscillations. An example is shown on Figure 10 for the case of CO oxidation. Negative currents cause a decrease in the frequency of oscillations, while I>0 leads to a frequency increase (2,3).

2.8. NEMCA effect on chemisorption

It has been recently confirmed that the strength of chemisorptive bonds is indeed varied under NEMCA conditions (2,18,19). To this end the chemisorption of oxygen on Ag was investigated both under open-circuit conditions and during application of positive and negative potentials. The key results are shown on Figure 11 which depicts the kinetics of dissociative oxygen adsorption and desorption under open-circuit conditions (I=0) and upon application of a negative current (I=-110 μA, i.e., removal of O^{2-} from the catalyst and thus decreasing eΦ).

As shown on Figure 9 decreasing eΦ causes a five-fold increase in the rate of atomic oxygen adsorption, or sticking coefficient, a six-fold decrease in the apparent desorption rate constant and a two-fold increase in the equilibrium atomic oxygen uptake. All these observations show a significant strengthening of the Ag-atomic oxygen bond. This is because decreasing eΦ is expected to cause an increase in the binding energy of electron-acceptor adsorbates, such as chemisorbed atomic oxygen.

2.9. Spectroscopic evidence for spillover ions as the cause of NEMCA

There has been so far only one in situ XPS investigation of catalyst surfaces subject to NEMCA conditions (28,29). The main results are shown on Figure 12 and provide direct evidence for the creation of spillover oxide ions on a Ag catalyst surface interfaced with YSZ and exposed to vacuum at 400°C (28,29).

Upon positive current application, chemisorbed atomic oxygen is immediately formed (O 1s at 532.6 eV) followed by the gradual evolution of anionic oxygen (O 1s at 529.2 eV) which eventually causes a decrease in the amount of atomic oxygen due to the weakening of the Ag=0 bond with increasing catalyst eΦ. This transient behaviour is in good qualitative agreement with rate transients during ethylene epoxidation on Ag under similar temperature and imposed current conditions at atmospheric pressure (19).

3. CONCLUSIONS

Solid electrolytes can be used as active catalyst supports to reversibly promote catalyst surfaces. The promoting effect of solid electrolytes is due to an electrochemically driven and controlled spillover of ions on the catalyst surface.

The study of NEMCA has revealed that over wide ranges of work function eΦ, catalytic rates depend exponentially on catalyst work function. Potential technological applications of NEMCA may emerge as more reactions are studied.

Acknowledgement

Sincere thanks are expressed to the EEC Non-nuclear Energy and JOULE programmes for financial support and to the Alexander von Humboldt Foundation of Germany for a fellowship.

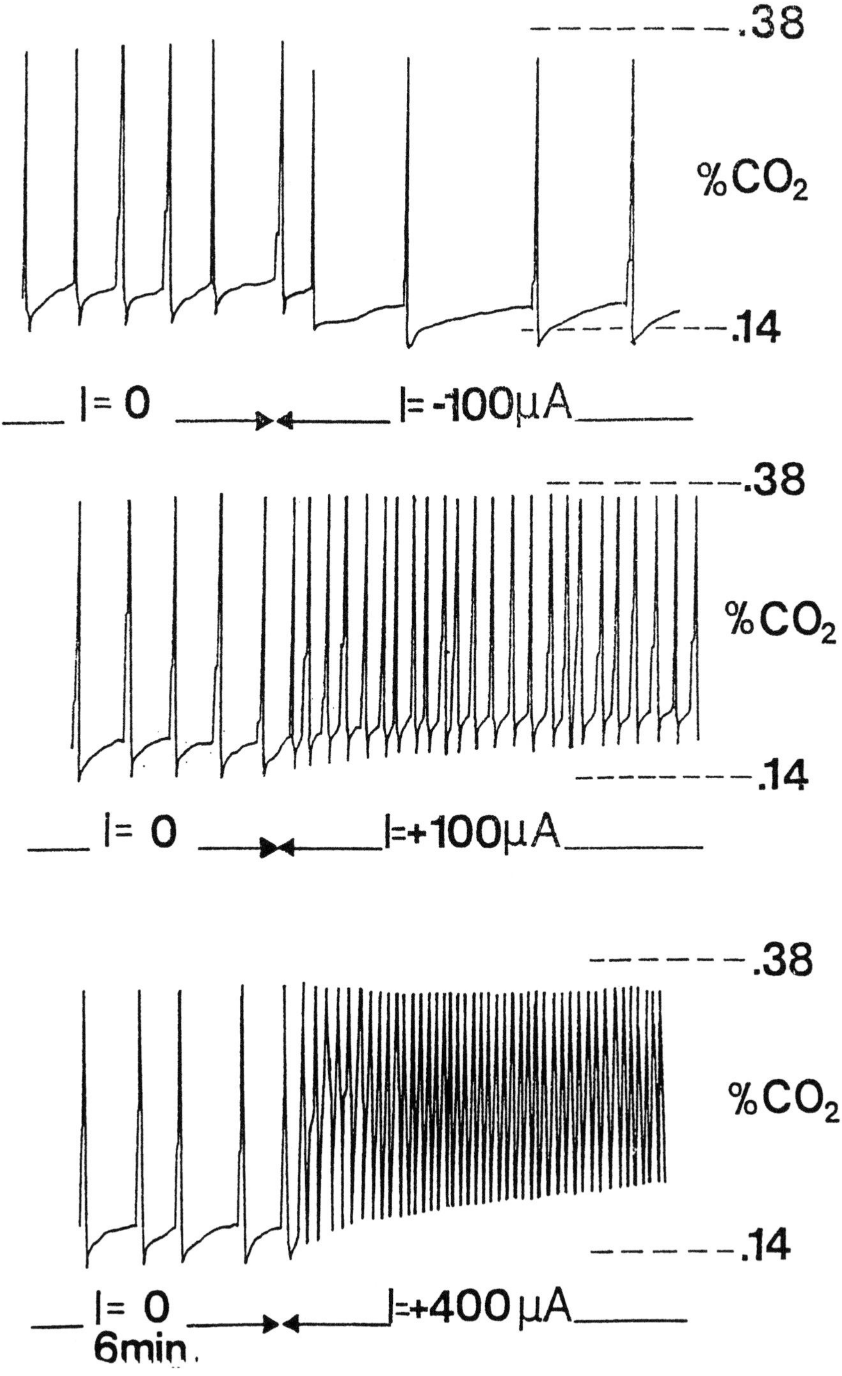

Figure 10. Effect of applied constant current on the frequency of the self-sustained rate oscillations during CO oxidation on Pt (ref. 3). Inlet composition : P_{CO}=0.47 kPa, P_{O_2}=16 kPa, T=332°C.

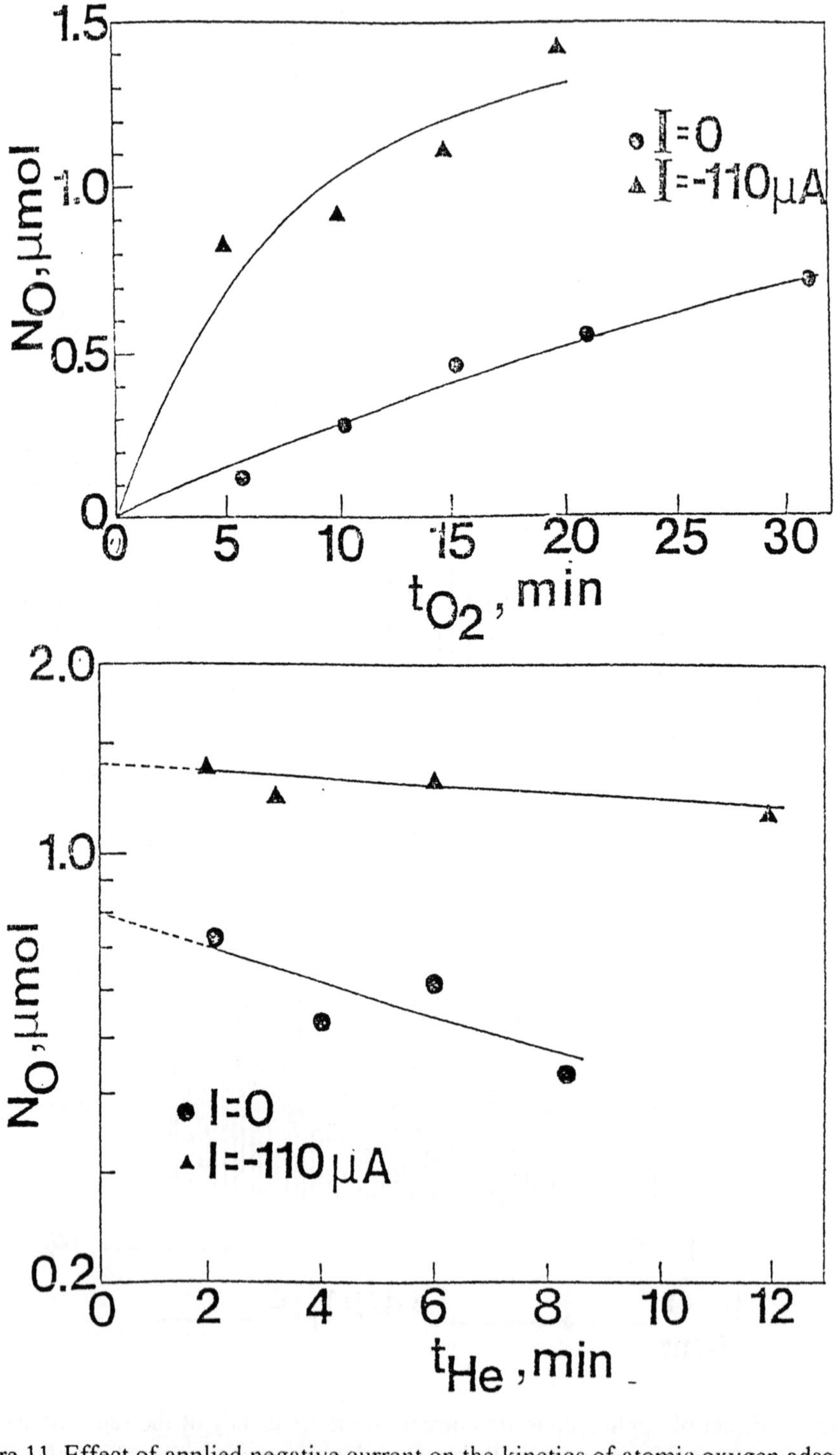

Figure 11. Effect of applied negative current on the kinetics of atomic oxygen adsorption (top) and desorption (bottom) T=435°C (ref. 18).

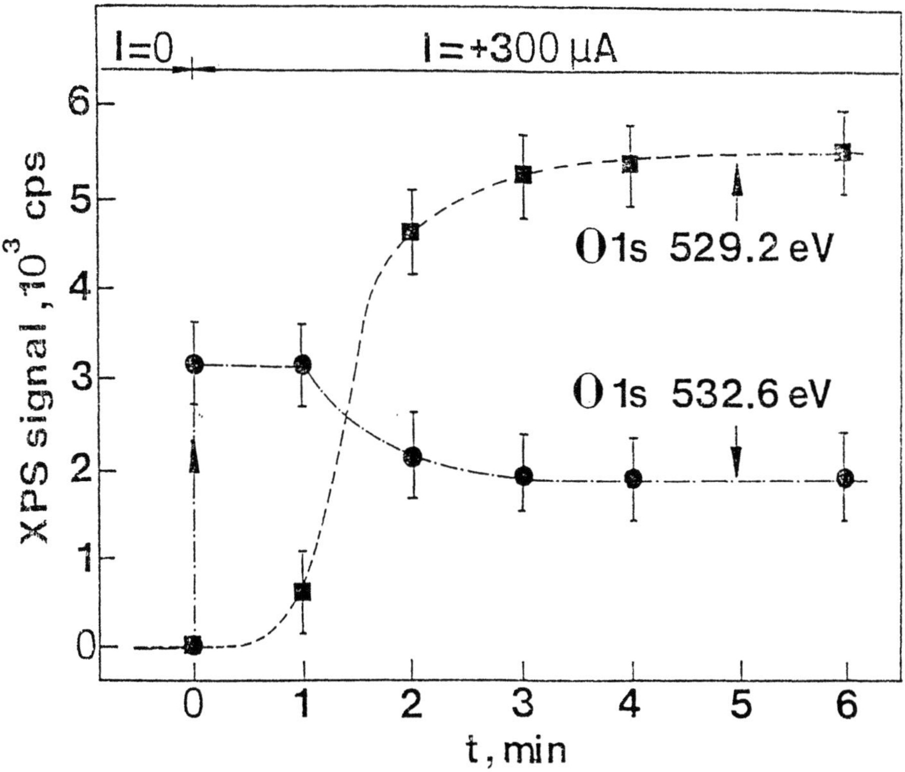

Figure 12. XPS O 1s signal response (peak height) to a step change in applied positive current. The peak at 532.6 eV is chemisorbed atomic oxygen and the peak at 529.2 eV is ionically bonded oxygen; Catalyst : Ag/YSZ; based on the XPS spectrum of Fig. 3 of Ref. 29 at 400°C; see text for discussion.

REFERENCES

1. Vayenas, C.G., Bebelis, S. and Ladas, S., Nature (London) 343 (6259), 625 (1990)
2. Vayenas, C.G., Bebelis, S., Yentekakis, I.V. and Lintz, H.-G., "Non-Faradaic Electrochemical Modification of Catalytic Activity: A status report", Catalysis Today, Elsevier, Amsterdam, 11(3), 303-442 (1992)
3. Yentekakis, I.V. and Vayenas, C.G., J. Catal. 111, 170 (1988)
4. Vayenas, C.G., Bebelis, S. and Neophytides, S., J. Phys. Chem. 92, 5083 (1988)
5. Bebelis, S. and Vayenas, C.G., J. Catal. 118, 125 (1989)
6. Neophytides, S. and Vayenas, C.G., J. Catal. 118, 147 (1989)
7. Vayenas, C.G., Bebelis, S., Neophytides, S. and Yentekakis, I.V., Appl. Phys. A 49, 95 (1989)

8. Vayenas, C.G., Bebelis, S., Yentekakis, I.V., Tsiakaras, P. and Karasali, H., Platinum Metals Rev. 34(3), 122 (1990)
9. Vayenas, C.G. and Neophytides, S., J. Catal. 127, 645 (1991)
10. Vayenas, C.G., Bebelis, S. and Despotopoulou, M., J. Catal. 128, 415 (1991)
11. Ladas, S., Bebelis, S. and Vayenas, C.G., Surf. Sci. 251/252, 1062 (1991)
12. Lintz, H-G. and Vayenas, C.G., Angew. Chem. 101, 725 (1989); Angew. Chem. Intl. Engl. Edn. 28, 708 (1989)
13. Vayenas, C.G., Bebelis, S. and Neophytides, S., in "New Developments in Selective Oxidation", G. Centi and P. Trifiro, Eds., Studies in Surface Science and Catalysis 55, pp. 643-652, Elsevier, Amsterdam, 1990
14. Vayenas, C.G., Bebelis, S., Yentekakis, I.V., Tsiakaras, P., Karasali, H. and Karavasilis, Ch., ISSI Lett. 2, 5 (1991)
15. Vayenas, C.G., Bebelis, S. and Kyriazis, C., Chemtech 21, 500 (1991)
16. Bebelis, S., Karavasilis, Ch., Karasali, H., Tsiakaras, P., Yentekakis, I.V. and Vayenas, C.G., in Proc. 2nd Intnl. Symp. on Solid Oxide Fuel Cells, Athens, Greece, pp. 353-360, EEC Publ., Luxembourg (1991)
17. Vayenas, C.G., Bebelis, S., Yentekakis, I.V., Tsiakaras, P., Karasali, H. and Karavasilis, Ch. in "Proc. 3d Intnl. Symp. on Systems with Fast Ionic Transport", Holzhau, Germany, 1991, Materials Science Forum 76, 141 (1991); P. Tsiakaras and C.G. Vayenas, ibid 76, 179 (1991); H. Karasali and C.G. Vayenas, ibid 76, 171 (1991); Ch. Karavasilis, S. Bebelis and C.G. Vayenas, ibid 76, 175 (1991); S. Bebelis and C.G. Vayenas, ibid 76, 221 (1991)
18. Bebelis, S. and Vayenas, C.G., J. Catal., 138, 000 (1992)
19. Bebelis, S. and Vayenas, C.G., J. Catal., 138, 000 (1992)
20. Eng, D. and Stoukides, M., J. Catal. 130, 306 (1991)
21. Alqahtany, H., Chiang, P., Eng, D. and Stoukides, M. Catal. Letters, in press (1992)
22. Mar'ina O.A. and Sobyanin, V.A., Catalysis Letters 13, 61 (1992)
23. Pritchard, J., Nature (London) 343 (6259), 592 (1990)
24. Tsiakaras, P. and Vayenas, C.G., J. Catalysis, in press (1992)
25. Yentekakis, I.V. and Vayenas, C.G., in preparation
26. Wolkenstein, Th., "The electronic Theory of Catalysis on Semiconductors", Pergamon, Oxford (1963)
27. Vayenas, C.G., Ioannides, A. and Bebelis, S., J. Catal. 129, 67 (1991)
28. Arakawa, T., Saito, A. and Shiokawa, J., Chem. Phys. Lett. 94, 250 (1983)
29. Arakawa, T., Saito, A. and Shiokawa, J., Appl. Surf. Sci. 16, 365 (1983)

SELECTIVE OXIDATION PATHWAYS AT OXIDE SURFACES: THE TRANSFORMATION OF ALKANES ON VANADYL PYROPHOSPHATE

G. CENTI
Dept. of Industrial Chemistry and Materials
V.le Risorgimento 4
40136 Bologna
Italy

ABSTRACT. The main aspects of the surface reactivity and reaction mechanisms in alkane selective oxidation on the vanadyl pyrophosphate active phase are discussed in relation to the results reported in the literature on the nature of active sites and mechanisms of reaction in alkene selective oxidation, in order to evidence which new aspects about the elementary steps, selective oxidation pathways and surface characteristics of oxide catalysts derive from the study of the surface structure, reaction dynamics and mechanism of reaction in C_3-C_5 alkane oxidation on $(VO)_2P_2O_7$. Particular attention is focused on the following topics: influence on the oxidation mechanism of the inhibition of acid sites, presence of competitive adsorption phenomena and role of the surface density of oxygen active sites in relation to the selective and unselective oxidation, relationship between number of available oxygen species in a surface limited environment and pathways of oxidation for different alkanes, nature of the surface changes during the reaction induced by the strong adsorption of some intermediates and dependence of the pathway of transformation on the relative rates of H- abstraction and O-insertion. In conclusion, the possible mechanism of activation and selective oxidation of alkanes on $(VO)_2P_2O_7$ is discussed.

1. Introduction

Selective oxidation reactions at solid surface are an important class of commercially applied transformations of hydrocarbons and today catalytic oxidation is the basis of the production of almost all monomers used in the manufacture of synthetic fibers, plastics and many other products. With the noticeable exception of silver for ethylene oxide production, almost all catalysts used in these reactions are based on transition metal mixed oxides or oxysalts. The combination of their acid-base and redox (with respect both to electron and atom transfer) surface properties determines the pathway for the selective transformation [1,2]. For example, depending on the choice of the oxide catalyst, it is possible to obtain acrolein, acrylic acid, acetone, propylene oxide, 1,5-hexadiene or benzene from propylene. The ultimate formation of carbon dioxide and water (the thermodynamically favoured products) always competes with these reactions. Therefore, all products of partial oxidation are derived by kinetic control of the reaction, a difficult task to achieve taking into consideration that usually the C-H bonds in the initial reactant are stronger than those in the intermediate products. For example, the C-H bond in butane is 98 kcal/mol and is 78 kcal/mol in butene, which makes the intermediates much more sensitive to further transformation than the starting hydrocarbon.

Two different problems, therefore, must be considered in the analysis of the selective

R. W. Joyner and R. A. van Santen (eds.), Elementary Reaction Steps in Heterogeneous Catalysis, 93–112.

pathways of oxidation at oxide surfaces: *i)* how to orient the surface reactivity towards the desired intermediate/product and *ii)* how to control the process of surface transformation selectively in order to stop the reaction to one intermediate/product avoiding overoxidation up to other molecules or carbon oxides. There is, however, an additional aspect which must be considered [3]. Selective oxidation surface transformations are usually complex multielectron, multistep and polyfunctional reactions [2], but, in addition, multiple pathways of transformation are possible. The relative rates in these possible pathways depend not only on the presence or not of some sites, but also can depend on the dynamics of surface transformation, eg. the lifetime and surface mobility of adsorbed species, the redox dynamics of the oxide transformation, the evolution of the surface characteristics during the catalytic reaction, the presence of competitive adsorption phenomena, the population and surface distribution of adsorbed species, and the *architecture* of the active surface. Some of these aspects are discussed in this contribution with special reference to the vanadyl pyrophosphate catalyst [4], an active phase with unique properties in the selective oxidation of the n-butane and n-pentane alkanes, which due to this peculiar reactivity, offers various opportunities for investigation of new aspects of the surface mechanisms in the selective oxidation of hydrocarbons with respect to the well known case of alkene selective transformation [1,5]. However, before analyzing which new suggestions regarding the relationship between surface structure, active oxygen species and selective pathways of reaction can derive from the study of the mechanisms of alkane oxidation on vanadyl pyrophosphate, it is useful to summarize the main concepts developed in the investigation of alkene selective oxidation on mixed oxides, because the reaction mechanism has been studied in detail and thus it is a good reference reaction.

1.1 ALKENE OXIDATION

When an alkene interacts with the surface of an oxide, weak hydrogen bonds may be initially formed with the surface Brønsted sites, and depending on the acid strength of these sites, their protons may form hydrogen bonds with the π-bonds of the alkene or can be transferred to form a carbocation. The π-bond of the alkene can also interact with surface Lewis acid sites due to coordinatively unsaturated transition metal cations to form a surface π-complex. Depending on the basic character of nearlying structural oxygens, a nucleophilic attack on the α-hydrogen atoms may occur with formation of an allylic species bonded to the transition metal ion either side-on (π- allyl), or end-on (σ-allyl) [5]. An equilibrium between these two forms exists on the surface, but these two surface complexes may evolve along different pathways. The latter species is an intermediate in the oxidative coupling to form products with a higher C atom number. The former may undergo a nucleophilic attack from surface oxygen atoms to give an aldehyde or ketone depending on the attack to a primary or secondary carbon atom. For example, only dimerization of propylene to hexadiene occurs on Bi_2O_3, whereas acrolein is formed on Bi-molybdates. The selective transformation of an alkene thus requires defined acid-base and redox properties and an optimal arrangement of the various functionalities necessary to perform the entire transformation. On Bi-molybdates, for example, the active site may be viewed as a multinuclear site containing both Bi and Mo cations (*ensemble site*). The bridging oxygen between Bi and Mo has been suggested as being responsible for α-H abstraction from propylene, Mo as being responsible for chemisorption of the alkene through the π-bond and formation of the π-allyl complex and responsible for the electron transfer, the terminal oxygen of molybdenum as being responsible for oxygen insertion and the Bi atoms for O_2 reduction and dissociative chemisorption [5]. However, the selective behavior does not depend only on the presence of a definite surface structure of the active site. In fact, the insertion of structural oxygen in the hydrocarbon results in a surface oxygen vacancy. In the presence of gaseous oxygen, a series of equilibria may be established between the point defect and O_2, with oxygen gradually enriched with electrons until the state of O^{2-} is reached, which may be considered as structural oxygen of the surface layer of the oxide. In this process of incorporation, adsorbed oxygen species formed in the first steps, such as $(O_2)^-$ and O^-, possess different reactivity characteristics than structural oxygen and, in particular, it has been suggested that their electrophilic nature is responsible for the attack to the region of higher electron density (C=C double bond) to form peroxo or epoxy- complexes that are intermediates in the degradation of the carbon skeleton [1,6] and formation of carbon oxides. If a site of activation of oxygen is present near to a chemisorbed alkene and the electrophilic oxygen species formed have time to react with it before being incorporated into the surface oxide layer, a decrease in the selectivity is expected. On the other hand, surface oxygen vacancies may be compensated for also via a change in linkage between the coordination polyhedra, and in this case, reoxidation may also occur through bulk oxygen diffusion with oxygen species generated far from the site responsible for the selective transformation of the alkene. For this reason components with enhanced redox properties such as iron or cerium are introduced into the oxide matrix (Bi-molybdate) in order to permit rapid reconstruction of the reduced sites and maintain the fully oxidized active and selective sites [7 and references therein]. The rate determining step in propylene oxidation is the α-H abstraction [5,7], but the formation of the σ-allyl complex and the rate of reoxidation of the reduced Mo sites determine the selectivity and therefore a fully oxidized surface is needed.

In other cases, however, a different situation may be required which depends also on the type of molecules to be oxidized. $Fe_2(MoO_4)_3$ is a well known very selective catalyst in methanol oxidation to formaldehyde and an excess of MoO_3 is usually necessary in order to avoid the excessive reduction of the catalyst that lowers the selectivity [8]. However, when 1-butene oxidation on this catalyst is studied, selectivities to products of partial oxidation (butadiene and maleic anhydride) are obtained only when the gas-phase oxygen partial pressure is decreased and a surface reduction to β-$FeMoO_4$ occurs [2 and references therein]. Other analogous molybdates such as $CoMoO_4$ and $CdMoO_4$ are also selective at low oxygen concentration in 1-butene oxidation, but form only butadiene similarly to Bi-molybdate, whereas $MnMoO_4$ is also able to form maleic anhydride [2 and references therein]. Similarly in 1-butene oxidation, the two cuprous molybdates $Cu_2Mo_3O_{10}$ and $Cu_6Mo_4O_{15}$ composed of the same chemical elements in the same valence state, behave in two completely different ways from the catalytic point of view: the former is able to form only butadiene, whereas the latter selectively gives crotonaldehyde [1,6 and references therein]. The surface geometry of the ensemble site and the bulk oxide characteristics are certainly important aspects that determine these differences in the reactivity, but even so it is difficult to rationalize why on some of these molybdates only H-abstraction (butadiene formation) occurs and in other molybdates different types of O-insertion reactions occur, on the terminal C atom with formation of crotonaldehyde, and on the 1,4 positions of the intermediate butadiene with cyclization to form furan (intermediate to maleic anhydride) with consecutive further insertion up to maleic anhydride. Clearly then, the following question remains to be answered: Does the oxidation of

1-butene, for example, to the various reaction products depend only on the presence of terminal oxygens of Mo and the structure of the ensemble active site or do the presence of different oxygen species as well as other additional properties play a role in determining the pathway of oxidation of the hydrocarbon? No clear answer can be derived from the literature data that are relatively detailed for propylene oxidation to acrolein or acrylonitrile, but much less so for other reactions of hydrocarbon selective oxidation on other mixed oxides. Further studies are thus necessary to unravel the mechanism of elementary steps involved in these reactions and to obtain a better understanding of which factors control the surface reactivity.

1.2 ALKANE OXIDATION

In alkane oxidation, as compared to the transformation of the corresponding alkenes, the activation of the reagent is much more difficult due to the absence of available lone pairs and of empty orbitals, the strong bonding between C and H, and the presence of deep lying σ bonding orbitals and high lying σ^* levels. In addition, the difference in the reactivity towards H-abstraction between an alkene and its partially oxidized product is much less than the difference between H abstraction from an alkane and its corresponding alkene and therefore it is much more difficult to control the further oxidation of the reaction products. A direct consequence of this fact is that very few selective catalytic systems are known for alkane oxidation and vanadyl pyrophosphate is almost the only mixed oxide selective even at high hydrocarbon conversion [4], in contrast to the relatively broad range of mixed oxides selective in alkene oxidation. Furthermore, different types of products of partial oxidation may be obtained selectively starting from alkenes, whereas up to now only anhydrides (maleic anhydride from butane or maleic and phthalic anhydrides from n-pentane [9]) using vanadyl pyrophosphate, or acrylonitrile from propane using a V- antimonate based catalyst can be formed with good yields starting from alkanes [10].

Different surface characteristics may thus be required to transform selectively an alkane as compared to an alkene, but the more general question that will be addressed in the following sections is whether or not similarities or analogies exist in the reaction pathways, elementary steps and nature of surface sites between the oxidation of alkenes (as described above) and of alkanes on vanadyl pyrophosphate and which new aspects of the surface reactivity of oxides in oxidation reactions may be derived from this comparison.

2. Surface Structure of the Active Sites in Vanadyl Pyrophosphate

There is a general consensus in the literature [2,11 and references therein] that vanadyl pyrophosphate is the active phase in n-butane selective oxidation to maleic anhydride and that the catalyst microstructure influences the catalytic behavior in this reaction. In particular, several authors agree in observing a relationship between an increase in the specific surface area of the (100) crystalline plane of $(VO)_2P_2O_7$ and the activity- selectivity in n-butane oxidation, and even though conclusive evidence does not exist due to the inherent difficulty in obtaining experimental proof, it may be reasonably assumed that the active sites are located on this plane.

The vanadyl pyrophosphate framework consists of sheets made with pairs of highly

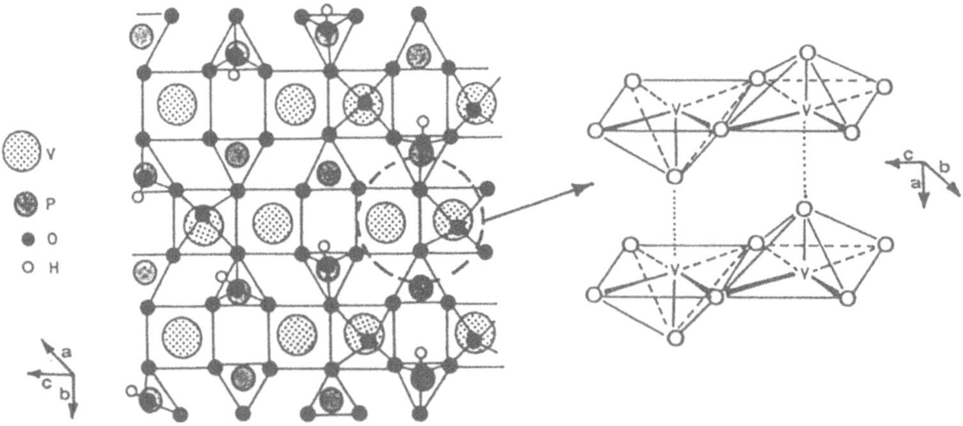

Fig. 1 Idealized crystal structure of the (100) plane of vanadyl pyrophosphate according to a simple truncation model of the bulk structure [15].

distorted VO_6 octahedra which share edges and have trans V=O bonds, equatorially linked to phosphate tetrahedra. These sheets are stacked upon each other, so as to form double columns of distorted VO_6 octahedra (chains O=V···O=V) parallel to bent pyrophosphate groups. Vanadium's sixth coordination site when directed inwardly weakly interacts with the underlying vanadyl group, and when directed at the surface is a Lewis acid site. Using infrared characterization of acetonitrile adsorption on vanadyl pyrophosphate [12], it is possible to show the presence of medium- strong Lewis acidity. In addition, a tentative correlation between acid strength of these sites and activity of the catalyst in n-butane selective conversion was found. Infrared characterization of the surface acidity of vanadyl pyrophosphate [12] also indicates the presence of Brønsted acid sites reasonably connected to the phosphorus of the surface terminating phosphate moieties (PO_3-OH). When the vanadyl pyrophosphate crystal is cleaved parallel to the (100) plane an idealized surface structure such as that shown in Fig. 1 may be assumed [13-15]: coupled vanadyl groups in trans positions are isolated from each other by pyrophosphate groups, half of which terminate as protonated orthophosphate moieties. Half of the surface V atoms is coordinatively unsatured and shows Lewis acidity, whereas the other vanadium atoms terminate in a V=O double bond. The formal oxidation state of vanadium is four. However, two results constrast with this idealized termination. The phosphorus to vanadium surface ratio, as determined by XPS or other surface sensitive techniques, is higher (by about two times) than the bulk value of 1.0 and the amount of Brønsted sites, as determined from surface titration with K^+ ions [14], is also about two times higher than that expected from the above idealized model.

An intriguing question about these problems has recently been suggested [13]: Is the (100) crystalline plane really terminated by the simple truncation of the bulk vanadyl pyrophosphate structure (Fig. 1) or, on the contrary, is it terminated by a phosphorus-rich layer, eg. not with protonated orthophosphates but with protonated pyrophosphate moie-

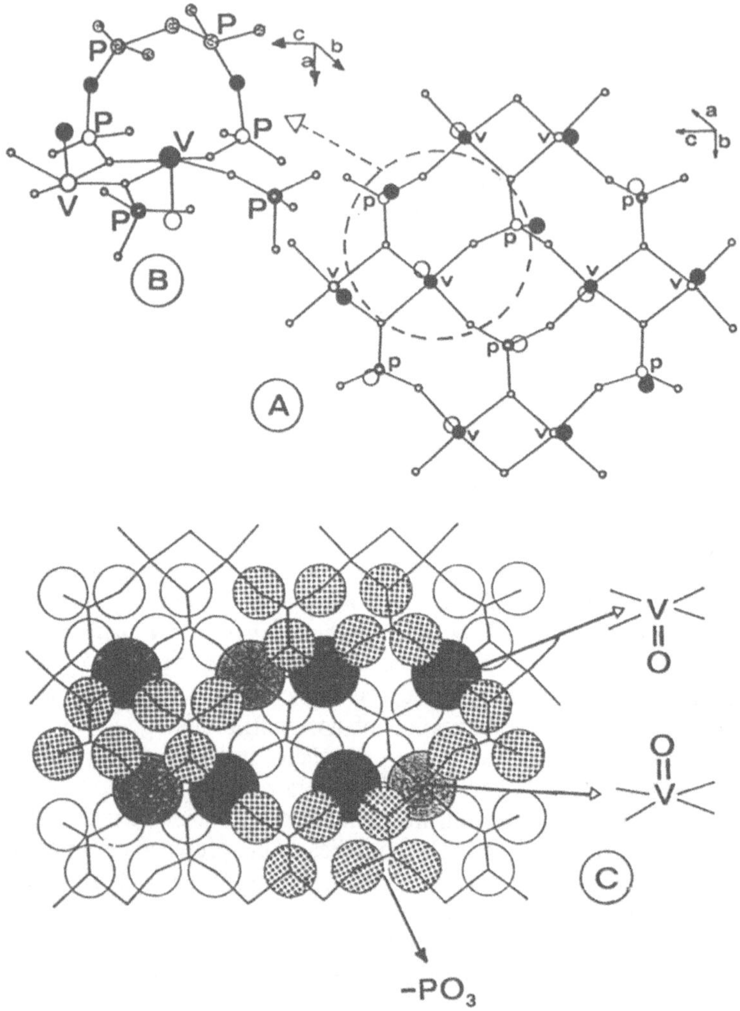

Fig. 2 Projection of the (100) plane of vanadyl pyrophosphate illustrating the vanadyl orientation according to polytype I [13] [with dark spheres are indicated the surface V, P or O atoms] (**A**), example of the pyrophosphate termination of this plane (with dotted spheres are indicated the surface phosphate groups omitted for clarity from part A of the scheme] (**B**) and model of the surface of the (100) plane of polytype I indicating the arrangement of surface pyrophosphate groups and the cavities creates from these groups (**C**) [13].

ties (Fig. 2 B)? In the second hypothesis was suggested [13] that a maximum bond valence for each atom under the constraint of generating a neutral surface is preserved and especially the surface model is in better agreement with the experimental evidence of P/V and P/OH surface ratios. An interesting consequence that derives from this hypothesis is that the pendant pyrophosphate groups provide steric isolation of the vanadium centers in clefts or cavities which limit the accessibility of gaseous reactants to vanadium

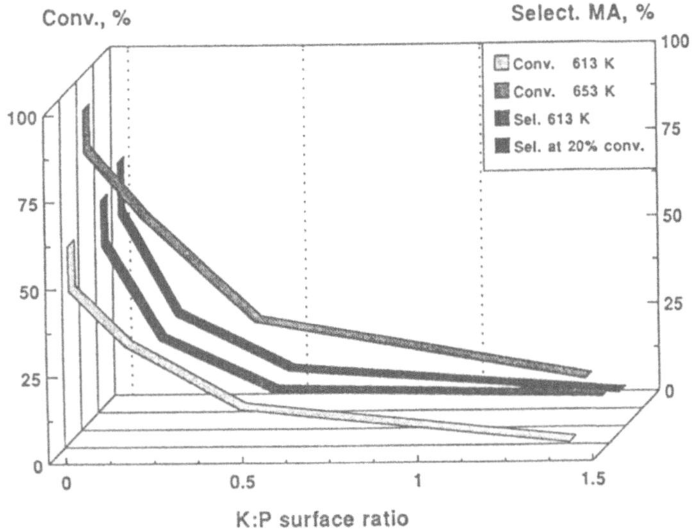

Fig. 3 Effect of the K/P surface ratio on the catalytic behavior in n-butane conversion to maleic anhydride of K-doped vanadyl pyrophosphate (n-butane conversion at 613 and 653 K, selectivity to maleic anhydride at 613 K and at 20% butane conversion) [14].

centers (Fig. 2 C) as well as can limit the surface mobility of adsorbed species [13]. In addition, the pendent pyrophosphate groups that overhang the ensemble of vanadium sites in the surface clefts present a higher number of hydrogen binding sites which can facilitate hydrogen acceptance and transport to sites of water formation and desorption [13] as well as can provide sites to orient the intermediate reaction product.

It should also be noted that structural results indicate that vanadyl pyrophosphate crystallizes with varying degrees of disorder in the vanadium positions [13] and especially with variability in the directional orientation of the vanadyl columns running perpendicular to the (100) surface. Various authors [2,11 and references therein] agree in suggesting a correlation between disorder in the structure of vanadyl pyrophosphate and its catalytic activity, even though the reason is not clear. However, the presence of disorder in the orientation of vanadyl columns also would result in a change in the symmetry and dimensions of surface cavities created from the pending pyrophosphate groups mentioned above with a consequent change in the accessible surface vanadium sites. The method of preparation and the transformations during thermal and catalytic treatments affect the degree of disorder in the vanadyl pyrophosphate [11] which consequently may be reflected in the surface structure according to the above mechanism.

3. Relationship between Surface Properties and Selectivity in Alkane Oxidation.

In the previous section evidence for the presence of both Brønsted (P-OH) and Lewis (coordinatively unsaturated V) acid sites on the surface of $(VO)_2P_2O_7$ was presented. Useful information on their role in the reaction mechanism derives from specific doping experiments.

3.1 DOPING OF ACID SITES AND ROLE ON THE SELECTIVE OXIDATION MECHANISM

Doping with K^+ ions in an anhydrous medium [14] leads to selective doping of the Brønsted acid sites of vanadyl pyrophosphate without affecting the Lewis acidity, as shown from infrared characterization of the surface acidity. The effect of doping on the catalytic behavior in n-butane oxidation to maleic anhydride is summarized in Fig. 3. The K/P surface ratio is calculated assuming all the K to have been added on the surface and a surface concentration of P atoms of $6.43 \cdot 10^8$ P surface atoms/m^2 of surface area as can be estimated from the surface model discussed above. Figure 3 illustrates the strong deactivation effect of K on the conversion and on the selectivity to maleic anhydride over the entire range of temperatures. A parallel increase in the selectivity to carbon oxides has been observed [14]. Even the presence of small amount of K on the surface (K/P = 0.004) may cause a considerable inhibition of the activity and of the selectivity. Similar strong deactivation of the selectivity to anhydrides (maleic and phthalic anhydrides) and on the rate of alkane depletion was also observed in the oxidation of n-pentane. This suggests a specific role of Brønsted acid sites both in the mechanism of alkane activation and in the selective pathway of oxygen insertion up to maleic anhydride. The study by infrared spectroscopy [14] of the reactivity of adsorbed species on vanadyl pyrophosphate doped or not with K shows that the anaerobic contact of 1-butene with undoped sample forms adsorbed species that, when in contact with O_2, easily transform to maleic anhydride. However, this second stage is inhibited when the sample is doped with K. Doping with K therefore specifically inhibits the stage of maleic anhydride synthesis from the already activated intermediate when gaseous oxygen is present, as well as to inhibits the step of alkane activation.

In order to understand the reasons for this effect, unexpected on the basis of the previous discussion on the pathways of alkene transformation on catalysts such as Bi-molybdate, it is useful to analyze the effect of preadsorbed ammonia on the surface reactivity of vanadyl pyrophosphate [16,17]. The reaction of propane conversion to acrylonitrile was studied in this case because it is simpler and provides better indications on the relationship between doping of surface sites and modification of selective mechanisms of oxidation. In addition, this reaction allows a more direct comparison with the mechanisms of propylene ammoxidation on Bi-molybdates and therefore a better analysis of the differences in the surface reactivity between vanadyl pyrophosphate, catalyst for alkane selective oxidation, and the catalysts for alkene selective transformation.

Combined thermogravimetric and infrared characterization of activated (400°C) ammonia adsorption on $(VO)_2P_2O_7$ [17] indicates that, in contrast to what happens with Bi-molybdate, no reaction occurs between V=O and ammonia to form the metal-imido species (V=NH) analogous to that indicated as responsible for N insertion on the π-allyl propylene complex intermediate on molybdate catalysts (Mo=NH) [5]. Ammonia on vanadyl pyrophosphate interacts with both the Brønsted and Lewis acid sites to form surface complexes relatively stable at 400°C and that relatively slowly decompose restoring the initial situation. This offers the opportunity to analyze how the surface reactivity on vanadyl pyrophosphate in alkane conversion changes as a function of the surface coverage with preadsorbed ammonia. The tests were done following the evolution with time-on-strem of the catalytic behavior in anaerobic or aerobic interaction of propane with the

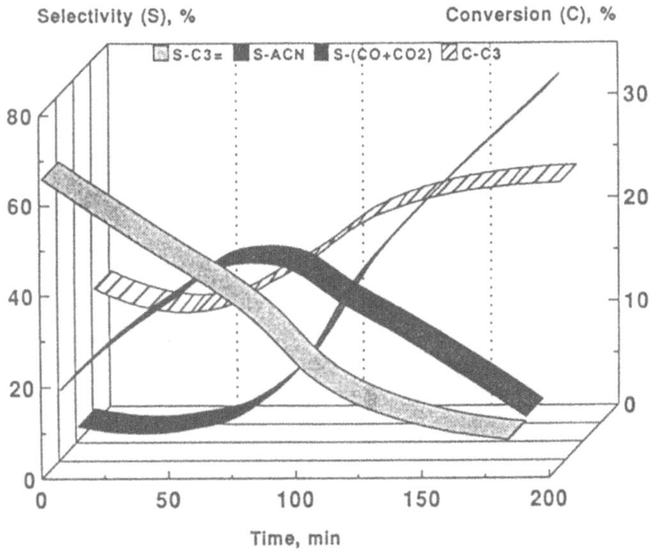

Fig. 4 Transient catalytic behavior in propane conversion on vanadyl pyrophosphate with preadsorbed ammonia. Exp. conditions: 403°C, preadsorption of ammonia for 30 min at the same temperature and then removal of gas-phase species with an He flow, flow of 6.6% propane and 7.6% oxygen in helium, 2 g of catalyst, total flow (STP) 6 L/h [16].

catalyst after preadsorption of ammonia and elimination of gas-phase NH_3. With increasing time-on-stream, due mainly to the incorporation of ammonia in the reaction products, the amount of adsorbed ammonia decreases and therefore the evolution of the catalytic reactivity with the time-on-stream is proportional to the change in the reactivity of $(VO)_2P_2O_7$ with its surface covered with decreasing amounts of chemisorbed ammonia. Figure 4 shows the results obtained at 400°C using a feed mixture of propane and O_2 with a 1:1 ratio [16].

Initially, for the higher surface coverage with chemisorbed ammonia, propane is oxidatively dehydrogenated to propylene, but is not further oxidized to acrylonitrile or to carbon oxides. For an intermediate surface coverage with adsorbed ammonia, the selectivity to acrylonitrile is maximum and for a very low surface coverage with adsorbed ammonia, the rate of propane depletion is higher, but mainly carbon oxides are formed. When the same type of experiments are carried out in the absence of gaseous oxygen, on the contrary, the rate of propane depletion is one order of magnitude lower and only propylene forms [16]. Similar results also were obtained using the catalyst after oxidation to the corresponding V^{5+}-phosphate ($VOPO_4$).

Infrared characterization of the interaction of C_3 hydrocarbons with $(VO)_2P_2O_7$ with preadsorbed ammonia [17] gives further indications about these effects. In particular, in the absence of oxygen, propane is not activated. Propylene, on the contrary, is adsorbed on the catalyst in the form of a π-bonded complex with vanadium Lewis acid sites that, in the absence of gaseous oxygen is slowly transformed to acrylonitrile through the intermediate formation of propylenamine. This mechanism therefore involves only H- abstraction steps. When gaseous oxygen is present, however, the evolution is completely different. Propane can be dehydrogenated to propylene and further transformed to surface acrylate species. The infrared analysis of the time-on- stream evolution of the surface species during the interaction of propane and oxygen with $(VO)_2P_2O_7$ with preadsorbed ammonia, furthermore, suggests that the surface acrylate species react with nearlying ammonium ions to form acrylonitrile and that this second pathway of reaction (possible only in the presence of O_2) is much more rapid than the first involving only H-abstraction steps (intermediate formation of propylenamine).

3.2 PRESENCE OF COMPETITIVE ADSORPTION PHENOMENA AND ROLE OF THE SURFACE DENSITY OF ACTIVE OXYGEN SITES

The analysis of propane interaction with vanadyl pyrophosphate with preadsorbed ammonia indicates some important considerations about the surface reactivity of this catalyst in alkane oxidation. The observed effects suggest competition between ammonia and O_2 chemisorption on vanadium Lewis acid sites. Only when ammonia is removed from these sites may oxygen be chemisorbed and this step is necessary to form the sites which can oxidatively dehydrogenate propane to propylene and insert oxygen to form a surface acrylate species. Tentatively, the sites responsible for these two properties may be attributed to chemisorbed activated oxygen species and not to structural oxygens related to the formation of some V^{5+}-phosphate surface islands due to local oxidation of $(VO)_2P_2O_7$ (no XRD detectable V^{5+} species is present). In fact, isothermal thermogravimetric curves of ammonia interaction with $(VO)_2P_2O_7$ do not suggest a preliminary reduction of the catalyst during ammonia preadsorption and the $(VO)_2P_2O_7$ used in these tests cannot be oxidized to a V^{5+}-phosphate at a reaction temperature of 400°C (a temperature about 150°C higher is required [14]). It should be noted, however, that probably the Brønsted sites nearlying to V Lewis sites are important in stabilizing these activated oxygen species forming an hydroperoxo-like species and this can explain why doping with K can inhibit both the rate of alkane transformation and the selective pathway of oxidation.

A further interesting aspect derives from the analysis of the evolution of the catalytic behavior of $(VO)_2P_2O_7$ as a function of the coverage with adsorbed ammonia. We can reasonably assume that ammonia blocks the sites for oxygen chemisorption and that the removal with the time-on-stream is nearly statistical. This indicates that initially few isolated activated oxygen species exist on the surface of vanadyl pyrophosphate and gradually the surface density of these species increases. When few isolated active oxygen sites are present propane can be transformed only to propylene. When the density of these sites is increased acrylate species can also be formed which are intermediate to acrylonitrile. Finally, when the density of these sites is too high, only carbon oxides can be formed. The comparison of these results with the previous conclusions on the selective and unselective pathways of alkene oxidation, for example on Bi-molybdate, thus indicates that the selectivity in one reaction cannot be associated only to different sites or different crystal faces, but also must be related to the surface density of sites which can be varied also with selective doping.

This concept may be further extended. In fact, activated oxygen species possess a surface mobility certainly higher than structural oxygens, because the bonds with the surface are weaker. There is no direct evidence in the literature about the possibility and rate of surface transport of activated oxygen species, but indirect indications are found, for example, from the synergic effects used as the basis of the theory of remote control in oxidation reactions [18]. Nevertheless, if this hypothesis is assumed to be realistic, it follows

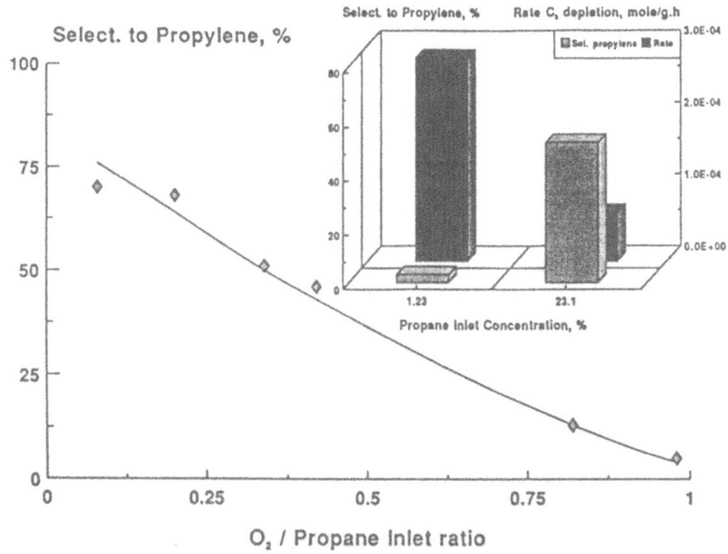

Fig. 5 Effect of the O_2/propane ratio on the selectivity to propylene on vanadyl pyrophosphate. Exp. conditions: 322°C, 23.1% propane, 5 g catalyst, total flow (STP) 40 cc/min. Compared in the inset are the selectivities to propylene and the rate of propane depletion at 322°C using a feed mixture containing in both cases 10.2% O_2, but 1.2% or 23.1% of propane.

that the selective formation of a product of oxidation as compared to complete oxidation to carbon oxides not only depends on the geometric surrounding of an adsorbed organic intermediate, but also on the rate of transformation of the intermediate (to a product which can desorb from the surface) with respect to the rate of surface transport of activated oxygen species. This rate of surface transport, in turn, depends on the surface characteristics of the active phase and therefore isolation of adsorbed intermediates in surface clefts or cavities, as discussed above in regard to the possible surface structure of vanadyl pyrophosphate [13], is fundamental to limit surface transport. It should be mentioned that $(VO)_2P_2O_7$ redox processes during the n-butane oxidation to maleic anhydride are limited to a near-surface region and no bulk structural oxygens partecipate in the reaction mechanism [19], in contrast to what occurs with Bi-molybdates. In addition, TAP reactor experiments [3,20] have shown that *i)* the surface lifetime of adsorbed intermediates in n-butane oxidation on $(VO)_2P_2O_7$ is quite long (in the order of some hundred milliseconds), *ii)* the surface lifetime influences the selectivity to maleic anhydride and *iii)* different species of oxygen are involved in the steps of alkane oxidative dehydrogenation and intermediate furan transformation to maleic anhydride and in the step of intermediate butadiene conversion to furan. The latter step involves the structural oxygen of the V=O double bond, whereas the results obtained tentatively suggest a role of activated oxygen species in the other two steps.

3.3 SELECTIVE AND UNSELECTIVE REACTION PATHWAYS

The relationship between surface lifetime of adsorbed species and selectivity can be shown also by the analysis of the behavior of vanadyl pyrophosphate in n-pentane oxidation which, in contrast to n-butane oxidation where only maleic anhydride forms, gives rise to the formation of both maleic and phthalic anhydrides. The main aspects of the reaction mechanism in this reaction and in particular of phthalic anhydride formation have been discussed elsewhere [9,21], but it is interesting to note that the kinetic analysis of n-pentane oxidation on vanadyl pyrophosphate [9] indicates that phthalic anhydride undergoes a higher rate of conversion to carbon oxides than does maleic anhydride, although the reasons on a chemical basis are not so evident. However, using a new microreactor apparatus (TAP) coupled to a mass quadrupole detector for a very fast response (sub-millesecond time resolution) [20] it is possible to clarify this aspect. In these transient experiments [21] in which n-pentane is pulsed over $(VO)_2P_2O_7$ the maximum in the phthalic anhydride transient response is consistently shifted to longer times and the pulse is consistently broader than for the maleic anhydride response. The analysis of the respective pulse shapes indicates that the rate of phthalic anhydride desorption is significatly lower than that of maleic anhydride, probably due to the presence of the aromatic ring. The increased surface lifetime of phthalic anhydride compared to maleic anhydride increases the possibility of further transformation to carbon oxides. Thus the difference in the desorption rates more than their intrinsic reactivity to oxidation products explains the difference in the stability of the two anhydrides towards consecutive oxidation.

A further significant aspect of the reactivity of $(VO)_2P_2O_7$ which provides further evidence for the role of the surface density of oxidizing sites on the selective and unselective oxidation pathways of alkane oxidation is related to the difference in the catalytic behavior between propane and n-butane. Using the same sample and reaction conditions for which maleic anhydride selectively forms from n-butane or maleic and phthalic anhydride are synthetized from n-pentane, propane is oxidized to carbon oxides [22]. Therefore, a small change in the nature of the alkane considerably alters the catalytic behavior and this effect may suggest that the selective pathway is determined by the geometry of adsorption of alkane or the possible nature of reaction intermediates (unlike the case of propane, cyclic products such as dihydrofuran and furan or their corresponding methylated molecules can be formed from both C_4 and C_5 hydrocarbons). However, if the propane transformation is studied on the same catalyst but using a high propane concentration (about 23%) and low oxygen concentration [22], a selective synthesis of propylene from propane is instead observed (Fig. 5). The increase in propane concentration induces deactivation of the surface reactivity. In the inset of Fig. 5, for example, the rate of propane depletion for an equal oxygen concentration (10.1%) in the feed, but a low or high propane concentration (1.2% and 23%, respectively), are compared. The corresponding selectivities to propylene also are reported.

In order to understand this effect it is useful to note that strongly adsorbed species form on $(VO)_2P_2O_7$ during the catalytic reaction of alkane oxidation, i.e. adspecies having a longer lifetime than the reactive intermediates due to their low rate of transformation and desorption. These species can be detected on catalysts which have been exctracted from a reactive environment and their amount can be determined by titration with oxygen in a thermobalance [14] or a pulse reactor [20]. In addition to these species, during reaction the catalyst surface is covered with other reactants, intermediates, products and coproducts. In the case of n-butane oxidation, these species include C_4 olefins, furan and maleic anhydride, carbon oxides and water, as well as other products not detected in steady-state reaction conditions including crotonaldehyde and phthalic anhydride [14]. The latter species can be determined by programmed heating of a catalyst after

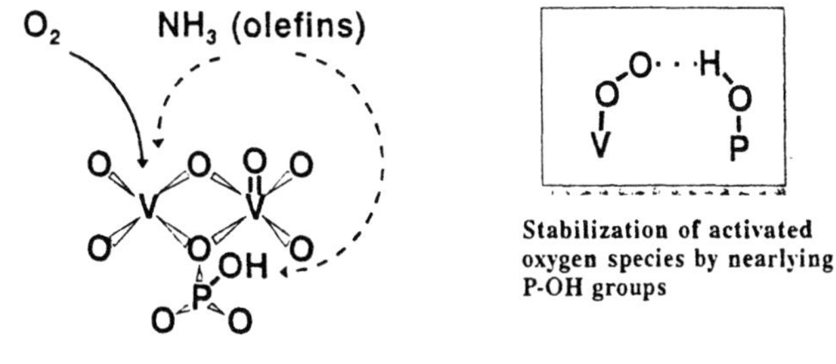

Fig. 6 Schematic diagram showing the presence of competitive adsorption phenomena between oxygen from one side and ammonia or olefins on the other side. Shown in the square is the possible effect of stabilization of chemisorbed activated oxygen species from nearlying Brønsted sites.

reaction. Infrared analysis [23] confirms these indications and shows that these species rapidly reform upon contact of a cleaned catalyst with an alkane-air flow. Their formation may be attributed to the strong interaction of alkenes or alkandienes with the surface acid sites of vanadyl pyrophosphate. Indeed, when the amount of strongly adsorbed species after n-butane or after butadiene steady state oxidation is compared, a much larger amount is found in the case of the olefin. Alkene intermediates may thus block some surface sites and this can explain why deactivation is observed when working with a high propane concentration (Fig. 5). However, in analogy with that discussed for ammonia adsorption, blockage of some surface sites in combination with a limited availability of oxygen for further conversion of the propylene allows the selective synthesis of propylene from propane.. In different working conditions, however, the same catalyst forms only carbon oxides. There is thus a strong similarity in the effect induced by chemisorption of ammonia and olefins. This analogy, shown schematically in Fig. 6, on the other hand, is quite reasonable taking into consideration the basic nature of the double bond. However, olefins can be chemisorbed more strongly than ammonia and indeed, when small amounts of olefins are cofed with the alkane-air mixture, complete deactivation of the surface reactivity of $(VO)_2O_2O_7$ occurs [22,23] with a considerable decrease both in the activity and in the selectivity. Similarly, on a selective $(VO)_2P_2O_7$ catalyst for n-butane or n-pentane oxidation, the oxidation of the corresponding C_4 or C_5 olefins gives rise to the formation mainly of carbon oxides [22] and after olefin oxidation the catalyst is found to be completely deactivated when consecutively tested in alkane oxidation. Only when these strongly adsorbed species are removed can the initial active behavior be restored.

This is a good example, therefore, of the influence of the surface changes during the catalytic reaction on the selective and unselective pathways of oxidation and the role of the surface density of sites of oxygen chemisorption on the selectivity.

3.4 SURFACE CAVITIES AND PATHWAYS OF ALKANE TRANFORMATION

The discussion in the previous section illustrated how it is possible to block some sites to obtain an improvment in the selectivity, but this does not explain why in the same conditions propane forms carbon oxides and n-butane is selectively transformed to maleic anhydride. This difference can be interpreted, however, taking into consideration the

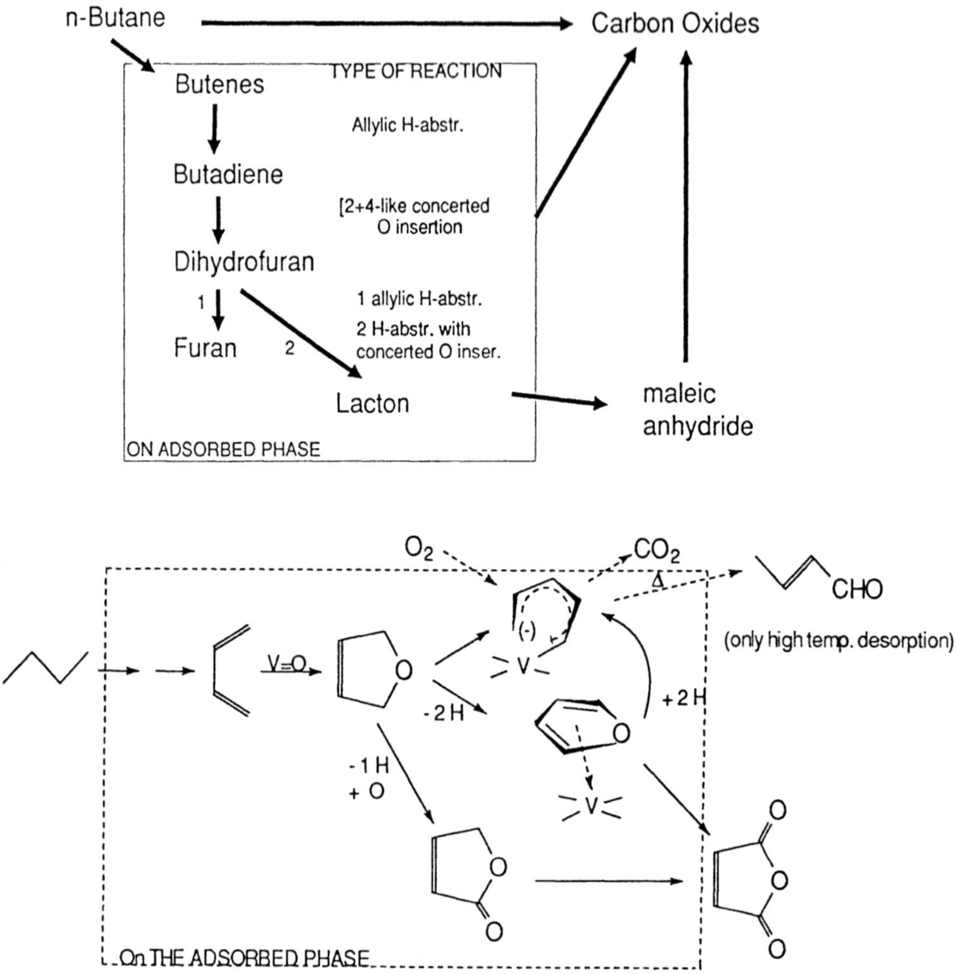

Fig. 7 Schematic reaction network in n-butane oxidation on vanadyl pyrophosphate where the presence of multiple pathways of transformation to maleic anhydride is shown (A) as well as the higher formation of products with reduced reactivity (strongly adsorbed species which desorb only at higher temperatures and react with oxygen forming carbon oxides) in the pathway involving the intermediate formation of furan (B).

possible presence of surface cavities on $(VO)_2P_2O_7$ created from pendent pyrophosphate groups (Fig. 2) and the different amount of oxygens required in the two cases. In fact, the amount of V sites for oxygen chemisorption and of V=O groups in these cavities must be limited, but the synthesis of maleic anhydride from n-butane, assuming all hydrogens transformed to water molecules, requires 7 O atoms as compared to one O atom required for propane to propylene transformation. In the latter case, therefore, more oxygens are available than necessary and oxidative degradation of the carbon skeleton may occur. In the case of n-pentane oxidation to maleic anhydride, assuming that the additional carbon as compared to butane is oxidized to CO_2, the transformation requires 10 O atoms which will be excessive for the number of oxygens available and therefore the rate of oxidation of intermediates decreases [14,24] allowing the occurrence of a parallel pathway of transformation involving a surface template reaction between adsorbed intermediates that leads to the formation of a product (phthalic anhydridse, C_8) with a higher number of carbon atoms than the starting alkane (n-pentane, C_5) together with maleic anhydride. This model, oversimplified for several reasons, such as the possibility of H-transfer to other sites able form water (infrared data, in agreement, indicate a high surface mobility of H-atoms which is enhanced in the presence of strongly adsorbed species [17]), nevertheless, may constitute a basis for the development of more complex and realistic models. Such models, based on the number of available oxygen species in restricted environments may well offer the key to the design of new catalytic applications of heterogeneous catalysts.

3.5 DEPENDENCE OF THE SURFACE PATHWAYS ON THE RELATIVE RATES OF H-ABSTRACTION AND O-INSERTION

The different behaviors of n-butane and n-pentane oxidation on vanadyl pyrophosphate may also be viewed from a slightly different perspective. Due to strong analogies, it is reasonable to hypothesize a very similar initial mechanism of transformation for n-butane and n-pentane. For the former it has been shown [2,4 and references therein] that the initial steps of the n-butane to maleic anhydride pathway involve the intermediate formation of adsorbed butenes and butadiene. Similarly, the formation of adsorbed pentadiene from n-pentane can be expected. However, at this stage a main difference distinguishes the two reaction mechanisms of n-butane and n-pentane oxidation, namely the presence of additional allylic H-atoms in pentadiene as compared to butadiene and thus the possibility of further easy H-abstraction to form adsorbed cyclopentadiene. A template addition between two adsorbed cyclopentadiene molecules may give rise to a hydrocarburic precursor that evolves to phthalic anhydride. The key step in the phthalic to maleic anhydride competitive synthesis from the common intermediate (adsorbed pentadiene) is thus the competition between O-insertion to form the precursor of maleic anhydride and H-abstraction to form the precursor of phthalic anhydride [24] as well as the surface concentration of near-lying unsaturated C_5 hydrocarbons which has been shown to be important in TAP reactor experiments [21]. In adsorbed butadiene the reaction of H-abstraction is more difficult due to the lower reactivity of the C-H bonds. Therefore, both the characteristics of the reactive intermediate itself and the relative rates of O-insertion and H-abstraction (which depend also on the local density of active oxygen sites) determine the pathway of transformation.

It is interesting to note that these concepts may be applied to rationalize the difference in the catalytic behavior in n-butane and n-pentane oxidation between $(VO)_2P_2O_7$ and $H_5PV_2Mo_{10}O_{40}$ heteropolyacid [25,26]. The latter catalyst also can oxidize n-butane to maleic anhydride, but with lower selectivity than $(VO)_2P_2O_7$ and forming a larger amount of acetic acid. In addition, in n-pentane oxidation, this catalyst, in contrast to vanadyl pyrophosphate, only forms maleic anhydride and not also phthalic anhydride. The oxidation of decalin is a useful test reaction to understand the difference in the surface reactivity of these two catalysts [25]. Decalin is oxidized on these catalysts to phthalic anhydride and naphthoquinone through the intermediate formation of tetralin and naphthalene. All these products may be detected in the gas-phase, unlike n-butane and n-pentane oxidation where the entire mechanism occurs on the adsorbed phase with desorption of only the final products (anhydrides, carbon oxide and water). Decalin thus offers the possibility of determining a relative index of the H-abstraction vs. O-insertion properties based on the ratio of the selectivities of tetralin plus naphthalene to phthalic anhydride plus naphthoquinone. In addition, since decalin is a saturated hydrocarbon, the phenomena of strong adsorption and surface deactivation mentioned before using unsaturated hydrocarbons such as olefins is not observed. The evaluation of this relative index shows the higher ratio of H-abstraction vs. O-insertion rates in $(VO)_2P_2O_7$ as compared to $H_5PV_2Mo_{10}O_{40}$ which can explain, according to that discussed above, why only maleic anhydride is found from n-pentane on this catalyst while both phthalic and maleic anhydrides are formed on vanadyl pyrophosphate. An analogous explanation may also be given for the lower selectivity to maleic anhydride from n-butane and the higher formation of acetic acid on the heteropolyacid [25].

This concept of relative rates of H-abstraction and O-insertion and their influence on the surface pathways of reaction and surface concentration of adspecies may be utilized to develop a surface microkinetic approach to the selectivity in alkane oxidation which, in turn, makes it possible to estimate how the selectivity and the relative formation of the various products change as a function of the surface density of active oxygen species in the O- insertion reaction [26].

In n-butane oxidation on vanadyl pyrophosphate, a step in the reaction network also exists which may evolve along two different routes depending on the relative rates of H-abstraction and O-insertion. The reaction of transformation of n-butane to maleic anhydride, as mentioned before, occurs entirely on the adsorbed phase and therefore only indirect evidence exists [2,4 and references therein], but there is good general agreement about the main steps. The reaction pattern is shown schematically in Fig. 7A together with the type of reactions involved in the various steps. n-Butane is converted through a series of H-abstraction reactions to the adsorbed butadiene intermediate which can react with a vanadyl (V=O) surface group to give a [2+4] cycloaddition forming a surface-bound 2,5-dihydrofuran. In agreement, the formation of dihydrofuran from butadiene on vanadium-phosphorus oxides (VPO) has been shown [27] and recently verified to be consistent with theoretical determinations based on the frontier orbital approach with extended Hückel calculations [28].

The intermediate dihydrofuran, however, can evolve in two main directions: 1) quite reactive allylic hydrogens are present which may be extracted to form furan (on VPO, 2,5-dihydrofuran is very easily transformed to furan [27] which then is converted to maleic anhydride) and 2) contemporaneous to one H- abstraction, an O-insertion may occur to form a lactone. Theoretical calculations [28] indicate that one possible reaction path is the interaction of adsorbed dihydrofuran with an adsorbed molecular oxygen species (possibly an η^2-peroxo species) activated on a nearlying coordinatively unsaturated vanadium. This oxygen species activates a carbon-hydrogen bond in the 2-position of 2,5-dihydrofuran leading to a transfer of a hydrogen to the dioxygen fragment. A surface bound hydroperoxide species and a radical-like 2,5-dihydrofuran derivative are formed. The surface bound hydroperoxide species then transfers an hydroxyl group back to the radical-like 2,5- dihydrofuran derivative forming the 2-hydroxy derivative. The remaining adsorbed atomic oxygen abstracts the two remaining hydrogens and is eliminated as water, with contemporaneous formation of the asymmetric lactone which is then further

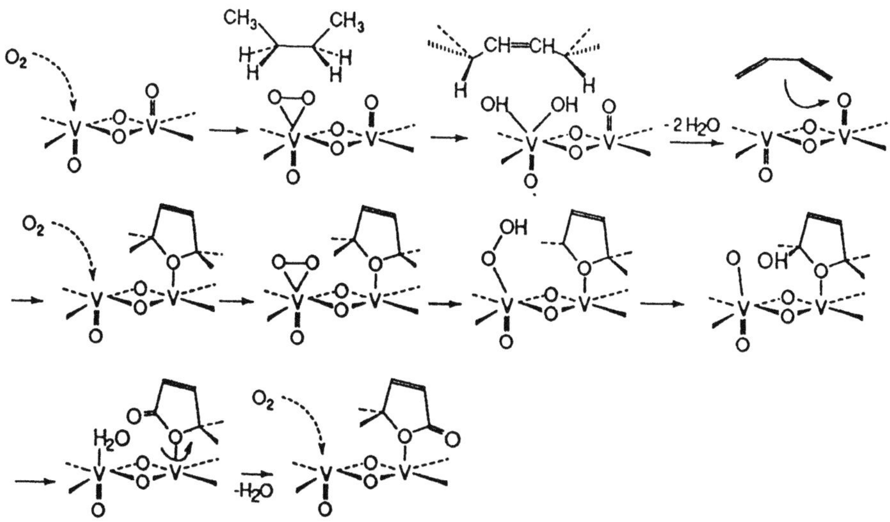

Fig. 8 Possible reaction mechanism of n-butane selective transformation on vanadyl pyrophosphate.

converted to maleic anhydride with a similar mechanism.

Infrared characterization of the adsorbed species on $(VO)_2P_2O_7$ after interaction of C_4 hydrocarbons is in agreement with this indication of the possibility of two pathways of reaction from butadiene to maleic anhydride involving the intermediate formation of furan or of the asymmetric lactone [24]. The evolution along the two pathways depends on the relative rates of (i) abstraction of the second hydrogen on the activated dihydrofuran to form furan and (ii) O attack to form the lactone according to the mechanism described above. However, the two pathways are not equivalent even though the same end product is formed, because furan interacts quite strongly with the surface V Lewis acid centres giving rise to the formation of surface by-products which may partially inhibit the surface reactivity with a lowering of the selectivity [24] (Fig. 7 B).

4. Mechanism of Alkane Selective Activation.

As noted above, the absence of direct proof about the elementary steps of the reaction due to the difficulty of studying very complex surface transformations (for example, n-butane oxidation is a 14 electron oxidation with abstraction of 8 H atoms and insertion of 3 O atoms; n-pentane transformation to phthalic anhydride is even more complex) which occur entirely on the adsorbed phase without desorption of any intermediate, makes any discussion about the mechanism of reaction rather speculative. Even so, an explanation must be sought for the results obtained regarding the reactivity of the system. In particular, the mechanism of n-butane selective transformation to the intermediate butadiene on $(VO)_2P_2O_7$ will be discussed, because the main aspects of the consecutive steps up to maleic anhydride have been commented on above as well as the influence on the selective pathways of transformation of changes in the nature of the hydrocarbon.

Various evidence [2,4 and references therein] has been obtained which indicates that the activation step of n-butane involves the abstraction of 2 H atoms to form the corresponding adsorbed butene molecule. In this step, using labelled compounds [19] it was shown that n-butane does not reversibly chemisorb with hydrogen exchange onto the surface of vanadyl pyrophosphate, but is only irreversibly adsorbed. In this step, nearly exclusive abstraction of 2 H atoms from the methylenic C-H bond occurs. This suggests that a simultaneous abstraction of the two hydrogens from the carbon atoms in the 2- and 3-positions of the n-butane molecule occurs, as confirmed also from the analysis of the rates of reaction of a series of alkanes on $(VO)_2P_2O_7$ [4]. The absence of formation of typical products of cracking or skeletal rearrangement in alkane oxidation on $(VO)_2P_2O_7$ indicating the absence of formation of carbocations should also be noted. A second aspect of the surface reactivity of vanadyl pyrophosphate which must be considered is that many data indicate the strong analogy of surface oxygen species involved in the step of alkane activation to the corresponding olefin and in the step of formation of the anhydride, whereas a different type of oxygen sites is required for the transformation of butadiene to furan [4]. The data on the doping of surface acid sites discussed previously also agree in suggesting this correlation.

The nature of the active species which is able to give this selective activation of n-butane is questionable. A possibility is based on the formation of the η^2-peroxo species by O_2 chemisorption on the V Lewis sites of vanadyl pyrophosphate. The formation of this species has been shown to be reasonable from theoretical calculations [28] and was hypothesized to be responsible for the insertion of oxygen in the 2 position of adsorbed 2,5-dihydrofuran to form the asymmetric lactone [28]. This species may also be responsible for contemporaneous 2 H abstraction from the methylenic groups of n-butane to form the corresponding olefin and a geminal dihydroxy vanadium species. The latter species easily rearranges eliminating water and leaving an adsorbed atomic oxygen. This oxygen can abstract one hydrogen from the methylic terminal group The further H-abstraction from the other terminal methyl group forms butadiene. Tentatively, either bridging oxygen connected to vanadium and/or phosphorus may be involved in this second H abstraction. H-transfer to the V-hydroxyl group and water desorption restore the initial situation. Butadiene may then react with the nearlying vanadyl site and the reaction can proceed according to that discussed above.

5. Conclusions

The analysis of the reactivity, surface characteristics, possible mechanism and elementary steps in C_3-C_5 alkanes selective transformation on vanadyl pyrophosphate illustrates the peculiarities of the $(VO)_2P_2O_7$ catalytic system as well as the considerable differences in comparison with the selective oxidation of alkenes on mixed oxides such as Bi- molybdates. Some new aspects on the chemistry of heterogeneous catalytic oxidation are suggested based on the study of alkane oxidation on $(VO)_2P_2O_7$, such as the problem of the surface mobility of active oxygen species and lifetime of organic intermediates, the control of selective pathways of oxidation when the surface reaction occurs in restricted environments, the role of adsorbed (not-structural) oxygens in selective oxidation steps and the changes in the surface reactivity due to strong chemisorption of reaction intermediates. However, there is no conclusive evidence to prove the suggestions and therefore the challenge of the molecular level identification of the elementary steps in this reaction requires greater effort, but can open new directions of research on catalytic heterogeneous oxidation reactions.

6. References

[1] Haber, J. (1992) 'Catalytic Oxidation. State of the Art and Prospects', Ruiz, P. and Delmon, B. (eds.), New Developments in Selective Oxidation by Heterogeneous Catalysis, Elsevier Science Pub., Amsterdam, pp. 259-304.

[2] Centi, G. and Trifirò, F. (1984) 'Some Aspects of the Control of Selectivity in Catalytic Oxidation on Mixed Oxides: a Review', Appl. Catal. 12, 1-21.

[3] Gleaves, J.T. and Centi, G. (1992) 'The Reaction Mechanism of Alkane Selective Oxidation on Vanadyl Pyrophosphate Catalyst. Features Gleaned from TAP Reactor Transient Response Studies', Catal. Today 16, in press.

[4] Centi, G., Trifirò, F., Ebner, J.R. and Franchetti, V.M. (1988) 'Mechanistic Aspects of Maleic Anhydride Synthesis from C_4 Hydrocarbons over Phosphorus Vanadium Oxide', Chem Rev. 88, 55-80.

[5] Burrington, J.D., Kartisek, C.T., Grasselli, R.K. (1984) 'Surface Intermediates in Selective Propylene Oxidation and Ammoxidation over Heterogeneous Molybdate and Antimonate Catalysts', J. Catal. 87, 363-380.

[6] Haber, J. (1984) 'Catalysis and Surface Chemistry of Oxides', Procecdings 8th International Congress on Catalysis, Dechema Pub., Frankfurth AM Germany, Vol. I, pp. 85-111.

[7] Grasselli, R.K., Brazdil, J.F., Burrington, J.D. (1984) 'Solid State and Surface Mechanisms in Selective Oxidation and Ammoxidation Catalysis', Proceedings 8th International Congress on Catalysis, Dechema Pub., Frankfurth AM Germany, Vol. V, pp. 369-380.

[8] Carbucicchio, M. and Trifirò, F. (1976) 'Surface and Bulk Redox Processes in Iron-Molybdate-Based Catalysts', J. Catal. 45, 77-85.

[9] Centi, G., Lopez, J.N., Pinelli, D. and Trifirò, F. (1989) 'Synthesis of Phthalic Anhydride and Maleic Anhydrides from n-Pentane. 1. Kinetic Analysis of the Reaction Network', Ind. Eng. Chem. Res. 28, 400-406.

[10] Catani, R., Centi, G., Trifirò, F. and Grasselli, R.K. (1992) 'Kinetics and Reaction Network in Propane Ammoxidation to Acrylonitrile on V- Sb-Al based Mixed Oxides', Ind. Eng. Chem. Research 31, 107-119.

[11] Centi G. (1992) 'Vanadyl Pyrophosphate - A Critical Overview', Catal. Today , Vol. 16,.

[12] Busca, G., Centi, G., Trifirò, F. and Lorenzelli, V. (1986) 'Surface Acidity of Vanadyl Pyrophosphate, Active Phase in n-Butane Selective Oxidation', J. Phys. Chem. 90, 1337-1344.

[13] Ebner, J.R. and Thompson, M.R. (1991) 'Key Structure-Activity Relationship in the Vanadium-Phosphorus Oxide Catalyst System', Grasselli, R.K. and Sleight A.W. (eds.), Structure-Activity and Selectivity Relationship in Heterogeneous Catalysis, Elsevier Science Pub., Amsterdam, pp. 31-42.

[14] Centi, G., Golinelli, G. and Busca, G. (1990) 'Modification of the Surface Pathways in Alkane Oxidation by Selective Doping of Brønsted Acid Sites of Vanadyl Pyrophosphate', J. Phys. Chem. 94, 6813-6819.

[15] Centi, G., Trifirò, F., Busca, G., Ebner, J. and Gleaves, J. (1989) 'Nature of Active Species of $(VO)_2P_2O_7$ for Selective Oxidation of n- Butane to Maleic Anhydride', Faraday Discuss. Chem. Soc. 87, 215-225.
[16] Centi, G. Tosarelli, T. and Trifirò, F. (1992) 'Acrylonitrile from Propane on $(VO)_2P_2O_7$ with Preadsorbed Ammonia. 1. Role of Competitive Adsorption Phenomena in Determining the Selectivity', J. Catal., in press.
[17] Centi, and Perathoner, S. (1992) 'Acrylonitrile from Propane on $(VO)_2P_2O_7$ with Preadsorbed Ammonia. 2. Study of the Mechanism of Ammonia Adsorption and Reaction with C_3 Hydrocarbons', J. Catal., in press.
[18] Weng, L.T., Ruiz, P. and Delmon, B. (1992) 'Classification of the Roles of oxides as Catalysts for Selective Oxidation of Olefins', Ruiz, P. and Delmon, B. (eds.), New Developments in Selective Oxidation by Heterogeneous Catalysis, Elsevier Science Pub., Amsterdam, pp. 399-413.
[19] Pepera, M.A., Callahan, J.L., Desmond, M.J., Milberger, E.C., Blum, P.R. and Bremer, M.J. (1985) 'Fundamental Study of the Oxidation of Butene over Vanadyl Pyrophosphate', J. Am. Chem. Soc. 107, 4883-4892.
[20] Gleaves, J.T., Ebner, J.R. and Kuechler, T.C. (1988) 'TemporalAnalysis of Products (TAP) - A Unique Catalyst Evaluation System with Submillisecond Time Resolution', Catal. Rev.-Sci. Eng. 30, 49-116.
[21] Centi, G., Gleaves, J.T., Golinelli, G. and Trifirò, F. (1992) 'Problems and Outlook for the Selective Heterogeneous Oxidation of C_5 Alkanes', Ruiz, P. and Delmon, B. (eds.), New Developments in Selective Oxidation by Heterogeneous Catalysis, Elsevier Science Pub., Amsterdam, pp. 231-246.
[22] Centi, G. (1992) 'Selective Heterogeneous Oxidation of Light Alkanes. What Differentiates Alkane from Alkene Feedstocks?', Chem. Letters, in press.
[23] Centi, G., Gleaves, J., Golinelli, G., Perathoner, S. and Trifirò, F. (1991) 'In-situ Control of Vanadyl Pyrophosphate Deactivation by Addition of SO_2 during C_4-C_5 Alkane Oxidation', Catalyst Deactivation 1991, Bartholomew, C.H. and Butt, J.B. (eds.), Elsevier Science Pub., Amsterdam, pp. 449-456.
[24] Busca, G. and Centi, G. (1989) 'Surface Dynamics of Adsorbed Species on Heterogeneous Catalysts: Evidence from the Oxidation of C_4 and C_5 Alkanes on Vanadyl Pyrophosphate', J. Am. Chem. Soc. 111, 46-54.
[25] Centi, G. and Trifirò, F. (1991) 'Selective Oxidation of Light Alkanes: Comparison between Vanadyl Pyrophosphate and V- Molybdophosphoric Acid', Catalysis Science and Technology, Yoshida, S., Takezawa, N. and Ono T. (eds.), VCH Verlags Basel Pub., Vol 1, pp. 225-230.
[26] Centi, G. and Trifirò, F. (1990) 'Surface Kinetics of Adsorbed Intermediates: Selective Oxidation of C_4-C_5 Alkanes', Chem. Eng. Science 45, 2589-2596.
[27] Centi, G. and Trifirò, F. (1986) 'Furan Production by Oxygen Insertion in the 1,4 position of Butadiene on VPO based Catalysts', J. Molec. Catal. 35, 255-265.
[28] Schiøtt, B., Jørgensen, A. and Hoffmann, R. (1991) 'Formation of Maleic Anhydride on a Vanadyl Pyrophosphate Surface. A Theoretical Study of the Mechanism', J. Phys. Chem. 95, 2297-2307.

Theory of Brønsted Acidity in Zeolites

R.A. van Santen*, G.J. Kramer+, W.P.J.H. Jacobs*
* *Schuit Institute of Catalysis, Eindhoven University of Technology, P.O. Box 513, 5600 MB Eindhoven The Netherlands.*

+ *Koninklijke/Shell-Laboratorium, Amsterdam, (Shell Research B.V.), P.O. Box 3003, 1003 AA Amsterdam The Netherlands*

ABSTRACT. The nature of the chemical bond of protons in a zeolite is analysed on the basis of theoretical and spectroscopic results. Of interest is the dependence on zeolite structure as well as composition. The zeolitic OH bond is mainly covalent. Proton attachment to the zeolite lattice causes a weakening of neighbouring Si-O and Al-O bonds. The effective increase in volume of the bridging oxygen atom causes a local deformation, that changes the strength of the lattice-chemical bonds over a few bond distances. Proton concentration effects as well as lattice-composition effects can be understood on the basis of the lattice-relaxation model. The energetics of proton transfer is controlled by the need to stabilize the resulting Zwitter-ion. The positive charge on the cation becomes stabilized by contact with basic lattice-oxygen atoms.

1. Introduction

How the reactivity of the acidic protons attached to the framework-oxygen atoms in the micropores of zeolites depends on composition, location and zeolite framework is of obvious importance to the application of zeolites in acid catalyzed reactions. Whereas significant progress has been made to determine this relationship, it is still not completely understood. One of the reasons is the complexity of the catalytic reaction cycle, that is composed of different reaction steps. Proton transfer between lattice and substrate is only one of them and not necessarily always the rate limiting step. Secondly the acidity of the proton, as probed in the catalytic event depends on its state before reaction as well as the stability of the "Zwitter-ion" state, generated

R. W. Joyner and R. A. van Santen (eds.), Elementary Reaction Steps in Heterogeneous Catalysis, 113–131.

upon proton transfer to the reacting molecule. The latter may vary dependent on substrate and, as we will see, depends also strongly on the interaction between ion and zeolite wall.

Here we will discuss recent results of theoretical as well as spectroscopic studies probing the details of the chemical bond of the proton with the zeolite and the response of the proton with reacting basic molecules.

Protons in zeolites have been extensively studied using many different physical techniques, especially infrared spectroscopy, NMR spectroscopy and neutron scattering. Because of their well-defined nature zeolites belong to the exceptional class of catalysts that allow for a definitive structural characterization of its catalytically active site. Computational chemical techniques require such detailed knowledge. The use of ab-initio calculations has contributed in an essential way to an understanding of the chemical features that control zeolite acidity. An important theoretical question we will address is the validity of the cluster approach to approximate the protonic site. A second important theoretical approach is the use of lattice-energy calculations and lattice-structure simulations based on force fields determined from accurate quantum-chemical calculations.

In the following two sections we will highlight our present understanding of the nature of the chemical bond of the proton. In a subsequent section we will present the results of theoretical studies on proton transfer to adsorbed basic molecules. It will appear that two properties of the solid state are very important to zeolite acidity:

- electrostatics, the positive charge developing on the proton has to be compensated by a negative charge generating on the zeolite lattice.
- lattice relaxation, changes in chemical bonding affect the forces acting on the lattice atoms involved.

In a final section the implications of the physical chemistry of proton transfer to acid catalysis will be shortly outlined.

2. The proton Bond

The zeolite lattice consists of a three dimensional network of tetrahedra connecting four valent or three valent metal cations such as Si or Al, each having four oxygen atoms as neighbours. One oxygen atom has two metal cations as nearest neighbours (figure 1). When all lattice ions are Si, the zeolite lattice has the composition SiO_2 and is a polymorph of quartz. Brønsted acidic sites are generated when silicon, which has a formal valency of four, is replaced by a metal atom with a lower valency. Most common is the replacement of silicon by aluminum with a formal valence of three. A proton is attached to the oxygen atom connecting a neighbouring silicon and aluminum atom, resulting in a chemically stable situation. Note that now the oxygen atom becomes three coordinated.

Quantum-chemical calculations have convincingly shown that the SiO, AlO as well as the OH bonds have considerable covalency, resulting in a relatively weak OH bond.

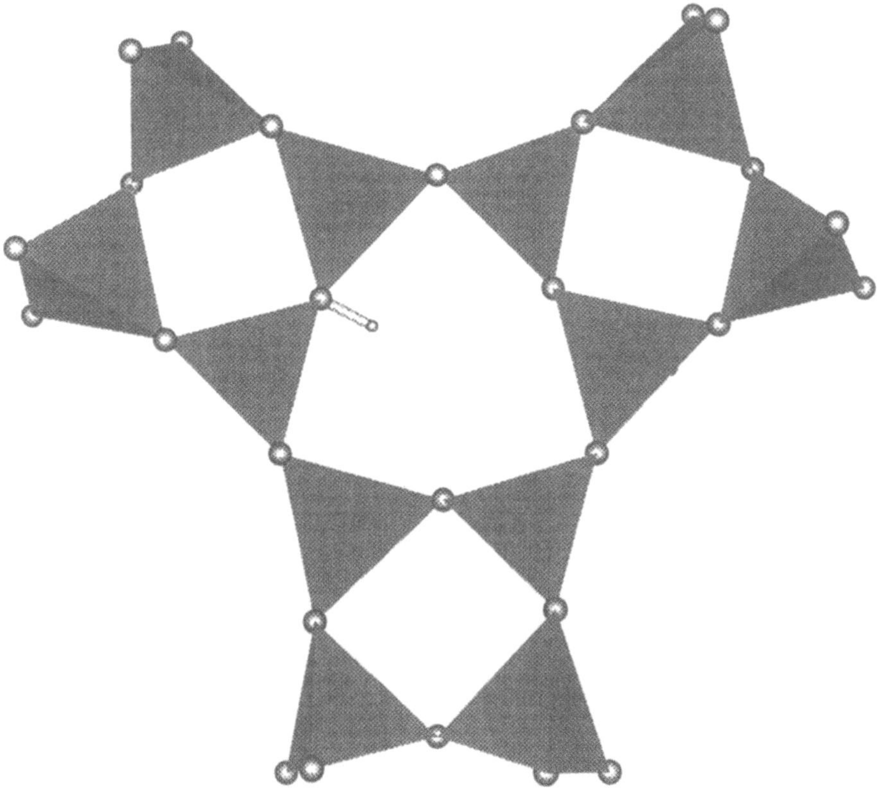

Figure 1. Connectivity of T-atoms. The T-atoms are in the centre of the tetrahedra formed by four oxygen atoms: when one silicon atom is replaced by an aluminum atom and a neighbouring oxygen atom becomes protonated.

The "onium" type coordination of oxygen is the fundamental reason for the high acidity of the attached proton. Cluster calculations on H or OH terminated rings in figure 2 give illustrations of such geometries providing detailed geometric information on the acidic site [1]. The results of ab-initio cluster calculations on protonated rings formed from four tetrahedra are shown. As we will see crystallographically different sites in a zeolite show small differences. One of the most interesting features is the long Al-O bond compared to the Si-O bond (0.188 nm and 0.172 nm respectively). The acidic site can also be seen as a silanol group promoted by a the Lewis acidic group. Whereas the OH bondlength changes little, the heterolytic bond energy changes from approximately 21.8 eV [2] to 13.7 eV [3].

Kazansky [4] has analyzed the OH interaction-potential by an analysis of the over-

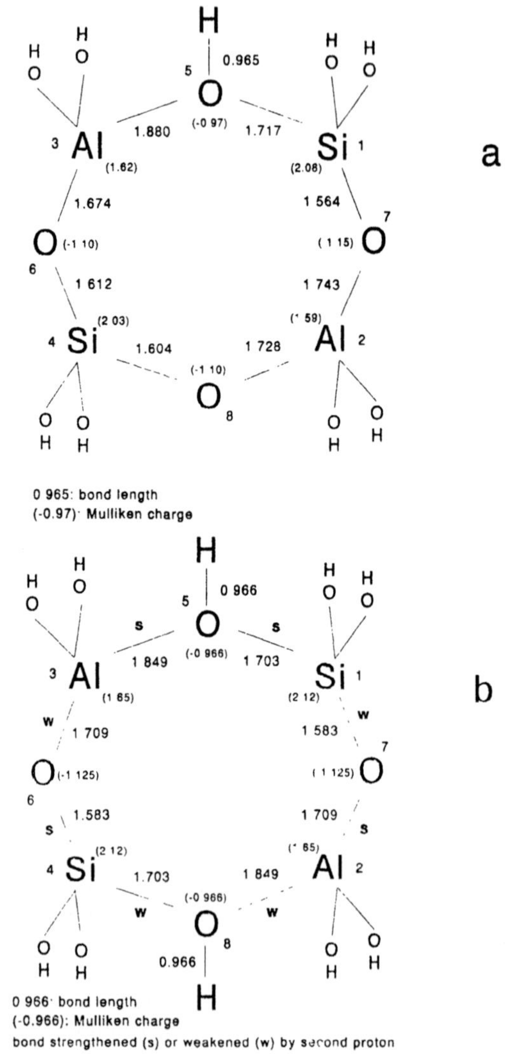

Figure 2. Four-ring cluster containing two silicon and two aluminum atoms and one (a) or two (b) acidic protons [1].

tones of the OH stretch frequencies measured in the near infrared spectroscopic range. He found that the covalent OH interaction-energy varies little with acidity. The constant homolytic bondstrength of OH indicates that the factor that contributes most to the strong acidity is mainly the increased stabilization of the negative charge left on the lattice with deprotonation.

1H MAS NMR spectra enable the experimental determination of the Al-H distance [5]. For zeolite Y values of 0.237 nm and 0.248 nm have been found for protons located

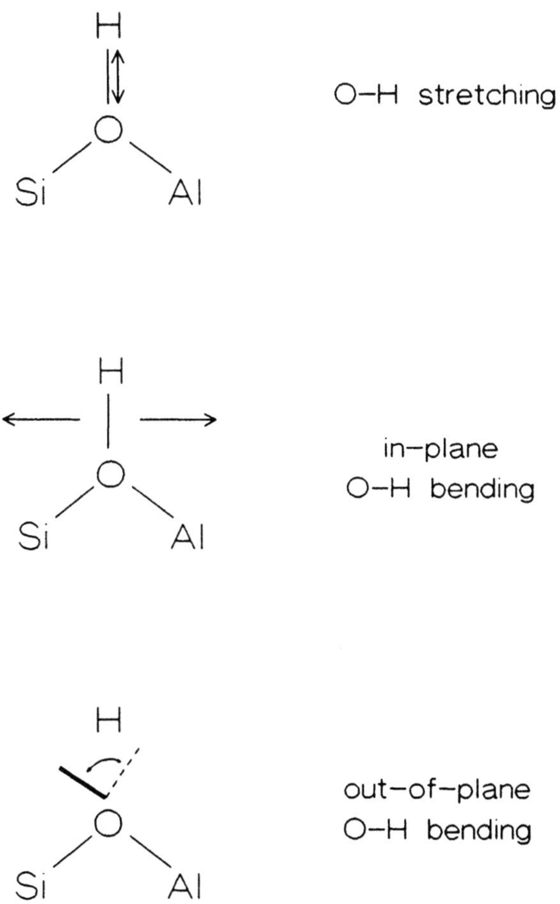

Figure 3. The three vibrational modes of the Brønsted acidic group.

in the supercages and sodalite cages respectively.

The cluster calculations indicate a large change in the equilibrium geometry of the ditetrahedral cluster upon deprotonation (compare figures 2a and 2b, atom-position 8). As is shown in figure 2b, the AlO and SiO distances shorten. These changes are a consequence of the covalent bonding and the Bond Order Conservation principle [6]. Since no electrons of the bridging oxygen atom are involved any more in bonding with the OH bond, they become available to additional bonding for AlO and SiO. The result is an increase in the SiO and AlO bondstrengths.

Ab-initio quantum-chemical calculations also provide information on the vibrational modes of the OH group. Infrared spectroscopy and inelastic neutrons scattering [7] can be used to compare predictions with experiments. The three vibrational modes of the Brønsted acidic group are sketched in figure 3. Typical values of the

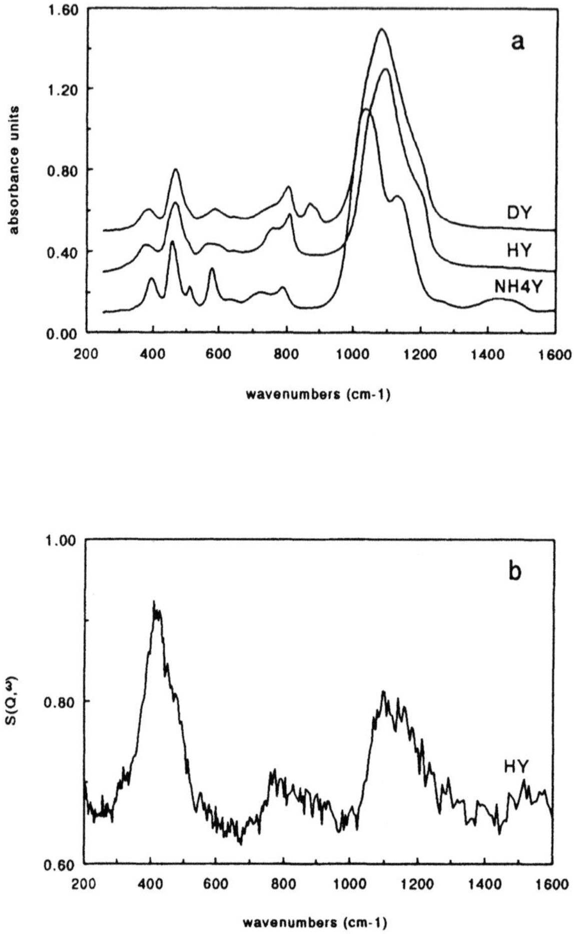

Figure 4. Infrared active lattice modes of NH_4Y, HY, DY (a) [11,12], and inelastic neutron scattering spectrum of HY (b) [7].

OH stretch frequencies of the Brønsted acidic protons vary between 3650 and 3550 cm^{-1}, to be compared with the silanol frequency at 3745 cm^{-1}. The relatively small decrease in stretch frequency corresponds to a slightly weakened OH force constant of the acidic protons. The in plane bending mode of the Brønsted acidic protons has a frequency of approximately 1050 cm^{-1} [7-10], the out-of plane bending mode of approximately 400 cm^{-1} [7,9,10]. Indirect measurement of the bending modes are available from infrared combination bands [7,8]. Direct measurements have been done using inelastic neutron scattering spectroscopy [7,10]. Experimental and theoretical values [9] are in close agreement.

Experimental indications of local structural relaxation upon deprotonation have been provided by infrared spectroscopic studies of the lattice-modes of zeolite Y, comparing the spectra of NH_4-Y with the H-Y form [11]. Figure 4a shows the corresponding infrared spectra. Also shown are the frequencies of the bending OH modes as measured by inelastic neutron scattering (figure 4b).

Protonation of the zeolite lattice results in significant changes of the lattice vibrational modes. One observes an upwards shift around 750 and 1100 cm^{-1}, decreases in intensity around 600 cm^{-1} and smaller changes at 400 cm^{-1}. Figure 4a shows also the infrared spectrum of deuterated zeolite Y. New peaks arise near 860 cm^{-1}[12]. The shifts around 750 and 1100 cm^{-1} and the losses in intensity below 600 cm^{-1} remain. Clearly in the region from 700 to 1100 cm^{-1} there is a strong coupling between the in-plane OH bending mode and the Si-O-Al stretching modes, that are resonant in energy. The existance of this coupling between the OH bending modes and the lattice modes is concluded from results from inelastic neutron scattering experiments [10]. Here, due to the coupling, the intensity of the lattice modes of H-Y zeolites is increased compared to those for Na-Y zeolites. The intensity of the modes between 600 - 300 cm^{-1} change very similarly for the protonated as well as deuterated system. The modes are mainly combinations of tetrahedral bending modes as well as Si-O-Al symmetric stretching modes, sensitive to deformation of the lattice [13]. These changes are due to the deformation of the lattice upon protonation.

Whereas the results of the ab-initio cluster calculations indicate the possibility of structural deformation upon protonation, embedding of the cluster in the three dimensional zeolite lattice is needed in order to evaluate to which degree it actually will occur in the solid. This can be conveniently done using energy minimization schemes applicable to the solid state. The equilibrium atomic geometries can be computed using atomic force fields [14]. Such force fields can be developed based on a fit of the parameters of classical interatomic potentials to potential energy surfaces computed quantum-chemically for small clusters [15]. One concludes from such studies that the changes in local chemical bonding due to proton attachment also lead to deformation of the lattice. However, the deformations are very local and disappear at a few atomic distances from the proton location [16]. This is illustrated in figure 5 for proton attachment to the O(1) atom of the double-six ring of faujasite. The geometry optimized structures of the protonated and non-protonated sites are compared. Atom positions before and after proton attachment are shown. One observes the longer SiO and AlO distances of the protonated system and the rapid decline of the bondlength changes at a larger distance from the proton.

Sites generated upon protonation require more space than the proton-free site. In zeolites atom distances of the lattice atoms vary for crystallographically different positions. One may expect to find a relation between bond angles and bondlengths around a site before protonation and the protonation energy. Structural and compositional effects will be discussed in the next section. The main conclusion of this section is that experimentally as well as theoretically local structural relaxation can be considered well established.

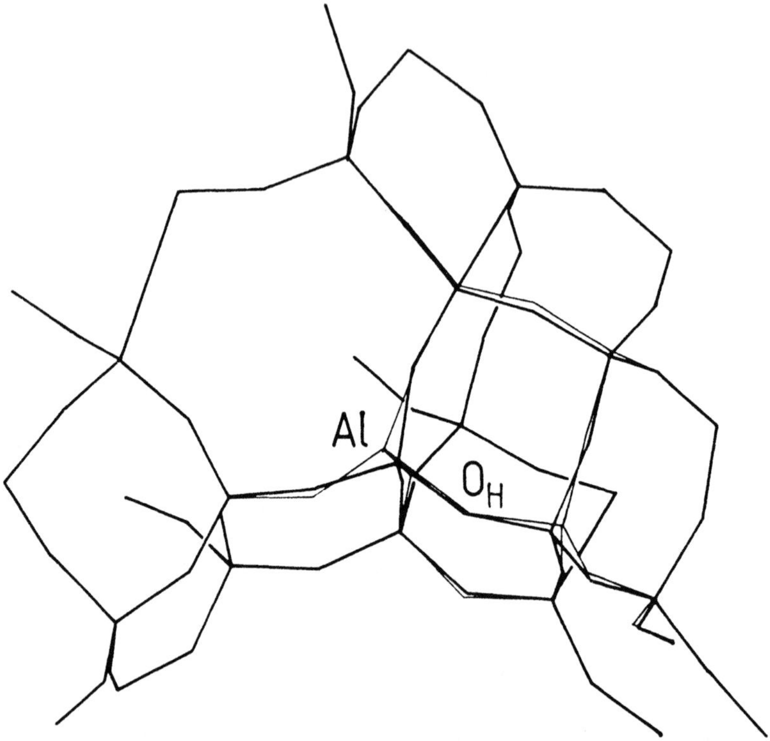

Figure 5. Lattice relaxation due to protonation of the double-six ring in the Faujasite lattice [16].

3. Proton affinity differences.

When one compares the stretch frequency of the OH bond in a zeolite in the same structure but varying the aluminum content, small changes in the infrared frequency are observed. Figure 6 shows the change in the infrared frequencies of high-frequency OH stretching modes in zeolite Y as a function of increasing Si/Al ratio [17]. Whereas at a ratio of Si/Al = 2.4, one observes only the silanol and two Brønsted acidic hydroxyl groups for zeolite Y (the high-frequency and low-frequency bands), upon an increase of the Si/Al ratio infrared HF OH peaks at lower frequencies appear, indicative of a stronger Brønsted acidity. This is in agreement with the shift to lower field of the corresponding lines in the 1H MAS NMR spectra (see figure 6).

In a zeolite according to the Löwenstein rule, aluminum ions cannot occupy neighbouring tetrahedra. This exclusion principle derives from the negative charge build up around the two tetrahedra leading to repulsive effects. At a low Si/Al ratio each [Si-O-Al] unit is connected to six tetrahedra of which a maximum of three is occupied by Al.

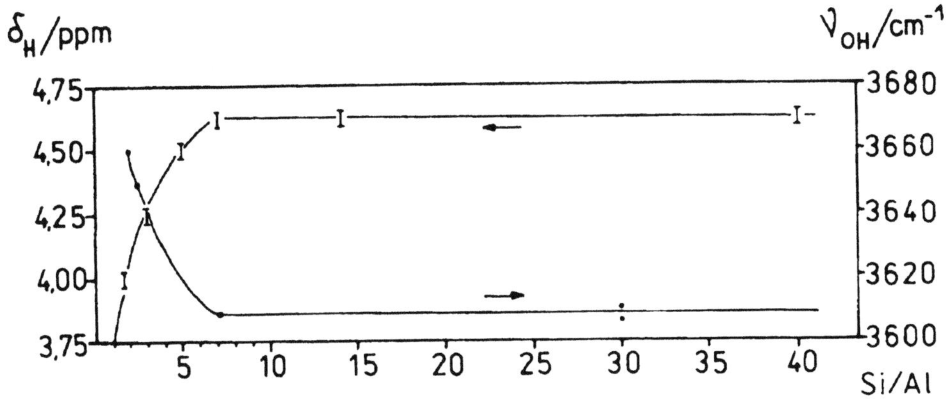

Figure 6. Infrared high-frequency OH stretch frequency (OH) as a function of the Si/Al ratio [17] and chemical shift (H) of the corresponding line in the 1H MAS NMR spectrum as a function of the Si/Al ratio.

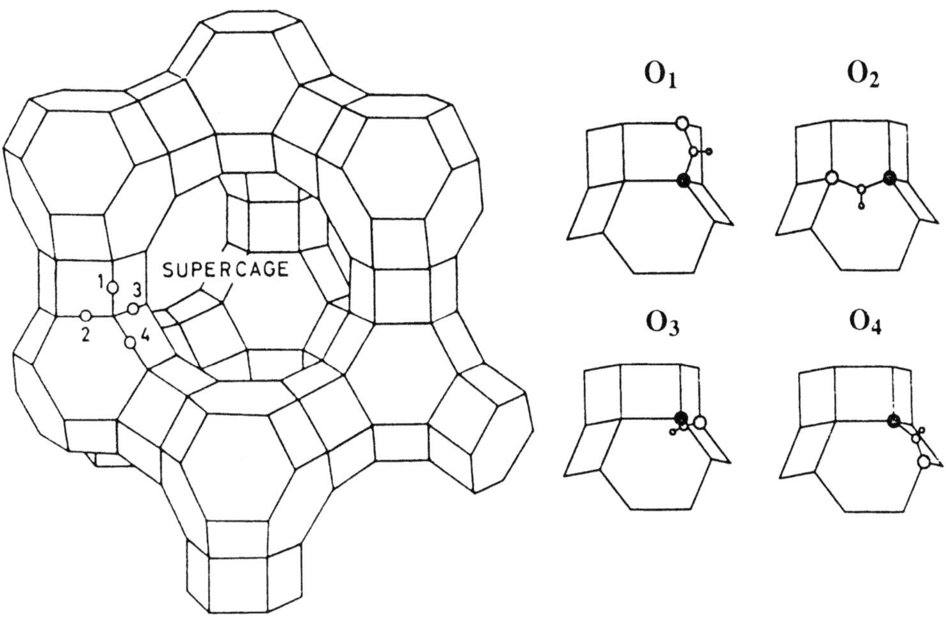

Figure 7. The four different oxygen atoms in the Faujasite lattice[21].

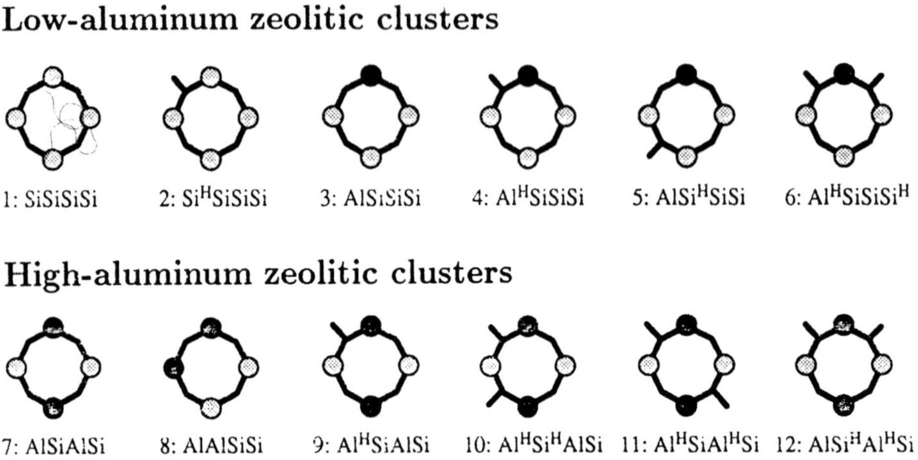

Figure 8. Four-ring clusters with different Si/Al ratios and different proton concentrations[1], see also table 1.

In the faujasite lattice positions are equivalent, but the four oxygen atoms forming a tetrahedron are non-equivalent (figure 7). Of these four oxygen atoms, protons are attached most strongly to O(1) and O(3) sites, giving rise to a HF (3650-3610 cm^{-1}) and LF (3550-3570 cm^{-1}) OH infrared absorption peak. The reasons for this will be discussed below. With a decrease of aluminum content, the probability for a [Si-OH-Al] unit to be surrounded by three tetrahedra with aluminum decreases. With a statistical distribution of aluminum over the tetrahedra (constrained by Löwenstein's rule) now different embedding possibilities of the [Si-OH-Al] unit exist. The frequency changes for the HF OH depend only on changes of composition in tetrahedra in the first coordination shell of the [Si-OH-Al] unit and converge to the same value at Si/Al>>10.

A decreasing heterolytic proton bond dissociation energy with increasing Si/Al ratio is also found from ab-initio calculations on clusters of varying Si/Al content. Figure 8 shows rings of four tetrahedra, terminated by end-on OH groups with varying Si/Al ratio. As follows from the corresponding table 1 the deprotonation energy is smallest for the tetrahedral four ring with one aluminum atom and the protons prefer attachment to an oxygen atom also coordinated to an aluminum atom. When the aluminum content increases, the deprotonation energy increases for a cluster with 2 protons as well as the negatively charged cluster, simulating compensation by cations. In figure 2 the corresponding geometries and charges are shown.

Experimentally the Brønsted acidity of zeolite Y appears to be a strong function of proton content. In figure 9 calorimetric measurements are presented of the ammonia adsorption on zeolite Y with three different compositions [18,19]. The interaction

Table 1. Total energies (in atomic units) and proton affinities (in eV) of 4-ring clusters. Complete geometry optimizations were performed at the STO-3G level; the 3-21G results refer to the STO3G-optimized geometry with an additional optimization of the proton position. Results marked by an asterisk refer to the abstraction of one proton only; the result in parentheses is influenced by a different configuration of the terminal hydrogen atoms [1].

		A:STO3G optimized geometry		B:3-21G STO3G geometry		C:3-21G optimized geometry	
	ring	E(a.u.)	PA(eV)	E(a.u.)	PA(eV)	E(a.u.)	PA(eV)
1	SiSiSiSi	-203.318558		-2048.522833			
2	$Si^{H}SiSiSi$	-203.779993	12.56	-2048.839850	8.63		
3	AlSiSiSi	-201.331399		-2001.777284		-2001.850643	
4	$Al^{H}SiSiSi$	-202.002739	18.27	-2002.279709	13.67	-2002.374317	14.25
5	$AlSi^{H}SiSi$	-201.925292	16.16	-2002.208117	11.72	-2002.309532	12.49
6	$Al^{H}SiSiSi^{H}$					-2002.749718	10.22*
7	AlSiAlSi					-2002.988098	
8	AlAlSiSi	-199.161940		-1954.885995			
9	$Al^{H}SiAlSi$	-200.024318	21.98	-1955.542784	16.77	-1955.626118	17.36
10	$Al^{H}Si^{H}AlSi$	-200.633693	16.58*	-1955.994844	12.29*	(-1956.120927	13.46*)
11	$Al^{H}SiAl^{H}Si$	-200.726481	19.10*	-1956.067643	14.27*	-1956.163893	14.63*
12	$AlSi^{H}Al^{H}Si$	-200.683936	17.94*	-1956.024463	13.11*	-1956.1256167	13.58*

Table 2. The frequencies of the experimental and calculated infrared spectra given in cm^{-1}. Experimental values are for NH_4^+ in: Mordenite (MOR), Faujasite (FAU), Beta (BET) and Erionite (ERI). The other spectra are calculated ones. 2H and 3H are the notation for the doubly and triply bonded NH_4^+. H is the structure in which the proton is attached to the zeolite, the peak given here thus is the OH stretching. COAD is a singly bonded NH_4^+ with one NH_3 coadsorbed (N-H stretching of the proton pointing towards the zeolite is shifted to 2123 cm^{-1}, but is not included in the table because it is mixed with a Si-H stretching).
(I) means that these peaks have a reasonable intensity, the other peaks have negligible intensity but are included for completeness [24].

MOR	FAU	BET	ERI	2H	3H	H	COAD
2780	2800	2970	2840	2623(I)	3103(I)	3142(I)	3153(I)
2930	3040		3068	2740(I)	3141(I)	3360	3363
3180	3270	3200	3260	3418	3478(I)	3478	3401
3400	3360	3460	3384	3495		3483	3473
							3476
							3483

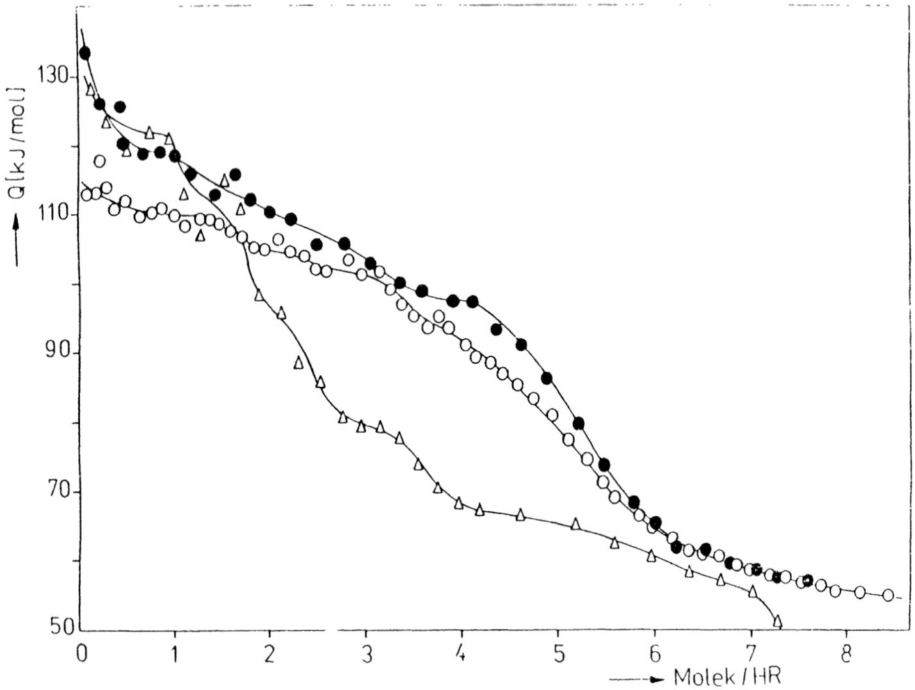

Figure 9. Differential heat of sorption (Q) versus the number of NH_3 molecules per supercage, measured at 423 K on Y type zeolites [19]: ∘ NaY(85), Si/Al=2.4; • HY(100), Si/Al=2.4; △ DY-3, Si/Al=4.1.

with zeolite Y (Si/Al = 2.4) is compared at 100% proton exchange and 85% proton exchange. One notes a significant increase in the initial heat of adsorption due to protonation of ammonia. The same increase in heat of adsorption is found for the dealuminated material.

Adding a proton to position O(8) results in a weakening of the O(5) - H bond, (enhancing the intrinsic Brønsted acid strength) and a strengthening of the neighbouring Si(3) - O(5) and Al(1) - O(5) bonds. This can be deduced from the bondlength differences in figure 2. Whereas proton attachment to position O(8) lengthens the Si(2) - O(8) and Al(4) - O(8) bonds, the next Si(2) - O(7) and Al(4) - O(6) bonds become stronger. As a function of distance to the proton on O(8) these changes alternate causing the O(5) - H bond to weaken. This again is a consequence of covalent bonding and the Bond Order Conservation principle. When one of the bonds to an atom weakens, the other will become stronger. The changes in atomic charges are very small hence one concludes that proton-strength differences are dominated by changes in covalent bondstrength.

For the LF OH peaks in zeolite Y a different behaviour upon dealumination is observed. With increasing Si/Al ratio the maximum for this band is shifted to higher

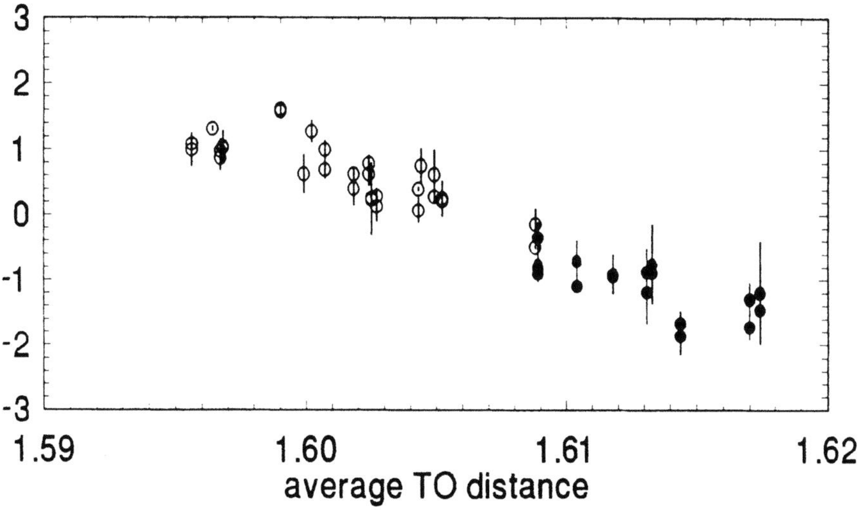

Figure 10. Relation between average T-O distance, O-O distance and T-O-T angle in the relaxed all-silica structure of MFI, and the (averaged) proton energy of the corresponding oxygen site [1].

frequencies. This behaviour cannot be explained by the small clusters discussed here, since the acidic protons, which are responsible for this infrared band, interact with two close oxygen neighbours at a distance of 0.26 nm [20].

To relate changes in OH bondstrength with composition to changes in lattice covalent bonding is one way of interpreting consequences of changes in electron density. As discussed elsewhere analysis in terms of HOMO-LUMO interactions leads to very similar conclusions [6]. It is important to realize that the changes in acidity are ascribed to electron-density differences and not to changes in atom charges.

Using force-fields, differences in proton affinity have been theoretically analyzed for two low aluminum content zeolite frameworks: zeolite Y which is a low density zeolite and silicalite of high density [1,21]. In zeolite Y the four oxygen-atom positions are non-equivalent. In silicalite 12 atom-positions are non-equivalent and 48 different oxygen atom positions are in principle possible. The deprotonation energy is calculated for fully relaxed bridging oxygen sites in the presence and absence of an added proton. These studies have demonstrated the importance of the zeolite lattice structure for the intrinsic Brønsted acidity of lattice protons. It is found that the deprotonation energies vary within an interval of 20 kcal/mole for different sites. For the MFI structure the amount of structural relaxation and corresponding energy gain is found to vary very little for the different sites, whereas there are large differences for the protonation energies. The parameter of the fully relaxed SiO_2 polymorph that correlates strongly with the OH interaction strength at a particular site is the SiO bondlength (figure 10). The larger the average SiO bondlength the stronger the OH bond. This is in line with the previously discussed bulkiness of the [Si-OH-Al] group compared to the Si-O-Si unit. The most strongly Brønsted acidic sites are found at sites that are spatially most constrained.

According to the results presented, the dominant effect controlling the OH bondstrength is the proton concentration (see table 1). When the proton concentration is low, the intrinsic Brønsted acidity per proton increases with decreasing Al-concentration. Structural diffences have a comparable effect as variation in Si/Al ratio.

4. Proton-transfer

In the acid - base reaction:

$$ZH + B \rightleftarrows Z^- + HB^+$$

charge is separated. In solution the cost of charge separation is overcome by solvation of the ions generated by solvent molecules. The dielectric constant of a zeolite is very small (~4), the energy of charge separation now has to be counteracted by the interaction between the positively charged protonated base and the negative charge of the zeolite wall.

Cluster-calculations on the protonation of NH_3 provide an interesting illustration of the importance Zwitter-ion stabilization [22,23]. Very accurate ab-initio calculations of the NH_3-[Si-OH-Al] dimer interaction with the ammonia nitrogen atom coordinating to the bridging OH site show that the OH distance remains 0.098 nm to be compared with the long HN distance of 0.277 nm (figure 11a). So proton transfer does not occur. The interaction between ammonia and OH shifts the OH frequency downwards by 380 cm^{-1} indicating a weakening of the OH bond. The computed interaction energy is 55 kJ/mole. The experimentally measured heat of adsorption of NH_3 is between 110 kJ/mole and 160 kJ/mole. Coordination as shown in figures 11b and 11c makes Zwitter-ion formation possible [23]. The NH_4^+ ion coordinates with two or three of its hydrogen atoms to the negatively charged oxygen atoms surrounding the Al atom. The computed interaction energies are very similar and equal to 114 kJ/mole. For the bidentate and tridentate bonded NH_4^+-ions, the infrared spectrum can be computed. Each coordination type is characterized by a different spectrum. Comparison with experimentally measured spectra enables an assignment of the two different coordination types(table 2). Also the frequency of the external vibration of NH_4^+ versus zeolite wall can be computed. For the bidentate and tridentate bonded NH_4^+-ions the obtained frequencies vary from 250 to 350 cm^{-1} [24]. These low frequencies compare very well with typical values of NH_4^+ (80-200 cm^{-1}) found for zeolite Y [25].

Comparison of the geometry of the aluminum-tetrahedral cluster before NH_3 interaction (figure 11d) and the interacting clusters (figure 11b and 11c) show that the geometry around the tetrahedra changes when the proton is transferred. In a rigid lattice this would be prevented and as a consequence no proton transfer would occur. This is what is found theorically if no geometric relaxation is allowed when computing proton transfer. So proton-transfer requires a flexible lattice, that allows for local deformations.

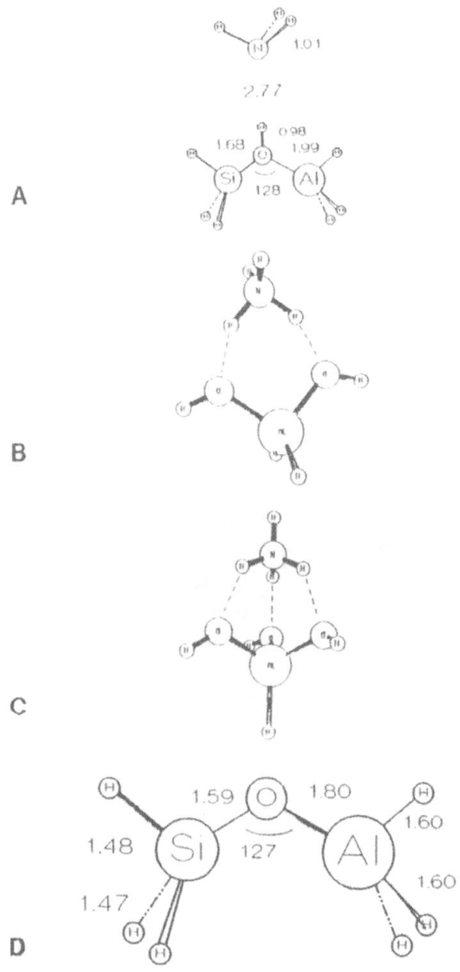

Figure 11. Clusters showing the interaction of NH_3 with the Brønsted site [23].

The equal interaction energies of NH_4^+ coordinated bidentate or tridentate implies that the NH_4^+ ion moves in a very flat potential. Neutron scattering energy loss data show that NH_4^+ is nearly freely rotating in the supercages of the Y-zeolite, but have a slightly more hindered rotation when in the smaller sodalite cage [26].

Whereas in water the basicity of ammonia is larger than that of pyridine, temperature programmed desorption experiments show always a higher temperature of desorption for adsorbed pyridine than ammonia [27,28]. The low basicity of pyridine compared to ammonia in water stems from the high heat of hydration of ammo-

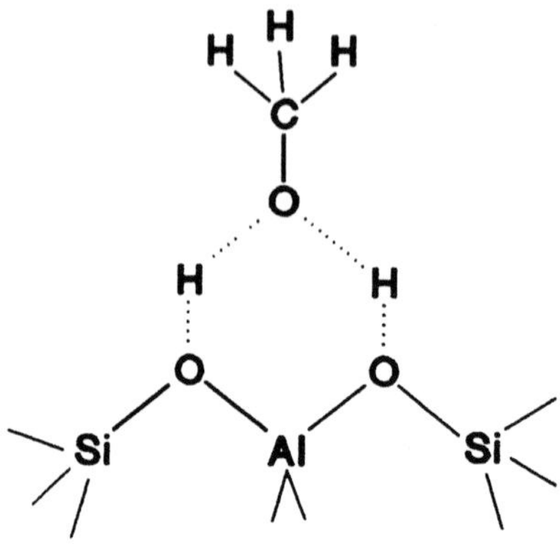

Figure 12. Transition state in the protonation of methanol [30].

nia compared to pyridine. The gas phase protonation energy of pyridine is 10.8 eV, whereas that of NH_3 is 9.9 eV [29]. The difference relates to the to the higher s-character of the pyridine lone pair compared to that of ammonia. The absence of hydration and the vacuum-type environment of the zeolite is responsible for the stronger basicity of pyridine in the zeolite environment.

Other quantum-chemical cluster calculations on the zeolite proton-polar molecule interaction show similar results. Protonation of H_2O leads to a hydronium ion that becomes adsorbed in bidentate coordination. Calculations [30] as well as infrared spectroscopic measurements [31] also find that protonated methanol becomes coordinated as a bidentate. This can be considered Brønsted acid-Lewis base assisted protonation. The Brønsted acid is the proton, the Lewis base is the other oxygen atom around aluminum to which the methanol proton attaches (figure 12). These results have an important consequence for the transition state of reactions that proceed via carbenium-ions as intermediates:

$$\mathrm{ZH + H2C = CH2 \longrightarrow Z^- \quad \underset{carbenium-ion}{H2C^+ - CH3} \longrightarrow propagation}$$

Carbenium-ion formation by ethylene has been studied quantum chemically [32,33]. Ethylene can interact in two ways with the proton. It can form initially a π-complex without proton transfer. In this state interaction is weak. When the proton becomes attached to ethylene (the transition state), the positive charge that develops at the other end of the carbenium ion becomes stabilized by interaction with the negative charge on an other atom around aluminum. This is the Brønsted acid-Lewis base assisted proton transfer mechanism. In the case of ethylene this leads to formation of a stable O-C bond, the σ-complex or surface alkoxide. The formation of σ-complexes has been confirmed by NMR [34,35].

5. Concluding remarks

In the previous section the quantum chemistry of an important intermediate to acid catalysis, the carbenium ion, was discussed. The relative stability depends on the deprotonation energy as well as the stabilization of the organic substrate by the negative charge generated on lattice oxygen atoms. The deprotonation energy is sensitive to lattice relaxation and depends on local geometric constraints due to long range structure of the zeolites as well as zeolite composition.

Acid catalysis is controlled by the chemistry of the protonation site as we discussed it here, as well as the adaption behaviour of reactant and products molecules.

6. References

1. G.J. Kramer and R.A. van Santen, submitted.

2. J. Sauer and W. Schirmer, In: "Innovation in Zeolite Materials Science (Studies in Surface Science and Catalysis 37), Eds. P.J. Grobet, W.J. Mortier, E.F. Vansant and G. Schulz-Eklhoff, Plenum, New York, 1986, p.323.

3. J. Sauer, J. Phys. Chem. 91 (1987) 2315.

4. V.B. Kazansky, Acc. Chem. Res. 24 (1991) 379.

5. D. Fenzke, M. Hunger, H. Pfeifer and J. Magn-Reson, 95 (1991) 477.

6. R.A. van Santen, Theoretical Heterogeneous Catalysis, World Scientific (Singapore) 1991, p.204, 320.

7. W.P.J.H. Jacobs, H. Jobic, J.H.M.C. van Wolput and R.A. van Santen, Zeolites 12 (1992) 315.

8. L.M. Kustov, V. Yu. Borovkov and V.B. Kazansky, J. Catal. 72 (1981) 149.

9. J. Sauer, J. Mol. Catal. 54 (1989) 312.

10. H. Jobic, J. Catal. 131 (1991) 289.

11. W.P.J.H. Jacobs, A.J.M. de Man, J.H.M.C. van Wolput and R.A. van Santen,

Proceedings, 9th Int. Zeol. Conf., July 5-10, 1992, Montreal.

12. W.P.J.H. Jacobs, H. Jobic, J.H.M.C. van Wolput and R.A. van Santen, in preparation.

13. R.A. van Santen and D.L. Vogel, Adv. Solid. State Chem. 1 (1989) 151.

14. C.R.A. Catlow and W.C. Mackrodt, Computer Simulations of Solids, Lecture Notes in Physics 66, Springer, Berlin, 1982.

15. G.J. Kramer, N.P. Farragher, B.H.W. van Beest and R.A. van Santen, Phys. Rev. B 43 (1991) 5068.

16. G.J. Kramer, A.J.M. de Man and R.A. van Santen, J. Am. Chem. Soc. 113 (1991) 6435.

17. D. Freude, M. Hunger, H. Pfeifer and W. Schwieger, Chem. Phys. Lett. 128 (1986) 62.

18. U. Lohse, B. Parlitz, V. Parlitz and V. Patzelova, J. Phys. Chem. 93 (1089) 3677.

19. U. Lohse and J. Jänchen, unpubl. results.

20. M. Czjzek, H. Jobic, A.N. Fitch and T. Vogt, J. Phys. Chem. 96 (1992) 1535.

21. K.P. Schröder, J. Sauer, M. Leslie, C.R.A. Catlow and J. Thomas, Chem. Phys. Lett. 188 (1992) 320.

22. E.H. Teunissen, F.B. Duijneveldt and R.A. van Santen, J. Phys. Chem. 96 (1992) 366.

23. E.H. Teunissen, R.A. van Santen, A.P.J. Jansen and F.B. van Duijneveldt, J. Phys. Chem., to appear.

24. E.H. Teunissen, W.P.J.H. Jacobs, A.P.J. Jansen and R.A. van Santen, Proceedings, 9th Int. Zeol. Conf., July 5-10, 1992, Montreal.

25. G.A. Ozin, M.D. Baker, J. Godber and C.J. Gil, J. Phys. Chem. 93 (1989) 2899.

26. W.P.J.H. Jacobs, R.A. van Santen and H. Jobic, submitted.

27. H.G. Karge and V. Dondur, J. Phys. Chem. 94 (1990) 765.

28. H.G. Karge, V. Dondur and J. Weitkamp, J. Phys. Chem. 95 (1991) 283.

29. W.J. Hehre, L. Radow, P.V.R. Schleyer and J.A. Pople, In: "Ab Initio Molecular

Orbital Theory", John Wiley & Sohns, (1986).

30. J. Sauer, C. Kölmel, F. Haase and R. Ahlrichs, Proceedings, 9th Int. Zeol. Conf., July 5-10, 1992, Montreal.

31. G. Mirth, A. Kogelbauer and J.A. Lercher, Proceedings, 9th Int. Zeol. Conf. July 5-10, 1992, Montreal.

32. V.B. Kazansky and I.N. Senchenya, J. Catal. 119 (1989) 108.

33. I.N. Senchenya and V.B. Kazansky, Catal. Lett. 8 (1991) 317.

34. M.T. Aronson, R.J. Gorte, W.E. Farneth and D. White, J. Am. Chem. Soc. 111 (1989) 840

35. J.F. Haw, B.R. Richardson, I.S. Oshiro, N.D. Lazo and J.A. Speed, J. Am. Chem. Soc. 111 (1989) 2052.

Session One, *Catalytic Reactivity*

Rapporteur P.B. WELLS
Department of Chemistry
University of Hull
Hull
HU6 7RX

The six lectures that make up this session have been very varied, but central to each, and to our discussions of catalysis in general, is the concept of the active site. Thus I have attempted to draw out the essence of the models used in the hope that it will both stimulate and underpin wider discussions.

The paper by Somorjai showed that defects and atomically rough surfaces are frequently associated with high activity (and that in consequence, our site models should be three-dimensional rather than two-dimensional), that adsorbates may induce restructuring (so we must beware having preconceived ideas about site geometry), that the surface may be mobile during reaction, co-adsorption effects may occur (ie. one adsorbate may regulate to activity of another so that adsorbate reactivity may not be solely determined by the adsorbate/site relationship), high reaction rates may be achieved at oxide/metal interfaces, and permanently held residues may be part of the total description of the active surface and may be vital to its good health.

By contrast, Rooney presented the approach of a physical chemist concerned to achieve the synthesis of complex organic compounds at oxide and metal surfaces. His premise was that site concentrations may be very low at both oxide and metal surfaces. Furthermore, he showed that the very simple process of σ-alkyl/π-alkene interconversion, sustainable at single metal atom sites, can be used to interpret a wealth of observations involving H/D exchange in cyclic and polycyclic molecules, syntheses of polycyclic compounds, carbon skeletal rearrangements and reactions in which radical species may be mobile at high temperatures. He also showed how small changes in polarity at a metal atom site might influence heterolytic as against homolytic bond fission - again at single metal atom sites.

A rather different philosophical approach was adopted in the paper by Basset in which catalytic activity was considered from the standpoint of the organometallic character of the active site. He demonstrated, for the case of a silica-supported zirconium hydride catalyst, how chemical evidence had been acquired for all of the elementary steps in the hydrogenolysis of neopentane at ambient temperature. Note that the sites again contained but one metal atom. A case was made that the thought processes, and terminology of the organometallic chemist are more precise, and more effective than those commonly used by workers in the field of heterogeneous catalysis. Thus he points out that one person's oxidative addition is another's dissociative chemisorption and that one person's ligand effect is another's promotion or poison effect. Most importantly, he claimed that the approach of organometallic chemistry can lead directly to creative site design, to the generation of new catalytic materials, and to catalysts in which all of the transition metal atoms are active sites.

In the fourth paper we moved from the potential glories of the tailored organometallic catalyst back to the more sober realm of supported metals, but now in the innovative form of thin metal films on ceramic electrolyte supports in which the active sites are subject to electrochemical promotion. Vayenas described the NEMCA (Non-Faradaic Electrochemical Modification of Chemical Activity) effect in which an ionic species (eg.O^{2-} or Na^{+}) is electrochemically pumped to a metal film surface (eg. Pt), under

R. W. Joyner and R. A. van Santen (eds.), Elementary Reaction Steps in Heterogeneous Catalysis, 133–136.

conditions such that high reactivity is generated on the three-phase boundary (ceramic/metal/reactive gas) and surface activation extends to the whole metal surface. For example, in Pt-catalysed ethene oxidation each O^{2-} so pumped to the three-phase boundary caused about 74,000 additional oxygen atoms to react with ethene. This was the fourth consecutive lecture to mention enhanced rates of catalysis at the oxide/metal/ reactant interface. In this case the NEMCA effect influences adsorption by increasing the sticking probability of oxygen, decreasing its desorption rate constant, and thus causing an increase in oxygen coverage. The systems show 'volcano relationships' of activity versus voltage, and XPS evidence was presented for the oxygen spill-over that is central to the interpretation of the NEMCA effect.

Centi moved the debate into the area of oxide catalysis, and in particular to considerations of butane oxidation to maleic anhydride and propane oxidation to a variety of products, each reaction catalysed by vanadyl pyrophosphate. Active sites are located in the surface region and consist of vanadium centres in clefts and cavities providing limited accessibility to reactants. (These are oxide analogues of the defects discussed by Somorjai and the shaped sites that provide selectivity by restricting the mode of adsorption, as proposed by Basset). The concentration of isolated but mobile oxygen species at the sites described regulates the selectivity in propane oxidation in the presence of ammonia, the possible products being propene, acrylate, and carbon dioxide. The presence of islands of inactive adsorbate enhance selectivity to alkane (again, we note parallel behaviour as between oxide and metal surfaces). This paper might well have been entitled 'Control of selectivity via restrictions of site environment.'

The final paper by van Santen was concerned with zeolite catalysis and the applications of theoretical calculations to an understanding of adsorption in these materials [1]. Acid catalysis in zeolites is controlled to a great extent by the chemistry of the protonation site. I mention but one of the many interesting facets of this paper: in proton transfer, such as in the conversion of an ammonia molecule to the ammonium ion, the transfer can occur only if there is a structural relaxation of the zeolite framework. Proton transfer requires lattice flexibility. This returned us again to the dynamic character of catalyst sites and to the realisation that all chemical steps at solid surfaces have the potential to regulate or reorganise the environment in which they occur. The advances that we achieve in heterogeneous catalysis over the next decade may well depend on, and be measured by, the extent to which we are able to develop and refine our concept of the active site.

Participants in the discussion highlighted the language difference used by various groups of chemists and physicists to describe common or similar problems in catalysis. Thus, heterogeneous catalysis attracts the interest of physical chemists, chemical physicists, organic chemists, organometallic chemists and chemical engineers. It was asserted that workers in these fields fail to communicate effectively, and the speakers were asked whether they perceived this to be a serious problem.

Rooney reminded the audience that the previous NATO ARW in catalysis, held in Sardinia some 20 years ago, was attended by workers in the fields of heterogeneous, homogeneous and enzymic catalysis [2], and that the difficulties of communication were all too apparent. Perhaps the environment in which to find solutions to the problem is in interdisciplinary collaborations and joint publications, or in joint investigations involving the use of new techniques.

The strongest driving force towards effective communication is interdisciplinary collaboration made necessary at the present time by the diminished funds available for research. Hardship will force us into this new mould. Examples can already be seen in areas where organic chemists have become acquainted with the language of the biologists. It was suggested that the theoretical chemist has it in his power to provide the common language, although we have no generalised theory for catalysis at the present time; when we do we shall find that we have a common language!

Activity patterns across the Periodic Table had been discussed in the session. Ponec pointed out, by reference to examples, that it was customary to attempt correlations of activity with single parameters, whereas in the real situation, many variables were relevant, the problem was multi-dimensional and multiple maxima would probably occur if a full evaluation were possible. The Panel adopted several approaches to this problem. Interpreting activity changes by reference to thermodynamic quantities such as heats of formation, tells only part of the story. We are fortunate in having some well-founded contradictory information available from surface science on the one hand and catalysis on the other, which may help us to identify missing links in our present arguments. A knowledge of the rate determining process which governs the observed activity is of the first importance. If the rate determining step remains the same under all of the experimental conditions being considered, then maybe correlation with one parameter may be valid; however, if the reaction path changes, then correlations, if they are possible at all, will be of higher dimensionality. While we must always be careful to compare reactions showing steady state behaviour, we must also take into account the fact that catalytic activity evolves to the steady state value. The number of factors that contribute to that evolution of activity are so great that steady state activity is seldom likely to correlate with only one parameter.

At this point the discussion widened into a general consideration of important objectives for the catalysis community over the next ten years. This was stimulated by Somorjai's view that catalysis must address the following problems. First, selectivity in organic reactions, eg. chiral catalysis, reactions involving biological molecules, and processes of interest to the fine chemicals industry. At the Catalysis 2000 Symposium in Budapest in July 1992 there was recognition of increasing environmental pressure to achieve the ultimate goal, namely 100% selectivity in 100% non-polluting processes. This may well involve the use of lower temperatures and liquid phase reactions. Second, active sites must be defined and their concentration determined. Surface science will need techniques with time resolution over may orders of magnitude in order to achieve this objective. Third, there is presently very little surface science applied to catalysis by solid acids and by oxides, and this situation must be remedied.

This analysis generated many vigorous responses. Industry concerned with the production of small scale speciality chemicals is unlikely to be in a position to fund work of generic significance, although, for example, steps to replace liquid acids by solid acids have been taken in Japan. Environmental pressures will lead to exploratory work and not fundamental research, to the ultimate detriment of scientific progress. The academic community has insights different from those of industry and the environmental lobby which are nonetheless valid. One of the basic tasks of the academic community is to develop the methodology for the solution of problems, and important areas to address include (i) the relationship between microscopic catalysis and macroscopic chemical engineering; (ii) catalysis at high surface coverages, and (iii) catalysis in the liquid phase including its relationship to electrochemistry.

Active site concentrations are currently determined under static conditions (eg. by use of CO chemisorption measurements), whereas what is required is the concentration under the conditions of reaction. In the reaction environment co-adsorbates may influence both surface site concentrations and selectivity, with the result that the concentration itself may be a dynamic quantity. Site concentrations are required for the determination of turnover frequency, but for truly effective catalytic systems (such as those involving Tolman cycles inside Tolman cycles, or reactions that are highly coupled in other ways) the very definition of site concentration presents philosophical difficulties.

The prospect of time resolved surface science studies stirred deep emotions! On the one hand, there was a view that the achievement of the required techniques lies too far in the future for serious consideration at present, which was opposed by the assertion that a commitment to state-of-the-art dynamics studies is one way ahead in the battle to obtain fundamental research funding, society's backing, and well motivated postgraduates; not to go down this path would be a mistake. Nevertheless, progress is currently being made; the time frame chosen should be appropriate for the studies in hand; good work with TAP reactors provides a suitable example. The approach should not be discarded simply because the whole range of 10^{-13} seconds to many hours may not be accessible for a few decades. Equally, however, the community would lose credibility if it were to shoot at inaccessible targets.

In the UK, funding authorities have recognised that the present surface science capability to study oxides is comparable to the position in metal catalysis 25 years ago; in consequence investigations of oxides has been a priority area.

The Panel also responded to questions of a more detailed nature. For example, will our recognition of surface restructuring lead to a complete understanding as to why some reactions are structure sensitive and some structure insensitive? Responses were varied. Some restructuring occurs with clean metal surfaces; some accompanies chemisorption and is induced and reversed on the chemisorption/desorption timescale; some restructuring of the adsorbate layer has been observed, and some structure sensitivity effects may be attributable to changes in hydrogen coverage and not to surface structure. This simple question does not admit of simple answers.

The application of results from surface science experiments at low pressures and coverages to catalytic reactions at relatively high pressures and coverages presents difficulties; how confident should we be about extrapolating such information? Somorjai replied that future surface science techniques must be high pressure friendly. The development of non-linear optical techniques, time-resolved STM, nmr studies and synchrotron-related techniques will all assist towards an understanding of chemisorption and catalytic reactivity at high surface coverage. Meanwhile, information obtained at high coverage but low pressure by use of low temperatures may be used, and may in due course be found to correlate with that obtained at higher pressures and temperatures.

In the special case of NEMCA reactions, is there a clear relation between the work function of the metal surface and the coverage achieved? If so, the volcano plot shown to the meeting would appear to represent a variation of activity with surface coverage. Vayenas replied that, for example with ß"-alumina as electrolyte, the effect is caused by the establishment of a surface coverage of sodium, which modified θ_{CO} and θ_O, and the sum of the changes provides the change in work function. He was asked whether oxygen coverage could not be changed by other means such as the use of semi-permeable membranes, but replied that very high pressure (kilobars) would be required to achieve the desired effect, and the oxygen under those conditions might not be in the right form. The chemistry of the spill-over process is as yet not well understood, and it might for example be useful to look for electrochemically induced reconstructions by STM.

1. The paper by van Santen *et al* was presented in this section for logistical reasons, although its subject matter is more appropriate to session 4.

2. Basolo, F. and Burwell, R.L. Jr. Eds., (1973) 'Catalysis: Progress in Research', Plenum, New York.

SURFACE REACTIVITY OF OXIDES

E. BORDES
Department of Chemical Engineering, University of Technology
B.P. 649, 60206 Compiègne Cedex
France

ABSTRACT. Considerations on the factors governing selectivity in different kinds of oxidation reactions of hydrocarbons and small oxygenated compounds are proposed. Among them the surface crystal field displayed by a catalyst, depending on its structure and on crystal morphology, is shown to be the primary factor. With the help of knowledge on methane coupling and butane oxidation, sites needed to activate and oxygenate molecules are proposed for alkanes. In this view the main assumptions obtained by a model (CMAS) describing the elementary steps of butane oxidation on vanadyl pyrophosphate are recalled.

1. Introduction

In heterogeneous catalysis, reactions performed on metals and oxides share the same complexity, particularly when solids are deposited on a support. The difficulty is to evaluate the real mechanisms which are involved at the surface. In the case of metals, the crystals take various forms as aggregates in which the coordination is different than in bulk metals, which give them a particular quality and favors the selectivity. Not only the support itself influences the reaction by means of acidic properties (for example), but also metal-support interactions can play a great role as for SMSI. However several parameters such as work function, atomic adsorption energies, etc, and even quantum chemistry calculations (SCF, Morse potential,...) allow to estimate the state of surface atoms as compared to the bulk and to draw models which are closer and closer from the reality.

Oxides of transition metals M are generally used in the selective oxidation catalysis of hydrocarbons and of small oxygenated molecules like methanol. In metallic solids, only one kind of bond, the metallic bond, is to be considered, and the spacial distribution of bonds is mainly isotropic at least in the bulk. On the contrary, inside an oxide like V_2O_5, V-O bonds range from covalent to ionic and their distribution around V is strongly anisotropic. Consequently the structural and electronic properties of MO_x oxides are also anisotropic even in the bulk, and the diversity of the M-O bonds is precisely related to their catalytic specificity. The methods used to characterize the surface give generally an average value of a given property because the results account for the whole grains. The contribution of the morphology of the crystallites should be taken into account because of the variety of bonds distributed among the various facets, some of them being responsible for selectivity. Happily the most well-known catalysts V_2O_5 and MoO_3 display a layered structure, so that their catalytic properties are known since a long time because the contribution of their cleavage plane is preponderant. This is less true for WO_3. In fact, it has been shown recently that, if MoO_3 is selective but non

R. W. Joyner and R. A. van Santen (eds.), Elementary Reaction Steps in Heterogeneous Catalysis, 137–153.

active in the oxidation of propene in acrolein, this is because the (010) cleavage plane is not the selective one.

In order to progress in the knowledge of the catalytic behavior of phases by which million tons per year of products are made industrially, we found necessary to "imagine" how can be an active and selective surface, with the help of electronic and structural criteria of selectivity. Without decreasing the importance of recent methods of characterization of oxide surface, it must be kept in mind indeed that it will be always very difficult to catch what happens exactly between a reacting molecule and several atoms of a solid surface in 10^{-9}-10^{-6} seconds. The aim of this paper is to recall the main characteristics of the oxidation of hydrocarbons, including the alkanes which are lately used and for which no mechanism is yet known. The role of the structure and of the reactivity of the catalytic crystals will be outlined. The main conclusions obtained with a model (CMAS), accounting for the transformation of a reactant (here butane) on an hypothetic perfect surface (vanadyl pyrophosphate) will be further developed.

2. Catalytic structure and reactivity in oxidation of unsaturated hydrocarbons.

In heterogeneous catalysis, the mild (amm)oxidation of hydrocarbons and the oxidative dehydrogenation (ODH) of oxygenated compounds (e.g. methanol) proceed on oxides and mixed oxides of transition metals. When reactants are not alkanes the mechanisms of reaction are fairly well established, at least for small molecules. They are based on some principles and/or experimental observations which are recalled below.

2.1. PARTICIPATION OF LATTICE OXYGEN.

The first point which has been demonstrated is the participation of the surface lattice oxygen(s) to the reaction, which is not made on adsorbed O species like on metals. These oxygens are further incorporated in the products of the reaction. The activation of, e.g., an olefin, proceeds by a concerted mechanism consisting in the abstraction by a surface O^{2-} of a hydrogen in α of the C=C bond, which is situated over a coordinatively unsaturated surface (c.u.s.) M^{n+} cation. For this purpose, the electronic configuration of M must be d^0 or d^{10} [1-7]. According to Haber [4], the activation of propene yielding the π-allylic intermediate is a nucleophilic process. This step is followed by the nucleophilic addition of O^{2-} and then by abstraction of a second hydrogen from this oxygenated species, yielding acrolein and water. However, when the electrophilic process of activation of gaseous oxygen leads to the formation of reactive O_2^{2-}, O_2^- or O^- forms instead of O^{2-}, these reactive specie can undergo an attack of C-H and/or of C-C bonds. This process is certainly responsible for the formation of carbon oxides in any reaction.

2.2. CONSIDERATIONS ON STRUCTURE AND REACTIVITY

MO_x oxides and their derivatives (mixed oxides and oxysalts) belong to the ReO_3 structural family in which the 3-D arrangement of M and O atoms is very particular. Taking V_2O_5 as an example, the small ionic radius and high charge of V allows an easy change of valence, from V^{5+} to V^{4+} and conversely. This $M^{n+}/M^{(n-1)+}$ redox system is involved in the well-known kinetic mechanism by which the oxidation of the hydrocarbon (-HC-) results in the formation of the oxidized product (-C-O) (and water) and the reduction of the catalyst (K), which is reoxidized by gaseous oxygen in a second step (KO):

$$(\text{-HC-}) + KO \rightarrow (\text{-C-O}) + K \qquad r_{red} \qquad (1)$$
$$K + 1/2\, O_2 \rightarrow KO \qquad r_{ox} \qquad (2)$$

A more detailed scheme states that reoxidation by dioxygen does not proceed at the site which is reduced after desorption of (-C-O), and that reoxidation of the surface can be made by diffusion of lattice oxygen. At the steady state the respective rates of reduction, r_{red}, and reoxidation, r_{ox}, of the catalytic solid are equal. However, until the steady state is reached, the solid facing the reactant (generally a mixture of (-HC-) and O_2) behaves according to its bulk **reactivity**. Indeed, if during the transitory state $r_{red} << r_{ox}$ or $r_{red} >> r_{ox}$, the resulting catalyst seems to be fully oxidized or reduced respectively. Molybdenum oxide and bismuth molybdates (facing propene) [2] and $(VO)_2P_2O_7$ (facing n-butane) follow respectively this behavior. Intermediate situations can be found when $r_{red} \approx r_{ox}$ with, e.g., V_2O_5 or $VOPO_4$ facing butene [8, 9]. In such cases the main difficulty is to stabilize a given proportion of the reduced and oxidized phases, keeping in mind that one of them can be inactive and/or unselective [8]. Another related point is that surface reoxidation can proceed mainly by diffusion of lattice O^{2-} from the bulk to the surface or with gaseous oxygen. Again bismuth molybdates and $(VO)_2P_2O_7$ are respectively representative of each behavior.

Another consequence of the small ionic radius and high charge of M is that its coordination is easily modified, from a (distorted) MO_6 octahedron to pentahedron and even MO_4 tetrahedron. Three consequences follow:

- Various kinds of metal-oxygen bonds exist, from the short (covalent) M=O found in all coordinations, to the long (more ionic) M...O bond (*trans* to M=O in octahedron), through all iono-covalent intermediates. Therefore a panel of M-O bonds with their specific length, strength and energy exists naturally in these oxides.
- The variety of bonds is responsible for the anisotropic character of crystal structures. Layered crystals result, as for V_2O_5 and MoO_3 which exhibit cleavage planes, because the long M...O bonds are cut at the surface.
- Owing to the possible change of coordination, several vacancies can be suffered by the solid (extended defects) without a structural collapse. When they are too much numerous, the reduction proceeds by aggregation of these vacancies, followed by nucleation of crystallographic shear planes due to the edge-sharing of octahedra earlier corner-shared. This mechanism does not affect the short M=O bonds which are perpendicular to the shearing plane, except in their length and in the coordination and oxidation state of M. Let us notice that crystallographic shear planes do not nucleate in bismuth molybdates, which tolerate only a small amount of surface vacancies, so that a deep reduction leads to decomposition.

To improve the catalytic properties a second cation M' (M' = V, Mo, W, Ti, Bi, P, Te,...) is generally added to M. M' can enter the lattice of MO_x as a solid solution, or give a mixed oxide (e.g. V_2MoO_8) or an oxysalt (e.g. $VOPO_4$). The result of course is that the structure and reactivity are more or less modified, and that supplementary bonds M'-O and M-O-M' are added to M=O, M-O-M, etc. Another possibility is that M' as a separated $M'O_y$ oxide exerts an effect through coherent interfaces between $M'O_y$ and MO_x, provided the structural misfits between them are small. The best example is V_2O_5/TiO_2 anatase for which a synergetic effect is observed in both catalytic (selectivity in phthalic anhydride from o-xylene enhanced) and not catalytic processes [5, 10].

2.3. CATALYTIC ANISOTROPY

It is not surprising therefore that oxidations are **structure-sensitive**, since a catalyst, with its own panel of bonds, structure and reactivity, is active and selective in one reaction. Once arrived on a surface, the molecule (e.g., unsaturated hydrocarbon) has to find at least one surface oxygen (O_s) to be activated while dehydrogenated (dehydrogenation), plus another one (at least) to be inserted (oxygenation). At least two kinds of O_s are therefore needed for mild oxidation [8].

In fact, the less complex the reaction, the more numerous the catalysts, as shown in Table 1 (reactions with alkanes are also presented and will be discussed later). In a simplified view, several catalysts can be used when water is the only oxygenated product of the reaction (ODH, oxidative coupling), except when the reactant is itself an oxidized compound. In such cases (methanol, isobutyric acid,...), another factor of selectivity is that this oxygen must remain in the product (as such or with another function). The oxidation of propene in acrolein is made on bismuth molybdates, but it is also possible to use tellurium-molybdenum or copper oxides with satisfactory results. This 4-electron reaction needs one O_s to be inserted in the intermediate and another O'_s to form water. On the contrary the oxidation of n-butane in maleic anhydride (MA) is a very demanding reaction since 14 electrons are exchanged, three oxygens are inserted while the C_4-chain must be cyclized, and four water

TABLE 1. Main reactions of oxidation and their catalysts.

Reactions	Number of electrons	Products	Catalysts
Dimerization			
2 CH_4	2	$C_2H_6 + H_2O$	alkaline earth and rare
	4	$C_2H_4 + 2H_2O$	earth oxides,
2 C_3H_6	6	$C_6H_6 + 3H_2O$	Bi-La-O
ODH			
C_4H_8	2	$C_4H_6 + H_2O$	molybdates
C_2H_6,	2	$C_2H_4 + H_2O$,	
C_3H_8, C_4H_{10}	2	C_3H_6, C_4H_8, + H_2O	Mg vanadates
ethylbenzene	2	styrene + H_2O	phosphates, molybdates,...
isobutyric acid	2	methacrylic acid + H_2O	iron phosphates, HPA*
CH_3OH	4	HCHO + H_2O	iron molybdates
Oxygenation			
C_2H_4	2	acetaldehyde	V_2O_5 (+ Pd)
C_3H_6	4	acrolein + H_2O	Bi-Mo-O, Te-Mo-O, U-Sb-O
C_4H_8	12	maleic anhydride + $3H_2O$	V-P-O, U-Mo-O
C_6H_6	20	maleic anhydride + $3H_2O$ + $2CO_2$	$(Mo_xV_{1-x})_2O_5$
o-xylene	12	phthalic anhydride + $3H_2O$	V_2O_5/TiO_2
CH_4	4	HCHO + H_2O	V_2O_5 or MoO_3/SiO_2
C_2H_6	6	$CH_3COOH + H_2O$	V-P-O or Mo-V-Nb-O
C_4H_{10}	14	maleic anhydride + $3H_2O$	$(VO)_2P_2O_7$

* HPA = heteropolymetallates.

molecules are formed. This process leads to the formation of at most seven vacancies, and even if they do not appear simultaneously, the lattice must accommodate several of them without structural collapse. One catalyst, $(VO)_2P_2O_7$, is known to provide yields close from 70 mol.% at high conversion [11]. Such high results were not obtained when we tried to find another selective catalyst, the multicomponent $CoMoO_4$-MoO_3/TiO_2 [12].

Since it has been shown that the atomic surface arrangements are primarily determined by the same structural principles than in the bulk, the chemisorption sites can be characterized by the same geometry [13]. By extension, M atoms and M-O bonds found in bulk catalysts can *a priori* be involved in a catalytic process. Taking a distorted MO_6 octahedron as example, covalent M=O bond will act as active sites, or at least as an electron reservoir, while at the opposite weak M...O bond will be either cut off, resulting in anionic vacancy Φ over M^{n+}, or restored with gaseous O_2, yielding weakly bonded O_s. A given crystal face is therefore characterized by a geometric and energetic set of oxygens linked with cations, some of them being unsaturated, which determines its **surface crystal field** [8].

We can add now that oxidations are crystal face-sensitive, which has been called **catalytic anisotropy** by Ziolkowski [14], and results in differences in reactivity of crystalline faces. Catalytic anisotropy was demonstrated in several cases [15], using single crystals of MoO_3 or V_2O_5 in presence of propene, butene, methanol, etc. In the case of MoO_3 it was found that (010) and (101) or (100) were selective in the formation of CO_2 and acrolein respectively by oxidation of propene. But on the contrary (010) which is not selective with propene allows methanol to be oxidized in HCHO and not in CO_2 (Fig. 1). Therefore it can be said that a given molecule facing a given surface crystal field is transformed accordingly in **a given** product, which results in selectivity.

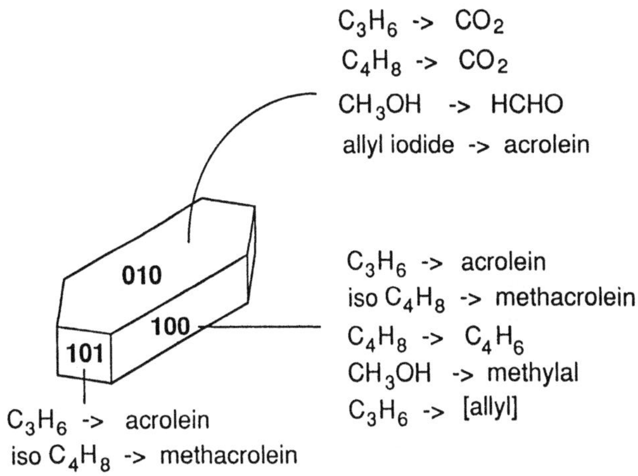

Figure 1. Catalytic anisotropy: reactions occuring on various faces of MoO_3 (after [15]).

Obviously, activity and/or selectivity are not observed when the morphology of the solid is not favorable, that is when the cleavage plane, which usually develops the largest area facing gaseous reactants, do not display the right sites. The poor activity and high selectivity exhibited by MoO_3 in propene oxidation has been attributed to the low area of the selective (101), (100) faces compared to (010), which is the cleavage plane in natural samples [15]. On the contrary, the cleavage plane (010) of V_2O_5 provides several sites able to (amm)-

oxidize more or less selectively various hydrocarbons which are linear C_2-C_4 olefins and C_6-C_8 aromatics, and even 3-picoline [16]. In turn, it is not easy to have the only one kind of sites which would be selective in one product, and this is the reason why generally selectivity is best achieved when a second cation M' is added. For examples, M' = P and Mo added to V-O system give rise to $VOPO_4$ and $(Mo_xV_{1-x})_2O_5$, which are active and selective in maleic anhydride by oxidation of butene and benzene respectively.

The crystal morphology, which determines the area and the nature of the faces which border the grains of the solid, has therefore a large influence on selectivity. The more demanding the reaction, the stronger is this influence. The fact that morphology depends at first on preparation enhances (if necessary) the importance of the methods used. A fruitful example is vanadyl pyrophosphate in the oxidation of n-butane [17-21], for which the best performance has been related by numerous authors to the occurence of (100) planes on the surface. Such plate-like crystals displaying (100) faces are obtained reproducibly by calcination of the layered precursor $VOHPO_4.0.5H_2O$, the dehydration of which is topotactic and yield pseudomorphic crystals of $(VO)_2P_2O_7$. The redox system is also modified because the γ form of $VOPO_4$ is obtained by oxidation of $(VO)_2P_2O_7$ instead of β [18, 19]. In turn, the crystal morphology of the precursor is related to preparation: several factors such as P/V ratio, concentrations of V and P of raw materials, duration of reflux, etc, among which the nature of alcohol used to reduce the vanadium-phosphate complex [22], have been identified [19-21].

3. Oxidation of Alkanes

Recently the use of alkanes as raw materials to offer an alternative to make olefins, or to obtain the same oxygenated products than with olefins, have been put in evidence owing to their cheapness. To date the most studied reactions in the field of functionalization of C_1-C_5 alkanes are the oxidative coupling of methane and the oxidation of n-butane (and pentane) in maleic anhydride. Although less developed for the moment, the oxidative dehydrogenation of ethane, propane, butane in the olefins and the oxidations of ethane in acetic acid or propane in acrolein are under investigation. Ammoxidation of propane in acrylonitrile on two kinds of catalysts, Bi-V-Mo-O and V-Al-Sb-O, is already fairly well patented (Sohio). Generally speaking, oxidation reactions are characterized by low conversions and low productivities and their development is still a challenge. Twenty-five mol.% of C_2-hydrocarbons seem to be the maximum yield obtained by coupling of methane [23]. Owing to the chemical inertness of alkanes the temperature can be very high, which favours homogeneous gas phase reactions hindering selectivity and difficult to control. An exception is the case of n-butane for which several commercial processes have been elaborated. With the active and selective $(VO)_2P_2O_7$ catalyst, high conversions of butane and selectivities in maleic anhydride of 60-72 mol.% are claimed according to the process. Academic works on this subject are continuously in progress. This reaction presents two advantages: the catalytic performances are high and the structural properties of the solid are well-known, although it belongs to the complicated VPO system [18, 19]. By studying this reaction models can be drawn and the main characteristics which could be valuable for oxidation of other alkanes can be determined by comparison.

Therefore, by combining what is known on the reactions with methane and butane, and by recalling the above selectivity criteria for oxidation of olefins, some directions can be proposed in order to find new catalysts or to improve the properties of others.

3.1. COUPLING AND DEHYDROGENATION OF ALKANES

Unlike for unsaturated hydrocarbons the mechanism of activation of alkanes in presence of oxygen[1] is not known. Olefins are weak donor of π-electrons which are accepted by metallic cations with empty d^0 (or full d^{10}) orbital: soft bases interact with hard (or soft) acid cations [24]. Oxidation catalysts are also said to be first dehydrogenating agents, and consequently these two functions are required for selective oxidation of unsaturated hydrocarbons [1]. But what about alkanes?

The main works on oxidative coupling of methane have been performed, at high temperatures and high methane to oxygen ratio, on oxides which are not of transition metals (Table 1). Catalysts are made of basic oxides like MgO, eventually doped with stronger bases like Li_2O; more recently rare earth oxides were also studied. Activation should therefore mainly proceed by withdrawing of a first hydrogen by a strongly basic surface O_s. It has long been recognized that $CH_3\bullet$ radical is the primary intermediate in the case of methane. Homogeneous recombination of radicals leads to the formation of ethane and ethylene. One part of the produced ethylene comes also from the direct ODH of ethane on the catalyst [23, 25]. The ODH of C_2-C_4 is indeed possible on alkaline earth oxides [26], and also on magnesium vanadates which are currently studied [27, 28]. It seems that for the latter performances depend more on operating C_n/O_2 ratio and on the presence of oxygens bonded to magnesium than on the kind of vanadate. The point common to methane coupling and ODH of C_2-C_4 is therefore the presence of hard acid cations (Li^+, Mg^{2+}, La^{3+},...) bonded to hard basic O_s^{2-} anions able to withdraw the first hydrogen from the molecule. Let us recall that in these reactions the surface lattice oxygen is not inserted in the hydrocarbon molecule and goes away in the form of water. The question of knowing whether a C-C bond is or not adsorbed over a cation (as found with unsaturated hydrocarbon), or over O-M, does not seem to be actual in these cases.

3.2. MILD OXIDATION OF ALKANES

It is well known that activity and selectivity in butane oxidation in maleic anhydride are strongly related to the occurence of $(VO)_2P_2O_7$ and to a mean oxidation state of bulk vanadium close from +4.0 (among other features) [17-21]. Inactive catalysts are obtained when vanadium (V) phosphates like $VOPO_4$ are used. The common point with the above reactions is that, once more, a basic oxygen bonded to a hard acid cation is necessary to activate n-butane. Accordingly, V^{4+} (or VO^{2+}), or P^{5+} acids to activate butane are less hard than Li^+, Mg^{2+}, La^{3+},..., used for methane [23]. However in the present case, once n-butane is activated, it must incorporate three surface oxygens to yield MA, and four other oxygens result in water. Notwithstanding the fact that V^{5+} are necessary to ensure these oxygenations, the successive modifications needed by the transformation of C_4H_{10} in $C_4H_2O_3$ can proceed only if the intermediate complex is strongly anchored on the surface. In the present case, enough basic oxygens (e.g. bonded to V^{4+} or P^{5+}) are present to catch hydrogens. Slightly less basic O_s^{2-} could be responsible for anchoring of the molecule by formation of -C...O_s bond. This hypothesis is supported by the model which will be developed further (*vide infra*).

Another example of demanding reaction, more recently examined, is the oxidation of methane in formaldehyde which is carried out between 450-650°C and in oxygen-lean feed

---

1 Oxidation by molecular oxygen will only be considered.

on vanadium or molybdenum oxides supported on silica [29, 30]. The presence of an ethoxide intermediate seems to be related with the selectivity in formaldehyde. The surface $-O_sCH_3$ fragment would be further dehydrogenated and then desorb in the form of CH_2O. Let us recall indeed that selectivity in acrolein from propene is obtained when the second H-abstraction occurs **after** insertion of the lattice oxygen. It has long been known that silica promotes the formation of Mo^{5+} and Mo^{4+}, the latter (or more exactly the oxygens bonded to) being responsible for the complete degradation of CH_2O in CO_2.

In oxidation reactions, the first H-abstraction by a basic oxygen $O\text{-}Mo^{5+}$ or $O\text{-}V^{4+}$ could be concerted with the formation of the ethoxide intermediate (or any O-containing fragment) following the $V\text{-}O_s...CH\text{-}$ anchoring. This anchoring would be done either by opening of the strongly covalent V=O (or Mo=O) bonds of oxocations, or rather on the opposite side $O_s...V(=O)$, electrons of these bonds being delocalized to help the anchoring. It cannot be ascertained whether these oxocations are VO^{2+}or VO^{3+} (MoO^{3+} or MoO_2^{2+}), although high oxidation states are necessary in order for the catalyst to be reduced when the oxidized product desorbs.

Selectivity criteria for the mild oxidation of alkanes can be summarized as (i), basic surface oxygen bonded to hard acid oxocation, (ii), hard acid cation able to undergo a redox mechanism and bonded to oxygen on which anchoring of carbon fragment occur, (iii), anisotropic structure allowing the formation of extended defects which favour the surface reoxidation. In the case of ODH, the first part of (ii) only would be necessary, enough basic oxygens ensuring the dehydrogenation. By using these criteria, we have found a catalyst able to oxidize ethane in acetic acid [31]: several formula containing VO^{2+} and/or MoO^{3+} were tested for activation at ambient pressure. All of them were active but their stability was shown to depend primarily on the relative amount of dioxygen as compared to ethane. With the most stable like $(VO)_2P_2O_7$ or Mo-V-Nb-O, at low temperature (200-300°C), acetic acid (S = 60-100 mol.%) was obtained with a low conversion of ethane. At higher temperature (300-500°C), conversion increases up to 15 mol.% and ethylene (+ CO, CO_2) is produced with a large selectivity. The difficulty is that two contradictory conditions are required, since $O\text{-}V^{4+}$ are needed to activate ethane while $O\text{-}V^{5+}$ are needed for oxygen to be further incorporated. Too much V^{4+} give a very active catalyst but poorly selective while too much V^{5+} lead to inactivity, which is the case when oxygen is in excess in the feed.

4. Oxidation of Butane and Butene in Maleic Anhydride: Model of Active Sites.

At first sight, trying to establish a model of active sites for a complicated reaction like butane oxidation to maleic anhydride, which is performed on vanadyl pyrophosphate, the structure of which has been solved with a poor resolution factor (R = 0.09) [32], is a very tentative task. However one examining the literature is easily convinced that this reaction on this catalyst is a good example for the following reasons:

1- The kinetics are simply described by a direct oxidation of butane, CO_2 being the by-product formed by oxidation of both reactant and MA,

2- Catalytic results obtained with the best catalysts are conversion 80-95 mol.%, MA selectivity 68-72 mol.%, by-products being acetic acid and CO_x. These results are good as compared with oxidation of n-butenes on the same kind of catalyst, MA selectivity being less than 52 mol.%, with 18 by-products at high conversion [9].

3- One catalyst, $(VO)_2P_2O_7$, is known. The best performances are related with this single phase (90 % at least needed), and also with its special morphology: selectivity is correlated with the presence of layered crystals displaying (100) planes. More complicated examples are found in the literature. High performances in the oxidation of propene in acrolein are obtained which multicomponent molybdates including at least six metallic cations.

4- Although the structure is complicated and solved with poor accuracy, the (100) plane of $(VO)_2P_2O_7$ is made from particular entities consisting in pairs of edge-sharing distorted octahedra (VO_6) surrounded by phosphates (PO_4) [32]. In each pair of (VO_6) short V=O bonds are in *trans* position. In the perpendicular plane (001) columns of pairs of (VO_6) extend parallel to straight and bent pyrophosphate (P_2O_7) groups. Owing to this strong structural anisotropy it is easy to have a picture of what could be the arrangement of atoms on a (100) surface.

5- At last, the set of V-O bond lengths in neighboring octahedra presents the peculiarity of being significantly different: focusing on V=O bonds only, it is seen that they are in the range 1.73-1.70 and 1.53-1.55 Å (Table 2). These values compare well with those found in V^{4+} and V^{5+} compounds respectively. In other words each pair of octahedra contains both the reduced and the oxidized cations. Strong electron delocalization inside octahedra and tetrahedra (P-O bonds are strongly covalent), and between them, must exist in order to compensate. It must be outlined that these values cannot be accounted for by the poor resolution factor. Several crystallographers examined this point and found roughly the same features. However a strong disordering, which would be the variability in the directional orientation of the vanadyl columns running perpendicular to the (100) surface, has been noticed in samples obtained by dehydration of the $VOHPO_4.0.5H_2O$ precursor [33].

TABLE 2. Examples of bond length R, energy e_{Ox} (kcal.mol^{-1}) of V-O bonds and type of site, as calculated by the relation $e = P / (R - Q)$, P and Q empirical constants [39].

Sites	R (Å)	e_{Ox}	Type (from inactive A to very active D
V1,O1 (V^{4+})	1.730	-117.1	B
V1,O1'	2.211	- 42.7	D
V4,O13 (V^{5+})	1.551	- 193.2	A
V4,O13'	2.367	- 34.5	D

All P -O bonds are C-type.

4.1. RECALL ON C.M.A.S.

The Crystallographic Model of Active Sites (CMAS), elaborated by J. Ziolkowski, was already described in several papers and applied to several oxidation reactions [14, 34-38]. One example is the oxidation of propene on MoO_3 [14], the application of the model accounting for the main observations made with single crystal [15]. This model provides the **geometric and energetic map** of each crystal face by using relations between bond length - bond strength - bond energy [39, 40], in relation with the energetics of the elementary steps of the reaction, for each face to be considered. Among them, steps such as adsorptions of molecular oxygen, of hydrogen (coming from the molecule), insertion of surface oxygen (in the molecule), formation and desorption of H_2O, as well as the mobility of these species can be completely characterized. The energetically-easiest pathway is that one for which the smallest endothermal barriers exist, and concerted exo- and endothermal steps are also considered over these sites. These catalysts working in presence of O_2, the first point is to consider the actual state of a face equilibrated with oxygen. In some cases the surface is mainly covered by oxygens (some faces of $(VO)_2P_2O_7$), in others mainly reduced (V_2O_5, MoO_3). The use of CMAS therefore allows, 1) to consider the surface structure on which active sites are considered as a multiplet, 2) to determine the enthalpies of all possible and necessary steps of the reaction, and 3) to indicate the energetically easiest pathway for the whole reaction.

4.2. APPLICATION TO BUTANE ON (100) $(VO)_2P_2O_7$

The model and its application to butane and butene on (100) $(VO)_2P_2O_7$ and on some other faces are fully described in our papers [36-38]. Here will be recalled only the most important points. In the structure four vanadium (V1-V4), four phosphorus (P1-P4) and 18 oxygens are crystallographically non-equivalent. For the sake of simplicity, the bond lengths and bond energies can be distributed in several ranges. So V1, V2 behave as V^{4+}, and V3, V4 as V^{5+}, and four kinds of O_S are bonded to V and P more or less strongly according to e_{Ox}. They are labelled A (inactive) to D (very active) (Table 2). Figure 2 shows the framework of (100), consisting in pairs of octahedra linked with tetrahedra, the top layer being constituted by oxygens A to D over V and P cations.

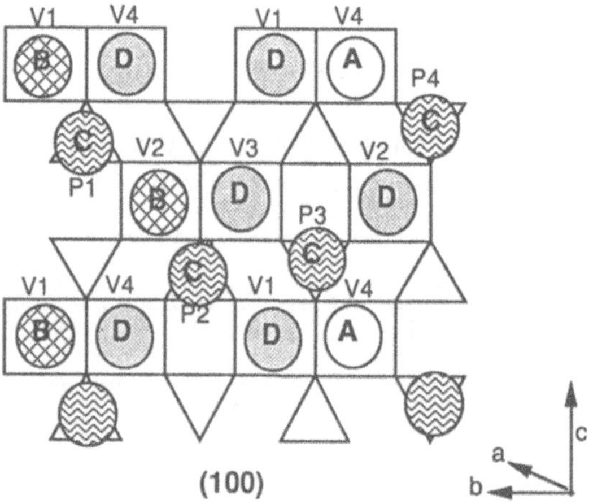

Figure 2. Framework of (100) $(VO)_2P_2O_7$ (solid lines): pairs of edge-sharing octahedra extending along b are linked along c by phosphate groups. Oxygens in the top layer (large circles) are labelled A to D according to their degree of unsaturation (see Table 3).

4.2.1. *Elementary steps involving oxygen.* Let us consider first the elementary steps with oxygen, i.e. its "mobility" on the surface, and the way of adsorption of O_2 when Φ vacancies are present. The (100) face of $(VO)_2P_2O_7$ in its equilibrium state is fully covered by oxygens, since all energies of binding, e_{Ox} are negative (Table 3). Once vacancies are created (after desorption of products, or because defects already exist), molecular oxygen is easily adsorbed on pairs of neighboring sites (V-V or V-P) ($q_{Ox} < 0$). The transfer of O from one site to a neighboring vacant one is not so easy ($q_{Otr} > 0$), and very difficult further. Oxygen spillover, which is invoked in the model of remote control [41], seems therefore unlikely here, since the long range transfer is defavorable.

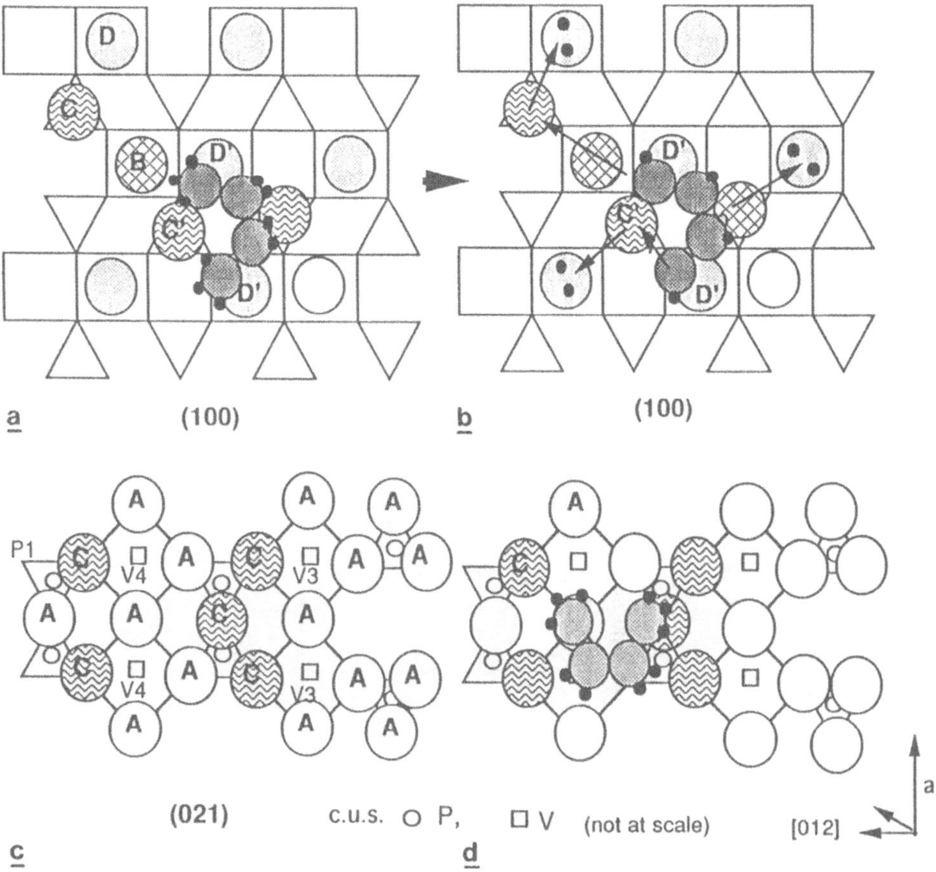

Figure 3. Adsorption of n-butane on (100) and (021) $(VO)_2P_2O_7$. a) Anchoring of butane over surface oxygens D', and b) dehydrogenation steps (arrows indicate paths followed by H) while butane is O-anchored on (100). c) Active sites , and d) one possible way of adsorption of butane on (021). Medium shaded and small black circles are C and H respectively (carbon, oxygens and lattice at scale),

4.2.2. *Elementary steps involving hydrogen.* Interactions between surface oxygens and hydrogens of the molecule are of several types (Table 3). During the reaction several hydrogens are withdrawn from the molecule (e_{Ha}), then adsorbed (e_H), and can move on the surface (q_{Htr}) as long as they are not in a stable state. For example H abstracted on B goes on C and stops on D where OH is more stable (see Fig. 3b). It can wait for another H in order to make water (q_W). Except for the desorption of H_2O which is endothermal, all these energies are negative, particularly e_H and e_{Htr}. This means that hydroxyls are very probably present, and that spillover of hydrogen is easy on this surface. To summarize this part, the surface of (100) should be covered mainly by O and OH, which is in accordance with experiments made by infrared spectroscopy [42].

4.2.3. *Adsorption and reaction of n-butane.* Butane arriving over (100) $(VO)_2P_2O_7$ is "chemically polarized" by this surface crystal field, as it meets with very active oxygens able to dehydrogenate it and displayed in a particular way. According to the energetically-easiest pathway found in former work [36], the adsorbed molecule bridges over two rows of pairs of

TABLE 3
Energy states $-e_{Ox}$ of surface oxygen, enthalpies of dissociative adsorption $-q_{Ox}$ of O_2 and of oxygen transfer q_{Otr} between adjacent sites ; energy states $-e_H$ of hydrogen adsorbed on A-D sites, enthalpies q_{Htr} of H-transfer between adjacent sites and q_W of desorption of water on (100) $(VO)_2P_2O_7$ (all energies in $kcal.mol^{-1}$) (after [37]).

Sites	$-e_{Ox}$*	Diss. Ads. of O_2 on	$-q_{Ox}$**	O transfer from → to	q_{Otr}	$-e_H$	H transfer from → to	q_{Htr}	Desorp. H_2O from → to	q_W
A	200	A-D	120 - 150	A → D	150 - 170	-	-	-	-	-
B	120	B-D	30 - 50	B → D	80 - 90	102 - 90	D → B	14 - 25	B → D	11 - 22
C	87	A-C	160 - 185	A → C	130	113	C → B	10 - 25	C → D	35 - 45
D	40	B-C	83 - 110	B → C	30 - 40	116	D → C	0 - 6	B → C	58
		D-C	0 - 20	C → D	40 - 50					

* : a low $-e_{Ox}$ means weakly bonded oxygen.
** : ranges are due to crystallographically non-equivalent V, P, O sites.

edge-sharing octahedra by means of C...O bonds (Fig. 3a). By a concerted mechanism involving each terminal carbon, anchoring of carbon on D and H-abstraction on B and/or C, D proceeds. The following steps consist in successive dehydrations leading to the formation of 3 H_2O (Fig. 3b), while C=C bonds are formed in the O-C anchored complex, etc (Fig. 4).

The cluster of active sites (multiplet) involved in the whole process (loss of 8 H in the form of water, insertion of 3 O) extends over a little more than one unit cell (Fig. 3a, b). It is situated on the cross of three routes by which hydrogens, once easily abstracted (high $-e_H$), move easily (to give further H_2O) (Tables 3 and 4). During the transformation, the anchoring C...O on both terminal carbons is very strong, these oxygens being inserted finally as the

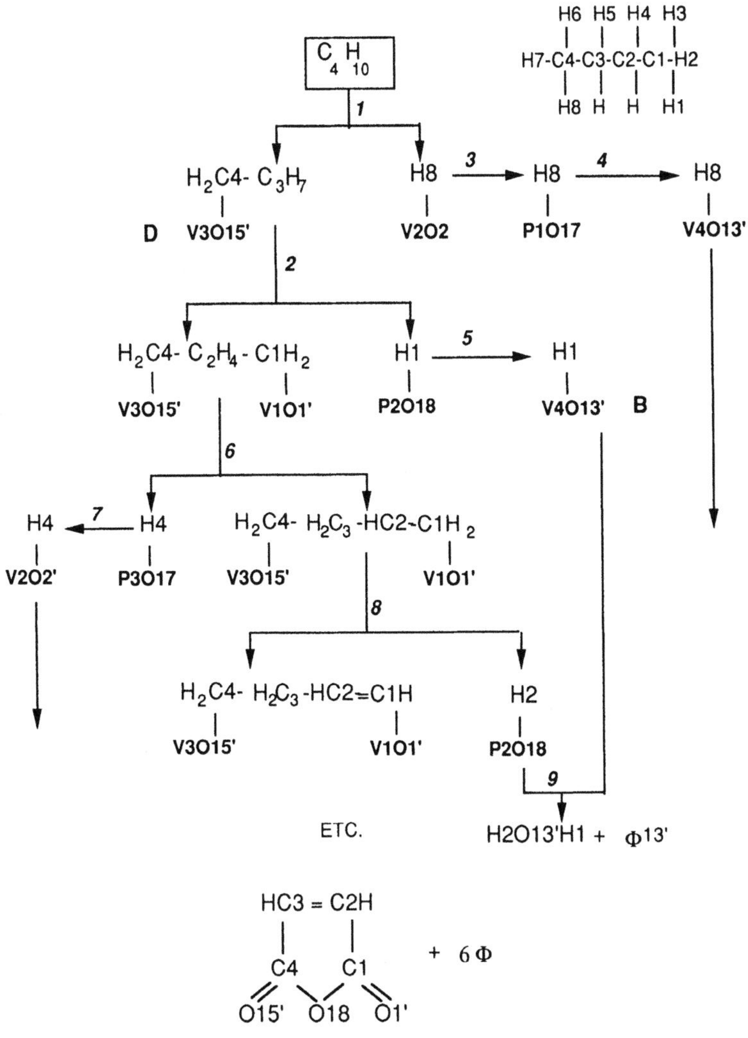

Figure 4. The first elementary steps of the oxidation of butane in maleic anhydride. C and H are labelled as in the upright; V, P, O are labelled as in the structure description [32]. Italic numbers refer to the successive steps (after [37]).

TABLE 4. Energies e_{Ox}, e_H of O and H states over the surface, q_{Otr}, q_{Htr} of transfer of O and H, and q_W of water, for various (hkl) planes of $(VO)_2P_2O_7$ (all heats in kcal.mol^{-1}) (after [37]).

Planes	$-e_{Ox}$	q_{Otr}	q_W	$-e_H$	q_{Htr}
(100)	V sites : 35-45 P sites : 87-91*	45	11-40	90-116	1-26
(021)	137-142	1-3	35-40	62-67	1-3
(001)	91-105	0-11	78-80	109-112	0-2
(010)	137-138	1	39	67	1

* only one P-site is used ; the heat is decreased in the concerted step.

C=O ketone groups of maleic anhydride. The furan-type oxygen is C', already linked with phosphorus (low e_{Ox}). The desorption of MA and 4 H_2O leaves six vacancies on the surface, the seventh being filled by oxygen before the end of the process. The reoxidation of the surface is easy as outlined above.

As such, this model accounts for several features noticed by workers (see [36] for details). One more point to stress is that there is no need to consider the temporary formation of butene, which is frequently considered as an intermediate. On the contrary it seems that oxygenation is prior to dehydrogenation in several cases: "butene" formed in step 8 (Fig. 4) is still O-anchored. The strong anchoring on both ends of the molecule is for us a key point to get selectivity.

4.3. APPLICATION TO OTHER FACES OF $(VO)_2P_2O_7$

Let us summarize the results obtained in the same way by considering (001), (010) and (021) faces [37, 38]. These faces are perpendicular to the (100) cleavage plane, and so O...V=O bonds do not protrude any more and are saturated A sites. Sometimes the cations are not masked by oxygen and could behave like Lewis acids.

Concerning the elementary steps and comparing with (100) (Table 4), all these faces have also a high power of H-abstraction (high $-e_H$), and H-spillover is as easy. The heat of adsorption of O is stronger (high $-e_{Ox}$) and vacancies are more easily refilled by surface O transfer (low q_{Otr}) than for (100) (very high $-e_{Ox}$): the result is a strong hindrance of O-abstraction limiting its further insertion in the molecule. The desorption of water is often more difficult (high q_W) which gives a high additional energy barrier limiting the rate. Moreover, the anchoring over two oxygens and concerted H-abstraction step during the activation of n-butane is no longer possible on these faces, on which a large proportion of oxygen sites are inactive. The example of a (021) face with A and C sites (Fig. 4a) shows that generally the number and the geometric disposition of the active O, to bridge the $C(H_3)$, or to be inserted, etc, are not favorable (Fig. 4b). These faces are therefore expected to be poorly active and to give by-products, including CO_2.

4.4. CASE OF BUTENE ON (100) $(VO)_2P_2O_7$

As (100) (like other faces) is covered by oxygen and/or OH, n-butene does not find unsaturated cations on which the usual π-complex can be established after abstraction of a first H. Instead of making a C...O bond on D, the C=C bond reacts with these highly unsaurated oxygens, which does not favor the former double anchoring. According to the position(s) of the molecule, few (less than 3) available active O are found in the vicinity [37]. Epoxybutanes (→ CO, CO_2), oxygenated C_2, C_3 by-products, eventually crotonaldehyde, furan, can be formed. (100) faces of $(VO)_2P_2O_7$ would not be favorable for selective butene oxidation in MA.

Let us recall that the selective VPO catalysts used in butene oxidation (P/V = 1/1) are not prepared like those devoted to butane oxidation [9]. The methods used aim at having a maximum of c.u.s. V^{5+} cations which are the active sites in the active $VOPO_4$ phase. However the reactivity of $VOPO_4$ is such that, even in the large excess of oxygen in the feed (1 vol.% of C_4 in air) $VOPO_4$ is reduced largely in $(VO)_2P_2O_7$, up to a mean stoichiometry $VPO_{4.7}$ [9]. If $(VO)_2P_2O_7$ itself is not able to activate butene selectively, this feature, together with the high reactivity of butene, could account for the numerous by-products found beside MA.

5. Conclusion.

In this paper have been exposed some considerations dealing with selectivity: the role of structure and surface reactivity of solids has been stressed for unsaturated and saturated reactants. Owing to the importance of the crystal morphology which has also been outlined, more and more experiments are needed to optimize the parameters of preparation, particularly in the field of crystallization science.

6. References

1 Germain, J.E., (1972), 'Catalytic Oxidation of Hydrocarbons and Related Reactions', Intra-Sci. Chem. Rept., 6, 101-134.

2 Grasselli, R.K., Burrington, J.D. and Brazdil, J.F., (1982), 'Mechanistic Features of Selective Oxidation and Ammoxidation Catalysis', J. Chem. Soc., Faraday Disc., 72, 203-223.

3 Hucknall, D.J., (1974), "Selective Oxidation of Hydrocarbons", Academic Press, London.

4 Haber, J., (1985), 'Catalysis by Transition Metal Oxides', in Solid State Chemistry in Catalysis, R.K. Grasselli and J.F. Brazdil Eds, ACS Symp. Series, 279, 3-21.

5 Courtine, P., (1985), 'Thermodynamic and Structural Aspects of Interfacial Effects in Mild Oxidation Catalysts', in "Solid State Chemistry in Catalysis", R.K. Grasselli and J.F. Brazdil Eds, ACS Symp. Series, 279, 37-56.

6 Sokolowski, V.D., (1990), 'Principles of Oxidative Catalysis on Solid Oxides', Catal. Rev.-Sci. Eng., 32, 1-49.

7 Kung, H.H., (1989), "Transition Metal Oxides", Stud. Surf. Sci. Catal., 45.

8 Bordes, E., (1991), 'Relative Influence of Structure and Reactivity of V- and Mo-containing Catalysts in Mild Oxidation of Hydrocarbons', in Structure-Activity and Selectivity Relationships in Heterogeneous Catalysis, R.K. Grasselli and A.W. Sleight Eds., Stud. Surf. Sci. Catal., 67, 21-30.

9 Bordes, E. and Courtine, P., (1979), 'Selectivity Criteria in the Mild Oxidation of Butene in Maleic Anhydride on VPO Catalysts', J. Catal., 57, 236-248.

10 Véjux, A. and Courtine, P., (1978), 'Interfacial Reactions between V_2O_5 and TiO_2 (Anatase): Role of the Structural Properties', J. Sol. State Chem., 23, 93-103.

11 Contractor, R.M., Ebner, J. and Mummey, M.J., (1990), 'Butane Oxidation in Transport Bed Reactor. Redox Characteristics of the Vanadium Phophorus Oxide Catalyst', in

"New Developments in Selective Oxidation" (G. Centi and F. Trifiro Eds.), Stud. Surf. Sci. Catal., 55, 553-562.
12 Jung, S.J., Bordes, E. and Courtine, P., (1985), 'Co-Mo-O-Ti, a New Catalyst Effective in Butane Oxidation in Maleic Anhydride', in "Adsorption and Catalysis on Oxide Surfaces", M. Che and G.C. Bond Eds., Stud. Surf. Sci. Catal., 21, 345-354.
13 Stone, F.S., (1975), 'The Significance for Oxide Catalysis of Electronic Properties and Structure', J. Sol. State Chem., 12, 271-281.
14 Ziolkowski, J., (1983), 'Catalytic Anisotropy of MoO_3 in Oxidation Reactions in the Light of Bond Strength Model of Active Sites', J. Catal., 80, 263-273.
15 Volta, J.C. and Portefaix, J.L., (1985), 'Structure Sensitivity of Mild Oxidation Reactions on Oxide Catalysts. A Review', Appl. Catal., 18, 1-32.
16 Andersson, A. and Lundin, S.T., (1979), 'Activities of Vanadium Oxides in Ammoxidation of 3-Picoline', J. Catal., 58, 383-92.
17 "Selective Catalytic Oxidation of C-4 Hydrocarbons to Maleic Anhydride", (1987), B.K. Hodnett Ed., Catal. Today, 1.
18 Bordes, E., (1987), 'Crystallochemistry of V-P-O Phases and Application to Catalysis', Catal. Today, 1, 499-526.
19 Bordes, E., (1988), 'Reactivity and Crystal Chemistry of V-P-O phases related to C_4-Hydrocarbon Catalytic Oxidation', Catal. Today, 3, 163-74.
20 Hodnett, B.K., 'Vanadium-Phosphorus Oxide Catalysts for the Selective Oxidation of C_4 Hydrocarbons to Maleic Anhydride', Catal. Rev.-Sci. Eng., 27 (1985) 373.
21 'Forum on Vanadyl Pyrophosphate Catalysts', (1992), G. Centi Ed., Catal. Today, in press.
22 Bordes, E., (1991), 'Influence of the Crystal Morphology of Active Vanadyl Pyrophosphate on Selectivity in the Oxidation of n-Butane in Maleic Anhydride', Fall Meeting, Materials Research Society, Dec. 2-6, Boston, preprint U4.2.
23 McCarty, J.G., McEwen, A.B. and Quinlan, M.A., (1990), 'Models of the Direct Catalytic Partial Oxidation of Light alkanes', in "New Developments in Selective Oxidation" (G. Centi and F. Trifiro Eds.), Stud. Surf. Sci. Catal., 55, 405-415.
24 Pearson, R.G., (1973), "Hard and Soft Acids and Bases", Dowden, Hutchinson and Ross Inc., Stroudsburg, Penn., USA.
25 Lee, J.S. and Oyama, S.T., (1988), 'Oxidative Coupling of Methane to Higher Hydrocarbons', Catal. Rev. Sci. Eng., 30, 249-280.
26 Seshan, K., Swann, H.M., Smits, R.H.H., Van Ommen, J.G. and Ross, J.R.H., (1990), 'Oxidative Dehydrogenation of Alkanes over Alkali and Alkaline Earth Orthovanadates', Stud. Surf. Sci. Catal., 55, 505-512.
27 Patel, D., Kung, M.C. and Kung, H.H., (1988), 'Selective Oxidative Dehydrogenation of Alkanes over Mg Vanadates', Proc. 9th Int. Cong. Catalysis, M.J. Phillips and M. Ternan Eds, Calgary, vol 4, 1554-1561.
28 Siew Hew Sam, D., Soenen, V. and Volta, J.C., (1990), 'Oxidative Dehydrogenation of Propane over V-Mg-O Catalysts', J. Catal., 123, 417-435.
29 Kung, H.H., (1989), in 'Transition Metal Oxides', Stud. Surf. Sci. Catal., 45, pp. 212-15.
30 Spencer, N.D., (1988), 'Partial Oxidation of Methane to Formaldehyde by Means of Molecular Oxygen', J. Catal., 109, 187-197.
31 Merzouki, M., Taouk, B., Monceaux, L., Bordes, E. and Courtine, P., (1991), Catalytic Properties of Promoted Vanadium Oxide in the Oxidation of Ethane in Acetic Acid', "New Developments in Selective Oxidation", 3rd Europ. Work. Meet., April 8-10, Louvain-la-Neuve, preprint, Vol. 1.

32 Gorbunova, Yu.E. and Linde, S.A., (1979), 'Crystallographic Structure of Vanadyl Pyrophosphate', Dokl. Akad. Nauk SSSR, 245, 5864-5872.

33 Ebner, J.R. and Thompson, M.R., (1991), 'Key Structure-Activity Relationships in the Vanadium Phosphorus Oxide Catalyst System', in "Structure-Activity and Selectivity Relationships in Heterogeneous Catalysis," R.K. Grasselli and A.W. Sleight Eds., Stud. Surf. Sci. Catal., 67, 31-42.

34 Ziolkowski, J., (1983) 'Catalytic Properties of Defective Brannerite-Type Vanadates', J. Catal., 81, 311-327.

35 Ziolkowski, J., (1986), 'Crystallochemical Model of Active Sites on Oxide Catalysts', J. Catal., 100, 45-58.

36 Ziolkowski, J., Bordes, E. and Courtine, P., (1990), 'Oxidation of Butane and Butene on the (100) face of $(VO)_2P_2O_7$: a Dynamic View in Terms of the Crystallochemical Model of Active Sites', J. Catal., 122, 126-150.

37 Ziolkowski, J., Bordes, E. and Courtine, P., (1991), 'A Dynamic Model of the Oxidation of n-Butane and 1-Butene on various Crystalline Faces of $(VO)_2P_2O_7$', "New Developments in Selective Oxidation", G. Centi and F. Trifiro Eds., Stud. Sci. Surf. Catal.,55, 625-632.

38 J. Ziolkowski, E. Bordes, (1992), 'A Dynamic Description of the Oxidation of n-Butane on Various Faces of $(VO)_2P_2O_7$ in Terms of the Crystallochemical Model of Active Sites', submitted to J. Molec. Catal.

39 Ziolkowski, J., and Dziembaj, L., (1985), 'Empirical Relationship Between Individual Cation-Oxygen Bond Length and Bond Energy in Crystals and in Molecules', J. Catal., 57, 291-299.

40 Ziolkowski, J., (1989), 'New Methods of Calculation of the Surface Enthalpy of Solids', Surf. Sci., 209, 536-561.

41 Weng, L.T., Ruiz, P. and Delmon, B., (1989), 'Evidence of the Migration of Oxygen Species from Sb_2O_4 to MoO_3 in MoO_3-Sb_2O_4 Selective Oxidation Catalysts', J. Molec. Catal., 52, 349-360.

42 Puttock, S.J. and Rochester, C.H., (1986), 'Infrared Study of Water and Pyridine Adsorption on the Surface of Anhydrous Vanadyl Pyrophosphate', J. Chem. Soc., Faraday Trans. I, 82, 2773-2779.

ADSORPTION AND REACTIONS OF CO, NO, H_2 AND O_2 ON GROUP VIII METAL SURFACES

B.E.Nieuwenhuys
Leiden University
Gorlaeus Laboratories
P.O.Box 9502, 2300 RA Leiden
The Netherlands

ABSTRACT. Group VIII metals are used as catalysts for a number of reactions involving the molecules H_2, CO, O_2, NO and NH_3. This paper will outline and discuss the differences in behaviour of the Group VIII metal surfaces in reactions of oxygen with hydrogen and carbon monoxide; nitric oxide with hydrogen and carbon monoxide; carbon monoxide with hydrogen; and ammonia with nitric oxide and oxygen. It will be shown that the differences in activity and selectivity of the various metals can be understood on the basis of the behaviour of these metals in the adsorption of the gases CO, NO, O_2, H_2 and NH_3.

1. Introduction

The purpose of this chapter is to review the current knowledge of the metal catalyzed reactions involving the following diatomic molecules: H_2, O_2, CO and N_2, viz.:

$$2H_2 + O_2 \rightarrow 2H_2O \quad (1)$$
$$2CO + O_2 \rightarrow 2CO_2 \quad (2)$$
$$2CO + 2NO \rightarrow 2CO_2 + N_2 \quad (3a)$$
$$CO + 2NO \rightarrow CO_2 + N_2O \quad (3b)$$
$$2H_2 + 2NO \rightarrow N_2 + 2H_2O \quad (4a)$$
$$5H_2 + 2NO \rightarrow 2NH_3 + 2H_2O \quad (4b)$$
$$H_2 + 2NO \rightarrow N_2O + H_2O \quad (4c)$$
$$\alpha H_2 + CO \rightarrow C_mH_xO_y + H_2O \quad (5)$$
$$\rightarrow CH_4 + H_2O \text{ (methanation)} \quad (5a)$$
$$\rightarrow C_nH_m + H_2O \text{ (n>1) (Fischer-Tropsch)} \quad (5b)$$
$$\rightarrow CH_3OH \quad (5c)$$
$$\rightarrow \text{various "oxygenates"} \quad (5d)$$
$$3H_2 + N_2 \rightarrow 2NH_3 \quad (6)$$

In addition to reactions 1 through 6 the following two reactions will be briefly discussed

$$4NH_3 + 5O_2 \rightarrow 4NO + 6H_2O \quad (7)$$

and

$$6NO + 4NH_3 \rightarrow 5N_2 + 6H_2O \quad (8)$$

Major emphasis of this review will be placed on the understanding of the differences in the behaviour of the various Group VIII metal surfaces towards the reactions (1) through (8). We shall correlate the catalytic properties of a metal with its adsorption behaviour. Therefore, it is appropriate to review some of the work describing the interaction of the individual molecules (reactants and reaction products) with metal surfaces as a preliminary to the discussion of the reactions. We shall discuss the adsorption on the basis of the nature of the adsorption states, the binding energy, the structure of the adsorption complex and the ordering of the adsorption complexes.

It should be noted here that it is not the aim of this chapter to review all the available relevant

R. W. Joyner and R. A. van Santen (eds.), Elementary Reaction Steps in Heterogeneous Catalysis, 155–177.

data. The data mentioned here and the papers cited here will be presented only to the extent that the findings illuminate the whole discussion.

The interest in these reactions has a very long history. Reaction (1) was already discussed in some detail by Faraday in 1849 [1] and almost a century later, Langmuir [1] proposed already a mechanism for the oxidation of H_2 and CO which appears to be essentially a correct description of the basic steps in the reaction. The methanation reaction has been reported already in 1902 by Sabatier [1]. In 1904 direct synthesis of ammonia was achieved when Haber passed a mixture of H_2 and N_2 over iron at 1000°C. Modern ammonia manufacturing plants are based on systems described by Haber in 1910 [2,3]. Already in 1908 Ostwald and Brauer operated the Pt catalyzed ammonia oxidation process on a commercial scale. The more recent interest in the catalytic reactions [1-8] has been greatly stimulated by the relevance of the reactions in connection with important industrial processes and environmental concern. Reactions 2,3,4 and 8 are used for diminishing air pollution. Reactions 5 are used for the production of synthetic fuels from, for example, coal and natural gas. Reaction (1) may become important because of the interest in hydrogen as a source of "clean" energy. Reactions 6 and 7 are applied for the production of nitric acid which is used in the manufacture of nitrate fertilizers, plastics, explosives and dyestuffs. In the author's opinion the use of modern surface science techniques, utilizing well defined surfaces, to the reactions (1-8) has been one of the most fruitful applications of the surface science techniques to the large field of heterogeneous catalysis.

2. Adsorption on metal surfaces

Many of the adsorption data presented in this paper are based on thermal desorption spectroscopy (TDS) measurements. This technique yields in a simple way information on the number of binding states and their bond strengths with the surface. In many cases adsorption of simple gases on polycrystalline Group VIII metal surfaces leads to TD spectra of the general type shown for these molecules adsorbed on a Rh filament in figs. 3,5-7. TDS measured on single crystal surfaces are often much more complicated. A polycrystalline surface exhibits all the adsorption sites of the crystal faces from which it is composed. Since these sites are simultaneously present, a TDS represents an average of the spectra of the different surface sites weighted according to the relative concentrations of these sites. Therefore, the measured TDS do not show the large structural variations that are observed for some gases using single crystal surfaces.

2.1 DISSOCIATIVE VERSUS MOLECULAR ADSORPTION

The adsorption of diatomic molecules on a metal surface may be considered as a competition between molecular and dissociative adsorption

$$AB_{gas} \quad \rightarrow \quad AB_{ads} \quad \rightarrow \quad A_{ads} + B_{ads}$$

Dissociative adsorption can occur when the bonds formed between the fragments of the dissociated molecule and the surface are much stronger than the bonds between the molecule and the surface. For example, molecular hydrogen, oxygen or nitrogen are only weakly adsorbed on the transition metals. Oxygen, hydrogen and nitrogen adatoms, on the other hand, are strongly bound on many metal surfaces. Therefore, dissociative adsorption is often thermodynamically possible as will be discussed below. Molecular adsorption of CO and NO is relatively strong on many metal surfaces. These adsorbates may undergo both dissociative and molecular adsorption on the same surface depending on the experimental conditions.

Figure 1 shows schematically three possible forms of potential energy curves for molecular and dissociative adsorption as the molecule approaches the surface. The molecularly adsorbed state can be considered as a precursor state for dissociation. The dotted curves represent molecular adsorption with a heat of adsorption Q_m at the equilibrium distance r_m. Molecular adsorption is usually a non-activated process. The solid curves represent adsorption of the atoms following dissociation of the molecule in the gas phase, the heat of adsorption is Q_d at the equilibrium distance $r_d < r_m$, corresponding to the short range chemical interaction. The cross-over point of the two curves

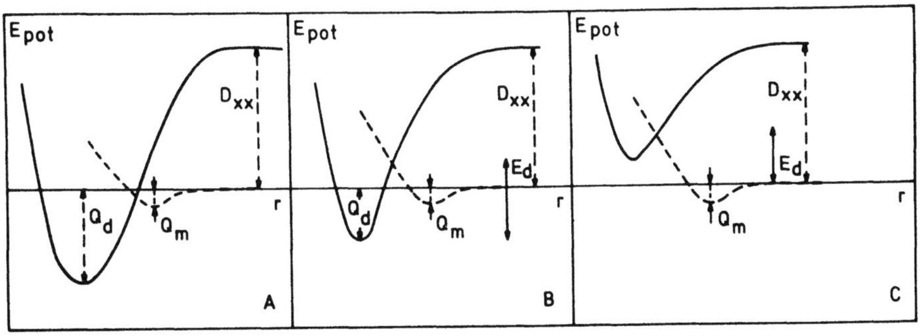

Figure 1.Potential energy plots for dissociative adsorption of a molecule XX
D_{xx} is the dissociation energy of the free molecule
E_d is the activation energy for dissociative adsorption
Q_m is the heat of adsorption in the molecular state
Q_d is the heat of dissociative adsorption
r is the reaction coordinate

determines the activation barrier of the dissociative adsorption. In figure 1A which applies, for example, to hydrogen on transition metals, the dissociative adsorption has a zero activation energy. In figure 1B dissociation results also in a strong chemisorption bond. However, the dissociative adsorption requires an activation energy which is smaller than in the gas phase. A typical example is hydrogen on a Cu(111) surface [4]. Figure 1C shows a typical example of an endothermic, dissociative adsorption. The potential energy of the two adsorbed atoms is higher than that of the free molecule, but lower than that of the two gaseous atoms. Therefore, dissociative adsorption can only take place when the molecule is pre-dissociated in the gas phase and at temperatures where the adsorbed atoms are immobile on the surface. At higher temperatures when the adsorbed atoms are sufficiently mobile over the surface, the atoms will recombine and desorb as a molecule. A typical example is hydrogen on gold.

It is often observed that molecular adsorption prevails at lower temperatures and that dissociative adsorption occurs at higher temperature. This may be caused by kinetics: the activation energy for dissociative adsorption is then too high for dissociation at lower temperature. It could also have a thermodynamic reason. If the number of surface sites where adsorption can take place is equal for molecular and dissociative adsorption, the surface can accomodate twice as many molecules in the molecular state than in the dissociative state. Hence, molecular adsorption will prevail if the heat of dissociative adsorption is not very much greater than the heat of molecular adsorption. The entropy change for adsorption is negative and, consequently, at sufficiently high temperature desorption will occur. In the case considered above only half the number of molecules can be adsorbed in the dissociative state than in the molecular state.As a result the entropy of the system will be lower for molecular adsorption and dissociation can occur at higher temperatures.

Dissociative adsorption requires a cluster of several free and adjacent metal atoms on the surface. Therefore, it is often found that dissociation occurs when the surface coverage is low and that molecular adsorption takes place above a certain coverage, provided that both dissociative and molecular adsorption can take place under the experimental conditions considered.

It is assumed that a diatomic molecule adsorbed parallel to the surface is a transition state for dissociation. The more favorable adsorption complex for molecules like CO on group VIII metal surface is that where the molecular axis is bonded perpendicular to the surface (or slightly tilted for NO). It has been demonstrated by ESDIAD (electron stimulated desorption ion angular distribution) that the molecular axis vibrates around the surface normal and that its amplitude increases with increasing temperature [5]. This is shown in figure 2.

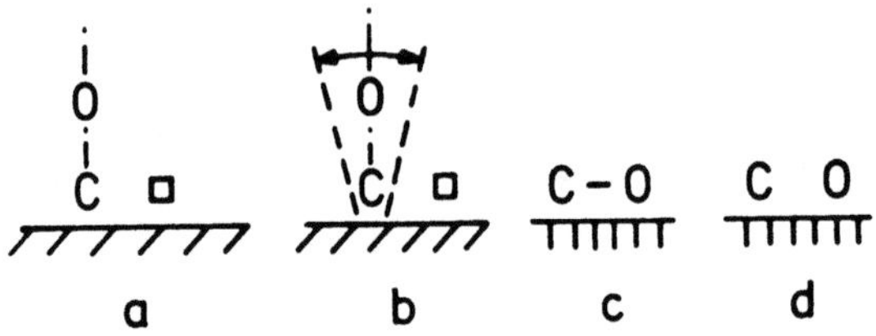

Figure 2. Molecules like CO are bonded on most of the Group VIII metal surfaces with the molecular axis perpendicular to the surface (fig.2a) or slightly tilted. The amplitude of the vibration around the surface normal increases with increasing temperature (2b). At a certain temperature an O-metal bond is formed (2c) and dissociation occurs immediately (2d)

At a certain temperature dissociation occurs. This temperature is probably that temperature where the vibration amplitude has become sufficiently large that a bond between the O-atom and the metal surface is formed resulting in dissociation.

Before we present a survey concerning the adsorption of each of the various reactants some general aspects of the dissociative adsorption will be treated. Table I tabulates the dissociation energies of the reactants in the gas phase. Table II shows some examples of the enthalpies of formation of metal (bulk) hydrides, nitrides, carbides and oxides.

Table I:

$D°_{298}$(kJ/mol)	
H_2	436
O_2	498
NO	631
N_2	945
CO	1076

Table II: Entalpies of formation of Ti, Mo, Pd and Ni hydrides, carbides, nitrides and oxides at 298K and 1 bar. Values in kJ/mol of H, C, N or O

Ti		Ni	
$TiH_{1.7}$	- 69*		
TiC	- 226	Ni_3C	+ 46
TiN	- 336	Ni_3N	+ 1
TiO_2	- 456	NiO	- 244

Mo		Pd	
Mo_2C	- 23		
Mo_2N	- 69		
MoO_2	- 272	PdO	- 85

* All values are taken from ref.6, except * (from ref.7)

The following relevant conclusions emerge:

(1) Dissociation in the gas phase becomes more difficult in the order $H_2 < O_2 < NO < CO$ and N_2.

(2) The order in strength of the metal-X bond (X=H,C,N and O) is M-H < M-C < M-N < M-O for the same metal.

(3) From (1) and (2) it follows, that dissociative adsorption on a certain metal surface becomes more favorable in the order N_2 and $CO < NO < O_2$.

(4) For the transition metals the M-X bond strength increases the further to the left in the Periodic Table the metal lies and it decreases on going down from 3d up to 5d metals. Brodén et al [8], and Nieuwenhuys [9], correlated the tendency for CO, N_2 and NO to dissociate with the position of the transition metal in the Periodic Table; the tendency to dissociate increases by going from Au to the direction of Ti. Benzinger [6] correlated this correlation with thermodynamic considerations: the dissociative heat of adsorption increases by going from Au to the direction of Ti in the Periodic Table because of the increasing bond strength M-X. It should be emphasized that this correlation is based on values of enthalpies of formation of (bulk) metal carbides, oxides, nitrides and hydrides. The M-X bond strength on the surface will differ from the bulk M-X bond strength. The effect of the surface structure on dissociation can be attributed to the variation of the M-X bond strength with the surface structure.

2.2 ADSORPTION OF THE REACTANTS

2.2.1 Adsorption of hydrogen. Hydrogen is known to be dissociatively adsorbed on transition metal surfaces and the initial sticking probability varies from about 0.05 to unity at $T \approx 100$-300K, depending on the metal and its surface structure. For example, the initial sticking probability for hydrogen on a Ni(111) surface is only 0.05, on the stepped Ni(S-)-8(111)x(100) surface it is 0.2 and on a Ni(110) surface it is essentially unity [1]. In general, the sticking probability is significantly smaller on perfectly flat surfaces than on surfaces with a high concentration of steps, surface defects, etc.

It appears that hydrogen prefers to be adsorbed on sites where it is coordinated with many metal atoms. On (111) surfaces of the fcc lattice, for example, hydrogen is adsorbed at low coverage on three-fold sites. There have been several attempts to correlate the heat of adsorption of small molecules on the various transition metals with the position of the metal across the Periodic Table. For several molecules the heat of adsorption decreases going from left to right across the table. For H, O and CO such a trend seems to be present [1], the variation of the heat of adsorption of H being much smaller than for O and CO. However, for the Group VIII metals, no systematic trend is apparent from the available data within the accuracy of the experiments [1].

For the Group VIII metals hydrogen desorbs around 350-400K, its initial heat of adsorption lies in the range 80-100 kJ mol^{-1} and is similar within 20% on similar surfaces of these metals.

Figure 3 shows an example of a TD spectrum.

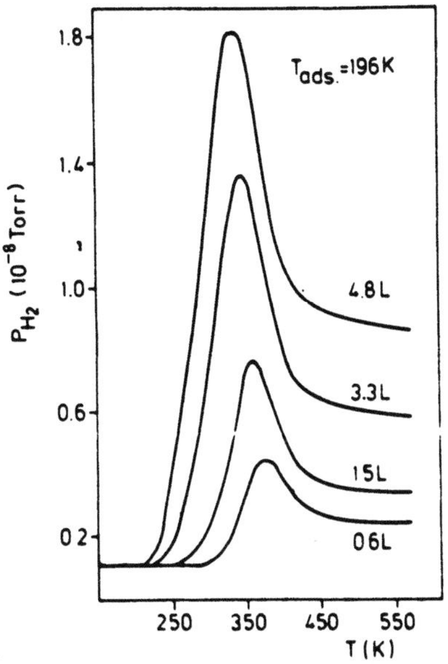

Figure 3. TDS of hydrogen from a Rh filament (from ref.1)

The bond strength is given by $B_{M-X} = (Q_X + D_{X2})/2$, where D is the dissociation energy of the molecule and Q the heat of adsorption. We can conclude that for the Group VIII metals the differences in the metal-hydrogen bond strength are very small, not more than about 5%. The heat of adsorption decreases significantly with surface coverage. The strong variation of Q with coverage is partly due to the heterogeneity of the surface. On single crystal surfaces, where Q also decreases strongly at higher coverage, repulsive interaction between the adsorbed H atoms is believed to play an important role [1].

The initial heat of adsorption varies considerably from crystal plane to crystal plane. This effect is especially large for hydrogen on Pt surfaces [1]: the initial activation energies for desorption differ about 29 kJ/mol, indicating differences in bond strength of 6% between the two densely packed surfaces (111) and (100).

In conclusion the adsorption of hydrogen is rather similar on Group VIII metals. The differences in initial Q and sticking probability (SP) for hydrogen on the various single crystal surfaces of a metal are larger than the differences in Q and SP among similar surfaces of the various Group VIII metals. Two recent reviews have been given in [10] and [11].

2.2.2 Adsorption of oxygen. The interaction of oxygen with Pt surfaces has been studied in detail with modern surface analytical techniques [1]. Since the TDS for oxygen on most of the other metals of Group VIII are qualitatively similar to those published for oxygen on Pt, we may expect essentially the same species on all these metals. At 100K oxygen is adsorbed on Pt(111) with a sticking probability of about unity in a molecularly adsorbed state (γ state) with a low heat of adsorption (37 kJ/mol) [12]. EELS combined with UPS [12] indicates an essentially single O-O bond with a significant electron transfer from the valence band of Pt into orbitals derived from the π_g antibonding oxygen levels and with the O-O bond axis parallel to the surface. Heating the surface above 170K results in the formation of adsorbed atomic oxygen (ß-O) and desorption of some oxygen (γ). The heat of adsorption of ß-O decreases rapidly with increasing coverage from 500 to 160 kJ/mol at $\Theta = 0.8\ \Theta_{max}$. On Pt(111) surfaces oxygen is adsorbed at 300K with a low sticking probability (< 0.1).

A third state is observed upon heating Pt(111) in the presence of oxygen in the 900K temperature range [12]. This "subsurface oxygen" which desorbs above 1200K has a very low reactivity towards CO and hydrogen. There is some controversy concerning the nature of this state [1]. Many authors believe that this state is a so-called surface oxide, wheras others attribute its appearance to oxide forming bulk impurities such as Si or Ca segregating to the surface during oxygen exposure at high temperature. On polycrystalline surfaces and on open surfaces the formation of an "oxide" layer at higher temperature is quite common, as is, for instance, nicely demonstrated by field emission microscopy [1].

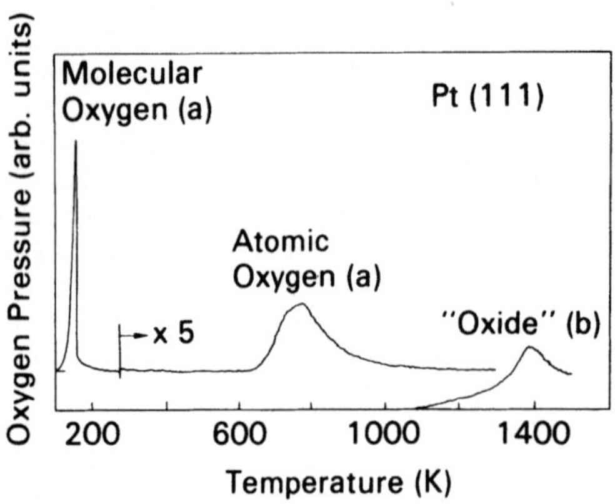

Figure 4. TDS of oxygen from Pt(111) (from ref.12, with permission)

In contrast to CO adsorption, the density of adatoms at saturation is not determined by the size of the adsorbate, but usually more open overlayer structures are formed. For example, (2x2) structures have been observed around room temperature for the (111) surfaces of Ni, Pd, Pt, Rh, Ir and the corresponding Ru(001) surface. However, it is not sure whether in all these cases the apparent (2x2) surface structures represent true (2x2) structures or structures consisting of (2x1) domains rotated 120°

to each other.

Kinetics of oxygen adsorption at ambient temperature vary greatly from metal to metal and from plane to plane. For example, oxygen is adsorbed with a sticking probability near unity on Ni surfaces, whereas on Pt(111) the sticking probability is below 0.1.
The initial heat of adsorption of ß-O is around 250kJ/mol on the Group VIII metals, corresponding to a metal-O bond strength of about 370kJ/mol. Because of the high desorption temperature, the appreciable decrease of Q with increasing coverage and the incorporation of oxygen into the bulk, it is not easy to find a correlation of $Q_{initial}$ with the position of the metal in the Periodic Table. A careful examination of reliable data suggests that the heat of adsorption increases in the following order: Pt(230 kJ/mol) = Pd<Ir<Rh<Ru<Ni(330kJ/mol). The variation in metal-O bond strength is about 55kJ/mol or 15%. The order in the heat of adsorption is similar to that in the heats of formation of the oxides. The tendency of the metal to form surface or bulk oxides also increases in this sequence.

We can conclude that, in contrast to the adsorption of hydrogen, the Group VIII metals exhibit significant differences in the metal-O bond strength.

2.2.3 Adsorption of CO. CO adsorption on metals is the most extensively studied adsorption system. A recent review has been given in [13]. We shall merely highlight here some of the features that are important for our discussion. TDS for CO on Group VIII metal surfaces show in general various peaks, which can be classified into three main peaks, as is shown schematically for CO on polycrystalline Rh in fig.5, corresponding to:

a) low T γ-states desorbing far below room temperature,
b) α states with a peak maximum in its TDS between 350 and 500K.
c) Sometimes, depending on the choice of the metal, its surface structure and the experimental conditions such as the temperature, additional states (ß) are observed with a maximum around 900K. These ß states arise from CO dissociation followed by recombination of C and O at the desorption temperature.

At room temperature adsorption normally takes place in the α state corresponding to molecularly adsorbed CO. Usually two or more α peaks are observed, depending on the surface structure and the CO coverage. These various peaks can be rationalized on the basis of distinct species adsorbed on different sites and in terms of lateral interactions.

A model for the α-CO-metal bond was first proposed by Blyholder [14]. According to this model the bond is formed by electron transfer from the highest filled molecular orbital of CO (5σ, which is essentially a non-bonding orbital with respect to the C-O bond) to unoccupied metal orbitals, implying a donation of electrons to the metal, accompanied by back-donation of electrons from occupied d orbitals into the lowest unfilled CO molecular orbital, the strongly antibonding 2π orbital.

The relative contributions of the σ- and the π-bonding to the metal-CO bond strength and to the net charge transfer, have been discussed in many papers. In the past ten years sufficient evidence has been found, based on theoretical calcuations [15,16] and inverse photoemission studies [17] that for Pd and Ni metal π-donation to CO is much more important for the CO-metal bond than the CO to metal σ-donation. However, the CO-Pt bond has a sizeable σ-donation component [17]. Also work function measurements indicate that the net electron trasfer from metal to CO is much larger for Rh, Pd and Ni than for Pt [9].

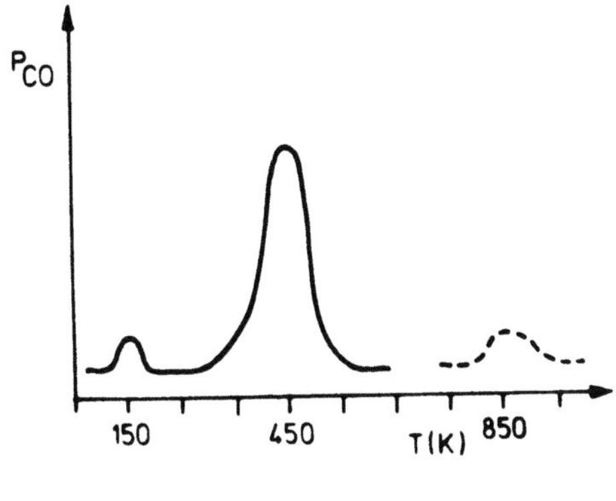

Figure 5. TDS of CO (from ref.1) from Rh

All the available experimental and theoretical data support the fact that the CO molecules are adsorbed on the densely packed surfaces of the group VIII metals with the C-O axis normal to the surface and with the carbon atom directed to the surface. Experimental evidence comes from angular resolved UPS, ESDIAD, ion scattering, LEED intensity analysis and EELS.

LEED, IR or EELS and TDS have been applied for CO on a large number of single crystal surfaces. The general features are quite similar. First a simple overlayer structure is formed up to medium coverage. For example, on the fcc(111) surfaces, $(\sqrt{3}x\sqrt{3})R30°$ structures are observed which are fully developed at $\Theta = 1/3$. The molecules are then located on most of the densely packed surfaces in identical sites. At higher coverage the overlayer unit cell is compressed,, new surface structures are observed and a part of the CO molecules forced to move into other sites. At the highest coverage a close-packed overlayer is formed which is largely determined by the CO-CO mutual repulsion and not by the substrate sites. The maximum density of adsorbed CO molecules is about $1x10^{15}$ molecules per cm^2. The sites where CO molecules are bound are reflected in the stretching vibration frequencies ν of the CO bond. In the pioneering work of Eischens et al.[18] bands below 2000 cm^{-1} were assigned to CO acting as a bridging ligand, $M_2C{=}O$, and bands between 2000 and 2100 cm^{-1} were attributed to linear CO species such as M=C=O. Numerous recent results have shown that Eischens' interpretation of the bands was essentially correct and more comprehensive correlations of ν with bonding sites have been suggested [19].

A substantial increase in ν occurs with increasing coverage. Both experimental observations based on the combined use of TDS, LEED and EELS/IR, and theoretical predictions suggest that the energy difference of CO bonded in "on-top" positions or in multi-fold sites is quite small. The only metal where bridged (or multi-fold) sites are the most favorable positions for CO is Pd. On Ir, Rh and Ru linear adsorbed CO is preferred. Also on Pt(111) "on-top" positions are preferred but the energy differences with the other positions are rather small, resulting in a high degree of disorder at room temperature as is observed by LEED [1].

The differences in heat of adsorption on the various Group VIII metals and for the various single crystal surfaces of a metal are about 20 kJ/mol^{-1} or 20% [1]. The highest values are found on Pd surfaces, which metal also distinguishes itself from the other Group VIII metals by its preference to adsorb CO in bridged or multi-fold positions. Since the behaviour of Pd in the relevant reactions is not particularly different from those of Pt and Rh, we can conclude that the position of CO is not of great importance for the catalytic behaviour of the metal.

Dissociative adsorption of CO has been found on a variety of transition metal surfaces, Brodén et al [8] and Nieuwenhuys [9] correlated the tendency for CO, N_2 and NO to dissociate with the position of the transition metal in the Periodic Table: the tendency to dissociation increases the further to the left the metal lies and it decreases on going from 3d up to 5d metals. Further, the borderline for dissociative or molecular adsorption moves to the right in the sequence CO, N_2, NO to O_2, being the same order as the bond strength in the free molecules. There is sufficient evidence for the proposed correlation. For example, W and Mo surfaces dissociate CO easily at room temperature, dissociative adsorption has not been reported for Pt, Ir and Pd (111) surfaces and CO dissociation has been reported to occur on Ni, Co and Ru at elevated temperatures. Benzinger [6] suggested that the state of adsorption (molecular or dissociative) is determined by thermodynamic criteria. In his paper the heat of dissociative adsorption was estimated from the heats of formation of metal carbides, nitrides and oxides. By this analysis, he was able to predict the correlation relating dissociative adsorption with position in the Periodic Table. The surface structure may have an additional influence on the dissociation. In general, close-packed surfaces are the least active and rough surfaces or surfaces with steps or kinks the most active in dissociation. The effect of surface structure on dissociation was attributed to the variation of metal-carbon etc., bond strength with surface structure.

For this paper it is relevant to conclude that for the Group VIII metal surfaces the variation in the initial heat of molecularly adsorbed CO is about 20 kJ/mol or 15%, and that the tendency to dissociation decreases by going from Fe to Pt in the Periodic Table.

2.2.4 *Adsorption of nitrogen.* In the temperature range 80-400K, nitrogen is only molecularly adsorbed on the Group VIII metal surfaces with the exception of Fe and Co where slow dissociation is observed on some surfaces above 300K [1]. The nature of the adsorption bond is rather similar to that of the CO metal bond . The heat of adsorption is, however, significantly lower for N_2 (40 kJ/mol) than for CO (130 kJ/mol) on the same surfaces, as is demonstrated by figure 6. For recent reviews the reader is referred to [20 and 21].

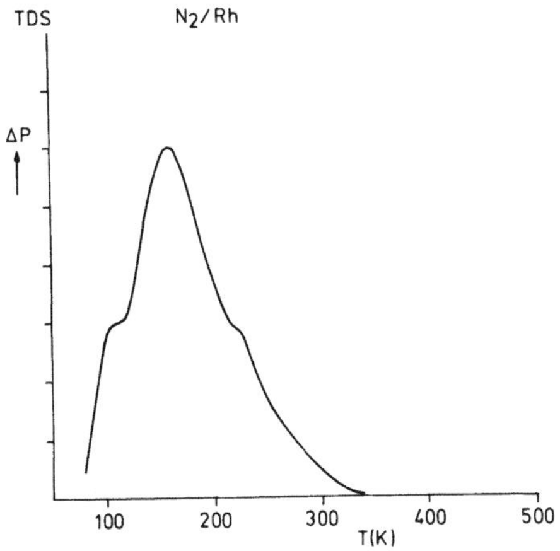

FEM studies point to a striking dependence on the surface structure for N_2 adsorption on Rh [1] and Ir [1]. IR reflectance [1], EELS [1] and angular resolved UPS [1] data for N_2 on Ni (110) and (100) clearly indicate that the N_2 molecular axis is perpendicular to the surface and that the molecule is bonded to a single Ni atom in a linear arrangement. On Ni(111) and Pd(111) the N_2 molecule is very weakly bound. The molecule is possibly adsorbed on this surface with its molecular axis parallel to the surface. The possibility for a lying-down geometry has also been put forward for N_2 on some other surfaces such as Fe(111). In conclusion molecular nitrogen is not adsorbed under the conditions of the NO reduction reactions as taking place on, for example, the automotive three-way catalyst.

Figure 6. TDS of nitrogen from a Rh filament (from ref.1)

2.2.5 *Adsorption of NO.* Figure 7 presents a TD spectrum for NO adsorbed on a Rh filament at 80 K [1] showing desorption peaks of NO, N_2 and O_2. The heat of adsorption of molecularly adsorbed NO is of the order of 110 kJ/mol^{-1} on the Group VIII metals.

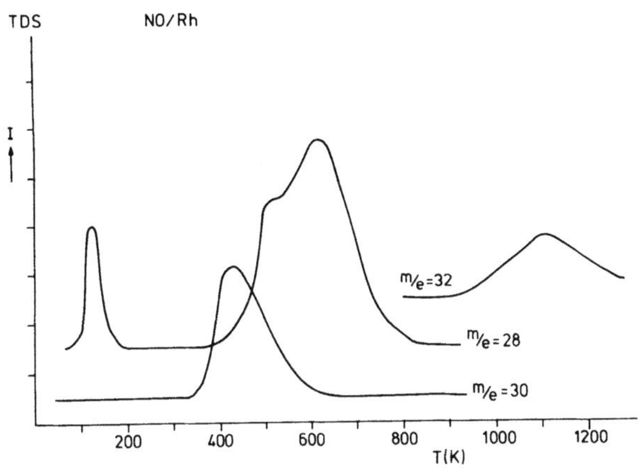

The energy barrier for NO dissociation is much lower for NO than for CO due to the presence of 1 electron in the antibonding 2π orbital. The temperature at which dissociation starts depends strongly on the NO coverage and is very sensitive to the surface structure. For example, when a NO covered Rh(331) surface is heated, dissociation starts already at 240K for a low NO coverage [22]. However, a temperature of 350K is required starting with a fully covered surface. These results show that some of the molecularly adsorbed NO must desorb in order to create free sites needed for dissociation.

Figure 7. TDS of NO, N_2 and O_2 following adsorption of NO on a Rh filament at 80K (from ref.1)

EELS spectra [1] for NO on Ru(001) at $T < 180K$ are consistent with a model in which NO is adsorbed in bridging or three-fold sites (state α_1, $\nu_{NO} = 1500\ cm^{-1}$) and in on-top sites (state α_2, $\nu_{NO} = 1800\ cm^{-1}$). At low coverage NO is adsorbed in the more strongly bound α_1 state and at coverages > 1/3 of saturation the on-top sites become populated. Exposure at ≈ 280K causes dissociation of α_1-NO. ESDIAD studies suggest that the N-O bond is predominantly normal to the surface in both molecular states. Ku et al. [1] concluded from their LEED observations following NO dissociation on a Ru(1010) surface that the N and O adatoms form separate islands on this surface. A number of studies suggest that molecularly adsorbed NO is highly inclined at low coverage whereas at high coverage the adsorbate has a perpendicular orientation. Examples are NO on Ni(111), Ni(100), Pt(100) and Rh(100) [23]. The tilted species is usually considered as a precursor for dissociation.

Somorjai et al [24], Gorte et al [25] and Masel [26] found that NO dissociation is negligible on perfect Pt(111) surfaces. Surface defects bind NO more tightly and induce dissociation. Gorte et al [25] reported that the Pt(100) surface has stronger bonding and much higher decomposition activity than the Pt(110) and (111) surfaces. The major tightly bound state on Pt(100) dissociates to yield 50% N_2 and O_2, whereas the fraction decomposed on the Pt(111) surface is less than 2%. The stepped Pt(410) surface with (100) terraces is much more active in NO dissociation than the (100) surface [26]. On Pd(111) the rate of NO dissociation becomes significant above 500K [1]. In addition to desorption of NO, N_2 and O_2, production of N_2O has been observed on Pt surfaces, probably by a reaction of adsorbed NO and adsorbed atomic N [1].

The available literature suggests that on Pd and Pt surfaces adsorption of NO is predominantly molecular at room temperature. Dissociation is observed at elevated temperatures and the surface structure has a significant influence on the extent of dissociation. In contrast to the adsorption of CO, NO dissociation can easily be detected on Ir, Rh, Ru and Ni surfaces at room temperature.

3. Adsorption of H_2O, CO_2 and NH_3

Various techniques such as EELS, UPS, XPS, TDS, FEM and ESDIAD have been used to study the interaction of water with clean metal surfaces [1]. Water is molecularly adsorbed at 80K and its molecular structure is preserved upon adsorption. The molecule is bonded to the surface via the O atoms, probably through the lone pair orbital. On Ru(001) adsorption occurs at low coverage without ordering and at higher coverages or for $T > 140K$, bilayer water islands are formed in registry with the substrate. Three H_2O TDS peaks have been observed on Ru(001): peaks at 215 and 180K due to different binding orientations of the H_2O molecules and a peak at 160K due to a multilayer(ice). On Ni(110) a higher temperature desorption peak (370K) was observed, which may indicate dissociative adsorption of H_2O. On Ni(111) no evidence was found for any decomposition product following thermal desorption of H_2O around 170K. Evidence was found for the existence of attractive lateral interactions consistent with clustering of H_2O molecules on Ni(111) even at fractional monolayer coverages.

We may conclude from the available data that molecular H_2O will desorb at temperatures at which the oxygen-hydrogen reaction occurs on the metals active in this reaction.

EELS, UPS and TDS studies on Pt(111) [27,28], unequivocally show the influence of preadsorbed O on the nature of the adsorption complex following exposure to H_2O. The spectra indicate that the following species are formed following exposure to water of a Pt(111) surface precovered with oxygen in the (2x2) structure:

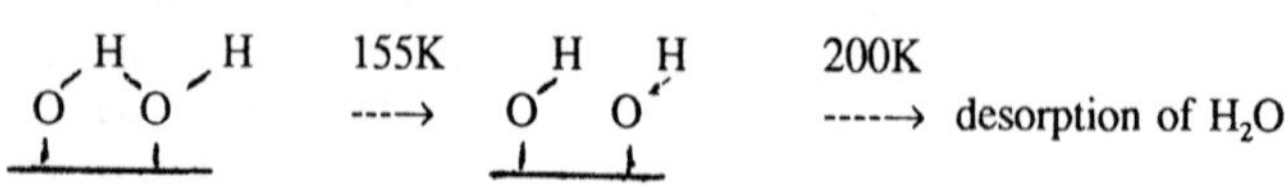

The hydroxyl groups decompose at a temperature of about 200K, resulting in the desorption of H_2O. A TDS and ESDIAD study of H_2O adsorption on O preadsorbed Ni(111) [1] shows a strong qualitative similarity with the behaviour on Pt(111): an additional H_2O desorption peak at 275K probably arising from the formation at surface OH species at low temperature and their subsequent disproportionation. Traces of preadsorbed O induce a high degree of azimuthal ordering in adsorbed water, probably by H bonding interactions. In contrast to Ni(111), preadsorption of oxygen on Ru(001) [1] inhibits the azimuthal ordering.

The interaction of CO_2 with the various Group VIII metals has been reviewed recently by Solymosi [29]. At 80K the adsorption on a Rh field emitter exhibits an interesting crystal face dependency [30]. In addition to molecular adsorption, dissociation occurs with an appreciable rate on the stepped surfaces around (111) and (100) at temperatures above 220K in agreement with measurements of Somorjai et al [31]. The latter authors used EELS, isotopic labelling, LEED and TDS and found that CO_2 is dissociatively adsorbed with a low sticking probability yielding adsorbed CO. The FEM patterns suggest that the surface structure of Rh has a striking influence on the ability of the metal to dissociate the CO_2 molecule. From Pt, CO_2 desorbs at a much lower temperature (90K corresponding with a heat of adsorption of about 20 kJ/mol) and dissociation was not detected [32].

Solymosi [29] found that the presence of B and alkali adatoms increases the binding energy of adsorbed CO_2 and promotes the dissociation. In the presence of preadsorbed oxygen carbonates are formed.

We can conclude that molecular CO_2 is not adsorbed under the usual conditions for the CO + O_2 reaction.

The interaction of NH_3 with metal surfaces has been examined on a number of metal surfaces with various techniques. Four adsorption states have been found on Ru(001) [33,35]. Two of these states, desorbing around 310 and 180K, were attributed to NH_3 chemisorbed in the first layer. The third state was attributed to 2nd layer NH_3 and desorbs around 140K, while the fourth state, desorbing around 115K, was attributed to multilayer NH_3.

Concerning the first two states it is well established that NH_3 is bound via the N atom through the lone pair orbital, with the H atoms pointed away from the surface. Some uncertainty exists concerning the adsorption sites. Some authors favor adsorption in threefold hollow sites while others favor atop sites. According to Zhou et al [34] it is most likely that the two states are both adsorbed with the threefold axis perpendicular towards the surface, one is adsorbed in threefold hollow sites and the other on atop sites.

Compared with Ru(001) ammonia adsorption on Pt(111) is only slightly different [36]. No evidence was found for significant dissociation of NH_3 on Pt(111) and Ru(001) under low pressure conditions. However, decomposition of NH_3 occurs at temperatures above 600K in a NH_3 atmosphere [37]. The effect of the surface structure on the dissociation is very large. For Pt the order in activity is (210)>(110)>(100>(111) [37].

4. Co-Adsorption

In order to understand the catalytic properties of the various Group VIII metal surfaces the interactions among neighbouring adsorbed species and their effects on the surface processes should be known since the surfaces may be covered with different kinds of adsorbates during the reaction.

Conrad et al [38] have studied in detail the mutual interaction of coadsorbed O and CO on a Pd(111) surface. Some of their relevant results are summarized here. O adsorption is inhibited by preadsorbed CO. At coverages below Θ_{CO} = 1/3 LEED patterns show that O and CO form separate domains on the surface. However, the behaviour is different when O is preadsorbed. CO can be adsorbed on the Pd(111) surface covered with O which is less densely packed than a saturated CO layer. The O adatom islands are then suppressed to domains of a ($\sqrt{3}$ x $\sqrt{3}$) R30° structure (Θ = 1/3) with a much larger local coverage than can be reached with O alone ((2x2) structure, Θ = 0.25). After further exposure the LEED pattern suggests the formation of mixed phases of O_{ads} and CO_{ads} (with local coverages of $\Theta_O = \Theta_{CO}$ = 0.5) which are embedded into CO domains. When these mixed phases are present CO_2 is already produced below room temperature. Coadsorption studies on other noble metal surfaces are consistent with this picture: preadsorbed CO inhibits the dissociative adsorption of oxygen whereas CO is adsorbed on a surface covered with O.

Lambert and Comrie [39] concluded from TDS on Pt(111) and (110) that gaseous CO displaces molecularly adsorbed NO (α_1) and also causes the conversion of this state to the other more weakly bound state α_2. The α_1 state was reported to be the only one which is reactive in the reaction with CO. Displacement of molecularly adsorbed NO by gaseous CO has also been observed on Pd surfaces. Thiel et al [40] have used EELS to observe directly the competition of CO adsorption with one of the two molecular states of NO on Ru(001). The stretching frequencies suggest that CO is linearly bonded to a single Ru atom, whereas NO prefers the bridging or multi-fold sites and that at higher coverage linear sites are also populated. The EELS spectra indicate that conversion from linear to multi-fold bonded NO occurs during adsorption of CO. Similar observations have been made recently for CO/NO mixtures on Pd(110) [41]. For each single-phase system the bridge site is preferred. When the two are coadsorbed, however, a mixed phase is formed in which the stable CO site is linear. In the presence of a CO-NO atmosphere Pt surfaces favor CO adsorption, as should be expected on the basis of the heats of adsorption [42]. Rh surfaces, however, adsorb both CO and NO [42]. Blocking of hydrogen adsorption by NO is observed on Rh surfaces [22]. The occurrence of strong, repulsive or attractive, interaction between coadsorbed H and NO has not been reported.

The interaction of coadsorbed H and CO has been studied on a variety of single crystal and polycrystalline surfaces under conditions where reaction does not occur [1]. The following phenomena have been observed: displacement of adsorbed H by CO, blocking of H adsorption by preadsorbed CO, segregation of the coadsorbed species, formation of mixed layers, a decrease in the H desorption temperature but also enhancement in the uptake of CO by the presence of preadsorbed H, and formation of new desorption peaks in the TDS. Preadsorbed CO strongly blocks hydrogen adsorption at 100K on Ru(001) [1] and Rh(111) [1]. The hydrogen TDS peaks shift to lower temperature in the coadsorbed layer on Ir(110) [1] and Ni(111) [1]. On Pd(110) [1], however, the data point to a weak attractive interaction between CO and H. The presence of CO on the Pd(111) surface reduces the uptake of hydrogen but no new CO or hydrogen desorption peaks occur [43]. Postdosing of CO to a hydrogen covered surface results in the formation of a high temperature desorption peak of hydrogen corresponding to subsurface hydrogen [43]. On Rh(100)[1] and Ni(100) [1] new low temperature desorption peaks are observed for hydrogen and on Ni(100) also for CO. These peaks with a maximum around 200K are found following exposure to CO at 100K of a surface precovered with H. Koel et al [44] confirmed these observations and extended them with UPS measurements. The data point to a single type of weakly held CO on a surface presaturated with H. Earlier EELS studies [1] also suggest that C-H or O-H bonds are not formed: the main effect is a displacement of CO from "on-top" sites into bridged sites. In summary, there is no evidence for the formation of directly bonded H-CO complexes.

In conclusion, the main effects found for coadsorption are blocking, islands formation, displacement by the component with higher heat of adsorption and site conversion. Strong attractive interactions between the two adsorbates do not take place. Hence, the catalytic reactions can be discussed on the basis of the adsorption of the individual reactants in combination with the coadsorption effects mentioned above.

5. Reactions

5.1 THE HYDROGEN-OXYGEN REACTION

Many studies [1] show that certain noble metal catalysts of Group VIII are the most effective catalysts for this reaction at low temperatures. From these studies we can conclude that the order in activity is: Ni<Ru, Rh, Os<Ir<Pd, Pt. Very effective multi-component catalysts contain both Pt and Rh or Ir.

An interesting observation is the occurrence of oscillations in the rate of water production [1]. The characteristics of these oscillations are similar to those reported for the other reactions discussed in this paper.

Fisher and Gland [27] studied the formation of H_2O from coadsorbed hydrogen and oxygen on Pt(111) by rapidly increasing the crystal temperature and monitoring the change in the H_2O pressure. H_2O desorption was observed at temperatures where hydroxyl groups decompose (200K) and from this similarity Fisher and Gland concluded that similar processes are rate limiting, indicating that hydroxyl groups might be intermediates in the reaction.

Norton et al [45] carried out reaction experiments with polycrystalline Pt, Pt(111) and Pt(100) precovered with O, using XPS, UPS and work function measurements. These authors did not find any evidence for the formation of OH groups under their experimental conditions. Water was the initial product. A factor of two difference was found between the activity of Pt(111) and that of Pt(100). In general, an induction period is observed, especially for high initial O coverages. Norton et al found that the adsorption of hydrogen is rate limiting. Hence, although OH species have been identified in coadsorption of H_2O and O at low temperatures on many surfaces, the first experiments failed to detect the possible OH intermediate during the $H_2 + O_2$ reaction. Recently, however, it has been found on Rh [46] and Pt(111) [47] at low temperature using EELS, and at high temperatures desorption of OH was observed by laser-induced fluorescence [48,49].

Two different mechanisms have been suggested for reactions like the oxidation of hydrogen and the oxydation of CO:

(a) An Eley-Rideal mechanism, at which one of the components reacts in the adsorbed state with the other molecule in the gaseous or in a physically adsorbed state: for example: $O_{ads} + H_{2gas} \rightarrow H_2O$.

(b) Langmuir-Hinshelwood mechanism, at which both components react with each other in the adsorbed state:

$$O_{ads} + H_{ads} \rightarrow (OH_{ads}) \xrightarrow{H_{ads}} H_2O \text{ or } O_{ads} + CO_{ads} \rightarrow CO_2$$

The available literature data suggest that the dominant reaction mechanism for all reactions discussed in this paper is the Langmuir-Hinshelwood mechanism. This is demonstrated in a very direct way by molecular beam experiments [50] as will be shown further on for the CO oxidation reaction.

Field emission studies [1] on Rh, Pt and Ir showed in a very direct way that the reaction proceeds by a Langmuir-Hinshelwood mechanism at the boundary between islands of immobile O atoms and mobile H. The direction of the movement of the reaction zone boundary with time indicates that the least active Rh surfaces are the surfaces with highest heat of adsorption of oxygen. The observed crystal face dependence suggests that the mobile H adatom chooses selectively the O adatom with low binding energy. The order in activity found in the FEM experiments is Rh < Ir < Pt, in agreement with the data for supported metals: Ni < Os, Rh, Ru < Ir < Pd, Pt. We have seen before that the heat of adsorption of oxygen and the heat of formation of oxides also decreases in this order. Apparently, both the crystal face dependence and the activity of the various Group VIII metals can be understood in terms of the metal-oxygen bond strength. On an active metal both O_2 and H_2 are dissociatively adsorbed, oxides are not formed under reaction conditions and the heat of adsorption for ß-O is relatively low. This suggests the existence of a vulcano shaped correlation between the reactivity of the metal and the binding energy of oxygen, as is shown schematically in figure 8. Such a vulcano shaped correlation is often observed for many catalytic reactions.

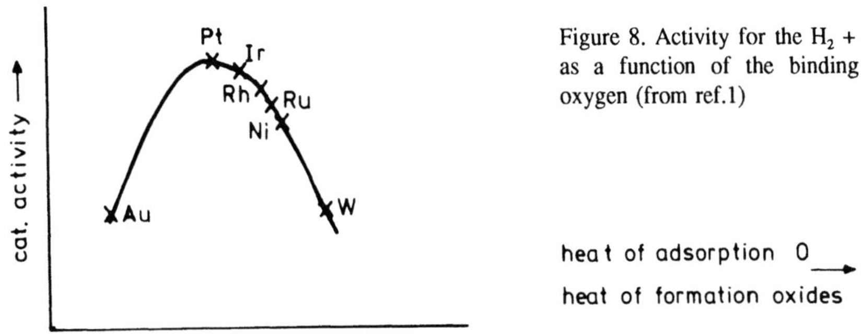

Figure 8. Activity for the $H_2 + O_2$ reaction as a function of the binding energy of oxygen (from ref.1)

5.2 THE OXIDATION OF CO

Usually, temperatures of 300-600K are required for this reaction over the active metal catalysts Pt, Pd, Ir, Rh, Ru and Ni. The main application of CO oxidation catalysts is for reducing automotive emission of CO. In addition, air purification devices for respiratory protection and CO gas sensors commonly employ CO oxidation catalysts. A novel application may be in sealed CO_2 lasers used for weather monitoring and in other remote sensing applications. For the development of long-life sealed CO_2 lasers novel catalysts must be developed which are able to oxidize CO near ambient temperature. A possible candidate is gold-manganese oxide [51]. The mechanisms of the processes responsible for the high activity are not fully understood. In the opinion of the present author the activity of this catalyst may be caused by spill-over of O atoms from the MnO_x to gold. Gold itself is not active in O-O bond scission. However, O adatoms are stable on gold at room temperature. In the presence of CO the O adatom can easily react with CO due to the low Au-O bond strength.

It has been established now that the dominant mechanism of the reaction is the Langmuir-Hinshelwood (LH) mechanism: $CO \rightarrow CO_{ads}$, $O_2 \rightarrow 2\ O_{ads}$, $O_{ads} + CO_{ads} \rightarrow CO_2$. The alternative Eley-Rideal mechanism fails to explain all the experimental data. Direct evidence in favour for the LH mechanism has been obtained from molecular beam experiments. Figure 9 [52] shows a typical example of the variation in rate of CO_2 production with time for a CO beam impinging an O adlayer on Pd(111) at a constant temperature of 374K, together with the variation of the surface concentration of O_{ads} and CO_{ads}.

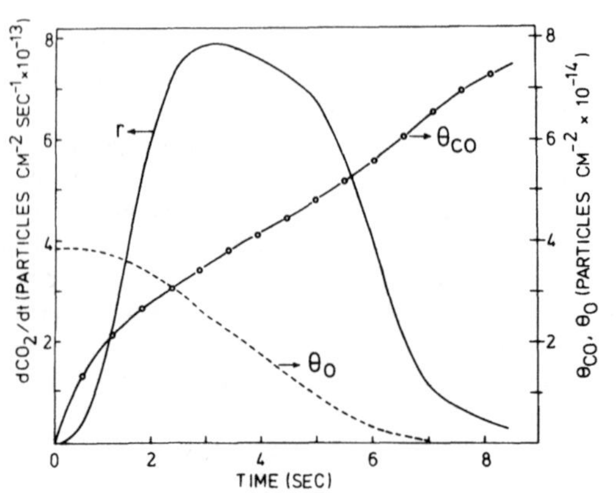

Figure 9. Rates of CO_2 formation on Pd(111) and O and CO coverages as a function of time at 374K (from ref.52, with permission)

The reaction exhibits an induction period and reaches its maximum rate after a few seconds. Obviously, the figure confirms that both reactants must be adsorbed in accord with the LH mechanism. Engel and Ertl [52] suggested that the adsorbed CO molecules first compress the O_{ads} layer resulting in a higher activity and leading to the onset of CO_2 formation.

Cant et al [53] studied the reaction over several supported metals and concluded that the activation energy is lower over Pt than over Ir, Rh, Pd and Ru catalysts.

Figure 10 shows the rate of CO_2 production over the (111), (100), (410) and (210) surfaces of a Pt0.25-Rh0.75 single crystal as a function of temperature using a 2:1 mixture of CO and O_2 at a total pressure of 2×10^{-7} mbar [54]. It shows that the reaction rate increases rapidly between 400 and 500K up to a temperature T_m where a maximum is reached after which it slowly decreases with increasing temperature. The temperature at which the maximum occurs increases with increasing CO pressure. Similar behaviour has been found on a large number of Pt, Pd, Ir, Rh and Ru surfaces. XPS measurements [1] show that the maximum rate is reached at a temperature at which the CO coverage is already small. All these and other observations indicate that adsorbed CO acts as inhibitor for O adsorption and, hence, for the reaction. The value of T_m is determined by the sticking probability, heat of adsorption of oxygen, and the heat of adsorption of CO. The relatively small CO inhibition on the (111) surface is consistent with the low heat of adsorption of CO on that surface.

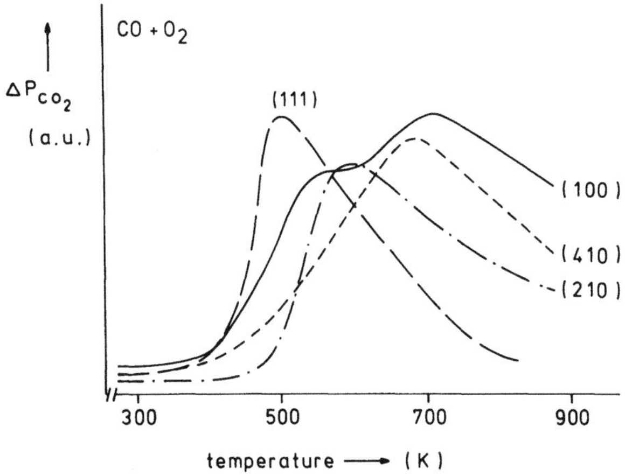

Figure 10. Steady state rates of CO_2 production for the CO + O_2 reaction over the (111), (100), (410) and (210) surface of a Pt0.25-Rh0.75 single crystal (from ref.54)

Oscillations in the rate of CO_2 production have been observed on a number of supported metal catalysts and single crystal surfaces. Similar oscillations have been observed for most of the reactions discussed in this chapter. An example is shown in figure 11 for the NO + NH_3 reaction [55]. Several models have been proposed to explain these oscillatory rates. Sales, Turner and Maple [56] associated the oscillation with a slow and reversible modification of the catalyst surface: slow oxidation and reduction of the metal surface inducing transitions between the two branches.

Ertl and his group [57] investigated in detail the oscillations on the (100) and (110) surfaces of Pt. The clean (100) and (110) surfaces reconstruct, i.e. the atomic arrangement in the topmost layer is not that of the corresponding bulk plane.

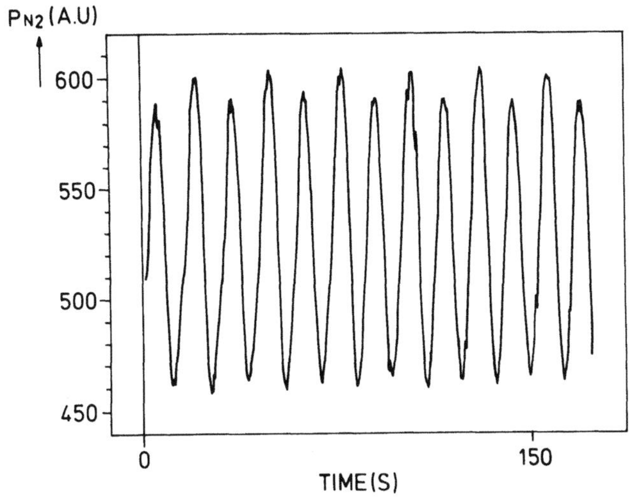

Figure 11. Double-peaked oscillations in the rate of N_2 production over Pt(100) observed for the NO + NH_3 reaction (from ref.55)

However, the reconstruction is lifted by adsorption of CO and NO when the coverage exceeds a certain critical coverage. The driving force is the higher heat of adsorption of CO on the non-reconstructed surface which overcompensates the energy required for altering the surface structure. The sticking probability is negligibly small on the reconstructed surface while it is much higher on the non-reconstructed surface. Ertl proposed the following model for the oscillations. The surface structure changes to the non-reconstructed surface due to CO-adsorption. Oxygen can then be adsorbed due to the relatively high sticking probability on this surface. The adsorbed O reacts with adsorbed CO molecules to yield CO_2 that is immediately released into the gas phase. As a result more sites are created for oxygen adsorption and the reaction rate increases. The CO coverage decreases because of its rapid reaction with O. At the critical CO coverage the surface structure transforms back into the reconstructed phase. Oxygen is then not adsorbed anymore and, hence, the reaction rate decreases. The CO coverage will then increase and beyond the critial coverage the surface structure transforms again to the non-reconstructed phase. This transformation completes the cycle of the oscillation. Recent studies suggest that rate oscillations under isothermal conditions can also occur on surfaces where the reconstruction model does not apply. A more general model was proposed involving the key role of vacant sites on the surface required for dissociation of molecules like NO and O_2 [55].

The CO oxidation is usually considered as a typical example of a structure insensitive reaction. For example, the catalytic activity is almost equal over Pt/Al_2O_3 catalysts with widely varying particle size at high CO concentrations [1]. However, the reaction became structure sensitive in excess oxygen. The observed temperature dependence of the steady state rates are almost similar on Pd(111), (100), (110) and (210) and polycrystalline Pd surfaces. Hence, a number of both single crystal and supported metal data indicate that the reaction is essentially surface insensitive. However, fig.10 illustrates that the reaction may be very sensitive to the surface structure under certain experimental conditions. Boudart suggested that lateral interactions between adsorbates erase the effect of different types of adsorption states, leading eventually to structure insensitivity [58]. A reactive mixture of CO and O_2 could also modify the catalyst particle shape in such a way that one type of surface structure dominates [48]. Another explanation could be that only the activity of one type of site is measured under certain experimental conditions [54]. Recent results of Yates et al [5] are consistent with this model. They found that the CO + O_2 reaction takes place at lower temperature on the terraces than on the steps of the Pt(211) surface using temperature programmed reaction conditions. However, at higher temperature the high diffusion rate of O adatoms over the surface obscured the differences of step and terrace sites.

We can conclude that for the CO oxidation reaction the reaction is often inhibited by CO adsorption. Both the CO + O_2 and the H_2 + O_2 reactions proceed via a LH mechanism and the active catalysts are the noble metals of Group VIII. Under conditions at which CO adsorption does not inhibit the reaction (large O_2/CO ratio or high T), the available data suggest that the order in activity is the same for both reactions: Pd, Pt > Ir> Ru. On an active catalyst CO is adsorbed in the molecular α state, CO dissociation and oxide formation do not occur under reaction conditions, and the metal-O bond strength is relatively weak.

5.3 REDUCTION OF NO BY CO AND H_2

Unfortunately, dinitrogen is just one of a variety of reaction products that can be formed in the catalytic converters used in automobiles, due to the presence of various gases in the exhaust gas. One of the undesirable reaction products is NH_3, which is formed especially under reducing conditions. Some of the processes that have been proposed involve adsorbed isocyanate(NCO) and adsorbed HNCO as intermediates [1]. The formation of NCO complexes has been observed on supported metals by means of IR. It has been established now that the presence and the stability of the NCO groups depend on the support [1].

Kobylinski and Taylor [59] have studied the NO + CO and NO + H_2 reactions on supported noble metals and found that the activity for the first reaction increases in the order Pt<Pd<Rh<Ru and for the second reaction Ru<Rh<Pt<Pd. The first reaction is slower than the second, only for Ru the

order was reversed. Ru is an excellent catalyst for the NO reduction with a minimum of NH_3 production. However, Ru forms volatile oxides under operating conditions resulting in an unacceptable catalyst loss. The most efficient catalyst appears to be Rh [60].

Lambert and Comrie [39] studied the CO + NO reaction on Pt(111) and (110) surfaces and concluded that the reaction proceeds by a LH mechanism between O and molecular CO. NO dissociation is also the prime step in the NO + H_2 reaction on a Pt foil at $p \approx 10^{-7}$ mbar [1]. NH_3 was found to be the major product below 600K and N_2 above 600K when the NO/H_2 ratio was $\approx 1/5$.

Siera et al [54] studied the CO + NO reaction over the same surfaces and under the same experimental conditions as those described earlier for the CO + O_2 reaction. At low temperatures (< 500K) the reaction rate decreases in the order Pt-Rh(111) >(100)>(410)>(210). This is the same order as was observed for the CO + O_2 reaction and it corresponds to the order in heat of adsorption of CO. NO dissociation decreases in the order (410) ~ (210) > (100) > (111). For both reactions the rate at low temperatures is controlled by CO desorption and not by NO dissociation.

Similar effects were found for the NO reduction by hydrogen over Rh single crystal surfaces [22]. The initiation process is NO desorption. Surface structural effects are related to differences in heat of adsorption of NO. Only at low NO coverages the intrinsic differences in NO dissociation activity determine the surface structural effects.

Hirano et al [61] studied in detail the NO + H_2 reactions on a number of pure Pt, pure Rh and Pt-Rh alloy surfaces. In particular, the mechanisms of the various processes resulting in the production of N_2, N_2O and NH_3 were investigated. For this purpose the authors studied the various reactions both in the 10^{-7} mbar range with access to the surface analytical techniques and in the higher pressure range. Isotopic labeled molecules were used in order to gather information concerning the mechanisms of the formation reactions of the N-containing products.

The main conclusions emerging from their studies are:

(1) The activities and the selectivities to N_2, N_2O and NH_3 depend strongly on the surface structure. This is illustrated by fig.12.

(2) Below 600K the main contribution to N_2 formation is via $NO_{ads} + N_{ads} \rightarrow N_2 + O_{ads}$. At higher temperatures the dominant mechanism is via $2N_{ads} \rightarrow N_2$.

(3) N_2O is formed via $NO_{ads} + N_{ads} \rightarrow N_2O$ whose contribution decreases with increasing temperature, because of the lower NO coverage.

(4) NH_3 is formed via dissociation of NO and subsequent hydrogenation: $N_{ads} + 3H \rightarrow NH_3$. Adsorbed NH was found as a reaction intermediate.

(5) The selectivities towards N_2, N_2O and NH_3 are determined by the relative concentrations of NO_{ads}, N_{ads} and H_{ads} on the surface.

(6) The selectivity towards N_2O and N_2. Both N_2 and N_2O can be formed by a reaction of adsorbed NO with adsorbed N. The availability of a vacant site near the NO_{ads} and N_{ads} may play a central role for the selectivity. A situation like the one sketched under b), where □ denotes a vacant site on the surface, may result in N_2 formation, whereas situation a) will result in N_2O formation.

a b

In summary, the surface science studies show that NO dissociation is a crucial step. Under steady state conditions the reaction is often inhibited by the presence of adsorbed CO or NO molecules. The data suggest a LH mechanism of O formed by dissociation of NO with adsorbed CO or H. Hence, an active catalyst is able to break the N-O bond, but not the C-O bond, under reaction conditions, and its metal-oxygen bond strength is rather small.

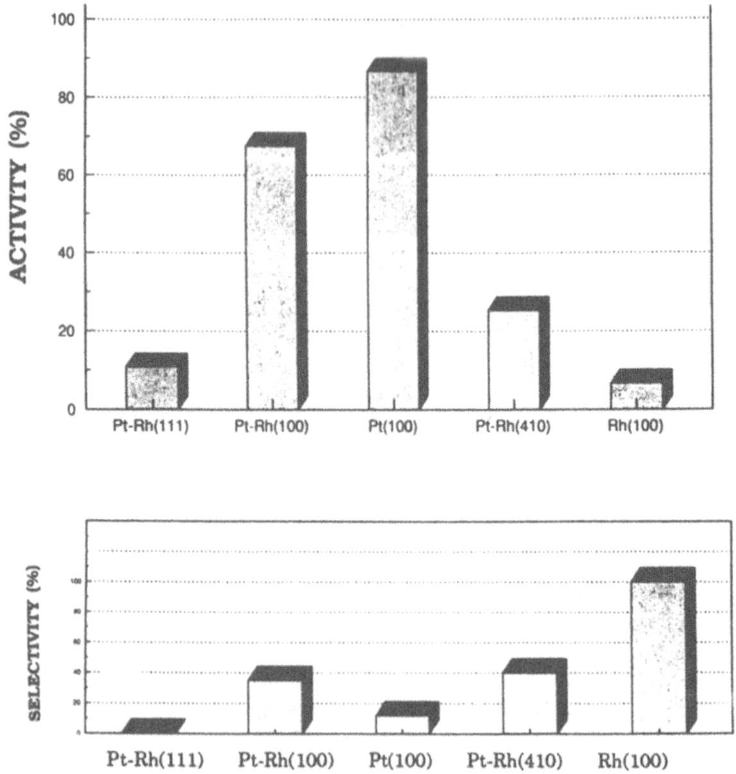

Figure 12. Product (N_2, NH_3) formation for the NO + H_2 reaction at constant temperature for Pt(100), Rh(100) and Pt-Rh(111), (100) and (410) surfaces (from ref.61). Activity (NO conversion) at 520K and selectivity to N_2 at 575K.

5.4. CATALYSIS IN AIR POLLUTION CONTROL-deNO$_x$ CATALYSTS

The automobile is identified as a major source of emission of hydrocarbons (hc), CO and nitrogen oxides (NO_x) to the atmosphere. Stationary sources such as conventional power plants are another major source of NO_x emission. Since 1967 the US federal emission standards for automobile emissions have become increasingly stringent.

Since 1975 a catalytic pollution control device is used in most new cars in order to meet the emission control requirements. The first generation of catalysts was based on Pt and Pd. This catalyst was capable to oxidize CO and hc into CO_2 and H_2O. Until 1983 the US federal standards for NO_x emission could be met by engine modifications. In 1983 increasing stringency of legislative standards necessitated the introduction of the so-called three-way catalyst which is capable to control CO, hc and NO_x emission simultaneously using a single catalyst. This catalyst is based on Pt and Rh, in line with the catalytic properties of these metals towards the relevant reactions 2,3 and 4, as has been discussed before.

The nitrogen oxides from automotive exhaust are converted into dinitrogen via reduction by hydrogen and CO. NO_x removal from the gases of stationary sources such as electricity plants and nitric acid plants is more complicated because the reducing agent reacts more easily with the excess oxygen than with NO_x.

Direct decomposition of NO_x is, at present, not an alternative for reduction, although it is thermodynamically feasible except at high temperatures. As we have seen the Pt group metals can easily decompose NO. Oxygen, however, is only released at high temperatures (see fig.7) and, hence, reducing agents are required. A catalyst suitable for direct decomposition must release O_2 at a much lower temperature than Pt. The best catalyst known for direct decomposition of NO is a Cu-ZSM-5 catalyst [62,63]. Unfortunately, sulfur-compounds poison this catalyst and its activity is not sufficiently large.

The only reducing agent known that can selectively reduce NO_x in the presence of O_2 is ammonia via process 8 and $4\ NH_3 + 4\ NO + O_2 \rightarrow 4\ N_2 + 6\ H_2O$ [64]. Again the Pt group metals are able to promote this selective reduction process [65]. Pt is already a very active catalyst at low temperature. Complete conversion of NO_x can be obtained at 180°C. However, above 200°C the conversion decreases with increasing temperature, ultimately giving apparently negative conversion [65]. The obvious explanation is reaction 7 for which reaction Pt is an excellent catalyst. V, W and Ti oxides or their combinations are less active than Pt and, therefore, preferred for the selective reduction. Commercial catalysts contain now one or more of these base metal oxide catalysts [64].

5.5 CO + H_2 REACTIONS

These reactions are among others important for the conversion of coal, natural gas or other C-containing materials into better transportable fuels. Synthesis gas (syngas) can be produced, for example, by steam gasification of coal. For the present review the interesting aspects are the differences in the activities and the selectivities of the Group VIII metals. Ni is selective in the methanation reaction, Fe, Co and Ru also have a good selectivity towards the Fischer-Tropsch synthesis of larger hydrocarbon molecules. The commercial catalyst for synthesis of methanol from syngas is based on Cu/ZnO. However, Pt, Ir, Rh and Pd are also selective at lower temperatures in the formation of methanol if the metals are combined with basic oxides like La_2O_3 or MgO. Rh catalysts promoted by V_2O_3 are able to produce ethanol and ethanal. Recent reviews have been given in [66].

Various reaction mechanisms have been proposed in the literature for the formation of CH_4 and higher hydrocarbons, including the following intermediates:

(1) C+O; active forms of carbon formed by dissociation of CO followed by hydrogenation, etc.

(2) A surface complex of CO and H_2, for example formation of H - C - OH followed by hydrogenolysis, dehydrocondensation, etc.

Using $^{12}CO/H_2$ mixtures, Ponec and coworkers [66] found that Ni, Co and Ru, that have been precovered with ^{13}C made by ^{13}CO dissociation, produce first $^{13}CH_4$. These observations show that methanation can proceed via hydrogenation of C, that surface C is the key-intermediate and CO dissociation an essential step.

Surface science studies performed on single crystal and polycrystalline surfaces of Ni, Ru, Fe and Rh and using AES and/or XPS and high pressure chambers support these conclusions [1]. CO dissociates under the reaction conditions (500-600K, $p \approx 1$ bar) and surface carbon appears to be a reaction intermediate. The reaction rate and the activation energy were found to be essentially similar for Ni(100), Ni(111) and supported Ni/Al_2O_3 catalysts. AES measurements [1] show that the active catalyst under steady-state conditions contains a submonolayer quantity (5% on Ni(100)) of a "carbidic"-like carbon species which is the precursor to CH_4 formation. Exposure of the Ni crystal to H_2/CO mixtures at T > 650K (depending on the hydrogen pressure) results in deactivation of the catalyst. This deactivation is directly related to the concentration of inactive "graphitic" carbon on the surface as shown by AES. Graphitic carbon overlayers do not react readily with hydrogen, a distinctly different behaviour than observed with carbidic carbon.

As we have seen before, the reaction of atomic H with atomic O is very rapid compared with the methanation reaction and the production of CO_2 from CO and O. For that reason H_2O is a principal reaction product whereas CO_2 is usually produced at a level of only a few percent of the CH_4 and H_2O.

The Fischer-Tropsch synthesis of larger hydrocarbon molecules is a polymerization reaction, initiated by CO dissociation. The propagation step, the growth of the chain, is the insertion of CH_x units [66].

It has been found that certain supported Pd catalysts [66] are active and selective catalysts in the synthesis of methanol. This behaviour of Pd has been attributed to the poor activity of Pd in the breaking of the C-O bond. However, it has already been mentioned, that the selectivity of precious metals for the production of methanol and of other oxygen containing compounds strongly depend on the support or promoter used. The mechanism of C_2 oxygenates formation probably includes insertion of CO in CH_x units and that of CH_3OH formation the presence of adsorbed formyl or formate as an intermediate [66]. The role of the promoter which is of essential importance for the production of O containing compounds is beyond the scope of this paper.

In summary, the data show that the activity and the selectivity of the metal in the methanation reaction is determined by:

(1) The metal-C and the metal-O bond strength
(2) The ability of the metal to dissociate CO. We have seen before that metals active in the dissociation of CO have also relatively large C- and O-metal bond strengths.
(3) The nature of surface C and O. Active metals dissociate CO easily under reaction conditions, but the metal-C (carbidic) bond is not very strong.

5.6 AMMONIA SYNTHESIS AND OXIDATION

The industrial catalyst used for ammonia synthesis, is made by fusing iron oxide with K_2NO_3 and Al_2O_3 and reduction which is usually performed inside the reactor. Often this "doubly-promoted" iron catalyst contains also CaO, MgO or SiO_2 in addition to the promoters Al_2O_3 and K_2O.

A plot of the catalytic activity of various metals versus the metal-N bond strength results in a vulcano shaped curve [67]. The most active metals Os and Fe have the optimum M-N bond strength. Metals like Pt and Rh bind N too weakly and metals like W too strongly. These metals are, therefore, poor catalysts for this particular reaction.

It has been accepted that NH_3 synthesis proceeds via dissociation of dinitrogen. Although there has been a long debate concerning the rate determining step, most researchers agree now that the N_2 dissociation is the rate determining step. In fact, this has been suggested already in the 1930's. Brill et al found some evidence from field emission microscopy, that the Fe(111) surface is more reactive than other Fe surfaces in N_2 dissociation and NH_3 synthesis [68]. Later studies by Ertl et al [69] and Somorjai et al [70] confirmed the enormous effect of the surface structure on N_2 dissociation and NH_3 synthesis with the open Fe(111) surface being more active than Fe(100) and Fe(110). Spencer et al [70] found the following relative rates for NH_3 formation on the (111), (100) and (110) surfaces of Fe: 418:25:1.

Ammonia is oxidized according to the stoichiometric reaction equation (7). The reaction is extremely exothermic and the heat of the reaction is usually sufficient to maintain the catalyst bed at the desired temperature. Undesired nitrogen is produced according to the following stoichiometric equation: $4\ NH_3 + 3\ O_2 \rightarrow 2\ N_2 + 6\ H_2O$. The catalysts requirement for the production of nitric acid is a high selectivity for reaction (7) under reaction conditions. The reaction must be carried out at a temperature above 900K under oxidizing conditions. Only a few catalysts can withstand these conditions and Pt was the obvious choice.

The catalyst is usually made up in the form of woven wire gauzes. The use of Pt-Rh alloys instead of pure Pt results in much better mechanical properties of the gauze and a significant reduction of Pt loss during the process.

Changes in surface structure and composition of the Pt-Rh alloy catalysts during heating in ammonia-air mixtures have been studied with a range of techniques including SEM, EPMA, XPS, AES, FIM and atom-probe FIM [71]. After oxidation a stable Rh-rich oxide layer is formed. However under some conditions the outer layers of the oxide may become enriched in Pt. This effect was ascribed to the redeposition of vapor-phase platinum oxide on the surface.

The mechanism of the oxidation reaction is still uncertain. It may be a radical reaction which initiates at the surface and possibly continues to play a role in the gas phase within the catalyst bed.

6. General remarks and conclusions

We have discussed that the metals active for the O_2 + H_2 reaction are the noble metals of Group VIII, expecially Pt with a low O-metal bond strength. For the CO + O_2 reaction the same metals are active. These metals do not easily break the C-O bond and do not form oxides under reaction conditions. Both reactions are LH reactions between adsorbed O and adsorbed H or CO. Active catalysts for the NO + CO and NO + H_2 reactions must be more active in bond breaking: NO must dissociate whereas the C-O bond must stay intact, followed by a LH reaction of O with adsorbed CO. For the methanation and NH_3 synthesis reaction the metals must be even more active in bond breaking: the CO and N_2 molecule must dissociate under reaction conditions. The reaction then proceeds by hydrogenation of C, N and O on the surface. Important factors that determine the performance of the metal in these reactions are the nature and bond strength of C, N and O on the surface.

7. References

1. B.E.Nieuwenhuys. Surf.Sci. 126 (1983) 307 and refs. therein.
   In order to limit the number of references the reader is referred to [1] for refs. to most of the papers published before 1983.
2. F.Haber, Z.Elektrochem. 16 (1910) 244
3. A.Mittasch, Adv.Catal. 2 (1950) 81
4. E.M.McCash, S.F.Parker, J.Pritchard and M.A.Chesters, Surf.Sci. 215 (1989) 363
5. R.D.Ramsier and J.T.Yates, Surf.Sci.Rep. 12 (1991) 243 and refs. therein
6. J.B.Benzinger, Appl.Sci.Sci., 6 (1980) 105
7. M.Th.Karapetyants and M.K.Karapetyants, Handbook of Thermodynamic Constants (Ann Arbor-Humphrey Science, Ann Arbor, 1970)
8. G.Brodén, T.N.Rhodin, C.Brucker, R.Bendow and Z.Hurych, Surf.Sci. 59 (1976) 593
9. B.E.Nieuwenhuys, Surf.Sci. 105 (1981) 505
10. K.Christmann, Surf.Sci.Rep. 9 (1988) 1
11. J.W.Davenport, P.J.Estrup, Chem.Phys.Solid State Surf.& Het.Catal. 3A (199) 39, eds. D.A.King and D.P.Woodruff, Elsevier/Amsterdam
12. J.L.Gland, B.A.Sexton and G.B.Fisher, Surf.Sci., 95 (1980) 587
13. J.P.Campuzano, Chem.Phys.Solid State Surf.& Het.Catal. 3A (1991) 389, eds. D.A.King and D.P.Woodruff, Elsevier/Amsterdam
14. G.Blyholder, J.Phys.Chem. 68 (1964) 2772
15. P.S.Bagus, C.J.Nelin, C.W.Bauschlicher, J.Vac.Sci.Technol. A2 (1984) 905
16. W.Muller, P.S.Bagus, J.Vac.Sci.Technol. A3 (1985) 1623
17. G.Rangelov, N.Memmel, E.Bertel, V.Dose, Surf.Sci. 251 (1991) 965
18. R.P.Eischens, W.A.Pliskin and S.A.Francis, J.Chem.Phys. 22 (1954) 1786
19. N.Sheppard and T.T.Nguyen in Adv.IR and Raman Spectr. Vol.5, eds.R.E.Hester and R.J.H.Clark, (Heyden, London, 1978)
20. D.A.King, Chem.Phys.Solid State Surf.& Het.Catal. 3A (1990) 39, eds.D.A.King and D.P.Woodruff, Elsevier/Amsterdam
21. C.N.R.Rao and P.Ranga Rao, Surf.Sci.Rep. 13 (1991) 221
22. R.M.Wolf, J.W.Bakker and B.E.Nieuwenhuys, Surf.Sci. 246 (1991) 135
23. A.Sandell, A.Nilsson and N.Martensson, Surf.Sci., 241 (1991) L1 and refs.therein
24. T.H.Lin and G.A.Somorjai, Surf.Sci., 107 (1981) 573
25. R.J.Gorte, L.D.Schmidt and J.L.Gland, Surf.Sci., 109 (1981) 367

26. R.I.Masel, Cat.Rev.Sci.Eng. 28 (1986) 335
27. G.B.Fisher and J.L.Gland, Surf.Sci. 94 (1980) 446
28. G.B.Fisher and B.A.Sexton, Phys.Rev.Lett. 44 (1980) 683
29. F.Solymosi, J.Molec.Catal. 65 (1991) 337
30. H.A.C.M.Hendrickx, A.P.J.M.Jongenelis and B.E.Nieuwenhuys, Surf.Sci. 154 (1985) 503
31. L.H.Dubois and G.A.Somorjai, Surf.Sci. 91 (1980) 514 and refs. therein
32. M.F.H.van Tol, A.Gielbert, R.M.Wolf, A.B.K.Lie and B.E.Nieuwenhuys, Surf.Sci., accepted for publication
33. C.Benndorf and T.E.Madey, Surf.Sci. 135 (1983) 164
34. Y.Zhou, S.Akhter and J.M.White, Surf.Sci. 202 (1988) 357
35. T.Sasaki, T.Aruga, H.Kuroda and Y.Iwasawa, Surf.Sci.Lett. 224 (1989) L969
36. B.A.Sexton and G.E.Mitchell, Surf.Sci. 99 (1980) 539
37. D.G.Löffler and L.D.Schmidt, Surf.Sci. 59 (1976) 195
38. H.Conrad, G.Ertl and J.Küppers, Surf.Sci. 76 (1978) 323
39. R.M.Lambert and C.M.Comrie, Surf.Sci. 46 (1974) 61
40. P.A.Thiel, W.H.Weinberg and J.T.Yates, J.Chem.Phys. 71 (1979) 1643
41. D.A.King, Faraday Disc.Chem.Soc. 87 (1989) 303
42. R.F.van Slooten and B.E.Nieuwenhuys, J.Catal. 122 (1990) 429
43. G.A.Kok, A.Noordermeer and B.E.Nieuwenhuys, Surf.Sci. 135 (1983) 65
44. B.E.Koel, D.E.Peebles and J.M.White, Surf.Sci. 107 (1981) L367
45. P.R.Norton in "Chem.Phys.Solid Surfaces and Heterogenous Catalysis", eds. D.A.King and D.P.Woodruff (Elsevier, Amsterdam, 1982), Vol.4, chapt.2
46. B.A.Gurney and W.Ho, J.Chem.Phys. 87 (1987) 5562
47. G.E.Mitchell and J.M.White, Chem.Phys.Lett. 135 (1987) 84
48. D.S.Y.Hsu, M.A.Hoffbauer and M.C.Lin, Surf.Sci. 184 (1987) 25
49. S.Ljunström, B.Kasemo, A.Rosén, T.Wahnström and E.Fridell, Surf.Sci. 216 (1989) 63
50. T.Engel and H.Kuipers, Surf.Sci. 90 (1979) 181
51. S.D.Gardner, G.B.Hoflund, B.T.Upchurch, D.R.Schryer, E.J.Kielin and J.Schryer, J.Catal. 129 (1991) 114
52. T.Engel and G.Ertl, J.Chem.Phys. 69 (1978) 1267
53. N.W.Cant, P.C.Hicks and B.S.Lennon, J.Catal. 54 (1978) 372
54. J.Siera, F.Rutten and B.E.Nieuwenhuys, Catal.Today, 10 (1991) 353
55. M.F.H.van Tol, J.Siera, P.D.Cobden and B.E.Nieuwenhuys, Surf.Sci. 274 (1992) 63 and M.F.H.van Tol, A.Gielbert and B.E.Nieuwenhuys, Catal.Lett. 16 (1992) 297
56. B.C.Sales, J.E.Turner and M.B.Maple, Surf.Sci. 114 (1982) 381
57. G.Ertl, Adv.Catal. 37 (1990) 213
58. M.Boudart and F.Rumpf, React.Kinet.Catal.Lett. 35 (1987) 95
59. T.P.Kobylinski and B.W.Taylor, J.Catal. 33 (1974) 376
60. K.C.Taylor, Catal.Sci.Techn., eds. J.R.Andersson and M.Boudart (Springer-Verlag-Berlin) 5 (1984) 119
61. H.Hirano, T.Yamada, K.Tanaka, J.Siera, P.Cobden and B.E.Nieuwenhuys, Surf.Sci. 262 (1992) 97 and proc.10th ICC, Budapest, 1992, paper 015
62. M.Iwamoto, H.Yahiro, K.Tanda, N.Mizuno, Y.Mine and S.Kagawa, J.Phys.Chem. 95 (1991) 3727
63. Y.Li and W.K.Hall, J.Catal. 129 (1991) 202
64. H.Bosch and F.Janssen, Catal.Today 4 (1988) 392
65. B.Harisson, A.F.Diwell and M.Wyatt, Plat.Met.Rev. 29 (1985) 50
66. V.Ponec, Stud.Surf.Sci.& Catal. 64 (1991) 117 and Catalysis Today 12 (1992) 227
67. A.Ozaki and K.Aika, Catal.Sci.Techn., eds. J.R.Andersson and M.Boudart (Springer-Verlag-Berlin) 1 (1981) 87
68. R.Brill, Ber.Bunsenges.Phys.Chem. 75 (1971) 455

69. G.Ertl, Catal.Rev.Sci.Eng. 21 (1980) 201
70. N.D.Spencer, R.C.Schoonmaker and G.A.Somorjai, J.Catal. 74 (1982) 129
71. F.C.M.J.M.van Delft, B.E.Nieuwenhuys, J.Siera and R.M.Wolf, Iron and Steel Inst.Japan Intern. 29 (1989) 550 and refs.therein

MODEL CATALYST STUDIES OF TITANIA SUPPORTED RHODIUM

J.EVANS, B.E.HAYDEN, J.F.W.MOSSELMANS and A.J.MURRAY
Department of Chemistry,
The University,
Southampton, SO9 5NH,
United Kingdom.

ABSTRACT A technique has been developed to deposit and characterise highly dispersed rhodium on titania (rutile) single crystal surfaces using MOCVD. The geminal dicarbonyl species can be synthesised directly on $TiO_2(110)$ from the pre-cursor $[Rh(CO)_2Cl]_2$, and is characterised by ν(C-O) bands at 2112 cm^{-1} and 2028 cm^{-1} observed in FT-RAIRS, with a $Rh(3d^{5/2})$ binding energy (309.1eV) close to that of the parent molecule (309.3eV). The frequencies vary by only ca. 9 cm^{-1} with coverage, and the gem-dicarbonyl is observed to undergo isotopic exchange at 300K. The gem-dicarbonyl is stable to 500K when decarbonylation takes place through CO desorption producing highly dispersed metallic rhodium particles. Decarbonylation of the gem-dicarbonyl is also induced by exposure to hydrogen at 300K, although this process is accompanied by a partial disproportionation of the CO resulting in adsorbed carbon. Subsequent adsorption of CO at 300K leads to a partial re-generation of the gem-dicarbonyl. In addition, a **slow** adsorption of linearly bound CO on metallic particles is observed, with ν(C-O) in the range 2071 cm^{-1}... 2064 cm^{-1}. Heating the rhodium overlayer to higher temperatures leads to the nucleation of larger metallic particles from which the gem-dicarbonyl cannot be re-formed by exposure to CO. Rhodium deposition by MVD at 300K leads to layer by layer growth of metallic rhodium. Exposure of MVD sub-monolayers of rhodium on $TiO_2(110)$ to CO does not result in the formation of the gem-dicarbonyl. A facile adsorption of linearly bound CO on rhodium is observed, with ν(C-O) in the range 2028 cm^{-1}... 2080 cm^{-1} for both sub-monolayer and multilayer MVD films deposited at 300K. The molecules desorb from the films in the temperature range 450-500K. The $TiO_2(110)$ substrate appears not to influence the CO adsorption behaviour even for the first atomic layer of rhodium. Heating the rhodium films to 500K and above produces rhodium particles which exhibit the **slow** adsorption behaviour which characterises particles produced from the MOCVD precursor.

1. INTRODUCTION

Finely dispersed rhodium particles supported on oxide surfaces act as catalysts to a number of important industrial processes **[1]**, including the hydrogenation of carbon monoxide, methane and higher mass hydrocarbons **[2]**, the reduction of NO_x **[3]** and the hydroformylation of olefins **[4]**. The extent of metal dispersion effects both the activity and selectivity of the catalyst **[5]**, in addition to having economic consequences. Not surprisingly, therefore, considerable effort has been directed towards a characterisation of high area oxide supported rhodium catalysts, and in particular their CO chemistry.

Following the original identification using IR spectroscopy of the gem-dicarbonyl species $Rh(CO)_2$ by Yang and Gartland **[6]** together with linear and bridged bound CO on

R. W. Joyner and R. A. van Santen (eds.), Elementary Reaction Steps in Heterogeneous Catalysis, 179–195.

metallic crystallites, it was suggested that the gem-dicarbonyl is CO adsorbed on an isolated (effectively oxidised) rhodium atom **[6-14]**, or is formed by CO adsorption at the edges of two dimensional rafts of metallic rhodium **[15,16]**. A **critical size (ca.10 atoms)** appeared to exist below which the gem-dicarbonyl could be observed following exposure of the surface to CO **[15,16]**. Subsequently it was shown using EXAFS **[17]** that **CO induced disruption of Rh-Rh bonds** takes place on highly dispersed rhodium on alumina. This disruption was shown to be slowest under dry conditions **[18]** and enhanced by the presence of surface hydroxyl groups **[19,20]** which appeared to be consumed during bond disruption. The behaviour of rhodium particles to CO exposure is, however, highly temperature dependent, and heating Rh/Al_2O_3 catalysts above 423K **[18,21]** or 473K **[22,23]** results in **reductive agglomeration**. Above 573K this process appears irreversible **[23]** implying the existence of larger crystallites that cannot be oxidatively disrupted to the gem-dicarbonyl by CO and OH. Although the majority of studies to date have been on Rh/Al_2O_3, a TiO_2 or SiO_2 support does not radically alter the surface chemistry. The clustering of rhodium particles on TiO_2 in the presence of CO is reported to take place at 473K **[24]**, and decarbonylation of the gem-dicarbonyl in the temperature range 373K **[25]** to 573K **[26]**.

More recently some considerable attention has been given to the generation of highly dispersed rhodium using organometallic precursors, in particular $[Rh(CO)_2Cl]_2$ **[27-35]**. We have used this route to produce highly dispersed rhodium on $TiO_2(110)$ in the form of the gem-dicarbonyl **[36]** in order to investigate its chemical activity and stability, and interconversions with the reduced phase **[37]**. We also demonstrate that such studies are inaccessible through rhodium deposition via direct MVD, although the chemistry of the larger metal crystallites is the same using either deposition technique. We also demonstrate the powerful combination of XPS and FT-RAIRS for studies of metal chemistry on oxide single crystal surfaces through the use of CO as the probe molecule. A more thorough discussions of the infra-red spectroscopy itself in a reflection mode on oxide surfaces will appear elsewhere [38].

2. EXPERIMENTAL

A schematic of the ultra high vacuum (UHV) system employed in these studies is shown in Figure 1. Facilities for low energy electron diffraction (LEED), x-ray photoelectron spectroscopy (XPS), Fourier transform reflection absorption infra-red spectroscopy (FT-RAIRS), ion bombardment, metal organic compound vapour deposition (MOCVD) and temperature programmed desorption (TPD) are incorporated in two working levels in the chamber. A quadrupole with mass separation to 1000 amu provides a monitor of the MOCVD process. The turbo-molecular pump is of the magnetic levitation design to minimise mechanical vibration during FR-RAIRS measurements, and when combined with a titanium sublimation pump provides a base pressure of 4×10^{-10} Torr.

The FT-RAIRS system is the Perkin Elmer 1720 using remote detection (MCT and InSb), and the optical path outside vacuum is flushed with dry nitrogen. The angle of incidence of the polarised radiation can be set at 83 degrees or 70 degrees. All measurements presented here have been made using p-polarised radiation at 83 degrees, more grazing than the Brewster angle of clean titania **[38-40]**. XPS data presented here was obtained using Mg Kα radiation at an electron pass energy of 50eV. Binding energies ($\pm$0.1eV) were referenced to $Ti(2p^{3/2})$ at 458.5eV **[41]**. Absolute coverages were estimated using the atomic sensitivity factors of Briggs and Seah **[42]**.

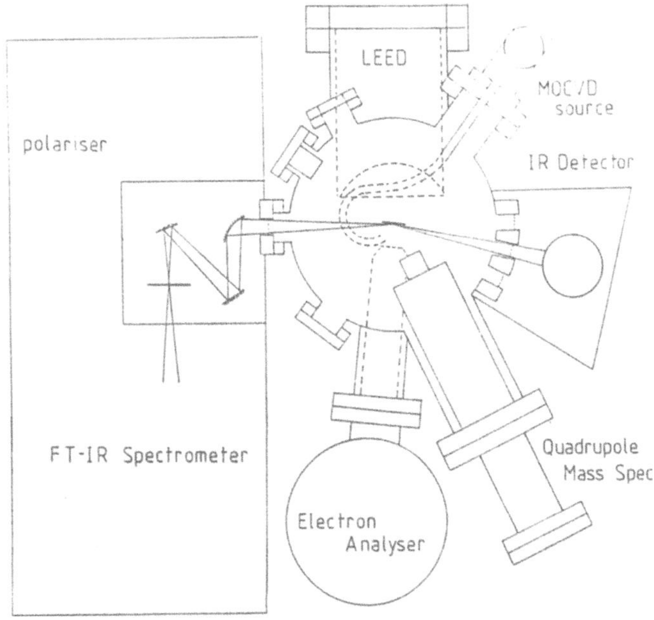

Figure 1. The experimental system used for the study of rhodium on $TiO_2(110)$

The sample temperature could be varied in the range 150 ... 1000K. The $TiO_2(110)$ single crystal was cleaned by cycles of Ne^+ or Ar^+ bombardment, annealing at 1000K and oxygen treatment at 400K following procedures described previously **[43,44]**.

3. RESULTS AND DISCUSSION

3.1 SYNTHESIS AND CHARACTERISATION OF RHODIUM GEM-DICARBONYL

Exposure of $TiO_2(110)$ at 300K to $[Rh(CO)_2Cl]_2$ results in the dissociative adsorption of the molecule in the form of the gem-dicarbonyl, $Rh(CO)_2$, and adsorbed chlorine **[36]**. The binding energy of the $Rh(3d^{5/2})$ in the gem-dicarbonyl (309.1eV) is only slightly shifted from that of the parent molecule (309.3eV), which we have obtained through the adsorption of multilayers at low temperature, and is in agreement with previous measurements **[1,5,34]**. The small shift in $Rh(3d^{5/2})$ binding energy demonstrates the difficulty in distinguishing the parent molecule from the gem-dicarbonyl by XPS, although the chlorine spectra give clear evidence that the $[Rh(CO)_2Cl]_2$ has dissociated to produce adsorbed chlorine. The binding energy $Cl(2p^{3/2})$ following adsorption of $[Rh(CO)_2Cl]_2$ at 300K is observed at 198.5eV, and is clearly shifted from that of the parent molecule (199.1eV). During the decarbonylation and nucleation of the rhodium species, the Cl($2p^{3/2}$) binding energy remains constant.

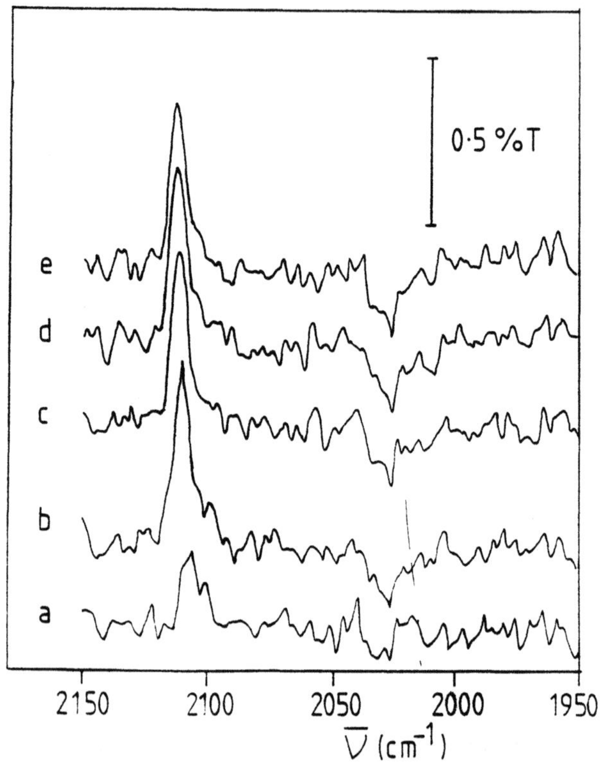

Figure 2. FT-RAIRS spectra of the gem-dicarbonyl formed on $TiO_2(110)$ during exposure to $[Rh(CO)_2Cl]_2$ at 300K: (a) 2.0L; (b) 4.0L; (c) 6.5L; (d) 9.3L; (e) 12.5L.

A series of FT-RAIRS spectra are shown in Figure 2 as a function of exposure, and yield a more unambiguous identification of $Rh(CO)_2$. As a result of the transparency of titania in the infra-red, absorption of radiation in the adsorbed overlayer can give rise to both an increase or decrease in reflectivity for experiments carried out at angles more grazing than the Brewster angle **[38-40,45]**. The band (increased reflectivity) at 2112 cm^{-1} is assigned to ν(C-O)$_{symm}$ of $Rh(CO)_2$ (coupling to the normal component of the p-polarised light). The band (decreased reflectivity) at 2028 cm^{-1} is assigned to ν(C-O)$_{assym}$ of $Rh(CO)_2$ (coupling to the tangential component of the p polarised light). These values are close **[46]** to those observed for this species on high area alumina **[7,8,14,15,20,22,26]**, silica **[8,9,26]** and titania **[9,25,26]** surfaces. A plot of the intensity of the two bands as a function of exposure is shown in Figure 3. This clearly demonstrates that the two bands originate from the same species, and that adsorption is Langmurian.

The ν(C-O) bands of the gem-dicarbonyl shift upwards no more than 9 cm^{-1} with increasing coverage. Dynamic dipole and chemical interactions between CO molecules adsorbing on the extended metal surface **[47]** give rise to the larger shifts characteristic of CO adsorption on Rh(111) **[46]**. The coverage independence of the $Rh(CO)_2$ infra-red vibrational bands has often been presented as evidence for an isolated species on high area powder supports **[6,7,12,13,14]**. Adsorption of the parent molecule takes place at temperatures below 250K, in agreement with previous UHV studies carried out on amorphous alumina substrates **[49-51]**, and give rise to a quite different FT-RAIRS spectra **[38]**. We note, however, that even during deposition at 200K, for initial (monolayer) adsorption of $[Rh(CO)_2Cl]_2$ on $TiO_2(110)$ we observe in FT-RAIRS the formation of the gem-dicarbonyl species. The saturation coverage of $Rh(CO)_2$ (Figure 2,3) is 0.35 ± 0.05 ML, estimated using XPS.

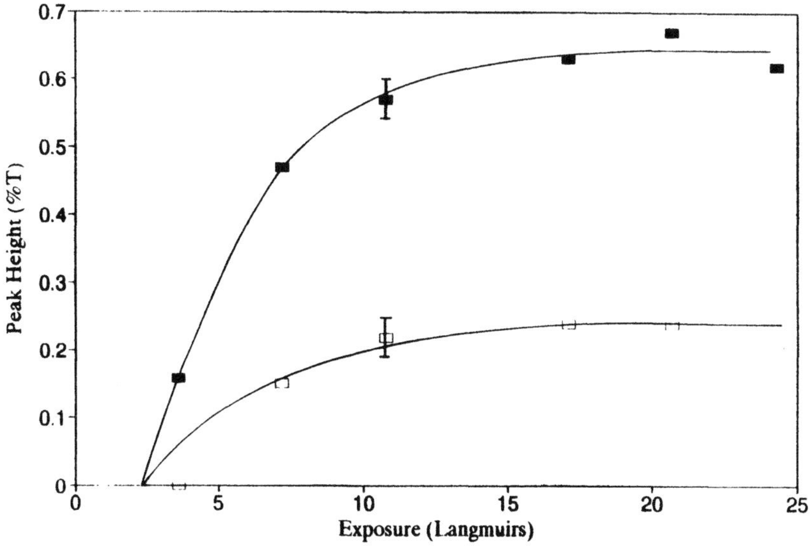

Figure 3. The peak heights of the symmetric and anti-symmetric carbonyl stretches of the gem-dicarbonyl as a function of $[Rh(CO)_2Cl]_2$ exposure.

3.2 STABILITY AND REACTIONS OF THE GEM-DICARBONYL

We have also been able to carry out CO isotopic exchange with this species. Exposure of the gem-dicarbonyl to pressures of ca. 1×10^{-3} Torr of $C^{13}O$ at 300K produces a sharp reduction in the intensity of both $Rh(C^{12}O)_2$ bands, and the $\nu(C\text{-}O)_{symm}$ band of $Rh(C^{13}O)_2$ appeared as expected at 2063 cm^{-1} (increase in reflectivity). The signal to noise was apparently insufficient to observe $\nu(C\text{-}O)_{asymm}$. The result is in agreement with previous observations **[7,12,33]** that $Rh(CO)_2$ is labile, undergoing isotopic exchange. Exposure of the $Rh(C^{13}O)_2$ to $C^{12}O$ did not result in complete recovery of all of the original $Rh(C^{12}O)_2$. There was also evidence from the XPS that some reduction of the rhodium had taken place following the exposure to CO at 1×10^{-3} Torr. This suggested that some reductive agglomeration of rhodium particles takes place at 300K on $TiO_2(110)$. Similar behaviour has been observed on powder oxide samples during exposure to CO at 373K, although no significant agglomeration was reported during isotopic exchange **[18,24]**.

The gem-dicarbonyl can be decarbonylated on $TiO_2(110)$ by heating the surface, or by exposure to hydrogen at 300K. Figure 4A shows a sequence of FT-RAIRS spectra taken as a function of temperature. The CO band intensity is reduced in the range 450 ... 550K, and

XPS reveals that CO desorbs around 500K from the surface cleanly, without the formation of residual carbon. Chlorine remains on the surface chemically unperturbed (constant 2p binding energy in XPS). Figure 4B shows the effect on the gem-dicarbonyl FT-RAIRS spectrum of exposure of the species to H_2 at 300K. A similar disappearance of the gem-dicarbonyl is observed. However XPS indicates that the reaction with H_2 at 300K results in a significant disproportionation (ca. 30%) of CO resulting in a residual of graphitic or carbidic carbon remaining on the surface.

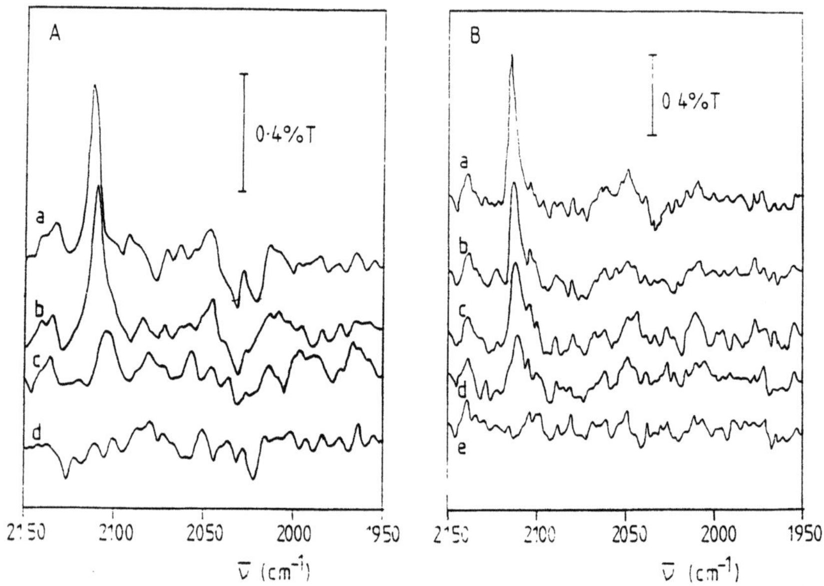

Figure 4. A. FT-RAIRS spectra of the gem-dicarbonyl species on $TiO_2(110)$ at (a) 300K, and following heat treatment to (b) 450K; (c)490K; (d) 500K.
B. FT-RAIRS spectra of the gem-dicarbonyl species on $TiO_2(110)$ at (a) 300K, and following a hydrogen exposure of (b) 600L; (c) 1200L; (d)3300L; (e)4500L; (f) $1x10^6$L.

A concomitant reduction of the rhodium takes place during both the heat treatment to 500K, and the H_2 chemical treatment at 300K. The XPS spectrum of the gem-dicarbonyl Rh(3d) region is shown in Figure 5A(a). Heating to 500K produces a phase of rhodium with an XPS spectrum shown in Figure 5A(b). The Rh($3d^{5/2}$) binding energy is shifted to ca. 307.4 in a broad peak, with no change observed in the rhodium peak intensity. The observed decomposition temperature is within the range of temperatures quoted in the literature for $Rh(CO)_2$ decomposition on the powdered oxides **[9,26]**. XPS also indicates that CO has desorbed from the surface. Clearly some reduction of the rhodium has taken place which is concomitant with CO removal. If some change in rhodium dispersion takes place during this process, it occurs only after CO has desorbed, since a simple reversal of the coverage dependent blue shift in $\nu(C\text{-}O)_{symm}$ (Figure 4A) is observed during the desorption.

XPS reveals a similar reduction of the rhodium following exposure of the gem-dicarbonyl to hydrogen (Figure 5B(b)). The binding energy of the Rh($3d^{5/2}$) level is shifted from 309.1 eV, that of the gem-dicarbonyl, to ca. 307.4 eV, again producing a slightly broadened peak. This change is concomitant with the disappearance of the carbonyl stretches in FT-RAIRS (Figure 4B).

Decarbonylation of the gem-dicarbonyl was also carried out by heating the surface to 800K. In this case the Rh($3d^{5/2}$) binding energy of the reduced species (307.0eV) was shifted further from that of the gem-dicarbonyl (Figure 5C(b)). Its value is the same as that expected for the extended metallic phase **[52]**. In addition, the peak width is smaller than that following the low temperature heating or hydrogen treatment, and similar to the gem-dicarbonyl.

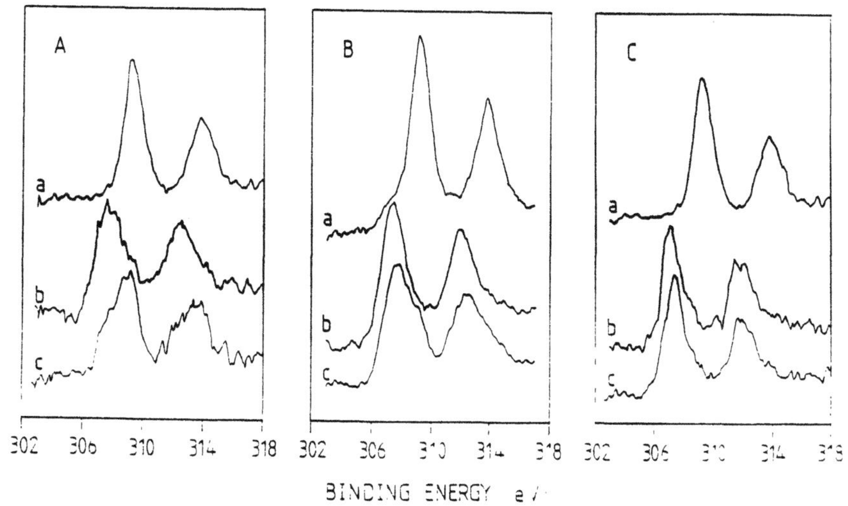

Figure 5. XPS spectra of the Rh(3d) core levels.

A. (a) the gem-dicarbonyl; (b) following heating (a) to 500K; and (c) following exposure of the heated surface (b) to 6 x 10^6L of CO at 300K.

B. (a) the gem-dicarbonyl; (b) following exposure of (a) to 1 x 10^6L of H_2 at 300K; and (c) following the exposure of (c) to 1 x 10^6L of CO at 300K.

C. (a) the gem-dicarbonyl; (b) following heating (a) to 800K; and (c) following the exposure of (b) to 3 x 10^7L of CO at 300K.

We ascribe the changes observed (Figure 4,5) during the thermal or chemical treatments to two distinct processes, namely a formal reduction of the rhodium oxidation state accompanying decarbonylation, and the aggregation of the reduced rhodium. Decarbonylation of the gem-dicarbonyl has clearly taken place at 500K, or following hydrogen exposure (Figure 4), and the intermediate Rh($3d^{5/2}$) binding energy, and broadness of XPS peaks, is associated with an inhomogeneous distribution of small metal particles or islands (Rh^0_x). Such phases can exhibit core levels with slightly higher binding energies **[53]** than found in the bulk metallic phase because of the dominance of initial state effects in photoemission. The effect of heating to a higher temperature (800K) is to cause a more extensive agglomeration of Rh^0_x, producing a more homogeneous distribution of larger particle sizes. This is consistent with the observation that the intensity of the Rh($3d^{5/2}$) peak is reduced following the 800K heat treatment (Figure 5C(b)). We considered the possibility that the presence of chlorine may be effecting the rhodium phases and binding energies, since it is present on the titania at 500K, but has largely desorbed by 800K. We have, however, carried out these experiments in the absence of the chlorine (removed by mild heat cycling with water **[37]**) and obtain the same results (section 3.3).

The reaction of hydrogen with chemisorbed CO on supported rhodium is of particular significance from the standpoint of a variety of catalytic reactions. For example, oxide supported rhodium exhibits a high activity towards CO and CO_2 hydrogenation, particularly on titania **[26]**. It is not surprising that a number of papers have focused on this aspect **[2,9,18,20,25]**. A similar effect to the facile hydrogen induced decarbonylation observed here on TiO_2(110) has been observed on Rh/Al_2O_3 **[18,20]**, however there are some subtle differences. The reaction is reported at a surface temperature of 373K, and is apparently absent at 300K **[18]**. On the other hand, it has also been reported that the gem-dicarbonyl reacts with H_2 to form Rh_x^0 at temperatures as low as 200K **[20]**. In the latter study, it was suggested that the supply of reactive hydrogen originates from Rh_x^0 on which hydrogen readily dissociates, and is subsequently supplied to the isolated gem-dicarbonyl via. spill-over onto the support. We clearly observe a facile reaction at 300K even in the absence of rhodium metal, leading us to speculate that either hydrogen is supplied directly from dissociation on the TiO_2(110) surface or a direct reaction of molecular hydrogen with the gem-dicarbonyl occurs.

We remark also on the partial disproportionation of the CO to carbon we observe following exposure of the gem-dicarbonyl to hydrogen at 300K. This route to carbon production does not involve CO dissociation on Rh_x^0, and we have never observed the latter process on TiO_2(110) supported particles of any size (see below). We therefore suggest that a competing reaction to the hydrogen induced displacement of CO takes place at the gem-dicarbonyl which can result in residual carbon. Since we are unable to identify at 300K any of the molecular intermediates that have been suggested for the reaction of $Rh(CO)_2$ with hydrogen, we are reluctant to speculate as to the detailed mechanism. Nevertheless, the facile reaction of the gem-dicarbonyl with hydrogen on TiO_2(110) suggests a possible direct mechanism to hydrogenated products involving this dispersed form of rhodium. Carbon is also a product of the hydrogenation **[see 26 and references therein]**, and reconversion through CO adsorption of the highly dispersed Rh_x^0 produced as a result of the hydrogen reaction (Figure 5B) can also take place (section 3.3).

3.3 THE REGENERATION OF THE GEM-DICARBONYL

If the surface which has been heated to 500K (section 3.2) is re-exposed to CO at 300K, CO is re-adsorbed at the surface. This is evidenced in the sequence of FT-RAIRS spectra shown in Figure 6. A band quickly re-appears at 2111 cm^{-1}, manifested as an increase in reflectivity. This peak has already appeared at an exposure of only ca. 100L. We associate its appearance with the re-generation of the rhodium gem-dicarbonyl from the dispersed rhodium particles. The lower signal to noise (resulting from temperature treatments to the sample) do not allow us to observe the associated anti-symmetric mode (Figure 2). The FT-RAIRS also suggests that only a partial re-conversion of the gem-dicarbonyl has taken place since the peak intensity of $\nu(C\text{-}O)_{symm}$ is lower than that observed for the original gem-dicarbonyl layer. This partial regeneration of the $Rh(CO)_2$ species is supported by the observed changes in the XPS spectrum of Rh(3d) resulting from CO exposure (Figure 5A(c)). This clearly reveals the partial regeneration of the oxidised form of rhodium with a Rh($3d^{5/2}$) binding energy (309.1eV) corresponding to the gem-dicarbonyl.

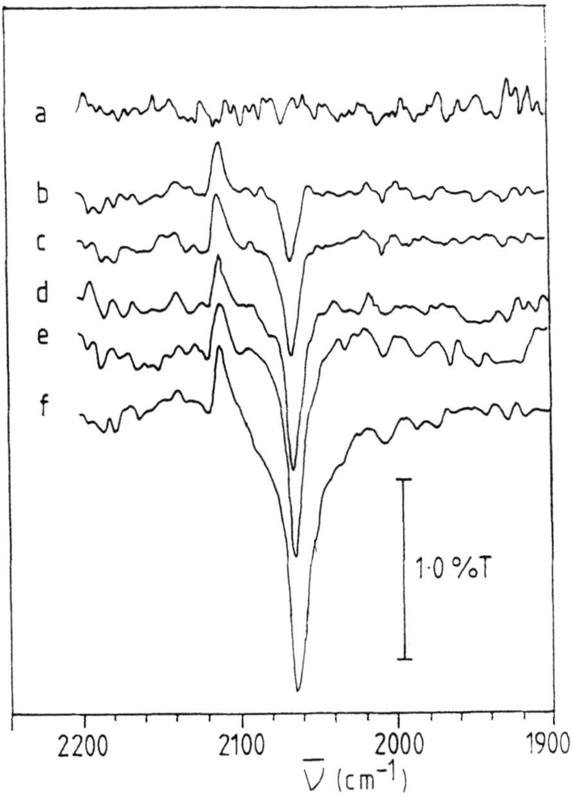

Figure 6. FT-RAIRS spectra taken during the re-adsorption of CO on a rhodium layer on TiO_2(110) produced by heating the gem-dicarbonyl to 500K. The CO exposures were (a) 0L; (b) $1.2x10^6$; (c) $1.9x10^6$; (d) $3.4x10^6$; (e) $4.6x10^6$; (f) $6.2x10^6$.

In addition to the re-appearance of $\nu(C\text{-}O)_{symm}$ of gem-dicarbonyl, a second band subsequently appears in the range 2071 cm^{-1} ... 2064 cm^{-1} at significantly higher CO exposures. This band manifests itself as a large absorption in the FT-RAIRS, and is associated with CO adsorbed in a linear configuration on Rh^0_x particles. This interpretation is supported by the XPS spectrum of the Rh(3d) core level (Figure 5A(c)) which reveals a proportion of Rh^0_x unperturbed by CO adsorption. Linearly adsorbed CO on such particles on high area supported samples is evidenced by a band in the range 2048-2067 cm^{-1} on alumina **19,20]**, 2030 cm^{-1}... 2075 cm^{-1} on titania **24]**, and is similar to range the observed frequency for linearly bound CO on Rh(111), 2020 ... 2070 cm^{-1} **[48]**. The FT-RAIRS characteristics on the particles is more akin to that obtained on the flat metal surfaces, exhibiting strong **absorption** behaviour, rather than the more complex behaviour exhibited by the oxide. This is an indication that the size of the particles is sufficient to exhibit the dielectric characteristics of the metallic phase. Some indication of the critical size is obtained from the experiments on thin rhodium films (section 3.4). It should be noted, however, that the initial frequency at which ν(C-O) is observed (2071 cm^{-1}), and the **small downward** frequency shift with coverage (to 2064 cm^{-1}) is subtly different to the characteristics of linear CO on supported metal particles **[19,20,24]**, Rh single crystals **[48]** and MVD films on TiO_2(110) (section 3.4). In these cases a more extensive upward chemical shift is exhibited.

It can be seen from Figure 6 that the re-formation of the gem-dicarbonyl is relatively facile, and is complete at 300K at the lowest ($1.2x10^6$L) CO exposure shown. Indeed the reconversion starts to take place at exposures of only 100L at a pressure of 2x 10^{-6} Torr. CO adsorption on the Rh^0_x clusters is significantly slower, and a more extensive adsorption takes place at higher exposures as judged by the IR band intensity. This is contrasted with the kinetics of $Rh(CO)_2$ formation at higher CO pressures on powder titania which is significantly slower than the adsorption of linearly bound CO on the metal particles. The rate of CO adsorption on the particles observed in this study (Figure 6) is significantly slower than on Rh(111) **[48]**. We do not consider that this effect originates from a chemical interaction of the metal with the substrate. Atomically thick rhodium layers on the TiO_2(110) surface exhibit CO adsorption rates, desorption temperatures and IR spectra very similar to the Rh(111) (section 3.4). The unusually small and downward coverage dependent frequency shift may provide the clue to the slow adsorption. We tentatively suggest that it is associated with oxygen which has migrated onto the rhodium surface from the titania. CO adsorption may take place slowly in islands on regions of the surface where oxygen has been reactively removed by CO. This aspect is the subject of continuing investigation.

The relative size of the peaks for ν(C-O) in the gem-dicarbonyl and on the Rh^0_x clusters (Figure 6) are deceiving. The XPS indicates that the amount of CO associated with the gem-dicarbonyl is significantly greater (even after the highest CO exposures) despite the relatively small size of the reflectivity change. This underlines the differences in the apparent sensitivities in FT-RAIRS. This is associated with the reflectivity differences of the metallic rhodium particles and the titania on which isolated gem-dicarbonyl species are adsorbed **[38]**. Heating the surface on which CO is adsorbed both in the form of the gem-dicarbonyl and the terminally bound species on Rh^0_x (Figure 6(f)) gives rise to a sequential loss of CO. The gem-dicarbonyl band at 2110cm^{-1} disappears at 450-500K (the same temperature as that found for the earlier thermally induced decarbonylation (Figure 4A)), leaving only the terminally bound species on Rh^0_x. The band of the terminally bound CO is slightly broadened, with its centre at 2050cm^{-1} (not shown). Further heating leads to the desorption of the terminally bound CO at 550-600K. The XPS also indicates the complete loss of the CO on the surface by 600K, and no residual carbon is detectable following desorption.

Exposure of the Rh^0_x particles produced as a result of hydrogen exposure at 300K (Figures 4B,5B) to CO also results in the re-generation of the gem-dicarbonyl, producing its characteristic FT-RAIRS spectrum (not shown). However, the re-generation is not as efficient as that observed from particles produced from the 500K thermolysis. This is reflected in the XPS spectrum taken following CO exposure (Figure 5B(c)) which reveals that only a small proportion of the highly dispersed phase Rh^0_x undergoes oxidation to the gem-dicarbonyl phase. This observation is somewhat surprising, since it will be shown that the extent of agglomeration is a function of surface temperature, and that the larger particles are unable to re-generate the gem-dicarbonyl. The above result suggests that the extent of agglomeration is not significantly different between 300K and 500K. It is also possible that the residual carbon formed during hydrogen exposure (section 3.2) inhibits the re-generation of the gem-dicarbonyl.

CO re-adsorption on the surface heated to 800K does not result in the re-formation of the gem-dicarbonyl species [37]. Only the linear bound species is re-adsorbed on Rh^0_x, with ν(C-O) initially at observed in FT-RAIRS at 2064cm^{-1}. there are some subtle differences to the spectra obtained following the 500K thermolysis. The peak is somewhat broader with a frequency more dependent on CO coverage **[37]**. In addition, shoulders develop at 2030cm^{-1} and 2010cm^{-1} at high coverages. We also observed a broad feature at 1870cm^{-1} which we initially attributed to the bridge bound species of CO on Rh^0_x **[48]**. Unfortunately, changes in the IR background during the annealing cycle can give rise to a similar broad structure, and a conclusive assignment awaits further investigation. The Rh(3d) XPS spectrum obtained following exposure of the surface to CO is shown in Figure 5C(c). It clearly confirms that the rhodium remains in the reduced form on the surface. We do observe at high coverages of CO a very small shift in Rh($3d^{5/2}$) binding energy to 307.4eV. We suggest that this is a result of is a small chemical shift induced by CO adsorption on the Rh^0_x clusters, associated with the electron withdrawing effect (Rh(3d) ---> CO(2π)) of the CO **[53]**.

It appears, therefore, that the gem-dicarbonyl can be re-generated by CO adsorption at 300K from Rh^0_x clusters following a mild heat treatment at 500K, but that this process is inhibited by heating the surface to 800K. Since adsorbed chlorine remains on the surface at 500K, but has desorbed at 800K, it is possible that the difference in behaviour is associated with the presence of the chlorine. We have therefore carried out experiments in which, following the formation of Rh^0_x clusters at 500K, the surface has been treated with water, and heat cycling to 500K, in order to remove the chlorine. XPS indicated that this process completely removes the adsorbed chlorine. This procedure was also found to be effective for the removal of adsorbed chlorine on amorphous alumina **[37]**. Exposure of the chlorine free surface to CO also produced the peak at 2111cm^{-1}, and it can be concluded that the gem-dicarbonyl formation was not promoted or inhibited by adsorbed chlorine.

We suggest that the difference in adsorption behaviour observed for the rhodium layers on surfaces heated to 500K and 800K is a result of the differing extent of nucleation of rhodium clusters as a function of temperature. Larger clusters produced at 800K are unable to undergo oxidative disruption by CO at 300K. This has also been suggested to be the case for rhodium particles on high area oxide surfaces. The decarbonylation of $Rh(CO)_2$ on rutile **[24]** at 523K is accompanied by structural re-arrangement with the formation of rhodium crystallites which could not be re-converted to gem-dicarbonyl at 300K. Decarbonylation of $Rh(CO)_2$ on alumina in the temperature range 473-573K also results in structural re-arrangement, and only at the lower temperatures of this range could the gem-dicarbonyl be re-formed by CO adsorption at 300K **[23]**. We suggest, therefore, that following thermolysis to 500K, or hydrogen exposure at 300K, a proportion of the reduced rhodium on TiO_2(110) (Figure 5A,5B) remains very dispersed, resulting in a facile re-conversion to gem-dicarbonyl on CO adsorption. It is at present unclear whether the cluster responsible for the re-conversion is a particular phase (eg. atomic sites or two dimensional islands) or simply a cluster with a size below some critical value which marks the onset of inactivity. Since CO adsorption on the cluster itself is shown to be relatively slow compared to gem-dicarbonyl formation at 300K (Figure 6), we speculate that the gem-dicarbonyl results from CO re-adsorption on a distinct phase of Rh^0_x on which CO is quickly adsorbed. It is not apparently generated as a result of the CO disruption of the larger particles which exhibit the adsorption of linear CO.

There has been some debate concerning the mechanism of the CO induced re-oxidation of Rh^0_x to Rh^I concomitant with the synthesis of $Rh(CO)_2$ (see **[24]** and references therein). Both a dissociative adsorption mechanism **[17]**, and one involving oxidation by surface hydroxyl groups **[18,20]**, have been suggested. We clearly see a reversible reduction/oxidation cycle on $TiO_2(110)$ induced by the desorption/re-adsorption of CO by the rhodium. The pressure regime in which the conversion is observed to take place on $TiO_2(110)$ ($2x10^{-6}$ Torr) is similar to the pressures at which the gem-dicarbonyl begins to form from rhodium on high area oxide surfaces. However, there is no evidence that surface hydroxyl groups are required for the regeneration of $Rh(CO)_2$ observed here on $TiO_2(110)$. No surface hydroxyl groups are adsorbed under the conditions of these experiments. In addition, during our experiments involving water treatment, for the removal of adsorbed chlorine, we have been unable to enhance gem-dicarbonyl formation. It is also unlikely that CO dissociation is responsible for the oxidation since we have been unable to detect any form of residual carbon (carbidic or graphitic) by XPS following conversions. We also fail to find any evidence from XPS **[54]** for the generation of OH groups as a result of the hydrogen induced decarbonylation of the gem-dicarbonyl (section 3.2), as might be expected on the basis of the results on alumina **[20]**

The rather facile reaction converting Rh^0 to Rh^I on the $TiO_2(110)$ surface is more likely to be associated with a small concentration of surface defects [37] on the titania surface. A local re-arrangement at these sites, probably involving the relatively mobile oxygen atoms and the redox behaviour of Ti^{4+}, could provide sites which could accomodate either Rh^I or Rh^0. A redox coupling involving the titanium has recently been suggested to be important in this type of conversion on titania **[17]**.

3.4 CO ADSORPTION ON RHODIUM FILMS.

We have also carried out a series of experiments in which the adsorption of CO is investigated on sub-monolayer and multilayer films of rhodium deposited directly using MVD techniques. The FT-RAIRS spectrum for CO adsorbed on a monolayer coverage of rhodium is shown in Figure 7. Similar spectra have been obtained in the rhodium coverage range 0.3 ... 1.0 ML for films deposited at 300K. A single band, exhibited as an increase in reflectivity, is observed in the range 2027 cm^{-1}... 2079 cm^{-1}. Adsorption is facile, and is complete following an exposure of 20L. This behaviour is similar to that observed on Rh(111) except that we have never observed the bridging band at high coverages. The fact that ν(C-O) manifests itself as an increase in reflectivity is a result of the layer by layer growth mode of the rhodium on $TiO_2(110)$ **[55]**. The rhodium overlayer is consequently in the form of an a single atom layer on the $TiO_2(110)$, and under these conditions the optical response in the region of the adsorbed CO molecule is not perturbed significantly from that of titania. Consequently, a coupling of a dipole to the normal component of the p-radiation results in increased reflectivity. A reversion to the metallic optical response takes place only after the growth of 10 ... 20 atomic layers **[38]**. This result suggests that the Rh^0_x particles on which linearly bound CO is observed have a diameter greater than about 10 atoms, since the FT-RAIRS appears to probe a CO molecule on a substrate with a metal dielectric response in the infra-red (section 3.3)

An important observation is that even after the exposure of sub-monolayer films to CO at pressures of 1 x 10^{-3} Torr for extended periods, we are unable to generate the gem-dicarbonyl species. However, heating the surface to 700K is accompanied by the desorption of CO at 500K, and subsequently a nucleation of the rhodium phase to produce three dimensional particles. This is evidenced by the series of FT-RAIRS spectra which are obtained following the re-exposure of such a surface to CO at 300K, shown in Figure 8.

Strong absorption bands are observed at 2064 cm^{-1}, producing spectra reminiscent of those obtained from the particles produced via the MOCVD route (Figure 6). We also note that adsorption on these particles is also significantly slower than on the newly deposited thin films, and similar to the behaviour observed on the MOCVD synthesised particles.

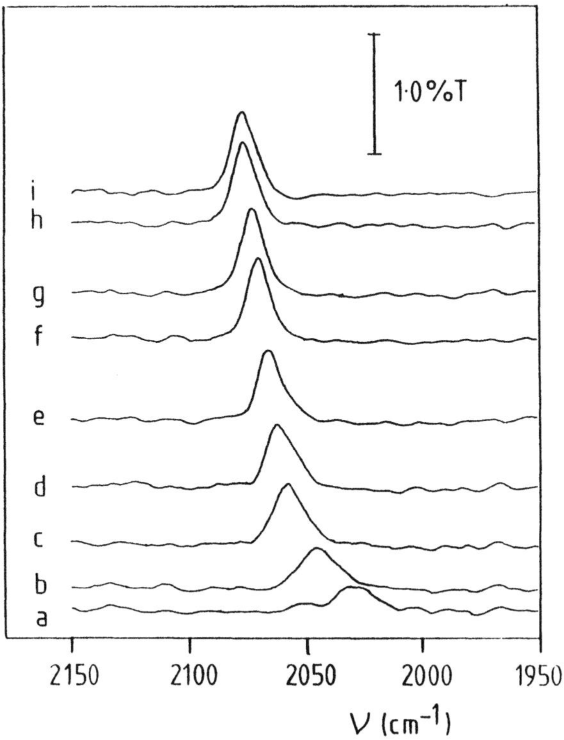

Figure 7. FT-RAIRS spectra of CO adsorption at 300K on a 1ML film of rhodium deposited my MVD at 300K on $TiO_2(110)$. The CO exposures were (a) 1.5L;(b) 3.0L;(c) 4.2L;d) 5.6L;(e) 7.0L;(f) 10.4L;(g) 14.3L;(h) 18.9L;(i) 25L.

4. CONCLUSIONS

The exposure of $TiO_2(110)$ to $[Rh(CO)_2Cl]_2$ at 300K results in its dissociative adsorption to produce the rhodium gem-dicarbonyl together with adsorbed chlorine. The gem-dicarbonyl is characterised in FT-RAIRS by bands at 2112 cm^{-1} and 2028 cm^{-1} which are associated with the symmetric and anti-symmetric carbonyl stretch vibrations respectively.

The gem-dicarbonyl undergoes thermal decarbonylation through the desorption of CO at 500K. Exposure to hydrogen at 300K also results in decarbonylation in a process which also results in the formation of residual carbon on the surface. The facile reaction of the gem-dicarbonyl with hydrogen on $TiO_2(110)$ does not require the presence of Rh^0_x as has been suggested to be the case for alumina supports. This route to adsorbed carbon may play an important role in low temperature CO hydrogenation on titania supported rhodium. This effect is reminiscent of the SMSI phenomenon.

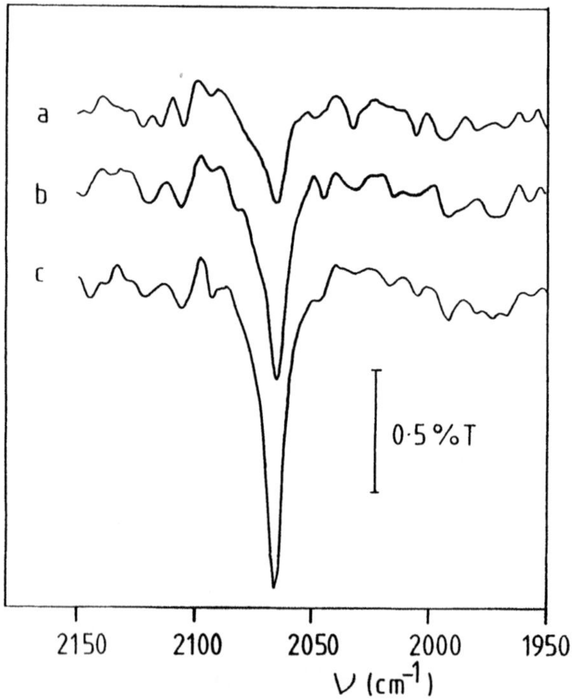

Figure 8. FT-RAIRS spectra of CO re-adsorption at 300K on the 1ML film of rhodium deposited by MVD which had subsequently been heated to 700K. The CO exposures were (a) $1x10^6$L;(b) $2x10^6$L;(c) $4x10^6$.

Decarbonylation through CO desorption at 500K, or reaction with hydrogen at 300K, is accompanied by the formation of a highly dispersed layer of Rh^0_x from which the gem-dicarbonyl can be partially re-generated through exposure to CO at 300K. CO adsorption in the form of the linearly bound species is observed on the remaining particles of Rh^0_x with ν(C-O) in the range 2071 cm^{-1}... 2064 cm^{-1}. CO adsorption on these particles is considerably slower than that observed on Rh(111) or on the newly deposited rhodium films on TiO_2(110). The slow adsorption of CO, and the unusually small and downward frequency shift, may be associated with adsorbed oxygen on the rhodium particles. A similar adsorption of linearly bound CO is observed on a more homogeneous dispersion of larger particles of Rh^0_x produced at 800K. However, unlike the highly dispersed layer produced at 500K or following hydrogen exposure at 300K, particles on a surface heated to 800K were unable to re-generate the gem-dicarbonyl through CO adsorption.

We suggest that since CO adsorption on the metallic particles is observed subsequent to gem-dicarbonyl formation, the latter is not derived from the CO induced disruption of these particular metal particles. Gem-dicarbonyl re-generation at 300K on $TiO_2(110)$ is a result of CO adsorption on particles below some critical size. This process seems to be uneffected by adsorbed chlorine, does not require surface OH groups, and does not involve the dissociation of CO. We cannot exclude the possibility that a disruption of the larger particles may take place under more extreme conditions of pressure and temperature.

Adsorption of CO on sub-monolayer films of rhodium appears unperturbed by the underlying $TiO_2(110)$ substrate, taking place quickly and producing a linear bound species with ν(C-O) in the range 2027 cm^{-1} ... 2079 cm^{-1}. Heating the films to 700K produce particles similar to those produced by the MOCVD route: CO is adsorbed on the particles only slowly, producing the linear bound species with ν(C-O) around 2064 cm^{-1}. The gem-dicarbonyl could not be synthesised through CO adsorption on any rhodium layers deposited by MVD.

5. ACKNOWLEDGEMENTS

This work has been supported by the S.E.R.C.(GR/F/52408), the University of Southampton and B.P.Research International and B.P.Chemicals Ltd. We also would like to acknowledge technical support from Perkin Elmer Ltd., and N.A.Williams and N.A.Cruise for the preparation of $[Rh(CO)_2Cl]_2$.

6. REFERENCES

1 Guczi, L. (1991) Ed. "Studies in Surface Science and Catalysis"; Elsevier, Amsterdam, Vol 64.
2 Solymosi, F., Tombacz, I., and Koscis M.J. (1982) J. Catal., 75, 78.
3 Solymosi, F. and Starkany, J. (1979) Appl. Surf. Sci., 3, 68.
4 Chuang, S.S. and Pien, S.I. (1992) J. Catal., 135, 618.
5 Yao, H.C., Yu Yao, Y.F. and Otto K. (1979) J. Catal. 56, 21.
6 Yang, A.C. and Garland, C.W. (1957) J. Phys. Chem. 61, 1504.
7 Yates, J.T. and Kolanski, K.J. (1983) J. Phys. Chem. 79, 1026.
8 Worley, S.D., Rice, C.A., Matteson, G.A., Curtis, C.W., Guin, J.A. and Tarrer A.R. (1982) J. Phys. Chem. 86, 2714.
9 Worley, S.D., Matteson, G.A., Caudill, R.J. (1983) J. Chem. Phys. 87, 1671.
10 Primet, M. (1979) J. Chem. Soc. Faraday Trans. 74, 2570.
11 Knözinger, H., Thornton, E.W. and Wolf, M. (1979) J. Chem. Soc. Faraday Trans. 75, 1888.
12 Yates, J.T., Duncan, T.M. and Worley, S.D., Vaughan, R.W. (1979) J. Chem. Phys. 70, 1219.
13 Yates, J.T., Duncan, T.M. and Vaughan, R.W. (1979) J. Chem. Phys. 71, 3908.
14 Wang, H.P. and Yates, J.T. (1984) J. Catal. 89, 79.
15 Yates, D.J.C., Murrell, L.L., and Prestridge, E.B. (1979) J. Catal., 57, 41.
16 Katzer, J.R., Sleight, A.W., Gajardo, P., Michel, J.B., Gleason, E.D. and Mcmillan, S. (1981) Faraday Discuss. Chem. Soc. 72, 121.

17 Van't Blik, H.F.J., van Zon, J.B., Huizinga, T., Vis, J.C., Koningsberger, D.C. and Prins, R (1985) J. Am. Chem. Soc. 107, 3139; (1983) J. Phys. Chem. 87, 2264.

18 Solymosi, F. and Pasztar, M. (1985) J. Phys. Chem. 89 4789; (1986) J. Phys. Chem. 90, 5312.

19 Basu, P., Panayotov, D. and Yates, J.T. (1987) J. Phys. Chem. 91, 3133.

20 Basu, P., Panayotov, D. and Yates, J.T.; (1988) J. Am. Chem. Soc. 110, 2074.

21 Anderson, J.A. and Rochester, C.H. (1991) J. Chem. Soc. Faraday Trans. 87, 1479; Anderson, J.A. (1991) J. Chem. Soc. Faraday Trans. 87, 3907.

22 Dictor, R. and Roberts, S. (1989) J. Phys. Chem. 93, 5846.

23 Solymosi, F. and Knözinger, H. (1990) J. Chem. Soc. Faraday Trans., 86, 389.

24 Buchanan, D.A., Hernandez, M.E., Solymosi, F. and White, J.M. (1990) J. Catal. 125, 456.

25 Conesa, J.C., Sainz, M.T., Soria, J., Munuera, G., Rives Arnau, V. and Munoz A.J. (1992) J. Mol. Catal. 17, 231.

26 Erdoheyl, A. and Solymosi, F. (1984) J. Catal. 84, 446.

27 Smith, G.C., Chojnacki, T.P., Dasgupta, S.R., Iwatate, K. and Watters K.L. (1975) Inorg. Chem. 14, 1419.

28 Smith, A.K., Hugues, F., Theolier, A., Basset, J.M., Ugo, R., Zanderighi, G.M., Bilhou, J.L., Bilou, V., Bougnol, T. and Graydon W.F. (1979) Inorg. Chem. 18, 3104.

29 Bowser, W.M. and Weinberg, W.H. (1980) J. Am. Chem. Soc. 103, 1453.

30 Basset, J.M., Theolier, A., Commereuc, D. and Chauvin, Y. (1985) J. Organomet. Chem. 279, 147.

31 McNulty, G.S., Cannon, K. and Schwartz, J. (1986) Inorg. Chem. 25, 2919.

32 Keyes, M.P. and Watters, K.L. (1986) J. Catal. 100, 477.

33 Robbins., J.L. (1986) J. Phys. Chem. 90, 3381.

34 Frederick, B.G., Apai, G. and Rhodin, T.N. (1987) J. Am. Chem. Soc. 109, 4797.

35 Keyes, M.P. and Wallers, K.L. (1988) J. Catal. 110, 96.

36 Evans, J., Hayden, B.E., Mosselmans, J.F.W. and Murray, A.J. (1992) J. Am. Chem. Soc. 114, 6912.

37 Evans, J., Hayden, B.E., Mosselmans, J.F.W. and Murray, A.J. Surf. Sci. Lett. in press.

38 Evans, J., Hayden, B.E., Mosselmans, J.F.W. and Murray, A.J.; to be submitted.

39 Chesters, M.A., Horn, A.B., Kellar, E.J.C., Parker, S.F. and Ravel, R. (1989) "Mechanism of Reactions of Organometallic Compounds with Surfaces", Cole-Hamilton, D.J. and Williams, J.O. (Eds.), NATO Advanced Studies Institute Series B, Plenum, 198.

40 Patel, H. and Pemble, M.E. (1991) J. Physique IV 1, 167.

41 Chang, T., Bernasek, S.L. and Schwartz, J. (1991) Langmuir 7, 1413.

42 Briggs, D. and Seah, M.P. (1983) "Practical Surface Analysis", John Wiley and Sons Ltd.

43 Kurtz, R.L., Stockbauer, R., Madey, T.E., Roman, E. and De Segovia, L. (1989) Surf. Sci. 218, 178.
44 Hayden, B.E. and Nicholson, G.P. (1992) Surf. Sci. 274, 277.
45 Chabal, Y.J. (1988) Surf. Sci. Rept. 8, 211; (1987) Springer Procs. in Phys. 22, 301.
46 Small upward frequency shifts are expected for the dipole coupling in the normal component of the p radiation in FT-RAIRS measurements under these conditions **[45,38]**.
47 Hayden, B.E. (1987) "Methods of Surface Characterisation", Ed. Yates,J.T. and Madey, T.E., Plenum Vol 4, p. 267.
48 Dubois, L.H. and Somorjai, G.A. (1980) Surf. Sci. 91, 514; Crowell, J.E. and Somorjai, G.A. (1984) Appl. Surf. Sci. 19, 252; Wagner F.T., Moylanand, T.E. and Schmieg, S.J. (1988) Surf. Sci. 195, 403.
49 Belton, D.N. and Schmieg, S.J. (1988) Appl. Surf. Sci. 32, 173.
50 Belton, D.N. and Schmieg, S.J. (1988) Surf. Sci. 199, 518.
51 Belton, D.N. and Schmieg, S.J. (1988) Surf. Sci. 202, 238.
52 DeLouise, L.A., White, E.J. and Winograd, N. (1984) Surf. Sci. 147, 252.
53 Apai, G., Lee, S.-T., Mason, M.G., Gerenser, L.J. and Gardner, S.A. (1979) J. Am. Chem. Soc. 101, 6880.
54 We have been able to monitor the presence of OH formed by adsorbing water on $TiO_2(110)$ at 300K by XPS, through its characteristic emission in the O 1s region.
55 XPS has been used to study the growth mode of the rhodium by monitoring the Rh and Ti core levels as a function of MVD deposition time.

THE DYNAMICS OF ALKANE ADSORPTION ON METALS

George W. Coulston
E.I. duPont de Nemours and Co.
Experimental Station
Wilmington, DE 19880-0262

Gary L. Haller
Yale University
Department of Chemical Engineering
New Haven, CT 06520

ABSTRACT. The dissociative adsorption of alkanes is the first and rate limiting step in several very important catalytic processes. In this review we summarize the current understanding of the dynamics of such adsorption for linear C_1-C_4 alkanes. The differences between precursor mediated adsorption, through both intrinsic and extrinsic states, and direct adsorption are discussed and a new model is presented which allows one to crudely predict when the precursor mediated process will dominate.

1. Introduction

In this review we will discuss the dynamics of alkane adsorption from an experimentalist's perspective. We have chosen to focus our discussion on alkanes since they are becoming of increasing importance in commercialized catalytic processes and because a comparison of results for different chain lengths allows for a test, perhaps only qualitative, of the generality of the statements to be made. Before going further, however, we should define what we mean by dynamics. Dynamics is often used in the catalysis and reaction engineering communities to refer to an important class of experiments which probe a reaction mechanism by measurements of unsteady-state kinetics. In this paper, dynamics is used in a different sense. As we use it here, dynamics distinguishes those experiments which probe the potential energy hypersurface (PES) from those experiments with outcomes predictable directly from conservation laws, i.e. kinematics. The connection to catalysis comes from the fact that the kinetics of a reacting system, steady-state or otherwise, are determined from the underlying dynamics and this connection is made through the relationship of the kinetic rate constant to the state-to-state cross-section. For example, consider the generic reaction A —> B. If the state-to-state cross-sections σ are known, then the kinetic rate constant can be determined from Equation (1),

$$k = \int \sum_{\varepsilon'} \rho(t,v,j,n)\, \sigma(t,v,j,n|t',v',j',n')\, d\varepsilon \qquad (1)$$

R. W. Joyner and R. A. van Santen (eds.), Elementary Reaction Steps in Heterogeneous Catalysis, 197–219.

where the state-to-state cross sections σ are summed over the accessible final states and averaged over the population distribution of the initial states ρ, with energies ε and ε', respectively [1].

As practiced in gas-phase dynamics experiments, one measures partially resolved quantum state-to-quantum state cross sections between entrance and exit channels (reactants and products) and these are then inverted (often by comparison with trajectories over a model potential) to give the PES, or at least that fraction of the hypersurface probed by the experiment. The situation is much worse (but not hopeless) for probes of the dynamics of heterogeneous reactions since one has to contend with increased excited state relaxation rates owing to the presence of the catalyst. Here, it is instructive to include some numbers. The relation between the lifetime of a state and the Fourier limited bandwidth of that state as probed by a suitable photon is $\Delta t=2\pi c\Delta v$. Assuming accommodation occurs on the time scale between 0.1ps and 100ps [2-4], each quantum state of the molecule probed will appear to be broadened by 0.05 to 53 cm^{-1}. This broadening is smaller than the spacing normally encountered between low lying vibrational levels and preserves the oscillator strength within a relatively narrow bandwidth. Thus, the states are reasonably well defined and one should be able to follow the vibrational relaxation of molecules adsorbed on surfaces; indeed, such experiments have been done. Therefore, surface reactions which involve reactants that can diffuse on a time scale short compared to the relaxation time ($D=10^{-1}$ to 10^{-4} cm^2/s for the molecule to randomly sample an area within 1 nm of its starting position) might, one day, be the subject of vibrational state-to-state experiments. The diffusion coefficient of CO has been measured to be $3x10^{-3}$ cm^2/s on Ni(111) [5] which suggests that CO oxidation on this surface might be probed in this way. The diffusivities of n-alkanes have been measured on Ru(001) and found to be on the order of 10^{-5} cm^2/s [6, 7] suggesting that they too might be studied in this way, although probably only for unimolecular reactions. Of course, in neither case would these be true state-to-state experiments, since one would have little control over the translational degrees of freedom of the adsorbates. However, they would provide valuable information regarding the role of molecular vibrations in surface catalyzed reactions.

Despite the difficulties alluded to above, progress has been made toward understanding the dynamics of adsorbate-surface interactions. Most of the experiments performed to date have probed barriers in the potential surface at either end of the reaction coordinate. Very detailed measurements have been performed of the final state energy partitioning in CO_2 produced by CO oxidation [8-15] and these experiments have served as indirect probes of the transition state in this reaction, but they have served more directly as probes of the barrier to dissociative CO_2 adsorption. We recently published our results for this system where we provided the first experimental evidence for the dependence of an adsorption cross-section on the relative phases of vibrational modes [8-10]. These studies are not discussed here since we have already reviewed them elsewhere [16]. Another system which has received much attention is the dissociative adsorption of H_2 on metals [16], particularly on copper. We will not discuss this system either, except to say that it has been and continues to be the source of many seminal ideas in adsorption dynamics. What we would like to discuss in this review is the dynamics of alkane adsorption. We are most interested, therefore, in reactions such as

$$C_nH_{2n+2}(g) \longrightarrow H(ads) + C_nH_{2n+1}(ads) \qquad (2)$$

although we do not mean to imply that Equation (2) represents an elementary step. Indeed, we have cause to write this review because in many cases it does not.

2. Dynamics of Dissociative Alkane Adsorption

2.1. METHANE

Because of the directional nature of their valence orbitals, the atoms of transition-metal surfaces can be considered unsaturated and, by analogy to homogeneous catalysis, one expects these surfaces to be active in both C-C and C-H bond activation [17]. While it is true that most molecules have a high dissociative chemisorption probability on clean transition metal surfaces, methane does not. This was demonstrated in the experiments performed by Winters [18, 19] who, working with W filaments, was the first to investigate the dynamics of dissociative methane chemisorption. By adjusting the filament diameter and the mean free path of the methane

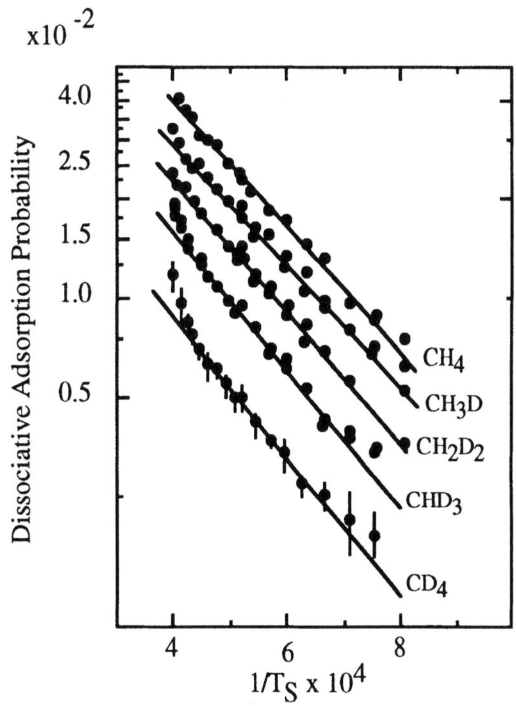

Figure 1: Dissociative adsorption probability of deuterated methane molecules on polycrystalline tungsten. Data extracted from ref [19].

molecule, he could effectively vary the surface temperature independently of the gas temperature. Thus, he could, in principle, have obtained Arrhenius type activation barriers for the surface temperature, T_S, and for the gas temperature, T_g, independently (this is accomplished in a bulb experiment by keeping the collision cross-section for molecules with the filament far below that for gas-gas and gas-bulb collisions). Of course, variation of T_g

implies an equilibrium of translational, rotational and vibrational energies (provided one has a sufficient number of gas-gas collisions). This can be contrasted with the use of free jet beam sources. The latter allows the translational and vibrational energies to be more or less varied independently (with the rotational energy always low, but varying somewhat with nozzle temperature and pressure [20, 21]). Winter's published data, however, was for a constant T_g=300K. In Figure 1, we show the dissociative sticking probability as a function of T_s plotted in Arrhenius form. As T_s is varied from 1200K to 2500K, the adsorption probability of CH_4 increases by nearly an order of magnitude, from $6x10^{-3}$ to $4x10^{-2}$. One can extract from Figure 1 an apparent activation energy equal to nearly 39 kJ/mol for all of the deuterated forms, although as pointed out by Winters, this value is somewhat subjective owing to the curvature in the Arrhenius plots. Based on this anomolous isotope effect (i.e. no change in the apparent activation energy), Winters suggested a model for dissociation based on tunneling from a vibrationally excited molecular species in thermal equilibrium with the surface. The implication for the involvement of a vibrationally excited molecular species was the strong temperature dependence of the sticking coefficient and Winters argued that this excitation occured by accommodation of CH_4 with the surface prior to dissociation. For adsorption via tunneling, the sticking coefficient can be written as in Equation (3)

$$S = \exp\left[-2\pi^2 L \frac{(V-E)}{h}\sqrt{\frac{\mu}{V}}\right] \tag{3}$$

where V is the height and L the width of the barrier, E is the initial energy of the tunneling particle in a mode directed perpendicular to the barrier and μ is the reduced mass of the system. Note that Equation (3) implicitly constrains the time development of the system in a single dimension. Vibrational excitation in the incident molecule would then serve, heuristically, to decrease the width of the barrier by increasing the average C-H bond length and also the relative collision energy of the H atom, i.e. E in Equation (3). Excitation of the phonon modes of the solid might also be expected to increase the value of E via coupling to the vibrational modes of the incident molecule, although to achieve the equivalent of one quantum of CH_4 vibrational excitation would require an nth order phonon process, where $n=\nu(CH_4)/\nu(W)$ and has a value typically of order 10. Since the Bose-Einstein distribution reduces to the Boltzmann distribution for $kT>>h\nu$, the apparent activation energy for the two processes would be the same, although the phonon assisted process can only be a dominating one when the sticking probability is very small since such high order phonon processes are unlikely events. It is not necessary, however, for the phonons to couple to the vibrational modes of the incident molecule for an enhancement in the adsorption probability to occur. Provided the incident mass is not significantly less than the surface atom mass, it is possible for the phonon modes to couple to the translational coordinate of the incident molecule during a nearly direct adsorption process. Such coupling is the basis for the phenomenon of thermally assisted tunneling [22, 23], which is to be adressed later.

The effect on the adsorption probability of vibrational excitation in the incident CH_4 was measured more directly in another early study of the dynamics of methane chemisorption on Rh field emitter tips (with T_s =245K). Stewart and Ehrlich [24] attributed a large increase in the sticking probability of methane from a heated effusive source (equivalent to a bulb experiment) to increased vibrational excitation of the incident molecules (600K $<$ Tg $<$

700K). From Arrhenius plots, they extracted an apparent activation barrier equal to 29 kJ/mol and estimated that fewer than 1 in 10^5 CH_4 molecules undergoes direct adsorption under the conditions of their experiment. As was the case for W, an isotope effect was observed, with CD_4 being ten times less likely to dissociatively adsorb than CH_4. However, contrary to Winter's work above, Stewart and Ehrlich did not invoke a tunneling mechanism to explain their data, but instead resorted to Slater's dynamical theory of unimolecular reactions [25]. In this model, the C-H bond is extended toward its dissociation limit by simultaneous occupation of those normal mode vibrations in CH_4 which combine to give the corresponding local mode representation of the C-H bond. The observed isotope effect would arise in this model from an increase in the density of vibrational states in the C-D bond. Based on this isotope effect, it was also claimed that the barrier to dissociative CH_4 adsorption was selective toward vibrational excitation, but not translational excitation, in the incident molecule. It is not possible, however, in our opinion, to extract conclusions about translational activation from the data presented by Stewart and Ehrlich. Subseqent work by the Ehrlich group on Rh [26] and W(211) [27] has also led them to the conclusion that a tunneling mechanism is not involved in the adsorption process and that vibrational excitation in the incident methane is most important for promoting dissociative adsorption.

Other early experiments performed by Schouten et al. [28-30] which probed the dissociative chemisorption of methane on the (110), (111) and (100) single crystal planes of Ni can also be used to address the importance of methane vibrational excitation in the adsorption process, particularly when their results are compared with the more recent results of Beebe et al. [31]. The most important difference between the two sets of experiments was the static gas pressure. Schouten et al. were working with a bulb pressure of 10^{-2} Torr, giving a mean free path large enough that the incident methane was equilibrated at the 300K bulb temperature. The pressure was higher in the experiments of Beebe et al. such that the gas was effectively in equilibrium with the surface and, thus, at the surface temperature. Thus, Schouten et al. could not measure chemisorption on Ni(111) and did not detect a temperature dependence of the chemisorption on Ni(100) (because the gas phase temperature did not vary with the temperature of the crystal). In contrast, Beebe et al. report the order of reactivity as Ni(111) < Ni(100) < Ni(110) above about 430K. They also compared their results for methane on Ni(111) to that reported for the steam reforming of methane. The rate-limiting step in this reaction is the dissociation of methane [32] so this should be a valid comparison. The rate estimated at 500K on the single crystal surfaces is about an order of magnitude higher than measured for small Ni particles. However, the single crystal rates are initial rates measured in the limit of zero coverage while the steady state rates measured by Rostrup-Nielsen will certainly involve finite carbon coverage. In this context, the agreement can be considered good. Finally, in recent investigations of methane adsorption on W(110) [33] and Ni(111) [34] where the vibrations were excited thermally and, thus, resulted in a Boltzmann distribution over all modes, vibrational energy has been found to be at least as efficient as translational energy for promoting dissociative chemisorption.

Nearly all of the work mentioned above has been performed using effusive sources and can thus be classified as bulb-type experiments. In contrast, free-jet expansion derived molecular beams have been the preferred reactant source for those probing the translational activation of CH_4 adsorption [33-38]. In all but one case [37] where the effect of translational energy on methane chemisorption has been investigated, the probability of chemisorption scales with the normal component of the kinetic energy as shown in Figure (2).

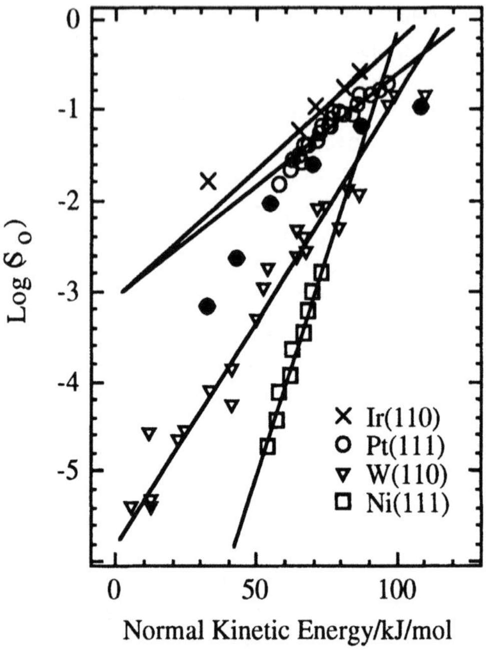

Figure 2: Initial dissociative adsorption probabilities for methane on Pt(111) [35], Ir(110) [39], Ni(111) [40], and W(110) [38]. The solid circles are the data of Luntz and Bethune on Pt(111) [36]. The solid lines correspond to the predictions of a one dimensional tunneling model. Reproduced with permission from ref [35]

Although there is not complete agreement as to its validity, the most popular model among the experimentalists to explain the scaling of the dissociative chemisorption of methane with the normal component of the kinetic energy has been a one dimensional tunneling model. The most persuasive experimental observation for quantum tunneling is the kinetic isotope effect where CD_4 has been found to have lower chemisorption probability than CH_4 (by about an order of magnitude on W(110) [33, 38] and Ni(111) [34]) but the same Arrhenius activation energy for translational energy. This was the same argument which led Winters to postulate a mechanism whereby the tunneling from a precursor state was enhanced by vibrational excitation. Figure (3) exhibits the initial dissociative chemisorption of methane on W (110) as a function of normal kinetic energy which has been varied both by changing the translational energy and the angle of incidence of the molecule measured relative to the surface normal. Note that the chemisorption probability increases by over four orders of magnitude as the normal component of translational energy is increased to 100 kJ/mol, i.e. the exponential increase in the chemisorption probability and the isotope effect is consistent with a tunneling mechanism.

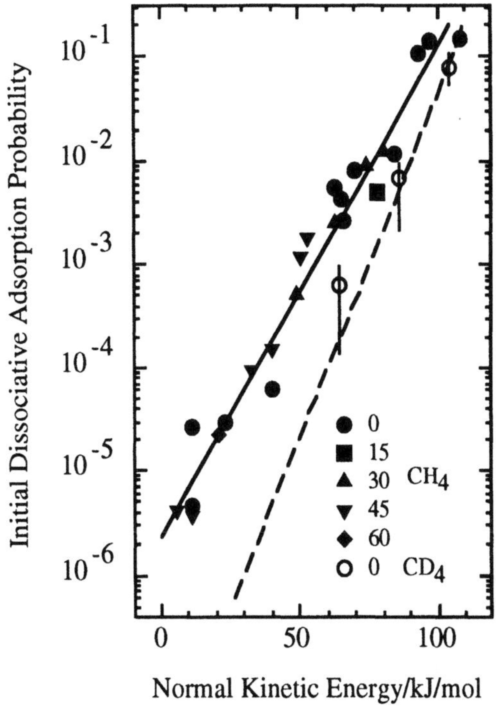

Figure 3: Initial dissociative adsorption probability of methane on W(110) as a function of beam energy and angle of incidence. Reproduced with permission from ref [38].

The tunneling parameters for the data in Figure (2) are tabulated below, without a correction for the differing degrees of vibrational excitation present in the incident CH_4 molecules [35]. Schoofs et al. [35] have calculated, on a point-by-point basis, the level of vibrational activation required in order to bring the data in Figure (2) for W(110) and Ni(111) into agreement with that for Pt(111). By defining a vibrational efficacy as

$$E = \frac{\Delta \ln(S_o)}{\Delta E_{vib}} \tag{4}$$

they find values for E ranging from 0.38-0.48 mol/kJ for adsorption on Ni(111) and 0.07-0.37 mol/kJ for adsorption on W(110). These values are, within experimental error, in agreement with those calculated from vibrational activation experiments carried out on the respective surfaces and, thus, one can conclude that the differences in Figure (2) are probably due largely to varying degrees of vibrational excitation in the incident CH_4. However, the differences in the vibrational efficacies between Ni(111) and W(110) and the coincidence of the data for Ir(110) with that of Pt(111) (apparently only at higher incident translational energies), even though the nozzle temperature varied over a large range in the case of Ir, indicates that the selectivity of the reaction coordinate toward vibrational energy depends on the metal [35]. It is also interesting to note that the calculated barrier thicknesses are largest for those cases where the incident vibrational energy was smallest, i.e. for W(110) and Ni(111), while no such correlation can be drawn from the barrier heights. In addition, the barrier thickness is larger for Ni(111) than for W(110), a behavior parralleled by the vibrational efficacies. As we stated earlier for the one dimensional tunneling model, vibrational excitation would be expected to increase the C-H bond length and, thereby, decrease the barrier width. By analogy to homogeneous chemistry, one does not expect to be able to extract a hydrogen from methane even at these hyperthermal translational energies without providing a means for stabilizing the resulting CH_3 fragment. Lee et al. [34] argue that when methane collides with a surface, it experiences a deformation from a tetrahedral to a pyramidal configuration which puts the C closer to the Ni surface and that this causes the barrier to H-atom tunneling to become narrow enough to break the C-H bond. An obvious extension and prediction of this picture is that a methane molecule resident on the Ni(111) surface can be distorted by a collision with another molecule, e.g. an Ar atom, and induced to dissociate. Indeed, Beckerle et al. [40] have demonstrated that such collision-induced dissociative chemisorption is possible.

The debate over the validity of the tunneling model to explain the dissociative adsorption of methane has been vigorous (e.g. refs. [41-44]). DePristo and Kara [45] do not find the one dimensional tunneling models convincing because, as they correctly argue, models that apriori restrict the motion of the system on the potential energy surface may significantly distort the dynamics. However, even though it is obvious that reduction of the dynamics to a single dimension warrants skepticism, one can also be sympathetic to its use because the correct multi-dimensional potentials are unknown and because the one dimensional models can have some predictive value.

Before going further we need to digress and make clear the distinction betweeen *direct* and *precursor mediated* adsorption. In direct adsorption, the energy required for activation is typically thought to be supplied by the incident molecule since there is essentially no time for energy to be transferred to the molecule from the surface. Such a description extends naturally from the trends observed in the beam experiments for activation by both incident kinetic and incident vibrational energies. In order to explain the activation observed with increases in surface temperature [18, 19, 36, 46], it is customary to invoke precursor mediated adsorption, where the adsorbate becomes trapped in a physisorption well long enough to become accommodated with the surface (section III). However, as pointed out by Luntz and

TABLE 1. Comparison of Tunneling Model for Ni, Ir, W, and Pt

Metal Surface	Barrier Height (kJ/mol)	Barrier thickness (A)
Ni(111)	99	0.47
Ir(110)	106	0.14
W(110)	115	0.25
Pt(111)	121	0.13

Bethune [36], precursor models are inconsistent with activation by incident kinetic and vibrational energy since excess incident energy tends to decrease the trapping probability (implicit in their argument, however, is the assumption that those molecules which do manage to trap become accommodated, although this needn't be the case for short lived precursors). Thus how does one explain the strong activation of methane by all three forms of energy, i.e. E_{trans}, E_{vib}, E_{surf}?

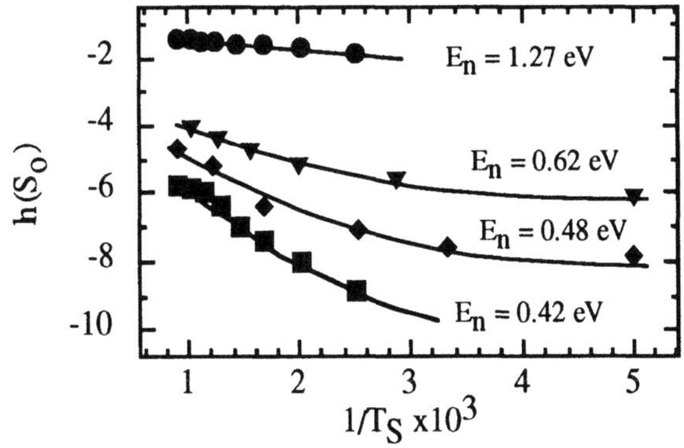

Figure 4a: Variation of the initial dissociative adsorption probability for methane on Pt(111). Data measured by Luntz and Bethune [36] and Harris et al. [22]. Reproduced with permission from ref [23].

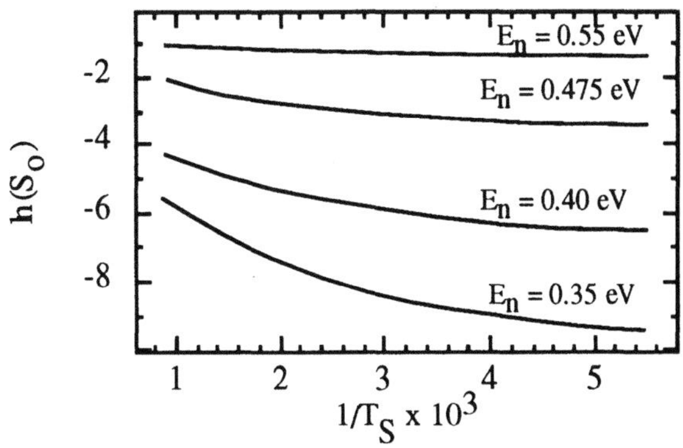

Figure 4b: Variation of the initial dissociative adsorption probability for methane on Pt(111). Predictions based on quantum dynamics model described in the text. Reproduced with permission from ref [23].

The answer may have come recently. Luntz et al. [22, 23] have proposed a model which can account *qualitatively* for all of the energy dependencies observed to date by invoking a single concerted dissociation process. The salient feature of the model is that phonons are allowed to modulate the tunneling barrier parameters, thereby, allowing these parameters to become sensitive to the surface temperature. This leads to the phenomenon of thermally assisted tunneling, which Luntz et al. claim is the dominant mechanism by which methane dissociates on metals. In simple terms, as the molecule impacts upon the surface, phonons of the proper phase serve to place the methane molecule closer to the top of the activation barrier, thereby, decreasing the distance through which the H atom must tunnel. It should be noted, however, that the model is only qualitative since, while it is able to reproduce the trends observed for changes in E_{trans}, E_{vib}, and E_{surf}, in all of the experiments discussed above, it frequently predicts dissociation probabilities in error by several orders of magnitude. This is illustrated above by comparison of Figures 4a and 4b.

To make the problem tractable, Luntz et al. assume that the orientation of CH_4 and its impact parameters with regard to the surface normal and surface unit cell are unimportant. Thus the surface was treated as flat, an assumption which is supported by the normal energy scaling of the dissociation probability. We anticipate that the authors can account for variations accross the surface unit cell by invoking a distribution of barriers, much like was done by Balooch et al. [47, 48] for the dissociative adsorption of H_2 on copper. Another important prediction of the model is the strong dependence on surface temperature of the kinetic isotope effect. This is shown in Figure (5) for several values of the normal kinetic energy.

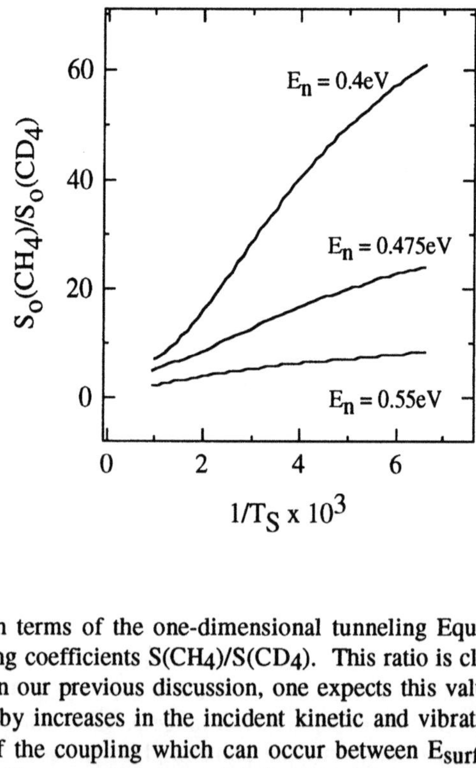

Figure 5: Variation of the kinetic isotope effect as a function of surface temperature based on predictions of dynamical calculations. Reproduced with permission from ref [23].

We can rationalize the predicted trends in terms of the one-dimensional tunneling Equation (3) simply by taking the ratio of the sticking coefficients $S(CH_4)/S(CD_4)$. This ratio is clearly sensitive to the quantity V-E and, based on our previous discussion, one expects this value to be reduced by increases in T_s as well as by increases in the incident kinetic and vibrational energies. Figure (5) is also suggestive of the coupling which can occur between E_{surf} and E_{trans}, and we expect also to lesser extent E_{vib}, and thus, as pointed out by Luntz and Harris [23], one should be careful when attempting to relate apparent activation barriers extracted from Arrhenius plots to barriers on the potential energy surface. This is another example of the difference between dynamics and kinetics.

The treatment of methane as a diatomic (R-H) in the model of Luntz et al. suggests that the same concept might be extended to the longer chain alkanes. It is the dissociative adsorption of these molecules we discuss next.

2.2. ETHANE, PROPANE AND BUTANE

The dynamics of the dissociative adsorption of ethane, propane and butane have been studied on Ir(110) [39, 49] and on Ni(100) [37]. On Ir(110), the dynamics are quite complex. (The actual surface is the reconstructed surface, Ir(100)—(1x2), which is unique in its low temperature dissociative adsorption of alkanes which occurs at 130K [50, 51].) Figures (6a)-(6c), which show the dissociative adsorption probabilities for ethane, propane and butane, respectively, exemplify this point. The reader will first note in Figure (6a) that for all but the lowest incident translational energies, the dissociative adsorption probability for ethane is temperature independent. Although Hamza et al.[39] suggested the slight temperature dependence for the 11 kJ/mol beam was within the error of their measurements, a subsequent study by Mullins and Weinberg [49] on the same surface convincingly shows the dependence is real. Furthermore, the temperature dependence becomes more pronounced as the chain length increases; e.g. see Figures 6b and 6c for adsorption of propane and butane, respectively. Whereas, ethane adsorption is temperature independent for incident energies above 11 kJ/mol (662K), the same becomes true for propane only for energies greater than 81 kJ/mol (4874K), and for butane only above 135 kJ/mol (8123K). The temperatures in parenthesis are those that would be required for a static gas to have the same mean incident energy on the surface, i.e. $\langle E \rangle = 2kT$. Above the respective critical energy of each alkane, the adsorption probability increases with increasing incident kinetic energy, just as it did for methane. Below this energy, however, the adsorption probabilities of propane and butane decrease with increasing incident energy. As mentioned in the previous section, this is consistent with precursor mediated adsorption being the dominant mechanism. The reason for the decrease will be addressed in Section III.

It is also interesting that, for incident energies below the critical value, the surface temperature at which the dissociative adsorption probability becomes temperature independent increases with increasing chain length. This trend might be expected because the residence time of an alkane in the precursor state should increase with the chain length due to both the increase in molecular mass and the increase in the quantity of interaction points between the molecule and the surface. (It is no coincidence that the activation energies for diffusion of normal alkanes also scale with increasing chain length [7].) However, as we point out in Section III, it is our belief that the trend is due to an increased stabilization of the transition state for precursor mediated adsorption compared to direct adsorption and that time in the precursor state plays a minor role. The dominance of the

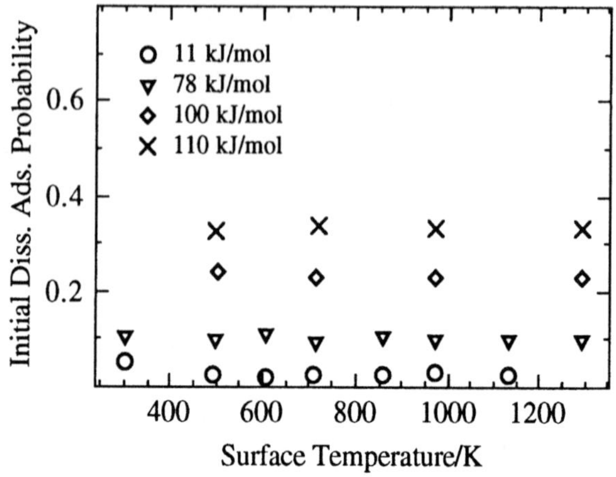

Figure 6a: Initial dissociative adsorption probability for ethane as a function of surface temperature measured at several incident kinetic energies. Reproduced with permission from ref [39].

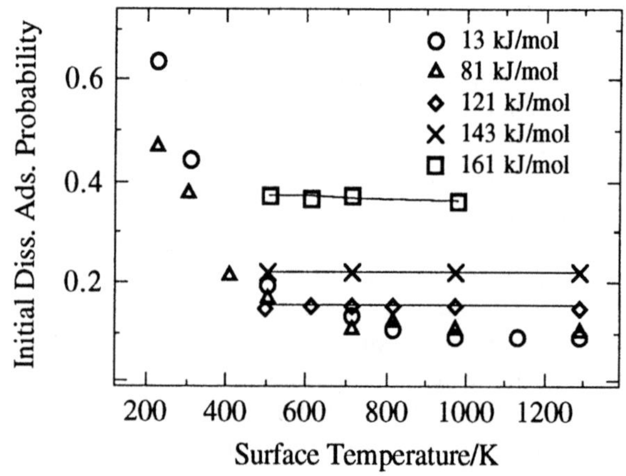

Figure 6b: Initial dissociative adsorption probability for propane as a function of surface temperature measured at several incident kinetic energies. Reproduced with permission from ref [39].

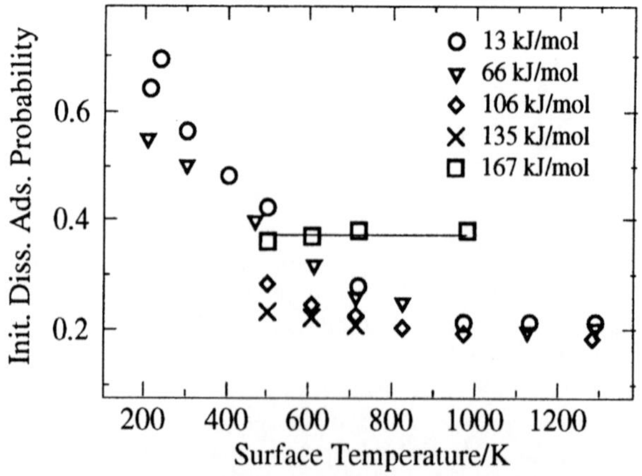

Figure 6c: Initial dissociative adsorption probability for butane as a function of surface temperature measured at several incident kinetic energies. Reproduced with permission from ref [39].

precursor mediated channel for propane and butane except for exceptionally high translational energies may be the reason for the apparent lack of a contribution from thermally assisted tunneling. In precursor mediated adsorption, the dominant effect of surface temperature is expected to be on the kinetic branching between dissociation and adsorption [52]. We note, however, that the data for ethane (and methane) on this surface also shows no evidence of this phenomenon.

The alkane chain length not only affects the precursor mediated channel, but also that for direct adsorption [39]. For surface temperatures above 970K, the dissociative adsorption probability for each alkane appears to have no contribution from a precursor mediated process. Furthermore, at high temperature, the dissociation probabilities converge to a limiting value as the kinetic energy is decreased, e.g. to 0.0 for methane, 0.03 for ethane, 0.09 for propane, and 0.21 for butane. Subtracting these constant values from the respective dissociation probabilities yields the results shown in Figure (7).

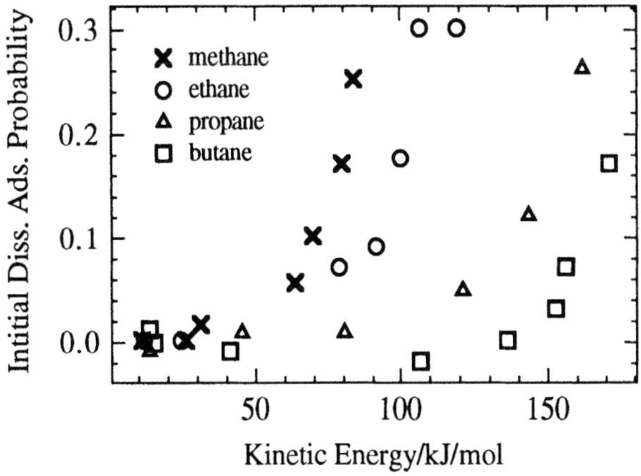

Figure 7: Initial dissociative adsorption probability as a function of incident kinetic energy for each alkane. The values have been adjusted as described in the text. Reproduced with permission from ref [39].

It seems that increasing chain length leads to increased translational activation. As we have already stated, increasing the mass of the incident molecule increases the coupling between the incident translational modes and the phonon modes of the solid and one thus expects the trapping probability to increase with increasing chain length. The curves in Figure (7) can be brought into near coincidence by accounting for this coupling within the framework of the hard cube model [39, 53] (see Section III). Since Ni has a much lower mass than Ir, one expects the differences to be even more pronounced in that system. Measurements by Hamza and Madix [37] confirm this.

The results for ethane, propane and butane adsorption on Ni (100) are qualitatively similar to those discussed for Ir (110) except, instead of having a minimum in the dissociation probabilities, there is a simple threshold before the onset of the direct translational activation [37]. That is, below a threshold of about 49.8 kJ/mol the dissociation probability is immeasurably small (less than 10^{-4}) and above the threshold the probabilty increases to one as a function of translational energy with a slope that depends on chain length. In order to estimate from the beam results the dissociation probability expected for a gas in equilibrium with the metal, it is necessary to extrapolate outside the range of translational energies that can be achieved in the molecular beam and to make some assumptions about what the effect of rotational and vibrational energies may be. Ignoring the effects of the internal energy of the molecules and simply convoluting the dissociative adsorption probability, as measured from the beam experiments, with an appropriate Maxwell-Boltzmann flux distribution results in the comparison shown in Figure (8). The dissociation probabilities from the beam experiments and those measured from an equilibrium Maxwell-Boltzmann distribution have little in common. The beam results predict a constant activation energy of 53 ± 4 kJ/mol, decreasing somewhat with chain length, and the activation energy measured in the equilibrium experiments is lower and decreases with increasing chain length while the absolute dissociation probability increases with chain length. Sault and Goodman [54] discuss the possible effects of rotational and vibrational energies (which were neglected in extrapolating the beam results), but suggest that the main effect must be due to dissociation of trapped, low energy molecules whose dissociation cannot be detected in the beam experiments but, because of their huge flux in the Maxwell-Boltzmann distribution, dominate in the equilibrium experiment. It is to this adsorption channel we turn next.

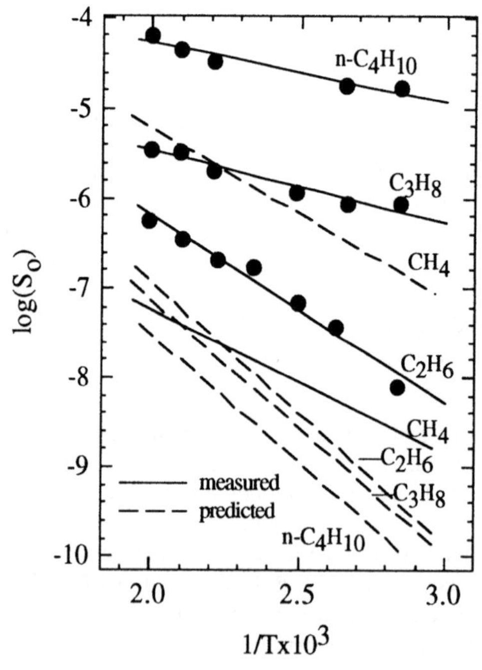

Figure 8: Comparison of dissociative adsorption probabilities measured in bulb experiments and predicted by beam experiments. Reproduced with permission from ref [54].

3. Dynamics of Trapping

Precursor states are characterized by weak interactions (typically dispersion forces) between the surface and the adsorbate and they have long been viewed as assisting in the adsorption process by confining the incident molecule to two dimensions. However, they can also gain influence by coupling energy from the surface into various internal modes of the adsorbate and can thus play a dominant role even when surface diffusion is not important. One speaks of intrinsic precursor states when the trapping occurs directly onto the bare surface and extrinsic states when the trapping is into a second layer state. However, at least for alkanes on metals and at the incident energies involved in typical catalytic reactions, the trapping probability is of order unity for both clean and saturated surfaces. In standard treatments the population of the precursor state is determined by balancing the trapping probability α with the rates of first order desorption from the trapped state and dissociative adsorption. In this way, the reaction step expressed by Equation (2) can be expanded into the mechanism shown below.

$$C_nH_{2n+2}\,(g) \rightleftharpoons C_nH_{2n+2}\,(t) \rightarrow H(a) + C_nH_{2n+1}\,(a)$$

Desorption from the chemisorbed state is easily incorporated into the above, providing one recognizes that desorption through both the precursor and direct channels should be considered in those cases where direct adsorption is also allowed.

3.1 INTRINSIC PRECURSOR STATES

Our primary focus here is on the dynamics of the trapping process and such data exists for ethane on Ir(110)-(1x2) [55, 56] and for methane through propane on Pt(111) [57, 58]. By far, the largest number of experiments have been performed for ethane, the smallest hydrocarbon for which trapping is expected to be significant. The experiments performed have been of the molecular beam type and have probed the effect of incident translational energy and angle of incidence on the trapping probability α. The above studies rely primarily on a technique first introduced by King and Wells [59] which allows α to be determined directly as that fraction of the molecular beam which is not reflected from the surface. The fraction reflected is proportional to the background pressure in the chamber and the total beam flux is determined by placing a flag in front of the surface which reflects all molecules. The measurements are performed under conditions where the probabilities for desorption and direct adsorption are low by working at low surface temperature (T_S) and low incident beam energies.

The trapping probabilites of methane-propane on Pt(111) are shown in Figure (9). The trapping probability decreases with increasing incident energy for each hydrocarbon and increases as a function of chain length at a given incident energy. This finding parallels that mentioned previously for the dissociative adsorption probability of alkanes on Ir(110) [39]. That discussion leads to the conclusion that trapping efficiency of an alkane should increase with increasing chain length and Figure (9) would appear to be a confirmation of that.

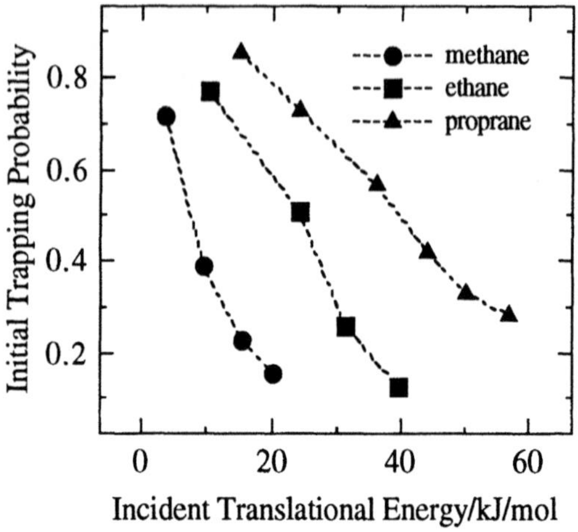

Figure 9: Initial trapping probability of methane, ethane, and propane on Pt(111) at 95K and normal incidence. Reproduced with permission from ref [57].

Note also that the trapping probabilities tend toward unity as the incidence energies approach zero. Rettner et al. [56] found similar trends for the trapping of ethane on Ir(110), as well as for the trapping of Ar and Xe on Pt(111), for Ar on 2H-W(100), and for N_2 on W(100). They were able, in all cases, to describe the trapping probability α as a function of energy and incidence angle with the relation given by Equation (5).

$$\alpha = \exp\left(\frac{-E\cos^n\theta}{E_e}\right) \tag{5}$$

where E_e is the energy at which the trapping probability has decreased to 1/e and n is an adjustable parameter. For ethane on Ir(110), n=0.5 and E_e=43kJ/mol. The data collected by Arumainayagam et al. for ethane on Pt(111) as a function incidence energy and angle can be correlated with $E\cos^{0.6}\theta$, although the data cannot be represented accurately by the exponential formula given by Equation (5). The values of E_e for methane, ethane and propane determined from Figure (9) are approximately 10, 30, and 50 kJ/mol, respectively, yielding a linear dependence on alkane chain length reminiscent of the activation energy for diffusion [7]. These trends are usually explained within the context of various kinematic models, the most popular of which is the simple hard cube model [53]. The model is based on conservation of energy and momentum in a purely impulsive collision between the incoming molecule of mass m and a surface atom of mass M. The model accounts for the attractive part of the interaction potential by increasing the collision energy by an amount equal to the well depth D. The conservation laws allow for the determination of the energy partitioning after the collision and, if the incident energy E (measured at infinite displacement) is less than

$$E_T = \frac{4D(m/M)}{[1-(m/M)]^2}, \tag{6}$$

then trapping is expected with unit probability. When E is greater than E_T, trapping cannot occur. The hard cube model thus predicts a trapping probability which resembles a step function and suggests why the precursor mediated adsorption probability decreases with increasing incident energy. The hard cube model can be modified to include the effect surface atom motion at elevated temperatures [60], with surface temperature serving to decrease the trapping probabilities of the molecules with energies just below E_T and to increase the trapping probabilities of those with energy slightly above E_T, i.e. the discontinuities of the step function are smoothed. Consistent with this, in cases where the effect of surface temperature on the trapping probability has been studied, only a very slight dependence has been found. While the hard cube model is useful for rationalizing data, it should be used with care and its predictive power should not be overestimated. For example, the model implicitly assumes normal energy scaling, i.e. n=2 in Equation (5), however, experiments show this is not the case. In addition, for alkane trapping on Pt(111), the model only works if M is allowed to be a decreasing linear function of energy [57], possibly accounting for the correlated motion of surface atoms during long lasting collisions [60]. Nevertheless, the model does provide a framework within which to correlate trapping data, even though it is essentially a kinematic model.

Before going on to consider the effect of translational energy on the trapping probability into an extrinsic precursor state, we should like to consider the conditions necessary for trapping mediated adsorption to dominate the direct adsorption. Weinberg has argued, based on transition state theory, that trapping mediated processes should nearly always dominate the direct adsorption process [61]. However, the conclusion was reached by analogy to the relative rates of a Langmuir-Hinshelwood reaction (trapping mediated adsorption) and an Rideal-Eley reaction (direct adsorption), which is what he actually estimated. Had he actually used a Langmuir-Hinshelwood mechanism in his calculation the analogy might have applied. However, what he considered to be a L-H mechanism was actually only a surface reaction since desorption of the reactant prior to reaction was not taken into account. That this consideration is important can be seen by the following simple (zero-order) argument: Consider that both direct adsorption and trapping mediated adsorption follow normal energy scaling and that the direct dissociative adsorption probability is unity for normal energies $E > E_D$ and zero otherwise and that the trapping probability is unity for normal energies $E < E_T$ and zero otherwise. Gas-kinetic theory then gives

$$\frac{F_T}{F_D} = \exp\left(\frac{E_D}{kT}\right)\left[1 - \exp\left(\frac{-E_T}{kT}\right)\right] \tag{7}$$

as the ratio of the flux of molecules that can trap to those that can dissociatively adsorb directly. This quantity is nearly always greater than unity indicating that more molecules can become trapped than can undergo direct dissociative adsorption. However, if one considers that trapped molecules only go on to dissociate with probability P given by

$$P = \left[1 + \frac{\nu_d}{\nu_r}\exp\left(\frac{E_r - E_d}{kT}\right)\right]^{-1} \tag{8}$$

then the ratio of trapping mediated dissociative adsorption to direct adsorption becomes

$$\frac{S_T}{S_D} = \exp\left(\frac{E_D}{kT}\right)\left[1 - \exp\left(\frac{-E_T}{kT}\right)\right]\left[1+\frac{\nu_d}{\nu_r}\exp\left(\frac{E_r - E_d}{kT}\right)\right]^{-1} \tag{9}$$

where the subscript d applies to desorption and r to dissociative adsorption from the precursor state. For cases where the reaction out of the precursor state is favored over desorption from it, the result in Equation (9) reduces back to that of Equation (7). An example of this is the dissociative adsorption of ethane on Ir(110)-(1x2) [49]. However, for most systems, one expects desorption to be favored from the precursor, both because $E_r > E_d$ and $\nu_d > \nu_r$. If one assumes that the transition state for the direct process lies Δ higher in energy than that for the precursor mediated process, then Equation (9) can be reduced to the following.

$$\frac{S_T}{S_D} = \frac{\nu_r}{\nu_d}\exp\left(\frac{\Delta}{kT}\right)\left[1 - \exp\left(\frac{-E_T}{kT}\right)\right] \tag{10}$$

Note that for $\Delta=0$ and $\nu_d > \nu_r$ this expression is less than unity indicating that the direct channel should dominate. Additionally, by considering E_T to be given by Equation (6), the quantity in square brackets will nearly always be approximately unity and equal contributions of both pathways to dissociative adsorption should only be expected for values of Δ given by the equality in Equation (11). Trapping becomes favored as Δ increases. In addition, this argument suggests that it is the trapped but unaccommodated molecules which will contribute more to the precursor mediated path than those molecules which have spent a long enough time in the well to become accommodated, i.e. the latter have a larger barrier to climb.

$$\Delta \geq kT \ln \frac{\nu_d}{\nu_r} \tag{11}$$

We can arrive at an order of magnitude estimate for the value of Δ required to give equal contributions for both trapping and direct dissociation if we take $\nu_d/\nu_r=100$ (this is a reasonable choice based on the data in ref [62]). At 500K, precursor mediated adsorption would be the favored pathway if the transition state for dissociation in this case is at least 4.5kcal/mol lower than that for the direct adsorption process. Since the amount of stabilization would not be expected to exceed some fraction of the trapping well depth (and we expect this fraction to be less than 0.5 for C-H bond breaking, i.e. a late barrier in the dissociation channel), one can rationalize why ethane adsorbs through a precursor on Pt(110)-(1x2) and methane does not, i.e. the physisorption well depth for methane is only 4.5kcal/mol while that for ethane is 7.7kcal/mol [62]. That ethane adsorption on Pt(111) does not seem to involve trapping [58] even though the trapping well depth (7.6kcal/mol) is similar to that on Pt(110) suggests that the dissociating C-H bond may be longer in the transition state on Pt(111) than on Pt(110) such that Δ is a smaller fraction of the trapping well depth on the (111) surface compared to the (110) surface. We realize that by making the assumption of normal energy scaling for both direct adsorption and precursor mediated adsorption, by taking the adsorption and trapping probabilities to be step functions of energy, and by neglecting the effect of internal energy on the direct adsorption and trapping processes we are ignoring many of the more interesting aspects of the dynamics of the chemisorption process. However, we expect our conclusions that (1) the transition states for the two pathways must be different for the precursor mediated process to dominate and (2) that the dominance of the precursor pathway may have little to do with the fact that many more molecules can trap than adsorb directly, may still be valid.

3.2 EXTRINSIC PRECURSOR STATES

We finish this review with a brief discussion of the results obtained for trapping into an adsorbed layer. Dynamics data currently exists only for ethane trapping on Pt(111) [58] and Ir(110)-(1x2) [55]. On both surfaces, the trapping probability increases with increasing surface coverage and a model originally due to Kisliuk [63] but modified to account for enhanced trapping into the second layer [55, 64] can be used to describe the kinetics. The salient features of the model are that it assumes that adsorption can occur either through Langmuirian adsorption (i.e. the molecules cannot diffuse on the surface prior to adsorption) or through an extrinsic precursor state, where adsorption occurs via trapping into a thermalized, weakly bound, second layer state. The sticking probability (ultimately into the first layer) is then given by

$$S(\theta) = S_0(1 - \theta) + \frac{\alpha\,(1 - \theta)\,q_m\theta}{1 - q_m\theta} \tag{12}$$

where S_0 is the initial direct sticking coefficient and α is that for trapping into second layer and q_m is given by

$$q_m = \frac{k_m'}{k_d' + k_m'} \tag{13}$$

where k_m' is the rate constant for adsorption into the first layer from the extrinsic precursor state and k_d' is the rate constant for desorption from the extrinsic precursor state.

Arumainayagam et al. used Equations (12) and (13) to fit data collected for trapping of ethane on Pt(111) at 95K at incidence angles ranging from 0 to 45° and at incident energies between 10 and 40 kJ/mol. In all cases, the best fits were achieved using q_m=0.999, suggesting that the time constant for migration of molecules in the second layer is much less than that for desorption. At a temperature of 95K, this implies that the energy for diffusion should be less than that for desorption by nearly 1 kcal/mol. Kang et al. [55] have used Monte Carlo simulations to estimate the barrier to diffusion and desorption of ethane from the extrinsic precursor on Ir(110)-(1x2). The values of 4.5kcal/mol and 3.7kcal/mol for desorpiton and diffusion, respectively, are in line with the 1kcal/mol estimate of the difference. We expect the comparison to be valid for ethane on different metals since the extrinsic precursor should be only weakly affected by the underlying surface. However, temperature programmed desorption studies of ethane from Pt(111) reveal that desorption from the second layer is approximately first order [58] suggesting that the extrinsic precursor state is somewhat influenced by the underlying surface (one expects zero order desorption from simple condensed phases [65]).

Perhaps most interesting from the point of view of dynamics are the results shown in Figures (10) and (11). The trapping probability of ethane on both clean and ethane saturated Pt(111) surfaces are shown in Figure (10) as a function of incident translational energy. Note that the trapping probability into the extrinsic state is larger than that into the intrinsic state for all incident energies.

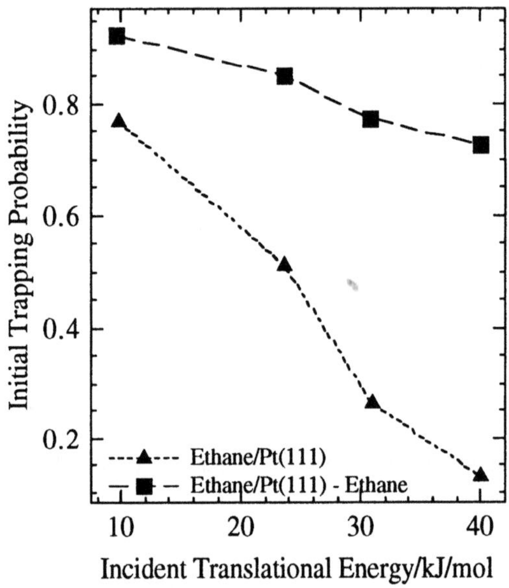

Figure 10: Initial trapping probability of ethane on a clean Pt(111) surface measured as a function of incident energy at 95K. Reproduced with permission from ref [66].

This behavior is easily rationalized within the context of the hard cube model, i.e. since the collision partner of an incident ethane is another ethane during trapping into the extrinsic state, the incident ethane dissipates a larger fraction of its energy into the adsorbed layer upon impact thus increasing the trapping probability. The data in Figure (11) demonstrate that the trapping probability is independent of the incidence angle, implying total energy scaling. Such scaling is typically ascribed to surface corrugation which effectively couples momentum components directed parrallel and perpendicular to the surface. Perhaps the best way to visualize this is as a collision between two spheres. If one sphere is embedded in a surface, the probability of a glancing collision of a second sphere with it increases with increasing angle from the surface normal. However, surface corrugation (both static and dynamic) serve to increase the probability that the embedded sphere suffers a more direct impact. Arumainayagam et al. [66] disuss the corrugation which produces Figure (11) in terms of the thermal motion normal to the surface of the first layer ethane molecules. These studies are important because they assess the dynamics of the adsorption process at high local coverage, i.e. under conditions likely to prevail during most commercial processes, particularly with respect to chemisorbed species. Although the results here are for like collision partners, the

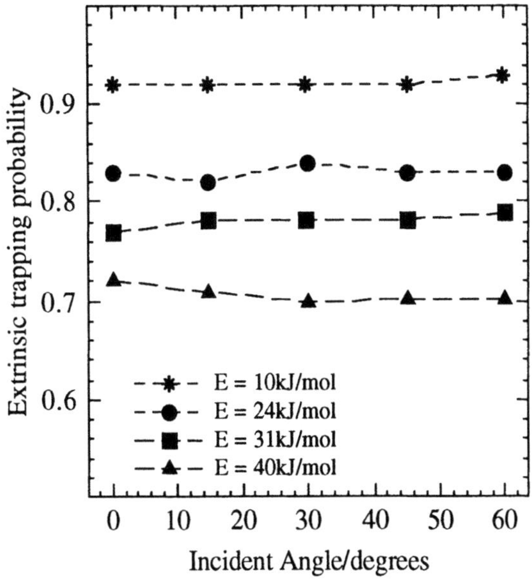

Figure 11: Initial trapping probability of ethane into the extrinsic precursor state measured at 95K on Pt(111). Reproduced with permission from ref [66].

kinematics of the trapping process, as embodied in the hard cube-model, will also apply for collision between unlike partners. Thus, these results can be considered more general.

4. References

[1]R. D. Levine and R. B. Bernstein (1987) Molecular Reaction Dynamics and Chemical Reactivity, Oxford University Press, New York.

[2]B. N. J. Persson and M. Persson (1980), Surf. Sci., **97**, 609.

[3]M. P. Casassa, E. J. Heilweil, J. C. Stephenson and R. R. Cavanagh (1986), J. Chem. Phys., **84**(4), 2361.

[4]J. D. Berkerle, M. P. Casassa, R. R. Cavanagh, E. J. Heilweil and J. C. Stephenson (1989), J. Chem. Phys., **90**(8), 4619.

[5]S. L. Tang, J. D. Beckerle, M. B. Lee and S. T. Ceyer (1986), J. Chem. Phys., **84**(11), 6488.

[6]M. V. Arena, E. D. Westre and S. M. George (1992), J. Chem. Phys., **96**(1), 808.

[7]J. L. Brand, M. V. Arena, A. A. Deckert and S. M. George (1990), J. Chem. Phys, **92**(8), 5136.

[8]G. W. Coulston (1990), Doctoral Dissertation, Yale University.

[9]G. W. Coulston and G. L. Haller (1990), J. Chem. Phys., **92**, 5752.

[10]G. W. Coulston and G. L. Haller (1991), J. Chem. Phys., **95**(9), 6932.

[11]D. A. Mantell, S. B. Ryali and G. L. Haller (1983), Chem. Phys. Lett., **102**(1), 37.

[12]D. A. Mantell, K. Kunimori, S. B. Ryali, G. L. Haller and J. B. Fenn (1986), Surf. Sci., **172**, 281.

[13]L. S. Brown and S. L. Bernasek (1985), J. Chem. Phys., **82**, 2110.

[14]L. S. Brown and S. J. Sibener (1989), J. Chem. Phys., **90**(5), 2807.

[15]S. L. Bernasek and S. Leone (1981), Chem. Phys. Lett., **84**, 401.

[16]G. L. Haller and G. W. Coulston (1991) 'Dynamics of Heterogeneously Catalyzed Reactions', in J. R. Anderson and M. Boudart (ed.), Catalysis: Science and Technology, Springer-Verlag, Heidelberg, vol. 9, pp. 131.

[17]R. H. Crabtree (1985), Chem. Rev., **85**, 245.

[18]H. Winters (1975), J. Chem. Phys., **62**, 2454.

[19]H. Winters (1976), J. Chem. Phys, **64**, 3495.

[20]J. B. Anderson and J. B. Fenn (1965), Phys. Fluids, **8**, 780.

[21]N. Abuaf, J. B. Anderson, R. P. Andres, J. B. Fenn and D. G. H. Marsden (1967), Science, **155**, 997.

[22]J. Harris, J. Simon, A. C. Luntz, C. B. Mullins and C. T. Rettner (1991), Phys. Rev. Lett., **67**(5), 652.

[23]A. C. Luntz and J. Harris (1991), Surf. Sci., **258**, 397.

[24]C. N. Stewart and G. Ehrlich (1975), J. Chem. Phys., **62**, 4672.

[25]N. B. Slater (1959) Theory of Unimolecular Reactions, Cornell University, Ithaca.

[26]S. G. Brass and G. Ehrlich (1987), J. Chem. Phys., **87**, 4285.

[27]T. Lo and G. Ehrlich (1987), Surf. Sci., **179**, L19.

[28]F. C. Schouten, E. W. Kaleveld and G. A. Bootsma (1977), Surface Sci., **63**, 460.

[29]F. C. Schouten, O. L. J. Gijzeman and G. A. Bootsma (1979), Bull. Soc. Chim. Belg., **88**, 541.

[30]F. C. Schouten, O. L. J. Gijzeman and G. A. Bootsma (1979), Surface Sci., **87**, 1.

[31]J. T. P. Beebe, D. W. Goodman, B. D. Kay and J. J. T. Yates (1987), J. Phys. Chem., **87**, 2305.

[32]J. R. Rostrup-Nielsen (1984) 'Catalytic Steam Reforming', in J. R. Anderson and M. Boudart (ed.), Catalysis: Science and Technology, Springer, Berlin, vol. 5, pp. 1.

[33]C. T. Rettner, H. E. Pfnur and D. J. Auerbach (1986), J. Chem. Phys, **84**, 4163.

[34]M. B. Lee, Q. Y. Yang and S. T. Ceyer (1987), J. Chem. Phys., **87**, 2724.

[35]G. R. Schoofs, C. R. Arumaninayagam, M. C. McMaster and R. J. Madix (1989), Surf. Sci., **215**, 1.

[36]A. C. Luntz and D. S. Bethune (1989), J. Phys. Chem, **90**, 1274.

[37]A. V. Hamza and R. J. Madix (1987), Surf. Sci., **179**, 25.

[38]C. T. Rettner, H. E. Pfnur and D. J. Auerbach (1985), Phys. Rev. Letts., **54**, 2716.

[39]A. V. Hamza, H.-P. Steinruck and R. J. Madix (1987), J. Chem. Phys., **86**, 6506.

[40]J. D. Beckerle, Q. Y. Yang, A. D. Johnson and S. T. Ceyer (1987), J. Chem. Phys., **86**, 7236.

[41]B. D. Kay and M. E. Coltrin (1988), Surf. Sci., **198**, L375.

[42]B. D. Kay and M. E. Coltrin (1988), Surf. Sci., **205**, L805.

[43]T. Lo and G. Ehrlich (1988), Surf. Sci., **198**, L380.

[44]T. Lo and G. Ehrlich (1988), Surf. Sci., **205**, L813.

[45]A. E. DePristo and A. Kara. (1989), Adv. Chem. Phys, , in press.

[46]S. G. Brass, D. A. Reed and G. Ehrlich (1979), J. Chem. Phys., **70**, 5244.

[47]M. Balooch, M. J. Cardillo, D. R. Miller and R. E. Stickney (1974), Surf. Sci., **46**, 358.

[48]M. Balooch and R. E. Stickney (1974), Surf. Sci., **44**, 310.

[49]C. B. Mullins and W. H. Weinberg (1990), J. Vac. Sci. Technol A, **8**(3), 2458.

[50]T. S. Wittrig, P. D. Szuromi and W. H. Weinberg (1982), J. Chem. Phys., **76**, 3305.

[51]P. D. Szuromi, J. R. Engstom and W. H. Weinberg (1984), J. Chem. Phys., **80**, 508.

[52]C. T. Rettner (1991) 'Molecular-Beam Studies of the Effect of Surface Temperature on the Probability of Dissociative Chemisorption', in D. J. Dwyer and F. M. Hoffmann (ed.), Surface Science of Catalysis: Insitu Probes and Reaction Kinetics, American Chemical Society, Washington, DC, vol. 482, pp.

[53]R. M. Logan and R. E. Stickney (1966), J. Chem. Phys., **44**, 195.

[54]A. G. Sault and D. W. Goodman (1988), J. Chem. Phys., **88**, 7232.

[55]H. C. Kang, C. B. Mullins and W. H. Weinberg (1990), J. Chem. Phys., **92**(2), 1399.

[56]C. T. Rettner, C. B. Mullins, D. S. Bethune, D. J. Auerbach, E. K. Schweizer and W. H. Weinberg (1990), J. Vac. Sci. Technol. A, **8**(3), 2699.

[57]C. R. Arumainayagam, M. C. McMaster and R. J. Madix (1991), J. Vac. Sci. Technol. A, **9**(3), 1581.

[58]C. R. Arumainayagam, G. R. Schoofs, M. C. McMaster and R. J. Madix (1991), J. Phys. Chem., **95**, 1041.

[59]D. A. King and M. G. Wells (1972), Surf. Sci., **29**, 454.

[60]E. K. Grimmelmann, J. C. Tully and M. J. Cardillo (1980), J. Chem. Phys., **72**(2), 1039.

[61]W. H. Weinberg (1991) 'Kinetics of Surface Reactions', in C. T. Rettner and M. N. R. Ashfold (ed.), Dynamics of Gas-Surface Interactions, The Royal Society of Chemistry, Cambridge, vol. pp. 171.

[62]W. H. Weinberg (1992), J. Vac. Sci. Technol A., **10**(4), 2271.

[63]P. Kisliuk (1957), J. Phys. Chem. Solids, **3**, 95.

[64]C. R. Arumainayagam, M. C. McMaster and R. J. Madix (1991), J. Phys. Chem., **95**, 2461.

[65]D. A. King (1979) in R. Vanselow (ed.), Chemistry and Physics of Solid Surfaces, CRC Press, Boca Raton, vol. 2, pp. 118.

[66]C. R. Arumainayagam, M. C. McMaster and R. J. Madix (1990), Surf. Sci. Lett., **237**, L424.

THE STUDY OF DYNAMICS AT CATALYTIC SURFACES WITH NUCLEAR MAGNETIC RESONANCE SPECTROSCOPY[1]

T. M. DUNCAN
School of Chemical Engineering
Cornell University
Ithaca, New York 14853
USA

ABSTRACT. The NMR methods to characterize molecular motions and chemical reaction are divided into three categories: relaxation methods, lineshape analysis, and spin-population labeling. The physical basis for each method is presented and a mathematical model of the NMR-observable parameters is derived. Each section concludes with examples from studies in heterogeneous catalysis.

1. Introduction

The capabilities of NMR for structural analysis are well-known; its utility for characterizing organic liquids is unrivaled and its applicability to solids is increasing. Although not as routine as structural analysis, NMR spectroscopy also has potential for characterizing molecular dynamics. This chapter describes three NMR methods for studies of molecular dynamics, with examples of applications to heterogeneous catalysis.

The three methods of NMR spectroscopy described in the following sections have unique capabilities to characterize molecular dynamics. These capabilities derive from three properties of NMR spectroscopy. (1) The coupling between the nuclear magnetic spins and the applied magnetic fields is stronger than the internal magnetic couplings of the material. One may excite specific chemical types and suppress specific internal couplings. (2) The internal magnetic couplings of the material (magnetic dipole-dipole, chemical shift) are known functions of the molecular structure and the spin operators, which allows one to translate a physical model into the language of magnetic resonance. Finally, (3) the non-equilibrium state of the nuclear magnetic spins excited by applied forces persists for long times relative to the excitation period and the internal couplings. Because lifetime broadening is small, there is potential for good resolution. These unique features of NMR spectroscopy, combined with the coincidence between NMR time scales and the times scales of key molecular phenomena, make NMR spectroscopy an ideal probe of dynamics at catalytic surfaces.

2. Relaxation Measurements

Molecular dynamics at catalytic surfaces may be characterized by measuring the rate of exchange of energy between the system of nuclear magnetic spins and the surrounding energy reservoir. The following discussion of this method is at an introductory level.

R. W. Joyner and R. A. van Santen (eds.), Elementary Reaction Steps in Heterogeneous Catalysis, 221–241.

Detailed discussions can be found in more advanced treatises on relaxation in nuclear magnetism [2-5] and on relaxation studies of interfacial systems [6-8].

2.1. BASIC THEORY

Relaxation phenomena are perhaps best couched in the spectroscopic description of nuclear magnetism as first put forward by Purcell, Torrey and Pound in 1945 [9,10]. A nucleus with spin quantum number I has a nuclear magnetic moment $m = \gamma \hbar I$ and has $2I + 1$ energy levels in a magnetic field. The energy splitting between levels is most conveniently expressed in units of frequency. The Larmor frequency in a magnetic field H_0 is defined as

$$\omega_0 = 2\pi\nu_0 = \gamma H_0 \, . \tag{1}$$

Typical gyromagnetic ratios (γ) and Larmor frequencies are given in Table 1.

TABLE 1. Nuclear Magnetic Parameters of Selected Isotopes

Nucleus	Gyromagnetic Ratio, γ (10^7 rad Tesla^{-1} s^{-1})	Larmor Frequency, ν_0 ($\gamma H_0/2\pi$ in MHz, at 2.35 Tesla)	Isotopic Abundance (%)	Spin, I
^1H	26.751	100.	99.99	1/2
^2H	4.106	15.4	0.015	1
^{13}C	6.726	25.2	1.11	1/2
^{27}Al	6.971	26.1	100.	5/2
^{29}Si	-5.314	19.9	4.7	1/2
^{103}Rh	-0.842	3.1	100.	1/2
^{195}Pt	5.751	21.5	33.8	1/2

An ensemble of spins at equilibrium with the energy reservoir of the surroundings (the lattice) has a Boltzmann distribution among the energy levels with a characteristic temperature equal to the macroscopic temperature of the sample. The population distribution is reflected in the net nuclear magnetism of the sample. To measure the rate of energy exchange, one may apply a short, intense rf pulse to perturb the nuclear spin system from equilibrium and then monitor the nuclear magnetism as a function of time. In a typical experiment, the rf irradiation is applied for a few μs and equilibration has a time constant of 10^{-2} to 1 s. This rate of exchange is characterized by a spin-lattice (or longitudinal) relaxation time, T_1. For homogeneous systems, the rate of equilibration is often exponential. Thus for a system irradiated with a $\pi/2$ which saturates the net nuclear magnetism, the nuclear magnetism at time t is

$$M(t) = M(\infty)\left\{1 - \exp\left[-t/T_1\right]\right\} , \tag{2}$$

such that $M(\infty)$ is the equilibrium magnetization. I shall now describe how motions are manifested in T_1.

Phenomenologically, energy exchange between a nuclear spin and the lattice requires the event to be energy conserving and of proper coupling. Proper coupling requires a fluctuating magnetic field to relax nuclei with magnetic dipoles ($I \geq \frac{1}{2}$), or an electric field gradient to relax nuclei with an electric quadrupole moment ($I > 1/2$). To conserve energy, the fluctuation most efficient for relaxation is at a rate equal to the frequency splitting between the nuclear magnetic energy levels. For example, an efficient, but usually uninteresting source of relaxation is a paramagnetic site; the electron spin (and its attendant magnetic field) fluctuates at $\sim 10^9$ sec^{-1}, which is comparable to typical Larmor frequencies, ω_0, as shown in Table 1. Note that molecular vibrations, which have frequencies of $\sim 10^{14}$ sec^{-1}, do not contribute to spin-lattice relaxation.

Two common sources of fluctuating magnetic fields caused by molecular motion are nuclear magnetic dipole-dipole coupling and chemical shift. The energy of dipole-dipole coupling between two magnetic moments μ_i and μ_j is classically

$$E_{dd_{i,j}} = \frac{\mu_i \cdot \mu_j}{r_{i,j}^3} - \frac{3(\mu_i \cdot \mathbf{r}_{i,j})(\mu_j \cdot \mathbf{r}_{i,j})}{r_{i,j}^5}, \tag{3}$$

such that $\mathbf{r}_{i,j}$ is the internuclear vector as shown in Figure 1. The dipolar coupling drops off quickly with internuclear distance; typically only nuclei within ~5 Å are significant. Note that the magnetic field corresponding to this coupling energy is $H_{dd_{i,j}} = E_{dd_{i,j}}/\hbar\gamma_i$. From equation (3), $\mu = \gamma\hbar I$, and the parameters in Table 1 (and $\hbar = 1.0544 \times 10^{-35}$ Tesla^2 cm^3 sec rad^{-1}) one calculates the magnetic field from a ^{1}H nucleus at 1 Å as ~0.002 Tesla.

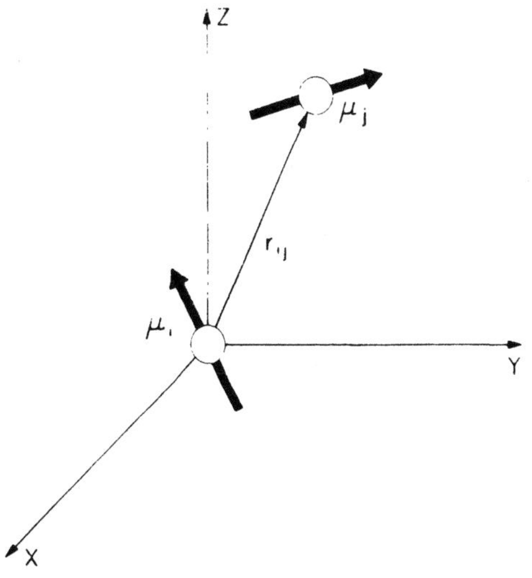

Figure 1. Coordinate system for the coupling between two nuclear magnetic moments μ_i and μ_j. The applied magnetic field, H_0, is oriented in the z direction.

The coupling between the nuclear magnetic moment and the magnetic dipolar field caused by nearby electrons is the chemical shift interaction. This coupling can be expressed similarly to equation (3). However, the parameter $r_{i,j}$ is ill-defined and does not appear explicitly. Retaining only the secular terms, the energy of the chemical shift is commonly simplified to

$$E_{cs} = \mu H_0 \delta(\theta_{cs}, \phi_{cs}), \tag{4}$$

$$= \mu H_0 \left[\delta_{iso} + \frac{\delta}{2}\left((3\cos^2\theta_{cs} - 1) + \eta \sin^2\theta_{cs} \cos 2\phi_{cs}\right)\right], \tag{5}$$

such that θ_{cs} and ϕ_{cs} are the orientation of the chemical shift principal axis system relative to H_0. For example, the z axis of the chemical shielding for molecular CO is along the C≡O bond. The parameters δ_{iso}, δ, and η are the isotropic shift, the chemical shift anisotropy and the chemical shift asymmetry which are characteristic of the chemical bonding. Some typical values are given in Table 2. Typically, for systems of ^1H the dipole-dipole coupling dominates. For ^{13}C, the dominant interaction depends on the chemical type; the dipole-dipole dominates for $-CH_3$ and the chemical shift dominates for CO.

TABLE 2. Typical Chemical Shift Parameters

Nucleus	Chemical Type	Chemical Shift Parameters δ_{iso} (ppm)	δ (ppm)	η
^1H	$-CH_3$	0	3	0.04
	$-COOH$	13	12	0.2
^{13}C	$-CH_3$	20	16	0.6
	$-C(O)-$	200	120	0.3
	aromatic	125	110	0.5
	CO	180	270	0.0

From "A Compilation of Chemical Shift Anisotropies," T. M. Duncan, (Farragut Press, Chicago, 1990).

With equations (3) and (5), and expressions for the Knight Shift and quadrupolar interaction, one can calculate the magnetic field from the spatial coordinates, *e.g.*, $\mathbf{r}_{i,j}$ or θ_{cs} and ϕ_{cs}. The expressions are separable into products of spin and spatial terms; the spatial terms are commonly represented as F(t). Calculating the fluctuation of magnetic fields amounts to calculating the correlation function

$$G(t) = \gamma^2 \hbar^2 \langle F(t') F(t'+t) \rangle_{t'} . \tag{6}$$

To determine the frequency distribution of the fluctuations a spectral density function may be calculated as

$$J(\omega) = \frac{1}{2\pi} \int_{-\infty}^{\infty} G(t) e^{i\omega t} dt . \tag{7}$$

Qualitatively, T_1 is proportional to the intensity of the spectral density function at the Larmor frequency, ω_0. For nuclear dipole-dipole coupling, the intensity at $2\omega_0$ also contributes. As first derived by Bloembergen, Purcell, and Pound (BPP) in 1948 [2,11], homonuclear dipole-dipole couplings cause a spin system to relax with a rate

$$\frac{1}{T_1} = \frac{3}{4} \gamma^4 \hbar^2 I(I+1)[J_1(\omega) + J_2(\omega)] , \tag{8}$$

where J_1 and J_2 are the spectral density functions of the couplings involving 1 and 2 quanta, respectively. Alternate formalisms were later offered by Kubo and Tomita (1954) [12] Hebel and Slichter (1959) [13] and Wangsness and Bloch (1953) [14], as extended by Redfield (1956), [15] which are described and contrasted in the monograph by Wolf [2].

To apply the theory, one devises a model for the molecular motion from which F(t) is obtained. One then calculates G(t), J(ω) and ultimately an expression for T_1. If the formalism chosen leads to non-exponential relaxation, one ultimately calculates an expression of M(t) different from the simple behavior of equation (2). Consider the BPP theory for a homogeneous system with isotropic motion. Furthermore, assume that the principal relaxation mechanism is fluctuating pairwise homonuclear dipole interactions. The correlation function is assumed to be exponential with a single correlation time τ_c.

$$G(t) = \gamma^2\hbar^2\langle F^2(t')\rangle \exp\left[-t/\tau_c\right] , \tag{9}$$

Inserting equation (9) into (7) obtains Lorentzian spectral density functions; examples are shown in Figure 2.

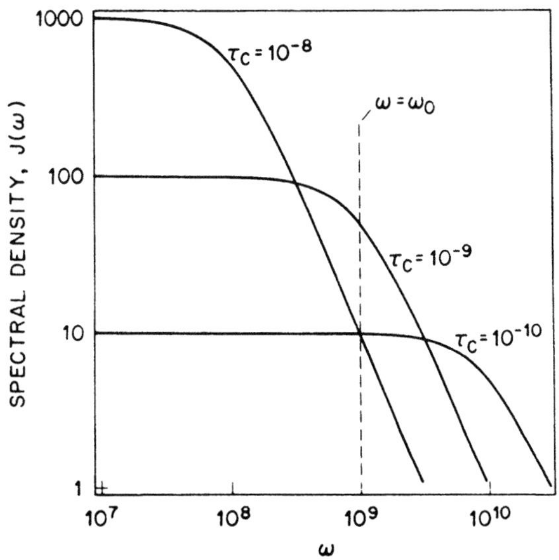

Figure 2. Lorentzian spectral density functions, $J(\omega) = t_c/\pi(1 + \omega^2 t_c^2)$, for three correlation times. For a given ω_0, the maximum J(ω) is at $\omega_0 = 1/t_c$, shown for $\omega_0 = 10^9$ sec^{-1}.

Substituting the spectral density functions into equation (8) yields

$$\frac{1}{T_1} = \frac{2}{5}\gamma^4\hbar^2 I(I+1)\frac{1}{r^6}\left[\frac{\tau_c}{1+\omega_0^2\tau_c^2} + \frac{4\tau_c}{1+4\omega_0^2\tau_c^2}\right], \tag{10}$$

where r is the pairwise internuclear separation [11]. The relation between T_1 and τ_c for a fixed ω_0 is plotted in Figure 3. The qualitative features shown in Figure 3 are also exhibited with more refined relaxation theories. The dependence of T_1 on τ_c, determined by measuring T_1 at different temperatures, will reveal whether the motion is in the slow regime (right side of Figure 3) or in the fast regime (left side). One may also calculate an activation energy, E_a, for example by assuming an Arrenhius behavior,

$$\tau_c = \tau_0 \exp\left[E_a/kT\right]. \tag{11}$$

Although effect of motion is largest at $\omega_0 = \tau_c^{-1}$, one can measure far away from the T_1 minimum. However, there are usually limits imposed by the sample. The most common limitation is the spin-lattice relaxation owing to ubiquitous paramagnetic impurities. In addition, conduction electrons may contribute to T_1 processes, as given by the Korringa relation, for species adsorbed on metallic particles. T_1 studies become ineffective in characterizing motion when the motional contribution becomes comparable to the other sources. Moreover, in any relaxation study it is imperative that each relaxation source be identified by its signature in magnetic field or temperature dependence. To this end, it is useful to measure T_1's of two different isotopes of the same species.

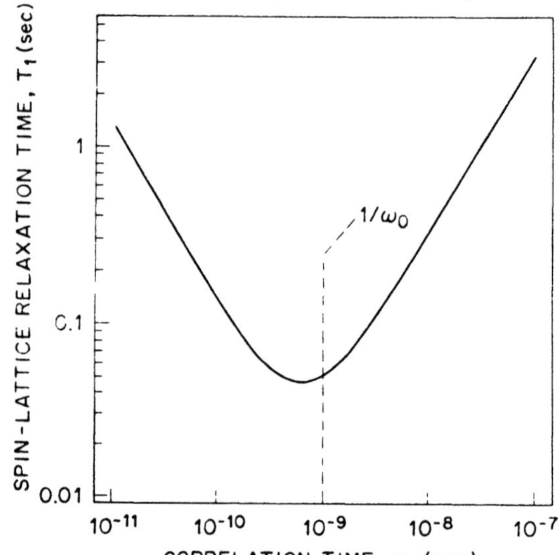

Figure 3. The spin-lattice relaxation time, T_1, as a function of the correlation time, τ_c, calculated for ^1H-^1H pairs at a distance of 1.5 Å (such as in water) at a Larmor frequency, ω_0, of 10^9 sec $^{-1}$ (ν_0 = 159 MHz). (see equation (10).)

Exponential relaxation as given in equation (2) is encountered in homogeneous systems characterized by isotropic motion and a uniform correlation time, τ_c. However, the very features that make catalytic surfaces chemically interesting also have potential to yield non-exponential recovery [6]. Potential causes include anisotropic motion and a distribution of correlation times. Anisotropic motion may be surface diffusion, or rotation about a fixed axis, such as a surface-bound methyl group. Because the samples are inherently powders, the orientational dependence of the motional effects may also cause non-exponential recovery. Correlation times will vary where there is variation in adsorption sites, or surface type (*i.e.* 100 versus 111). However, with this complexity comes the opportunity to learn more about the catalytic system and identify a specific model from a choice of many possibilities.

The range of relaxation methods may be extended by measuring other relaxation parameters, notably $T_{1\rho}$ (Tee-1-Ro). $T_{1\rho}$ is the time constant for the loss of coherence for nuclear magnetism locked perpendicular to H_0 with an rf field rotating at ω_0. Thus, it is analogous to T_1, but in a rotating frame, thus the name $T_{1\rho}$. Similar concepts apply and because the rotating magnetic fields are about 10^3 to 10^4 smaller than H_0, the technique is sensitive to correlation times proportionally longer.

2.2. EXAMPLES OF RELAXATION STUDIES ON CATALYSTS

Nearly all relaxation studies of adsorbates have concerned ^1H, and among this class, studies of water were the first and have dominated, beginning with the study of water on silica by Zimmerman, Hohnes and Lasater in 1956. [16] The results are summarized in several reviews [6-8,17,18]; the studies on the well-defined zeolites provide the best illustrations of the utility of the method. In a study of ^1H and ^2H relaxation of water in the supercages of a Na-X zeolite, Resing observed that the minimum T_1 occurred at the same temperature for both isotopes. [19] Furthermore, by varying H_0 (and thus ω_0), it was found that the T_1's were coincident at a given H_0 over a temperature range greater than 100 K. Whereas ^1H relaxation is caused by both diffusion and reorientation, ^2H has much smaller dipole-dipole couplings (γ is smaller) and consequently its relaxation is insensitive to diffusion effects. Thus, the equivalence of the T_1 minima implied that diffusion and reorientation had practically the same correlation time. That is, rotation was accomplished only by desorption.

Studies of ^{13}C T_1's to probe motions on surfaces are rare. In one example, the ^{13}C T_1 was measured to characterize the surface diffusion of isolated carbon atoms on alumina-supported Pt formed by decomposition of C_2H_4 and C_2H_2. [20] The carbon atoms are immobile on the Pt particles below 450 K, as revealed by a characteristic Korringa T_1 behavior; the product T_1T was independent of temperature below 400 K. Above 450 K, a plot of T_1^{-1} versus temperature is non-linear, which is interpreted as motion-induced relaxation. [4] In this model, T_1 is related to the time constant for the motion, τ_c, the Larmor frequency, ω_0, and the change in chemical shift upon moving from site to site, $\Delta\omega$, as

$$\frac{1}{T_1} = 2(\Delta\omega)^2 \frac{\tau_c}{1 + \omega_0^2\tau_c^2} \tag{12}$$

Compare equation (12) to the analogous equation resulting from fluctuations in dipole-dipole couplings, equation (10). The prefactor to the Lorentzian in equation (12) for chemical shift relaxation is the change in chemical shift, whereas the prefactor in equation (10) is the strength of the dipolar coupling. The chemical shift relation has no $2\omega_0$ term; the single nuclear spin of the chemical shift cannot absorb two quanta of energy. Returning to the carbon atoms on Pt, the authors extracted the motional contribution to the relaxation as a function of temperature by subtracting the relaxation effect of the conduction electrons and calculated an activation energy of 7 ± 1 kcal/mole. [20]

3. Lineshape Analysis

Although in principle relaxation methods might be employed to study motions with rates much slower than ω_0, the relaxation effect of such motions decreases rapidly as the difference from ω_0 increases, as shown in Figure 3. In heterogeneous catalysts, other phenomena may overwhelm the contributions from motions if the time constants for the motions are longer than $1/\omega_0$. Relaxation methods are thus rendered ineffective. For motions with rates slower than ω_0, lineshape analysis is more useful. To discuss the methods of lineshape analysis, it is more productive to describe the NMR technique in the context of nuclear magnetic induction. The nuclear magnetic induction description of NMR was first offered by Bloch, Hansen, and Packard [21], simultaneous with the spectroscopic description of NMR by Purcell, Torrey and Pound [9,10].

3.1. OVERVIEW

At equilibrium, the net nuclear magnetism is aligned parallel to H_0. An rf pulse at the Larmor frequency is applied to rotate the magnetization to lie perpendicular to H_0. The net magnetization then precesses about the z axis at the Larmor frequency. The net magnetization is the sum of the individual nuclear magnetic moments, each of which precesses at an angular velocity dictated by its local field. The distribution of fields in the sample causes a distribution of angular velocities and thus the magnetic moments lose coherence; the signal decays. For a sample with a distribution of frequencies described by a halfwidth $\Delta\omega$, the nuclear induction decay will lose coherence with a time constant $T_2 = 1/\Delta\omega$. A Fourier transform of the nuclear induction decay yields the NMR lineshape. A time decay with a time constant T_2 will yield a frequency spectrum with a linewidth $\Delta\omega$.

Consider a species reorienting or exchanging rapidly on the time scale of the NMR linewidth, $\tau_c \ll 1/\Delta\omega$. (The linewidth $\Delta\omega$ is taken as a measure of the distribution of magnetic fields in the rigid lattice limit, such as the separation between two peaks or the width of a dipolar-broadened Gaussian.) An atom with rapid, isotropic motion visits many environments (e.g., adsorption sites, geometric configurations, or chemical forms) during the observation a magnetic induction decay. A nucleus' lifetime in any of the many environments is too short to reveal the details of any individual environment. Rather, the NMR spectrum reflects an average environment. For example, the NMR spectrum of a specific site on a molecule in a pure liquid is a single sharp line, although the bond lengths and angles are temporal, owing to molecular vibrations which have time constants of approximately 10^{-14} s. Motion fast compared to $\Delta\omega$ yields only a sharp line and lineshape analysis reveals nothing about the details of the motion. One only knows that the rate is fast compared to $\Delta\omega$. In the extreme of motion slow compared to $\Delta\omega$, the rigid lattice limit, the species are immobile during the NMR decay and the resulting spectrum reveals the distribution of individual sites in the material; the spectrum is usually a broad line. But for motion with a rate comparable to the linewidth or for motion that is anisotropic, departures from the rigid lattice lineshape will reveal the details of the motion.

The motion of a nuclear magnetic moment between two environments with different magnetic fields illustrates the qualitative features of lineshape effects. Consider the exchange between two species, A and B,

$$A \underset{k_B}{\overset{k_A}{\rightleftarrows}} B \,. \tag{13}$$

Assume that in the limit of no exchange ($k_A \rightarrow 0$ and $k_B \rightarrow 0$) the NMR spectrum of each species is a Lorentzian peak. The spectrum for species A is therefore

$$I_A(\omega) = \frac{1}{\pi(\delta\omega)_A}\left[\frac{1}{1+\left[(\omega_A - \omega)/(\delta\omega)_A\right]^2}\right], \quad (14)$$

such that the peak is centered at ω_A and has a half-width at half-height of $(\delta\omega)_A$. The expression for the spectrum of species B is analogous. Thus the nuclear magnetic induction will lose coherence owing to two effects: the natural linewidths of each site ($\delta\omega_A$ and $\delta\omega_B$) and the exchange between environments with different magnetic fields ($\omega_A \neq \omega_B$). The time evolution is thus a pair of differential equations formed by augmenting the Bloch Equations with the effects of exchange

$$\frac{dM_A}{dt} = \delta\omega_A M_A + i(\omega_A - \omega)M_A - k_A M_A + k_B M_B \quad \text{and} \quad (15)$$

$$\frac{dM_B}{dt} = \delta\omega_B M_B + i(\omega_B - \omega)M_B - k_B M_B + k_A M_A , \quad (16)$$

such that M_A is the magnetization of A in the plane perpendicular to the *z* axis. Expressions for the NMR spectrum of this simple coupled system, obtained by the Fourier transform of $M_A + M_B$, are complex and can be found elsewhere. [22,23] A sequence of spectra for the simple case are shown in Figure 4. In the fast exchange limit, $k \gg (\omega_A - \omega_B) = \Delta\omega$, a single peak is observed. The characteristics of this single peak are insensitive to the exchange rate in this limit; to measure the rate one must appeal to spin-lattice relaxation methods such as described in the previous section. For smaller rate constants, such that $k \approx \Delta\omega$, the single peak broadens, then splits. The lineshape is a sensitive measure of k in this intermediate regime. Ultimately, the NMR spectrum becomes two distinct peaks in the slow exchange limit, $k \ll \Delta\omega$.

Lineshape analysis may be useful for determining mechanisms. Consider the slightly more complex case of ten sites which yields an NMR spectrum of ten resolved peaks in the slow exchange limit. One wishes to know which (if any) of the species exchange in this idealized system as the temperature is increased. The result will be obvious; as the temperature is varied, some peaks will remain unchanged whereas exchanging pairs will evolve in a sequence similar to the spectra in Figure 4.

For more complicated exchange processes, formulating the problem in terms of coupled differential equations becomes unwieldy. It is more productive to assume the system is a stationary Markov process [24] and adopt a density matrix formalism [23-25].
Simulating lineshapes is a profitable exercise for situations where the NMR spectrum has structure that show site distribution (such as the hypothetical two peak case above) or show orientational distribution, such as a chemical shift or quadrupolar powder pattern (vide infra). But frequently the NMR lineshape lacks structure. For spin systems dominated by many-body dipole-dipole couplings, as is usually the case for ^{1}H spectra, the NMR spectrum is a broad lump. Most often a dipolar-broadened ^{1}H NMR lineshape is a Gaussian in the rigid lattice limit,

$$I(\omega) = \frac{1}{(2\pi)^{1/2}\Delta\omega} \exp\left[-\frac{(\omega - \omega_0)^2}{2(\Delta\omega)^2}\right]. \quad (17)$$

The salient feature of the Gaussian lineshape is the linewidth, $\Delta\omega$, which can be calculated from the Van Vleck formula for the second moment,

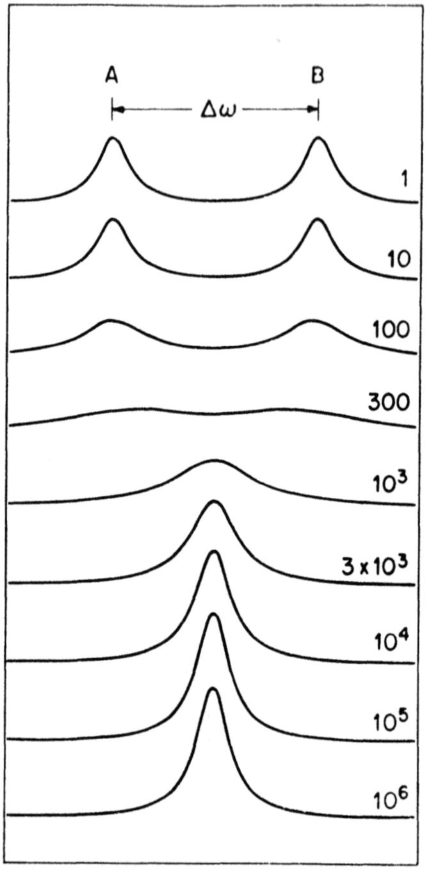

Figure 4. A sequence of NMR spectra for two exchanging species. The two peaks corresponding to the two species are of equal intensity, are separated by a nominal (ω_A - ω_B) = 10^3 rad/sec (= $10^3/2\pi$ = 159 Hz), and have equal linewidths, $\delta\omega_A = \delta\omega_B = 10^2$ rad/sec. At a lower temperature, the rate constant for the exchange, k (= $k_A = k_B$) is smaller than $\Delta\omega$ and consequently two peaks are resolved. As the rate constant becomes progressively larger, the individual peaks broaden and decrease, and a new peak appears at the center of mass, which is midway for the two peaks of equal intensity in this example.

$$\left\langle(\Delta\omega)^2\right\rangle = \frac{3}{4}\hbar^2 I(I+1)\left[\sum_j \frac{\left(1 - \cos^2\theta_{i,j}\right)^2}{r_{i,j}^6}\right] \tag{18}$$

Or, in the time domain, a Gaussian lineshape obtains from a Gaussian decay with time constant $T_2 = 1/\Delta\omega$, the rigid lattice limit. Increasing the temperature shortens the correlation time. At temperatures such that the correlation time is comparable to T_2, the spectrum narrows from a Gaussian line. At higher temperatures, the spectrum eventually becomes Lorentzian. Calculation of T_2 under motional narrowing parallels that of T_1 described previously; there are many methods and the best choice depends on the situation. For example, consider the result for a dipolar-coupled system assuming an exponential correlation function as in equation (9) [6], which yields,

$$\frac{1}{T_1} = (\Delta\omega)^2\left[\tau_c + \frac{5}{3}\frac{\tau_c}{1+\omega_0^2\tau_c^2} + \frac{2}{3}\frac{\tau_c}{1+4\omega_0^2\tau_c^2}\right], \text{ for } \tau_c < T_2 . \qquad (19)$$

Compare with equation (10), and recalling that $\langle(1 - 3\cos^2\theta)^2\rangle = 4/5$, note that for short τ_c, ($\tau_c < 1/\omega_0$), $T_1 \sim T_2$. The dependence of T_2 on τ_c is plotted in Figure 5.

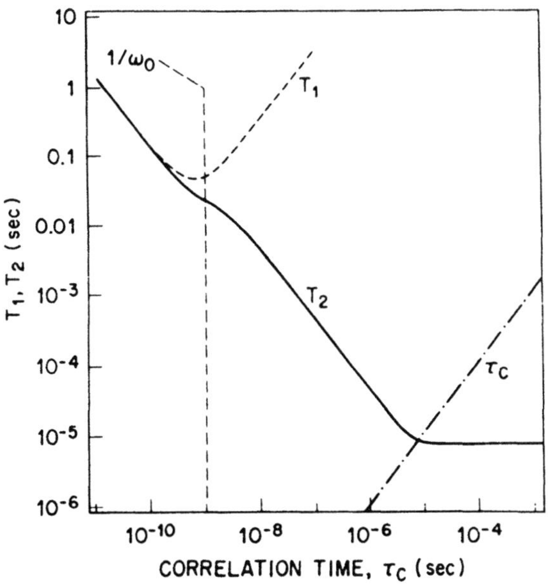

Figure 5. The spin-lattice and spin-spin relaxation times (T_1 and T_2) as a function of the correlation time τ_c for ^1H - ^1H pairs, as in Figure 3. For correlation times longer than $1/\omega_0$ (10^{-9} sec here), and shorter than the rigid-lattice limit for T_2 (7.5×10^{-6} sec here), the linewidth $1/T_2$ is a sensitive indicator of the motion.

3.2. EXAMPLES OF LINESHAPE STUDIES ON CATALYSTS

As an example of the use of T_1 (relaxation) and T_2 (lineshape) in concert to characterize a motional process, consider the hypothetical case of a sphere-like molecule (*i.e.*, SF_6 or CH_4) that rotates at a site and diffuses between sites. [6] Further, it is assumed that the activation energy for rotation is much less than that of diffusion. The calculated result [6], shown in Figure 6, is a combination of the characteristic curves shown in Figures 3 and 5.

As mentioned above, NMR spectra dominated by the chemical shift or the quadrupolar interaction reveal the details of the orientational distribution of the adsorbates. Consider the ^{13}C NMR spectrum of CO_2. In the rigid lattice limit ($\tau_c > 10^{-3}$ sec) the ^{13}C NMR spectrum

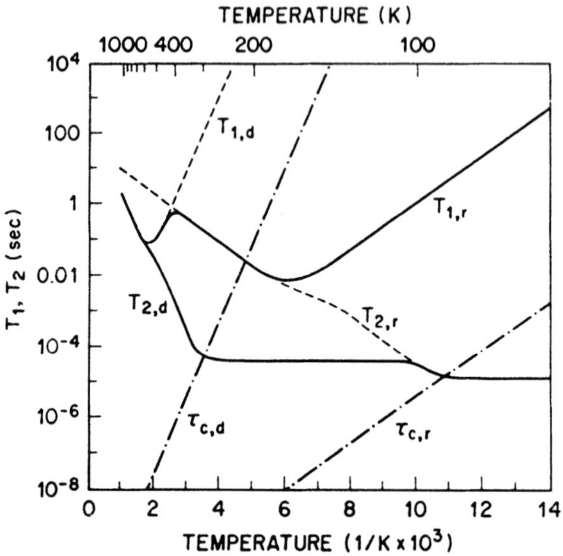

Figure 6. The calculated spin-lattice and spin-spin relaxation times (T_1 and T_2) for a hypothetical sphere-like molecule that rotates (subscript 'r') and diffuses (subscript 'd'). The correlation times $\tau_{c,d}$ and $\tau_{c,r}$ are calculated with equation (11) assuming $\tau_0 = 10^{-12}$ sec, and activation energies $E_{a,d}$ = 10 kcal/mole, and $E_{a,r}$ = 3 kcal/mole. The Larmor frequency is taken as $\omega_0 = 6.28 \times 10^7$ rad/sec, (ν_0 = 10 MHz), and $\langle(\Delta\omega)^2\rangle^{1/2} = 10^5$ rad/sec. The predicted values of T_1 and T_2 are shown by the solid lines. (From Reference [5].)

is a archetypal chemical shift powder pattern, as shown in the lower left of Figure 7 [26,27]. The basis for the lineshape can be derived from equation (5). The isotropic shift (δ_{iso}) is arbitrarily set to zero, the asymmetry parameter (η) is zero owing to the axial symmetry of CO_2, and the anisotropy (δ) is ~150 ppm. In a powder, there is an isotropic distribution of orientations. Those CO_2 molecules oriented with the symmetry axis parallel to H_0 (θ_{cs} = 0°) are chemically shifted ~100 ppm to lower frequency relative to the isotropic shift. Those CO_2 molecules oriented perpendicular to H_0 (θ_{cs} = 90°) are chemically shifted ~50 ppm to higher frequency. All intermediate orientations lie between these two extremes, as given by equation (5). The intensity distribution in the powder pattern derives from the statistics of the distribution; there are more CO_2 molecules oriented perpendicular than parallel.

CO_2 molecules rotating about an axis perpendicular to the O=C=O axis will exhibit the narrowed spectrum shown above the powder pattern on the left of Figure 7. Spectra intermediate to the rigid lattice powder pattern and the pattern for anisotropic rotation will reveal the correlation times for the motion. Such spectra are revealed by increasing the temperature from the rigid lattice limit, or by decreasing the CO_2 loading. Finally, in the limit of isotropic motion, a single sharp peak is observed. The observed ^{13}C NMR spectra for CO_2 on molecular sieve as a function of loading are shown on the right of Figure 7. The spectra progress from that of rigid CO_2 at low coverage, to spectra that include evidence of rotating CO_2, and finally to the spectrum of physisorbed CO_2 at higher coverage.

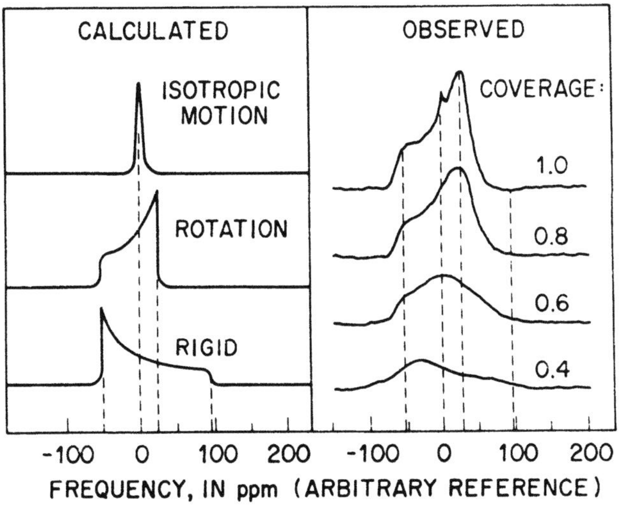

Figure 7. Calculated and experimentally observed ^{13}C NMR spectra of CO_2 adsorbed on mordenite. The three calculated spectra on the left comprise the lineshapes for three types of motion: rigid lattice, rotation about an axis perpendicular to the O=C=O bond, and isotropic motion. Each motion appears progressively as the loading of CO_2 increases.

The quadrupolar interaction, which dominates the spectra of nuclei with spin > ½, has an orientational dependence similar to that of the chemical shift interaction. Indeed, a quadrupolar powder pattern may be decomposed into a superposition of chemical shift-like powder patterns. Consider the case of ^2H in a C-H bond. To a good approximation, it may be assumed that the asymmetry parameter for the quadrupolar coupling in a C-H bond is zero. As such, the ^2H rigid lattice lineshape is the superposition of two chemical shift powder patterns such as shown for CO_2 in Figure 7; the second powder pattern is inverted through the center of mass. The spectrum formed by this sum is indicated in the lower left of Figure 8. Motion produces narrowed spectra articulated with many details such as peaks and shoulders [28]. These specific features provide for detailed descriptions of the motions, as well as activation energies, as is illustrated in Figure 8.

4. Spin-Population Labeling

Lineshape analysis is ineffective for characterizing motions with correlation times slower than the inverse linewidth. That is, systems whose NMR spectra are in the rigid lattice limit. In this limit, the method of spin-population labeling may be employed.

4.1. OVERVIEW

Spin-population labeling is analogous to the well-established method of isotopic labeling which involves tagging an appropriate reactant. One then observes the progression of the label as the tagged molecule adsorbs, reacts to form intermediates and products, then

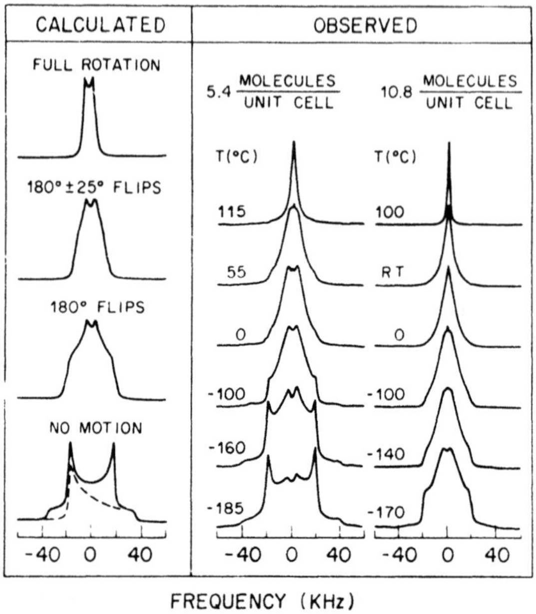

Figure 8. Calculated and experimentally observed ^2H NMR spectra of dimethyl ether (CD_3OCD_3) adsorbed on zeolite H-rho. The observed ^2H NMR spectra (right) show two trends: the spectra narrow with increasing temperature at a given loading, and are narrow with increasing loading at a given temperature. The observed spectra can be modeled as a superposition of five lineshapes, as follows. The calculated spectrum in the lower left corresponds to no molecular motion, only the rotation of the $-CD_3$ groups. (The lineshape for no methyl group rotation would be the same shape as the one at the bottom, but 3 times as wide.) 180° flips about the C_{2v} axis will narrow the spectrum as shown. Further narrowing is caused by fourfold jumps between orientations with azimuthal angles ± 25 ° and 180 ± 25 °, culminating with the spectrum of full rotation about the C_{2v} axis shown at the top. Finally, rapid isotropic reorientation yields a single sharp line (not shown.)

leaves the catalyst. The progress of the label may be monitored in situ, for example with infrared or NMR spectroscopy, or ex situ, for example by analysis of the reactor effluent. Typical isotopic labels are stable isotopes of elements key to the catalytic reactions: $^1H/^2H$, $^{12}C/^{13}C$, $^{14}N/^{15}N$, and $^{16}O/^{18}O$.

Spin-population labeling with NMR spectroscopy, like isotopic labeling, labels a particular nucleus on the probe molecule. Unlike isotopic labeling which requires chemical methods, the spin-population may be labeled with rf pulses. Labeling with NMR spectroscopy employs the spin-state of the nuclear magnetic dipole (*e.g.*, parallel or antiparallel with H_0) as follows.

The spin-population labeling experiment can be illustrated with the hypothetical two-peak spectrum used to describe lineshape analysis. Assume that the system is in the rigid lattice limit (relative to the NMR lineshape) such that the spectrum consists of two well-defined peaks, such as shown at the top of Figure 4. Recall that the NMR spectrum is obtained from a Fourier transform of a nuclear induction decay. The decay commenced after a detection rf pulse was applied to rotate the equilibrium magnetization from the z axis to the x axis. Consider the effect of preceding the detection rf pulse with an rf pulse tailored to selectively invert a portion of the equilibrium magnetization. For example, the selective pulse might rotate the magnetization of the A species from the parallel to antiparallel with the z axis without affecting the magnetization of the B species. A spectrum measured immediately after this selective inversion would consist of a normal B peak and an inverted A peak, as shown in Figure 9. Evidence of exchange may be detected by measuring spectra at various delays after the selective irradiation. If species A exchanges with species B, peak A will recover and *peak B will decrease in intensity*. Note that peak A is in a non-equilibrium state and will ultimately recover at a rate dictated by T_1, with or without exchange. Recovery of peak A faster than T_1 indicates exchange. An even stronger indication of exchange is a decrease of the intensity of peak B. The negative magnetization required to decrease peak B can come only from the inverted peak A. An idealized sequence of spectra for which $1/\Delta\omega < \tau_c < T_1$ are shown in Figure 9.

The use of spin-population labeling to determine mechanisms can be illustrated by expanding the simple two-peak system to a hypothetical ten-peak system, as used above to illustrate lineshape analysis. Again, one wishes to know which (if any) of the species in this ten-component system are exchanging. One selectively irradiates the first peak and acquires a series of spectra. If the species corresponding to the first peak is exchanging with any of the others, one of the other peaks will decrease in intensity. In this simple example, one might continue the experiment by inverting the remaining peaks, one at a time.

The advantages of spin-population labeling can be summarized by comparison with isotopic labeling. Isotopic labeling typically entails introducing a tagged species via a fluid phase. Owing to mass transport limitations, isotopic labeling is slow. NMR labeling is considerably faster. One of a pair of peaks may be labeled in a time $1/\Delta\omega$. For example, in the two-peak system exhibited by the ^1H NMR spectrum of OH groups of the silica support and hydrogen on Rh particles, the hydrogen peak was selectively inverted in 23 μsec [29]. To selectively burn a hole in a broad spectrum takes longer; ~1 msec was required to burn holes in the ^{13}C spectra of formic acid on titania [30], CO on Rh/silica [31], and CO on Ru/silica [32]. The spin-population labeling also has advantages in selectivity. Any species with a distinct NMR peak can be labeled. NMR-distinct includes species chemically shifted from each other, such as the ^{13}C spectra of carbonyls versus carbides, or a portion of the carbonyls (described below). Or, NMR-distinct can mean two overlapping species with different dipolar couplings. For example, the ^1H NMR peaks of CH_x species overlap those of surface hydroxyls, but the aliphatic hydrogen might be selectively labeled via cross-polarization from ^{13}C. A limitation of spin-population labeling with NMR is the lifetime of the transient label. The motional process to be studied must have time constant shorter than T_1, which typically requires a time constant shorter than 0.1 to 10 sec.

Spin-population labeling has application only to inhomogeneously broadened spectra, such as chemical shift or quadrupolar powder patterns. The technique is ineffective for homogeneously broadened spectra, such as the NMR spectra of systems dominated by dipole-dipole couplings. The spin-spin couplings that give the spectrum its width also cause an exchange of spin states unaccompanied by molecular motion, called spin diffusion. An attempt to selectively invert a portion of a homogeneously broadened line causes the intensity of the entire line to saturate to zero.

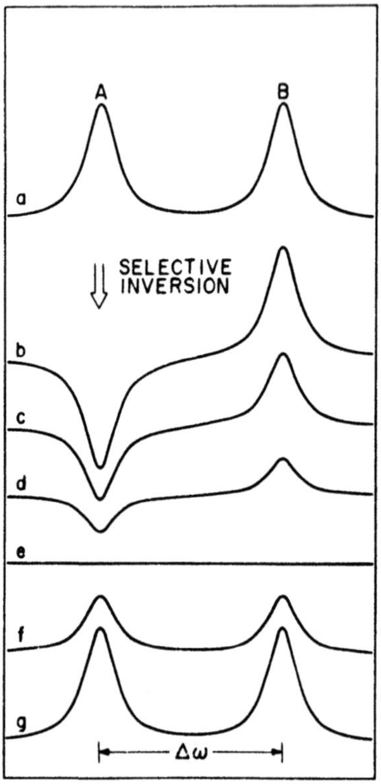

Figure 9. The sequence of spectra in an idealized selective inversion experiment sample with two species of equal amount exchanging with a time constant τ_c such that $1/\Delta\omega \ll \tau_c \ll T_1$, where $\Delta\omega$ is the splitting between the peaks. At top (a) is the NMR spectrum of the system at equilibrium. Spectrum (b) is observed immediately after selective irradiation to invert the magnetization of species A. After a delay $t \approx \tau_c$, the inverted peak partially recovers *and* the unirradiated peak decreases in intensity by an equal amount. Spectra (d) and (e) show the continued results of exchange. After a delay $\tau_c \ll t \approx T_1$, the system partially recovers, as represented by spectrum (f). Finally, after a delay $t \gg T_1$, the system recovers to the equilibrium state and the experiment may be repeated.

4.2. EXAMPLES OF SPIN-POPULATION LABELING

The method of spin-population labeling has been demonstrated for CO on Rh/silica [31] and Ru/silica [32], and is summarized in a recent review [33]. The ^{13}C NMR spectra of CO adsorbed on the metal particles of these samples are well-defined chemical shielding powder patterns, as shown for linearly-bonded CO at the top of Figure 10. Thus, the CO is static, or is moving with a the time constant longer than the reciprocal linewidth ($\tau_c >$ 100 μs). As described above for CO_2, the linewidth is due to the orientational dependence

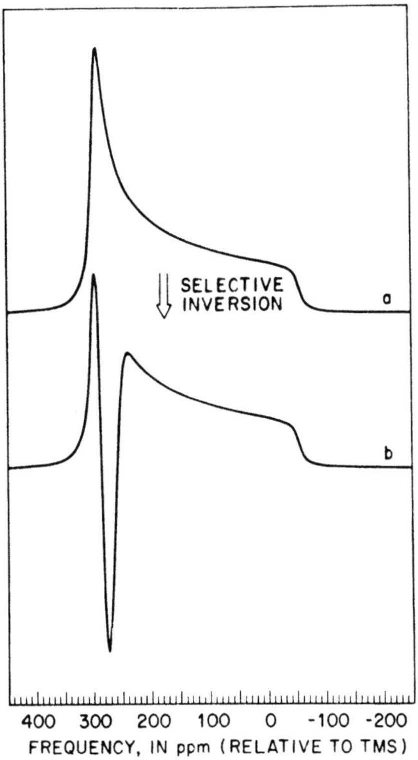

Fig. 10. The ^{13}C NMR spectrum of linearly-bonded CO on Rh at equilibrium (top) and immediately after selective inversion at 272 ppm, which selectively labels those CO molecules oriented at ~74° relative to H_0.

of the chemical shielding anisotropy, as given in equation (5). For species with pseudo-axial-symmetry ($\eta = 0$), such as for CO_2 or linearly-bonded CO, there is a one-to-one correspondence between the chemical shift and the orientation of the symmetry axis relative to H_0. Thus, selective inversion of a narrow portion of the powder pattern labels only those CO molecules bonded to surfaces with a specific orientation relative to H_0. Since the powder pattern of bridge-bonded CO completely overlaps that of linearly-bonded CO, there exists a surface orientation on which both types of CO have the same chemical shift. For CO on Rh, this corresponds to surfaces with normals at 74° relative to H_0. The appearance of the spectrum of linearly-bonded CO immediately after selective inversion is shown in Figure 10. It was observed that the rate of recovery of the inverted hole was faster than T_1 which indicated that CO's with selectively inverted spins are exchanging with unlabeled CO's. The labeled CO exchanged with unlabeled CO's at many different orientations, and thus no new holes were produced as the labeled hole recovered - only an overall decrease in the unlabeled portion was observed. The time dependence of the hole recovery was modeled by surface diffusion from labeled surfaces to adjacent unlabeled surfaces. The temperature dependence of the recovery yielded activation energies of 7±1 kcal/mole for CO on Rh [31] and 5.4±1 for CO on Ru. [32]

The selectivity of the spin-population labeling was demonstrated by labeling linearly-bonded CO molecules exclusive of adjacent bridge-bonded CO [31]. Although the ^{13}C NMR spectrum of bridge-bonded CO is entirely overlapped by that of linearly-bonded CO, the powder pattern of bridge-bonded CO is narrower than that of linearly-bonded CO, so it is possible to selectively invert a portion of the linearly-bonded CO without inverting any bridge-bonded CO. For CO on Rh, the linearly-bonded CO does not overlap the bridged-bonded CO for angles $\theta < 45°$. Selectively labeling the linearly-bonded CO was achieved by irradiating at 0 ppm, which corresponds to CO on surfaces with normals at 27° relative to H_0. The recovery of the hole at 0 ppm exhibited a plateau not observed when linear- and bridged-bonded CO were inverted simultaneously. This feature was interpreted as exchange between these two CO species on the labeled surface before the species exchanged with CO's on other surfaces. However, the poor sensitivity of the data prevented an accurate measurement of this process; an activation energy of 6±6 kcal/mole was calculated.

5. Summary

In heterogeneous catalysis, the key processes are adsorption/desorption, surface diffusion, and reaction. The characteristic time constants for these phenomena can be estimated from typical properties. For typical heats of adsorption between 10 and 30 kcal/mole, the residence time at 25° C ranges from 10^{-6} to 10^{9} sec; at 300° C the range is from 10^{-10} to 10^{-2} sec. For a diffusion coefficient of $10^{-8 \pm 2}$ cm^2 sec^{-1}, an adsorbate will travel a root-mean-square distance of 100 Å in about 10^{-6} to 10^{-2} sec. Assuming a turn-over number of $10^{3 \pm 3}$ $site^{-1}$ sec^{-1} gives reactant lifetimes that span the range 10^{-6} to 1 s.

The ranges of applicability of relaxation studies, lineshape analysis, and spin-population labeling are shown in Figure 11. Also shown is the temperature dependence of the time constant for a few activation energies. This schematic illustrates that within the experimentally accessible range of most NMR spectrometers (77 K to 400 K), the dynamic process will have a time constant compatible with one of the NMR methods.

It is emphasized that the ranges of applicability depicted in Figure 11 are arbitrary and may vary by many orders of magnitude between systems. There may also exist pathological conditions that prohibit the use of some methods. For example, extraneous relaxation processes (paramagnetic impurities, Korringa relaxation) may overwhelm motional contributions to relaxation, or homogeneous broadening may frustrate attempts to selectively label the nuclear magnetization. For these and other reasons, it is profitable to study two different isotopes in the material, such as ^{1}H and ^{2}H. The isotopes have gyromagnetic ratios that differ by a factor of 6.5, and the ^{1}H spectrum is usually dominated by dipole-dipole couplings, whereas the ^{2}H spectrum is dominated by quadrupolar couplings.

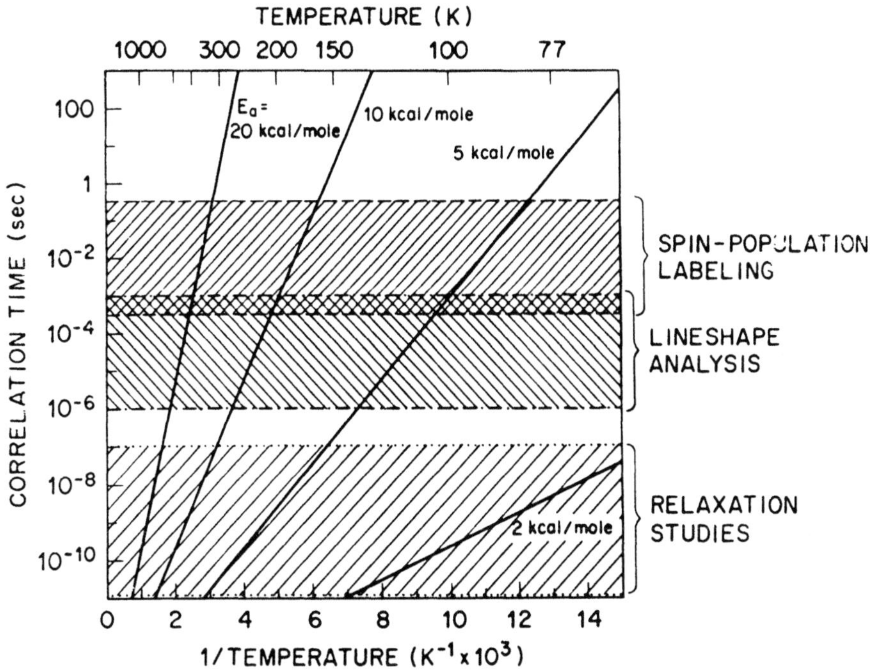

Fig. 11. The correlation time, τ_c as a function of temperature for various activation energies, E_a. The lines were calculated with equation (11) using $\tau_0 = 10^{-14}$ sec. Also shown are representative bands of τ_c for which the methods of relaxation studies, lineshape analysis, and spin-population labeling are applicable. These bands are estimates for illustration; the ranges of applicability with depend on many factors such as the dominant interaction (dipolar, chemical shift, or quadrupolar), the type of motion, and the level of paramagnetic impurities.

6. References

1. Adapted from Duncan, T. M. (1990) 'NMR Spectroscopy of Dynamics at Catalytic Surfaces', Japanese J. of Catalysis 32, 510-521.
2. Wolf, D. (1979) Spin-Temperature and Nuclear-Spin Relaxation in Matter: Basic Principles and Applications, Clarendon, New York.
3. Redfield, A. G. (1976) 'A Guide to Relaxation Theory,' NMR: Basic Principles and Progress 13, 1-12.
4. Slichter, C. P. (1989) Principles of Magnetic Resonance, Springer-Verlag, New York, 3rd ed. chapters 5 and 6.
5. Abragam, A. (1961) The Principles of Nuclear Magnetism, Oxford, London, chapter VIII.

6. Resing, H. A. (1967) 'Nuclear Magnetic Resonance Relaxation of Molecules Adsorbed on Surfaces', Adv. Molec. Relax. Proc. 1, 109.
7. Packer, K. J. (1967) 'Nuclear Spin Relaxation Studies of Molecules Adsorbed on Surfaces', in J. W. Emsley, J. Feeney and L H. Sutcliffe, (eds.), Progr. Nucl. Magn. Reson. Spectrosc. 3, 87-128.
8. Pfeifer, H. (1972) 'Nuclear Magnetic Relaxation and Relaxation of Molecules Adsorbed on Solids', NMR Basic Principles and Progr. 7, 53.
9. Purcell, E. M., Torrey, H. C., Pound, R. V. (1946) 'Resonance Absorption by Nuclear Magnetic Moments in a Solid', Phys. Rev. 69, 37.
10. Rigden, J. S. (1986) 'Quantum States and Precession: The Two Discoveries of NMR', Rev. Modern Phys. 58, 433.
11. Bloembergen, N.; Purcell, E. M., and Pound, R. V. (1948) 'Relaxation Effects in Nuclear Magnetic Resonance Absorption', Phys. Rev. 73, 679.
12. Kubo, R. and Tomita, K. (1954) 'A General Theory of Magnetic Resonance Absorption', J. Phys. Soc. Japan 9, 888.
13. Hebel, L. C. and Slichter, C. P. (1959) 'Nuclear Spin Relaxation in Normal and Superconducting Aluminum', Phys. Rev. 113, 1504.
14. Wangsness, R. K., and Bloch, F. (1953) 'The Dynamical Theory of Nuclear Induction', Phys. Rev. 89, 728; Bloch, F. (1956) 'Dynamical Theory of Nuclear Induction. II', Phys. Rev. 102, 104.
15. Redfield, A. G. (1957) 'On the Theory of Relaxation Processes'. *IBM J. Res. Develop.* 1, 19.
16. Zimmerman, J. R.; Holmes, B. G., and Lasater, J. A. (1956) 'A Study of Adsorbed Water on Silica Gel by Nuclear Resonance Techniques', J. Phys. Chem. 60, 1157.
17. Klinowski, J. (1984) 'Nuclear Magnetic Resonance Studies of Zeolites', Progr. NMR Spectrosc. 16, 237.
18. Lechert, H., and Basler, W. D. (1989) 'Molecular Motion in Zeolite Sorbents, Studied by Various NMR Methods', J. Phys. Chem. Solids 50, 497.
19. Resing, H. A. (1976) 'Quadrupole Coupling Constants in Adsorbed Water. Effects of Chemical Exchange', J. Phys. Chem. 80, 186.
20. Wang, P. -K; Ansermet, J. -P; Slichter, C. P., and Sinfelt, J. H. (1985) 'Observation of Isolated Carbon Atoms and the Study of Their Mobility on Pt Clusters by NMR', Phys. Rev. Lett. 55, 2731.
21. Bloch, F., Hansen, W. W., and Packard, M. (1946) 'Nuclear Induction', Phys. Rev. 69, 127; (1946) 'Nuclear Induction', Phys. Rev. 69, 680; (1946) 'The Nuclear Induction Experiment', Phys. Rev. 70, 474.
22. Rogers, M. T. and Woodbrey, J. C. (1962) 'A Proton Magnetic Resonance Study of Hindered Internal Rotation in some Substituted N,N-Dimethylamides,' J. Phys. Chem. 66, 540.
23. Sandström, J. (1982) Dynamic NMR Spectroscopy, Academic, New York.
24. Reference 4, pp. 447-451.
25. Kaplan, J. I. and Fraenkel, G. (1980) NMR of Chemically Exchanging Systems, Academic, New York.
26. (a) Sefcik, M. D.; Schaefer, J., and Stejskal, E. O. (1976) 'Characterization of the Small-Port Mordenite Adsorption Sites by Carbon-13 NMR', in H. A. Resing and C. G. Wade (eds.), Magnetic Resonance in Colloid and Interface Science, ACS Symp. Ser. 34 Amer. Chem. Soc., Washington, DC, 1976), p.109; (b) Sefcik, M. D.;

Schaefer, J., and Stejskal, E. O. (1977) 'Characterization of the Mordenite Sorption Sites by Carbon-13 NMR', in J. R. Katzer (ed.), Molecular Sieves II, ACS Symp. Ser. 40, p. 344.

27. Duncan, T. M., and Dybowski, C. (1981) 'Chemisorption and Surfaces Studied by Nuclear Magnetic Resonance Spectroscopy', Surf. Sci. Reports 1, 157.

28. Luz, Z., and Vega, A. J. (1986) 'Sorption of Dimethyl Ether in Zeolite H-rho Studies by Deuterium NMR', J. Phys. Chem. 90, 4903.

29. Root, T. W., and Duncan, T. M. (1987) 'Characterization of Hydrogen on Silica-Supported Rhodium with ^{1}H NMR Spectroscopy', Chem. Phys. Lett. 137, 57.

30. Root, T. W. and Duncan, T. M. (1986) 'Formic Acid on Titania: Motional Properties and Desorption of Chemisorbed Species', J. Catalysis 102,109.

31. Duncan, T. M.; Thayer, A. M., and Root, T. W. (1989) '^{13}C NMR Study of CO Surface Diffusion and Site Exchange', Phys. Rev. Lett. 63, 62.

32. Duncan, T. M.; Thayer, A. M., and Root, T. W. (1990) 'Surface Diffusion of CO on Silica-Supported Ru Particles: ^{13}C Nuclear Magnetic Resonance Studies', J. Chem. Phys. 92, 2663.

33. Duncan, T. M. (1990) 'Studies of Adsorbate Structure and Dynamics on Catalytic Surfaces with NMR Spectroscopy: CO on Metals', Colloids and Surfaces 45, 11.

Session Two, *Surface Science Studies in Catalysis*

Rapporteur H.P.C.E. KUIPERS
Koninklijke/Shell-Laboratorium, Amsterdam (Shell Research B.V.)
PO Box 3003
1003 AA Amsterdam
The Netherlands

Mme Bordes opened the session with an account of the surface reactivity of oxides. In the general discussion she returned to the question of language differences between the various branches of heterogeneous catalysis [1]. On the basis the presentations given up to that point she proposed a rough subdivision into three groups employing not only different languages but also concepts, namely those working on metal surfaces, oxide surfaces and zeolites. She illustrated her point by giving some examples of terms used in one field which had different names in the others. For instance the use of the term 'adsorbate induced reconstruction' in metal catalysis was claimed to resemble in some aspects the concept behind 'surface reactivity' in oxide catalysis, which could be a source of misunderstanding. Mme Bordes continued by stating that, contrary to what one might expect from the well-defined structures of zeolites and their good accessibility to analysis, they are much less well-defined, mainly since the inside and the outside of zeolite crystals may be very different and may contribute to unknown relative extents to catalysis. The impression gained from this discussion was that it is not so much the different languages used in the different disciplines which hinder proper communication, but rather the differences in the tacit assumptions made. These assumptions are based on a wealth of heuristic experience derived from observations made over a long period of time, but it appears difficult for workers in a different area to judge their significance and hence to develop a common starting point for discussion.

Centi added a remark to the discussion on language differences. He stated that the timescale at which surface reconstructions take place on an oxide surface is usually much longer than typical reaction times of the adsorbates, whereas in metal catalysis the two timescales are about the same. King supported this by mentioning as an example his observation that the CO induced reconstruction of Pd(110) can be completely suppressed by carbon atoms 'pinning' the surface, yielding more weakly adsorbed CO. Coulston expressed doubts as to the validity of Centi's statement that adsorbate induced reconstructions on oxide surfaces are much slower than on metal surfaces. Centi clarified this point by stressing that he had relative timescales in mind for surface reaction versus surface reconstruction. Since reactions on oxide surfaces are usually assumed to be very fast, on an absolute scale reconstructions of oxide and metal surfaces could be similar.

Nieuwenhuys in his presentation considered a number of catalytic reactions involving small molecules, CO, NO, O_2, H_2 and NH_2, relating catalytic activity and chemisorption properties. King commented that the proposed simple relationship between bond energy of adsorbates and reactivity could be misleading in some cases. As an example he mentioned that N_2 cannot be adsorbed dissociatively on Pt despite the fact that nitrogen adatoms on Pt are quite stable when prepared via atomic nitrogen. The barrier to dissociative adsorption appears to be extremely high, although the reason for this is unclear. Additionally he remarked that microcalorimetry will be able to provide the proper bond energies in future. It was suggested that the surface states close to the Fermi level should be important to adsorption and that these should be correlated with adsorption behaviour. King agreed but added that geometry is equally important.

R. W. Joyner and R. A. van Santen (eds.), Elementary Reaction Steps in Heterogeneous Catalysis, 243–247.

The session continued with Hayden's contribution 'Molecular Beam Studies of Surface Chemical Reactivity.' He used controlled reactive deposition of the rhodium precursor $(Rh(CO)_2Cl)_2$ to arrive at different densities of defined sites and configurations of sites on a TiO_2 (rutile) single crystal surface. These species' chemistry could subsequently be studied by controlled exposure to reactant gases (eg. H_2 and CO) and monitored by surface sensitive techniques (eg. XPS and RAIRS).

Joyner suggested that residual chlorine stemming from the decomposition of the Rh-precursors might have influenced the chemistry observed. Hayden replied that he had not found any indication for chlorine on the surface whatsoever and that the water added should have removed all chlorine. Furthermore, it was found remarkable that the Rh atoms preferred to stay arranged in two-dimensional rafts up to coverages approximating a monolayer rather than growing into three dimensions earlier. Hayden stated that nevertheless this was the case and that no indications could be found by the techniques available that deviations from two dimensional growth occurred at moderate coverages. Ponec added that the observed redispersion of small Rh particles to rafts upon exposure to CO is a well-known phenomenon for supported catalysts. Hayden commented that his work was done on single crystal surfaces, which allow a much more direct study with less ambiguity. King wondered whether there were any indications of CO bound to the oxide/metal interface. The answer was negative, all chemistry found could be explained with the metal and/or oxide surface. Somorjai asked whether any phenomena reminiscent of the strong metal support interaction, (SMSI), had been observed with the model catalysts prepared in this way. As yet no studies have been done under conditions (high temperature, reducing atmosphere) usually required for SMSI to occur. Basset mentioned that he was astonished by the ease with which the Rh precursor reacted with the surface. The reason appears to be the use of water. Along similar lines the question was raised whether titania was reduced during precursor decomposition. Indeed this is believed a possibility.

Nieuwenhuys commented on Hayden's presentation that this type of work represented a nice route towards preparing isolated 'active sites' which then can be studied in molecular detail. One particularly intriguing question is posed by Hayden's observation that the geminal dicarbonyl rhodium species could react so easily with H_2 in the absence of sites usually assumed to be required for the dissociation of hydrogen. Hayden commented that this was indeed so and that he would seek advice from experts in homogeneous catalysis in order to understand this phenomenon.

King presented his lecture 'The Role of Surface Structure in Adsorption on Metal Surfaces' [2]. With his studies on the adsorption of N_2 on tungsten surfaces he presented a nice example of extreme structure sensitivity. Dissociative adsorption of N_2 requires pairs (ensembles) of tungsten surface atoms arranged as in the W(110) surface. He could furthermore show that adsorbed nitrogen atoms reconstruct the surface, in a sense pulling the metal surface atoms into positions required for strong chemisorption. Also two-dimensional nucleation phenomena could be observed and analysed. Multiple peaks in TDS can be due to a number of phenomena like steps in the surface, configuration changes, molecular integrity, lateral interactions and desorption reconstruction. The frequently made assumption that preexponential factors are more or less constant during thermal desorption was rejected. As an example he mentioned the case of H_2 desorption from tungsten where not only the activation energy for desorption changes with coverage, but also the preexponential factor varies by more than seven orders of magnitude, due to reordering in the adsorbate layer. Furthermore, he showed that small quantities of strongly adsorbed species (eg. surface carbide) may 'pin' a metal surface into a structure with different chemistry. Overall this lecture presented a wealth of evidence for the fact that the dynamics/kinetics of surface reactions can be very strongly influenced, if not dominated by readjustment of the metal surface atoms upon chemisorption. Clearly the notion that metal surfaces may be assumed as a 'passive' well-behaved substrate must be abandoned.

Holloway asked how far the results for the N/W system matched the STM results of Behm *et al* (University of Munich) on the same system. Indeed the results appear to agree well. Somorjai recalled Langmuir's historical work on N/W in relation to tungsten light bulbs, filled with nitrogen in those days. He wondered in how far the results presented could be extrapolated to 1800 C. Joyner stressed the importance of knowing whether a molecule adsorbs directly or via a precursor state. In particular he referred to methane activation. King responded that direct adsorption is most likely to occur with light molecules, eg. methane. 'Heavy' molecules tend to be trapped first. Van Santen addressed the observed large changes in preexponential factor for hydrogen desorption from tungsten as a function of coverage. He proposed the delocalisation of protons in the surface potential as a possible further cause for this observation and King agreed that this might play a role. Finally Zhdanov remarked that if the reported nucleation phenomena for CO adsorption on Pt(110) represented a first order phase transition, then there should be a very strong temperature dependence. King confirmed that this was observed, but added that this can be very directly understood in terms of his finding that it takes an 'ensemble' of 5 CO molecules to initiate reconstruction of the surface.

Somorjai asked if the determination of sticking probabilities represents a worthwhile approach to understanding reactivity. His doubts were based on the fact that the sticking itself would always intermingle with surface diffusion and hence dynamic phenomena in the adsorbate layer and the surface underneath. King admitted that the latter is undoubtedly the case, but that nevertheless sticking probabilities represent a good approach to reactivity. Although being a 'lumped parameter', sticking probabilities are very good observables and descriptors of a system. Unravelling the various contributions would be the next step. Somorjai worried that the kinetic models used in this unravelling procedure contained too many parameters, which could not be determined in isolation, hence rendering the fitting to observations ambiguous. Haller added that from a practical point of view the concept of sticking probabilities was very useful for counting the number of 'active sites.' Ponec want to know whether the dynamics of surface diffusion, reconstruction and relaxation induced by a sticking species would have a large influence on the sticking process itself. This view was supported by Haller stating that these influences can be large in cases.

Haller continued with an overview lecture on 'The Dynamics of Alkane Adsorption on Metals.' He gave a summary of ways to probe barriers to adsorption of reactants, to reactions of intermediates and to desorption of products. In particular he addressed the question where the energy comes from that is required to overcome these barriers. Depending on the system, translational and/or vibrational and/or rotational energy, but also surface temperature can be very important. He summarised methods on how information was obtained in particular for small hydrocarbon molecules and derivatives (alkanes, alkenes, alcohols) on a variety of surfaces (metals and oxides).

In the discussion the influence of the presence of hydrogen was addressed. At present little is known in that respect, since the proper experiments still have to be performed. As yet the most detailed experiments have been done in the low coverage regime, hence remote from conditions used in catalytic practice. Joyner wondered whether the observed dissociative adsorption mechanisms for light molecules would require the consideration of more crucial aspects than 'direct tunnelling' alone. Haller responded that direct tunnelling was real and that looking at the precursor well alone would not reveal relevant information on the chemisorption process for these cases.

The last lecture of the session was presented by Duncan, on 'NMR studies of Metal Catalysts.' He gave an overview of the type of information NMR can provide on chemisorbed species and the limitations this technique has. Most work done so far involves relaxation studies, line shape analysis and spin-population labelling. However, the sensitivity of NMR precludes single crystal studies so far, hence supported metal particles have to be used in order to have a sufficient number of chemisorbed species in the sampled volume. Van Santen inquired whether the studies on the exchange between CO in on-top and bridging configurations had provided information on the coverage dependence of the rate of this phenomenon. This information is not accessible, because all studies done so far have been performed at full coverage. King wanted to know whether the activation energies for diffusion on metal surfaces found by NMR for supported metal catalysts were in line with the values found by surface science studies on single crystals. The background of his question was that in single crystal studies adsorbed species need not cross grain boundaries, whereas grain boundaries should be present in supported metal particles. Duncan responded that his results were in reasonable agreement with those of surface science studies. Apparently the influence of grain boundaries is not very important in the systems studies so far.

Joyner raised the question whether surface science data would allow an estimation of the critical particle size below which the chemistry of the bulk metal surface changed into a different chemistry. King commented that as yet one cannot give a single answer to that question. He mentioned recent observations Japanese workers, who found by use of TEM that the structures of small metal particles may be completely different from the known bulk structure. Ponec added that recent theoretical work of Norskov *et al* (University of Lyngby) has shown that small metal particles may be described as essentially consisting of a 'solid core' surrounded by a 'liquid shell.'

Duncan was asked about the sensitivity of NMR, more specifically what number of protons would represent the detection limits of this technique. The world record appears to be 10^{14} protons, albeit at 10 K. 10^{18} protons was considered a more representative number, while 10^{16} could be achieved with some struggle. Hence single crystal studies with NMR appear out of reach. Optical techniques providing information with similar specificity in combination with much higher sensitivity are however developing at a rapid pace.

In more general discussion Basset reopened the question of culture differences in the various fields of catalysis. He distinguished between the 'isolated atom' and 'metal surface' studies. In his opinion molecular chemistry and surface science are very different, particularly in that surface science so far can only handle very simple systems. Molecular chemistry should be further ahead since the consideration of phenomena like geometries, dynamics, energies, relaxation etc, are the rule rather than the exception. However, on the other hand he admitted that molecular chemistry lags behind surface science with respect to direct reactivity studies. He proposed to emphasise the study of the activity of chemisorbed and metalorganic molecules. Finally he recommended that surface scientists try to count the electrons during hypothetical reaction pathways, as molecular chemists do.

In summary, it is the Rapporteur's view that surface science continues to contribute detailed quantitative knowledge on the atomic scale to the field of heterogeneous catalysis. Surfaces and surface configurations can be prepared in a controlled fashion. The characterisation of these surfaces can be done by an ever increasing wealth of techniques down to atomic detail. Elementary steps of reactions with and on these surfaces provide quantitative information on the energetics, kinetics and even dynamics involved. However, the main weakness of surface science in relation to heterogeneous catalysis remains, namely that these studies have to be done at low pressures and hence with states of the surface which might or might not be representative or even relevant to catalytic practice. Truly surface sensitive *in-situ* techniques would be the obvious way out, but only very few of such techniques exist. Optical and NMR techniques show promise, although the latter appear to lack sufficient sensitivity to allow the study of properly prepared and characterised systems (eg. single crystals). A further approach, which is increasingly applied, involves the preparation and characterisation of model catalyst surfaces under 'controlled surface science conditions' followed by catalytic testing under conditions resembling those encountered in practice. In such a way the activity, selectivity and stability of model surfaces can be determined under relevant conditions. At the end of the test run reactants and products can be pumped off rapidly and the surface analysed. Although the surface will certainly be different in UHV than under reaction conditions relevant information can still be obtained like for instance the state of promoters (eg. valences, coverages, atomic compositions) or whether the surface is reconstructed or even completely roughened.

1. See the Report on session 1 for the origin of this discussion.
2. King's lecture is not included in these proceedings.

APPLICATIONS OF SYNCHROTRON RADIATION IN CATALYSIS

Richard W. Joyner,
Leverhulme Centre for Innovative Catalysis,
Department of Chemistry,
University of Liverpool,
PO Box 147, Liverpool, L69 3BX, UK.

ABSTRACT. Synchrotron radiation is finding increasing application in catalytic studies. The properties of synchrotron radiation which make it useful in catalytic studies are first discussed and the nature of the interactions between X-rays and matter reviewed. A number of applications are then described, including structural characterisation of metal containing zeolite and supported catalysts, alloying, and modification of catalyst structure by reaction.

1. Introduction

Synchrotron based techniques are making increasingly important contributions to the study of catalysts and catalysis. The synchrotron can provide intense, white, electromagnetic radiation extending from the far infra red almost to the gamma ray region of the spectrum. It is therefore potentially suitable for diffraction techniques and for a wide range of spectroscopies. In this presentation the most important properties of synchrotron radiation are first described. X-ray based techniques have so far been the most valuable in studies of heterogeneous catalysis, so the interactions of X-rays with matter are briefly discussed. Applications in catalysis are then considered in more detail, including structure determination, studies of metal - support interactions, chemisorption, reactivity and alloying. Lastly, but in keeping with the forward looking nature of this ARW, requirements for rapid progress in the future are assessed.

2. Synchrotron Radiation

It has long been recognised that when charged particles travelling at relativistic velocities are made to traverse a circular orbit, electromagnetic radiation will be emitted [1]. What is now called synchrotron radiation was first observed at the General Electric 70 MeV synchrotron in 1947 [2]. The intensity of the radiation varies monotonically with wavelength and depends, inter alia, on the mass and charge of the particle, its velocity and the radius of the curved orbit. For many years this radiation was regarded as a curiosity and a largely unwanted feature of accelerators built to carry out particle physics experiments. It became recognised only gradually that the synchrotron radiation possesses properties of value in many branches of science, from molecular biology, through chemistry and materials science, to physics. Use of synchrotron radiation became increasingly popular in the 1970s, and in 1982 the UK synchrotron radiation source at Daresbury became the first facility in the world to be

R. W. Joyner and R. A. van Santen (eds.), Elementary Reaction Steps in Heterogeneous Catalysis, 249–265.

built purely for experiments which use the synchrotron radiation. [Daresbury is also the birthplace of Lewis Carroll, and synchrotron studies sometimes seem to have an *Alice in Wonderland* dimension.] There are now about ten synchrotron radiation facilities operational world-wide, with the intense, bright European Synchrotron Facility at Grenoble being the most recent addition. Developments in synchrotron hardware and experimentation are reviewed in the journal Synchrotron Radiation News [3].

Electrons are the most convenient particles and are used in the majority of synchrotron sources. An operating synchrotron typically has many bunches of electrons circulating at once, each separated from each other by perhaps 2 nanoseconds. The synchrotron orbit is not circular, but consists of a number of bending magnets connected to each other by straight sections. The electron pathway is held at ultra-high vacuum to conserve the electron current. In the Daresbury source typical operating conditions involve 160 bunches of electrons, adding up to a current of 200 mA. The energy is 2 GeV and the current decays with a half-life of ca 30 h.

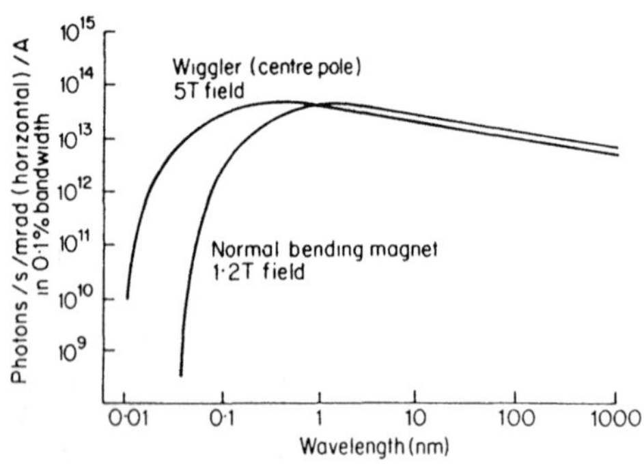

Figure 1. Flux produced by the synchrotron radiation source at Daresbury (UK)

The emission spectrum of the Daresbury SRS, representing a typical synchrotron radiation source is shown in Figure 1. The intensity increases gradually from the far infra red region and reaches a maximum typically in the far ultra-violet or X-ray region. At higher energies the intensity falls off rapidly. The radiation spectrum contains no characteristic lines and so offers distinct advantages over laboratory X-ray generators. The radiation intensity is high and may approach or exceed that of a characteristic X-ray line even after monochromatisation. The radiation also has other useful properties, it is highly collimated and plane polarised. Because of the bunched nature of the electron flux, it is also highly time resolved, although this property has not yet been much used in catalysis. Figure 1 shows the typical spectrum generated at a bending magnet in the synchrotron ring. The spectrum may be changed, and its maximum shifted to higher or lower wavelengths, through the use of *insertion devices* such as wigglers or undulators, which are inserted into straight sections of the electron pathway. Fig. 1 also shows the flux produced by a wiggler (magnetic field 5 Tesla) at the Daresbury source.

The natural ingenuity of scientists has meant that all of the synchrotron spectrum has been put to use. In catalysis the far UV/X-ray region is particularly valuable and will take up most of this chapter. The infra red region presents some advantages over laboratory sources for studying low frequency vibrational modes, such as the metal - carbon stretches in chemisorbed carbon monoxide [4]. These are normally invisible with laboratory IR sources, but are accessible through high resolution electron energy loss spectroscopy [5]. Because of the importance of the X-ray region of the spectrum we now briefly review the interactions of X-rays with matter and the information that can be obtained.

3. Interactions of X-rays with Matter

The interaction of X-rays with matter is relatively weak. The main processes which provide useful information are absorption, scattering and diffraction, as shown diagramatically in Fig. 2. X-ray scattering can be used to study the size distribution of small particles [6], but has so far been very little used in catalytic studies. The technique of X-ray diffraction is too well known to require introduction, but the use of a synchrotron as the X-ray source is novel. Synchrotron sources have the advantage of high *brightness*, ie they can deliver a high flux over a very small area, and this can be valuable for the study of very small crystals. Use of a synchrotron source may also allow X-ray diffraction to be used in conjunction with other techniques, for example

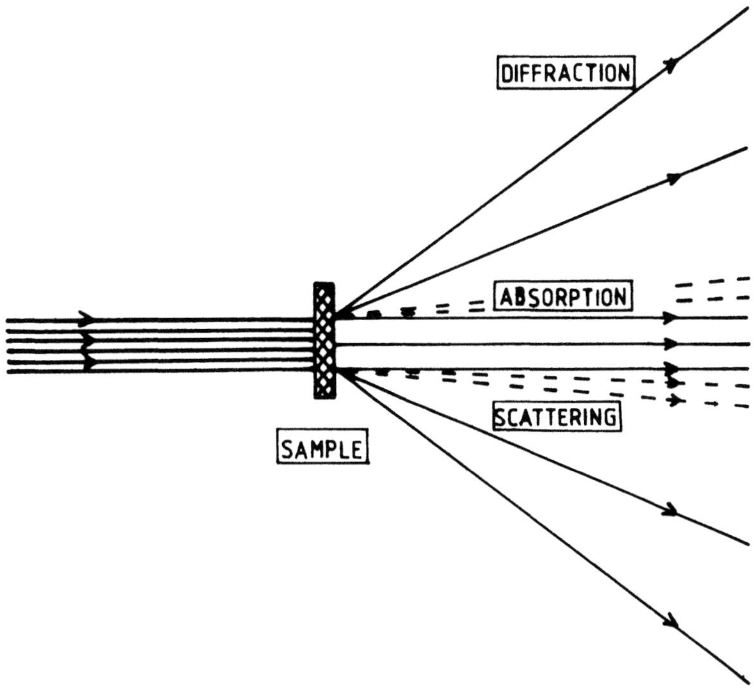

Figure 2 Schematic illustration of the ways in which X-rays interact with matter.

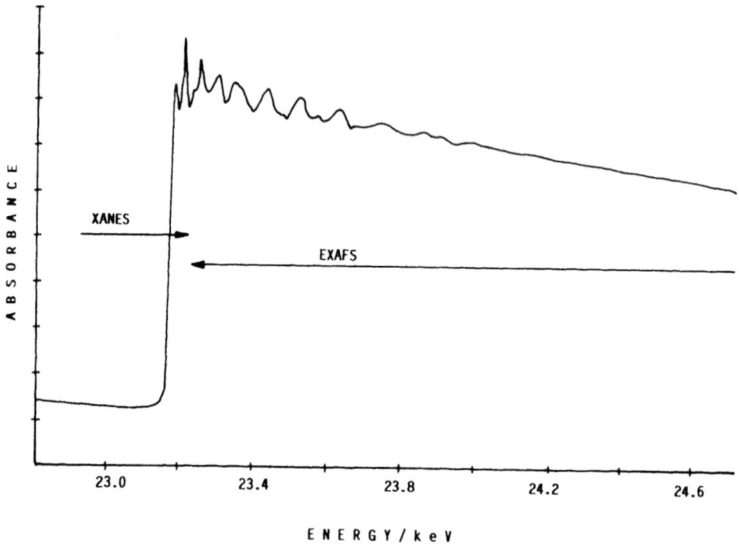

Figure 3. The X-ray absorption spectrum of a rhodium foil, indicating the near-edge region, (XANES or NEXAFS) and the extended fine structure or EXAFS region.

EXAFS provides information on the ***local environment*** of a ***specific element***

> ***Interatomic distances*** can be determined with an accuracy of *at best* **± 0.01Å**
>
> ***Coordination numbers*** can be determined to *at best* **± 10%,** (often much worse)
>
> ***Debye Waller factors:*** NB These are different to those measured in X-ray diffraction

EXAFS is therefore **less accurate** than X-ray diffraction and is most useful for poorly ordered samples

X-ray absorption, and it will be seen that this can be advantageous.

X-ray absorption spectroscopy has made the biggest impact in catalysis to date, and will be the main focus of this chapter. Of most interest are the absorption edges and their associated fine structure. The absorption coefficient of a solid sample, μ, is defined by:

$$\mu = \mathrm{Ln}(I_0/I) \qquad (1)$$

where I_0 and I are respectively the incident and transmitted X-ray flux at the wavelength of interest. μ varies smoothly with photon energy until this becomes just sufficient to excite electrons from a given inner shell to the Fermi level of the solid. When this happens there is a sudden, rapid increase in the X-ray absorption and the jump in the absorption spectrum, which is referred to as the absorption edge. As can be seen from Fig. 3, there is often considerable *fine structure* in the absorption spectrum, and this may extend for more than 1 keV above the edge. This fine structure may conveniently be divided into two regimes, the *near edge structure* or XANES is defined as the region extending up to ca 50 eV above the absorption edge, while oscillations beyond this are referred to as the *extended fine structure* or EXAFS. The physics of X-ray absorption is described elsewhere [7-9], and will not be presented in any detail here. X-ray absorption produces a photoelectron and the shape of the absorption spectrum reflects the interaction of this electron with its local environment. XANES provides fingerprint information from which the oxidation state and coordination environment of the absorbing element may be identified in favourable cases [10]. XANES has close parallels with electronic, uv/visible spectroscopy. The photoelectron undergoes extensive multiple scattering so that *ab initio* calculation of the XANES spectral region is normally not possible. Above the XANES range interference effects are dominant: as the X-ray photon energy changes so does the wavelength of the photoelectron and the result is an oscillatory change in the absorption cross section that is determined by the structural environment of the absorbing atom. EXAFS is thus a structural technique and the information that can be obtained is summarised in the box on the previous page. Because the method is much less accurate than X-ray diffraction, EXAFS is at its best when studying samples containing only short range order, where XRD is not useful. The study of small particles in catalysts has been and continues to be one of the most important fields of application of the EXAFS method.

EXAFS is at its most powerful when studying metal containing catalysts. Metals are strong absorbers of X-rays and metal particles are usually highly ordered and have high coordination numbers. Other solids of catalytic interest, such as oxides, zeolites and sulphides, are often much less ordered, for example as a result of Jahn Teller distortion, and coordination numbers are usually lower. The application of EXAFS will be illustrated by a wide range of examples, including the structure, reactivity and alloying of metal particles on conventional supports and in zeolites.

The synchrotron source is ideal for X-ray absorption spectroscopy. Bremsstrahlung radiation from a laboratory X-ray source may be used, but is much less satisfactory. The experiment is otherwise simple and is described elsewhere, together with details of data analysis [7,9]. The weak nature of the interaction between X-rays and matter means that the method is well adapted for *in situ* studies [11,12] The X-rays are introduced to the experimental environment through windows fabricated from a low atomic weight material such as beryllium, boron nitride or a suitable polymer. Experiments up to at least 20 bar pressure are possible [see 12 and discussion of this paper reported by Ponec].

4. Applications in Catalysis

4.1. STRUCTURAL CHARACTERISATION STUDIES

X-ray diffraction studies can usefully employ synchrotron radiation where only very small single crystals are available, a common situation for microporous solids such as zeolites and zeotypes. Interpretable data can be obtained from crystals with a cube side of only 20 μ, compared with a minimum requirement of 50 μ using conventional X-ray sources [13]. Examples include the structure of the zeolite *theta-1* [14] and the cobalt substituted aluminium phosphate CoAPO-21 [15]. Although similar results can be obtained from powder diffraction studies [16], particularly using profile analysis techniques such as that developed by Rietveld [17], the single crystal method is expected to give more reliable results.

A study of zeolite structure by Dooryhee et al. provides an excellent example of the use together of high resolution powder X-ray diffraction and EXAFS [18]. Zeolites typically contain alkali metal cations, to supply the charge deficit caused by the substitution of silicon by aluminium. These ions can be exchanged for transition metal ions which, on removal of zeolitic water become located at fixed, crystallographically distinct sites in the zeolite framework. Dooryhee et al. examined the environment of the nickel ion in the S1 site in zeolite Y, where it is octahedrally coordinated to 3 oxygen atoms from each of two parallel 6 rings. Using both EXAFS and synchrotron based X-ray diffraction they showed that the 6-ring undergoes considerable distortion,

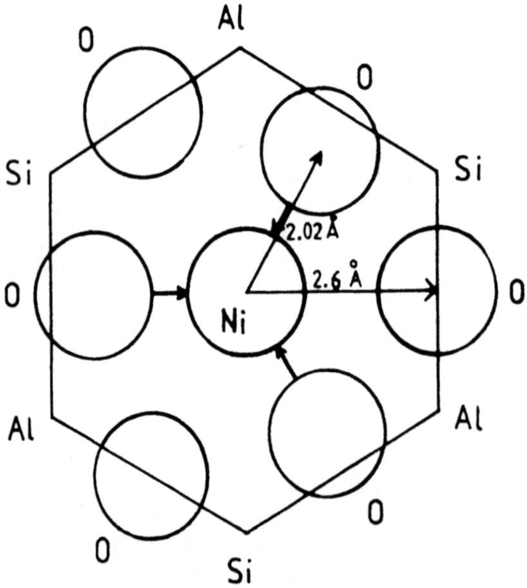

Figure 4. The distortion introduced into a 6-ring of zeolite Y by bonding to Ni^{2+}. The six ring is in the plane of the paper with the nickel ion above the plane. Nickel is also coordinated to three oxygen atoms of another sic ring (not shown). The Ni-O distance is that deduced from the EXAFS measurement. (*From Ref. 18, with permission*).

as shown in Fig. 4, with three of the oxygens in each 6-ring being pulled inward to the nickel ion. This study shows the advantage of EXAFS in focusing on the local structural environment around the absorber. EXAFS provides a reliable estimate of the nearest neighbour, short Ni - O bonding distance of 2.02 ± 0.02 Å, while from the X-ray diffraction data it was possible to determine only the average of short and long Ni - O distances, 2.25 Å.

EXAFS has been very widely used to study small metal particles in supported catalysts. Attention has focused on their size, structure and shape and on changes on interatomic distances compared to the bulk metal. Estimates of particle size may be obtained from the nearest neighbour coordination number (CN). This has the value 12 in the bulk, face-centred cubic structure and decreases with particle size: a 13 atom fcc particle has a nearest neighbour coordination number of 5.5. The variation in CN is most sensitive for very small particles, so EXAFS is at its most useful studying particles of < 30 Å in diameter. Size estimates must be treated with some care, since the accuracy with which CNs can be determined is poor (see above), and EXAFS provides only an **average** value. Where EXAFS estimates have been compared with more detailed studies using electron microscopy, however, agreement is reasonable [19].

EXAFS may be used to probe the crystal structure adopted by small metal particles, through studying the *non-nearest neighbour* distances in the RDF. Such studies are of interest as a large number of theoretical investigations now suggest that the preferred structure for small particles is *icosahedral* rather than close packed [20]. The icosahedron is a structure with five fold symmetry, which exposes hexagonally close packed faces, and so minimises surface energy compared to close packed structures. Icosahedral structures are not space filling, and so become less stable as their size increases. Calculations estimate that close packing becomes preferred above ca 55 atoms in the particle, corresponding to a diameter of typically 10-12 Å. We have studied small particles in supported rhodium [21] and platinum catalysts [19] and have shown that these adopt the fcc structure, since the $\sqrt{2}\mathbf{a}$ and $\sqrt{3}\mathbf{a}$ peaks expected from the fcc structure are indeed present (**a** is the nearest neighbour distance). Statistical significance tests [22] have been used to show that these features, which make a relatively small contribution to the spectrum, are meaningfully present. In contrast, Russian workers have claimed that EXAFS studies demonstrate the presence of icosahedral particles in giant palladium clusters, containing ca 555 atoms [23].

EXAFS is not a very good method for studying particle shape, although it was pointed out many years ago that the ratios of non-nearest neighbour CNs to the nearest neighbour CN vary with particle shape [24] and this has been quantified by Greegor et al. [25].

EXAFS studies consistently suggest that the nearest neighbour metal-metal interatomic distance is smaller in small particles than in the bulk, with contractions in the range 0.03 - 0.1 Å [19]. Theoretical analysis by Norskov et al. [26] has alternatively suggested that their is no reduction in interatomic distance, but that the observation is caused by *anharmonicity* in the metal-metal bonding. Almost all EXAFS interpretation methods assume that the vibration of near neighbours is well described by a harmonic oscillator and it is recognised that departure from this situation poses problems in analysis. It is nonetheless not clear whether anharmonicity is sufficient to explain the full extent of the contractions in interatomic distances that have been reported.

The power of combining X-ray absorption and diffraction has been demonstrated in a study of catalyst genesis by Couves et al. [27]. Catalysts for the industrial synthesis of methanol contain copper and zinc oxide, are prepared by coprecipitation.

The result after drying is a poorly ordered mixed phase resembling the minerals aurichalcite [28] or gerhardite [29]. When this material is heated in air or aged in solution [28] it decomposes into zinc oxide and copper oxide. Couves et al. have used XRD to monitor the phase changes during the decomposition of aurichalcite. EXAFS, performed simultaneously has shown that the local structural and electronic environment of Cu^{2+} is unchanged, during heating to 720 K, despite marked changes in the crystallinity of the solid phases present.

4.1.1. *Metal - Support Interactions* . It was originally hoped that X-ray absorption, and in particular EXAFS would provide valuable information on the structure of the metal/oxide interface in supported catalysts. Many EXAFS studies of supported metal catalysts have indeed detected metal - oxygen distances. The bond lengths observed are usually about 2Å, similar to those found in metal oxides. Our study of platinum in H-ZSM-5 is typical, since Pt-O distances of 2.02 $\pm$ 0.03 Å were found [19]. Platinum - oxygen coordination numbers of ca 2 were deduced and these were present even after the catalyst had been thoroughly reduced, at up to 800 K. It appears that the metal particles, average diameter 12 Å, are encapsulated by the distorted zeolite lattice and are in close contact with framework oxygen atoms.

Van Zon, Koningsberger et al. have studied the structure of iridium catalysts on γ-Al_2O_3, prepared by decomposition of $Ir_4(CO)_{12}$. [30]. Very small iridium particles are found, with an Ir-Ir coordination number of 4.7 implying about 10 atoms per particle. The iridium-oxygen coordination number reported is 2.9, which it is argued reflects an epitaxial relationship with the (111) plane of the support. The Ir-O distance is found to be 2.58 Å, considerably longer than that in any bulk oxide. Long metal oxygen distances are a feature of the results from this group which, despite considerable effort and care, have not been confirmed by other workers [31].

It therefore seems unlikely that X-ray absorption will provide much new information on the structural interaction between small metal particles and their support. It may be that progress could be made by considering all of the available data together, since it seems that a new approach to the problem of the structure of the metal support interface is required. Is there any correlation between the observation of metal - oxygen distances and the degree of crystallographic mismatch between the metal and the support? Is a fractal analysis of the structural problem appropriate?

There have been a number of attempts to study the *electronic* state of the catalyst by examining the shape of the absorption edge and in particular of the 'white line,' which is a characteristic feature of many edges. Lytle [32] has used this approach to study the reduction of rhenium on alumina catalysts, providing an early example of *time resolved* EXAFS. The size of the white line decreased steadily on reduction, starting with ReO^{4-} and ending, after reduction at 1023 K, with a line shape very close to that of metallic rhenium. Detailed analysis of the white line intensity has also been considered as a possible was to probe the vexed problem of electronic charge transfer between the support and small metal particles. Koningsberger has, however, recently been very pessimistic that unambiguous information can be obtained from white line analysis [33].

4.1.2. *Alloys* . The potential of EXAFS for studying alloy and bimetallic catalysts has been clear since the classic studies of Sinfelt et al, many of which are described in his book [34]. A range of copper containing bimetallic systems were studied initially, and it was shown, both by analysis of EXAFS data, and by changes in catalyst structure on exposure to oxygen, that copper forms an external 'skin' on a core of the more noble catalyst component [35]. Via, Sinfelt et al. have recently shown that more reliable analysis of their data can be obtained if the EXAFS data from both alloying components are analysed simultaneously, with a program which allows the constraint of parameters within physically acceptable limits [36]. Despite this impressive demonstration of the potential of EXAFS, there have been relatively few other alloying studies, particularly of the very small particles. The particle diameter is still small, at ca 12 Å, but is large enough to distort the channel structure of the ZSM-5 zeolite support.

We have been interested in Pt/Cr/H-ZSM-5 catalysts which are of potential interest for gasoline reforming [10,37]. Addition of chromium reduces the activity of these materials in propane hydrogenolysis by about an order of magnitude, as shown in Fig. 5. On reduction in hydrogen at 623 K platinum particles are formed and XPS shows the presence only of the chromium +3 oxidation state. When the catalyst is reduced at the temperature of gasoline reforming, 823 K, XPS indicates the appearance of some metallic chromium. Changes in the Pt L_{III} EXAFS also occur, which we have shown are due to a new contribution from Pt - Cr bonding with an interatomic distance of 2.65. This interatomic distance is characteristic of a Pt - Cr alloy, and its contribution to the EXAFS has been validated by statistical techniques [22] with a confidence level that exceeds 99%.

EXAFS and XPS together thus show the formation of a Pt/Cr alloy, but the reason for the decrease in hydrogenolysis activity is unclear. Although no results are available on the hydrogenolysis activity of chromium, comparison with that of iron, cobalt and nickel suggests that it should be much higher than that of platinum [38]. Relying on evidence from surface science studies of similar alloys [39] we have proposed the 'shell' model for the alloy particles shown in Fig. 6. The outer monolayer is pure platinum, with a weakened chemisorption capacity as a result of chromium enrichment in the next layer of the structure. The catalytic properties of the catalyst are consistent with such a structural and electronic proposal.

4.1.3. *Promoters and Poisons* . X-ray absorption spectroscopy should, in principle be able to make major contributions to our understanding of the structural changes associated with the presence of catalyst promoters and poisons. One example where there has been considerable progress concerns the location and environment of the cobalt promoter in $Co/Mo/Al_2O_3$ catalysts for hydrodesulphurisation. There appears to be a consensus that cobalt atoms are located at edge sites at the surface of MoS_2 slabs [40].

There continues to be considerable interest in the role of alkali metals as catalytic promoters for a wide range of reactions. Despite this no EXAFS studies have been reported on the structural environment of alkali metals in catalysts, although there have been successful adsorption studies on single crystals, by *surface* EXAFS [41]. In our group we have made a number of attempts to determine the alkali metal environment in carbon supported catalysts, but the EXAFS has been too weak to interpret reliably despite making measurements at 77 K. It seems that the promoter environment lacks even short range order.

EXAFS should be a good method to study the location of catalytic poisons and we plan to study sulphur on supported catalysts in the near future. A wide range of SEXAFS and NEXAFS studies on single crystals prove the viability of the approach [42,43].

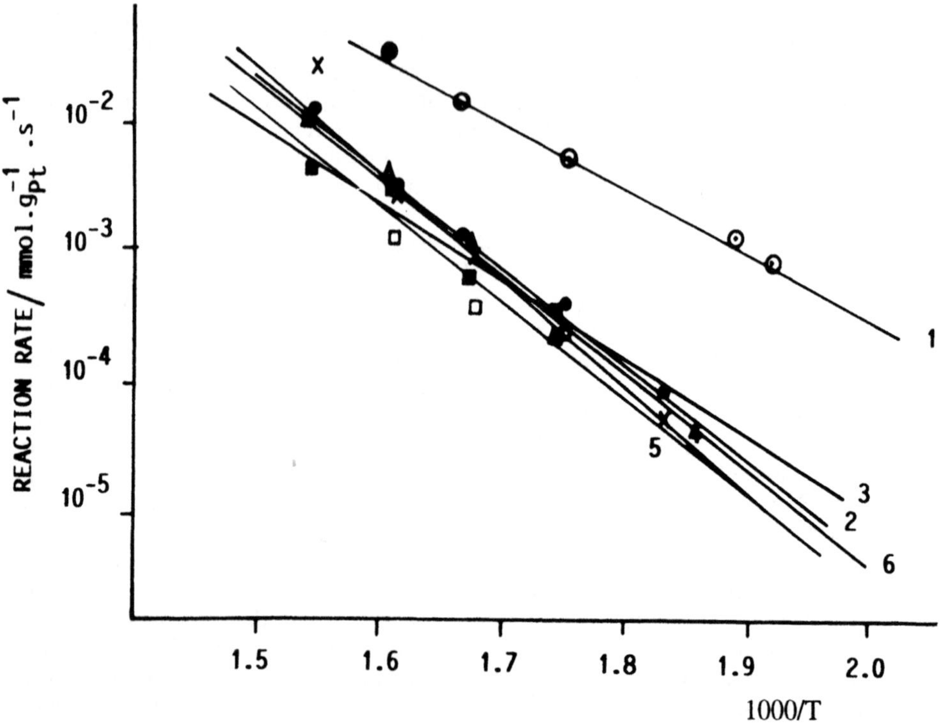

Figure 5. Arrhenius plots for propane hydrogenolysis over Pt/Cr/H-ZSM-5 catalysts. Catalyst No 1 does not contain chromium while the chromium content increases from 0.5% to 2.25% for catalysts 2-6.

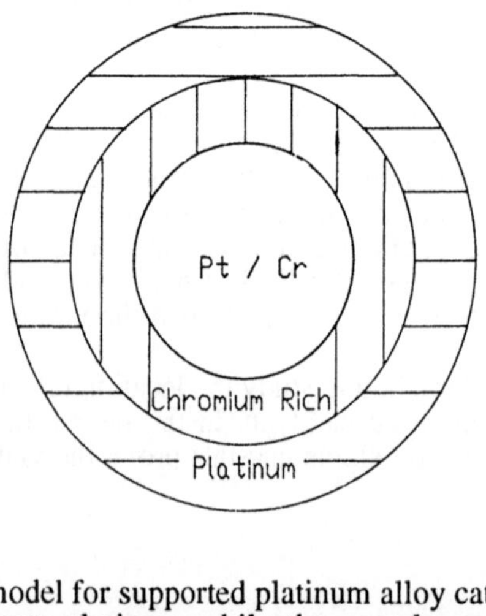

Figure 6. The 'shell' model for supported platinum alloy catalysts. The outer monolayer is suggested to be pure platinum while the next layer is enriched in the alloying component, in this case chromium. The core has the bulk composition.

4.2. CATALYTIC REACTIVITY STUDIES

4.2.1.*Chemisorption.* X-ray absorption spectroscopy has no inherent surface sensitivity, its successful application in catalysis relying on the circumstance that the surface often comprises a significant fraction of the whole. Rayment's group have used an interesting means of imparting surface sensitivity to the EXAFS measurement, through electron yield detection [44]. By monitoring electrons ejected from the solid, rather than X-ray absorption, a high degree of surface sensitivity is obtained. The relationship between normal EXAFS and that using electron yield detection parallels that between X-ray fluorescence analysis, which is not surface sensitive, and X-ray photoelectron spectroscopy, where most of the detected signal originates from the outermost 10 Å of the solid.

The experimental arrangement used is shown in Fig. 7. The authors have studied supported copper catalysts, where the copper particles are ca 100 Å in diameter. This size is too large for the transmission EXAFS experiment to provide reliable information on surface structure or composition since the dispersion ic ca 10%. Using electron yield detection the authors area able to report changes in the Cu K shell radial distribution function, resulting from heating metallic copper particles in CO_2 at 523 K, as shown in Fig. 8, and claim that the presence of ca 0.5 monolayer of oxygen on their copper particles can be detected.

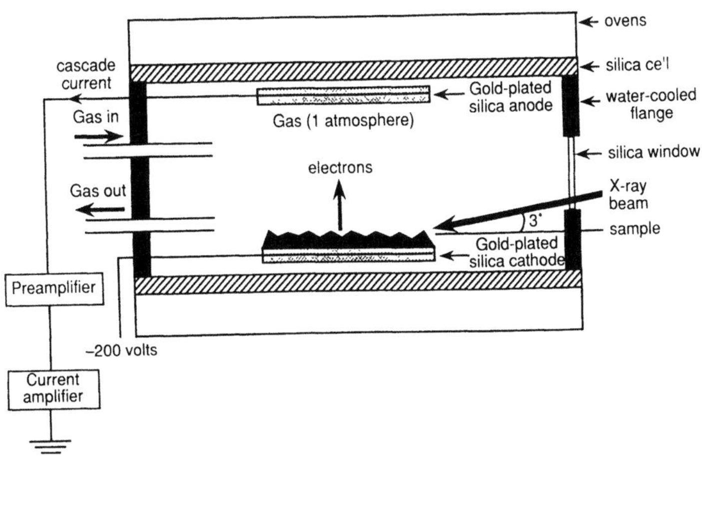

Figure 7. The experimental arrangement used to enhance surface sensitivity in EXAFS measurements using electron yield detection. (*From Reference 44, with permission.*)

For very small metal particles, the signature of any chemisorbed species may be detectable in the EXAFS spectrum. It will often be weak due to the low backscattering power of low Z adsorbates, but it should be visible in favourable cases and with good data. In our studies of rhodium/alumina catalysts interacting with synthesis gas, we have noted evidence of CO adsorption in the Rh EXAFS. The Rh-C bond length is

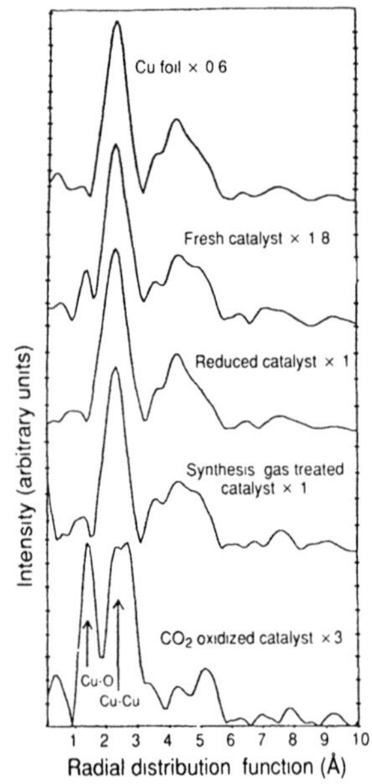

Figure 8. Radial distribution functions obtained from copper catalysts treated in various environments. (*From Reference 44, with permission.*)

1.87 ± 0.04 Å, typical of that found in metal carbonyls, and the Rh-C coordination number is 2.0 ± 0.4. Evidence of Rh - O bonding is less clear, since it is masked by the dominant feature in the spectrum, due to the Rh - Rh bonding distance, at ca 2.7 Å.

4.2.2. *Catalyst Reactivity* . Single crystal studies have demonstrated the importance of adsorbate induced restructuring of the active surface in catalysis (see G.A. Somorjai, pp3, this volume). It is thus important that studies of high area catalysts turn their attention to the dynamic structural changes that accompany reactivity. EXAFS is the most useful and direct technique for probing structural changes induced in small metal particles by adsorbates. The study of the decomposition of small rhodium particles by exposure to CO [45] represented an important early triumph for EXAFS in catalytic studies. Starting from rhodium trichloride solution, rhodium particles containing on average less than 20 atoms were formed on an alumina support after reduction in hydrogen. On exposure to CO at room temperature these undergo almost complete decomposition, to give species often represented as *gem*$Rh(I)(CO)_2$.

We have reexamined this interaction and have shown that the structural situation is more complexthan originally supposed. It is well known that chlorine from the precursor is retained on the γAl_2O_3 catalyst support [45,46]. Detailed analysis of the Rh K edge EXAFS, shown in Fig. 9 indicates that chlorine is associated with and stabilises the Rh(I) *gem* dicarbonyl species. In addition to features due to Rh-C, Rh-O and Rh-Rh bonding, there is a contribution from a Rh-Cl shell with a coordination number of 0.7 ± 0.2 and an interatomic distance of 2.34 ± 0.03 Å, close to the Rh-Cl distance in the crystalline dimer $Rh_2Cl_2(CO)_4$, 2.33 Å. In confirmation of the role of chlorine in stabilising the gem-dicarbonyl species at the catalyst surface, we have further shown that small particles prepared from a rhodium nitrate precursor have

A)

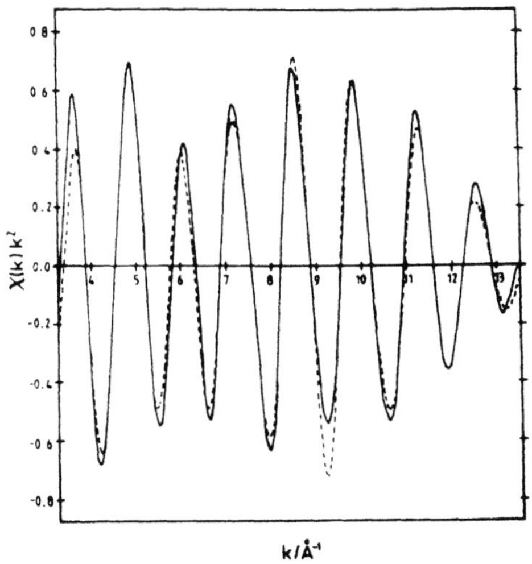
0.8
0.6
0.4
0.2
0.0
-0.2
-0.4
-0.6
-0.8
χ(k).k²
4
5
6
7
8
9
10
11
12
13
k/Å⁻¹

B)

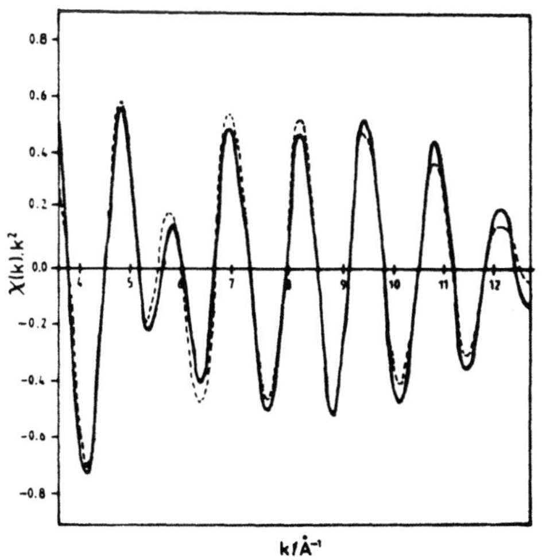
0.8
0.6
0.4
0.2
0.0
-0.2
-0.4
-0.6
-0.8
χ(k).k²
4
5
6
7
8
9
10
11
12
k/Å⁻¹

C)

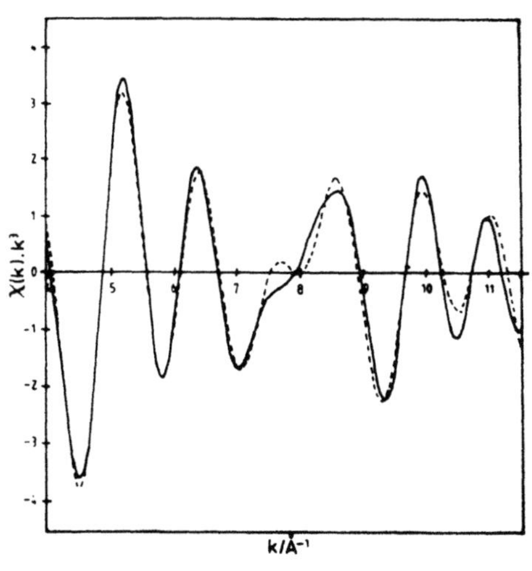
3
2
1
0
-1
-2
-3
-4
χ(k).k³
5
6
7
8
9
10
11
k/Å⁻¹

D)

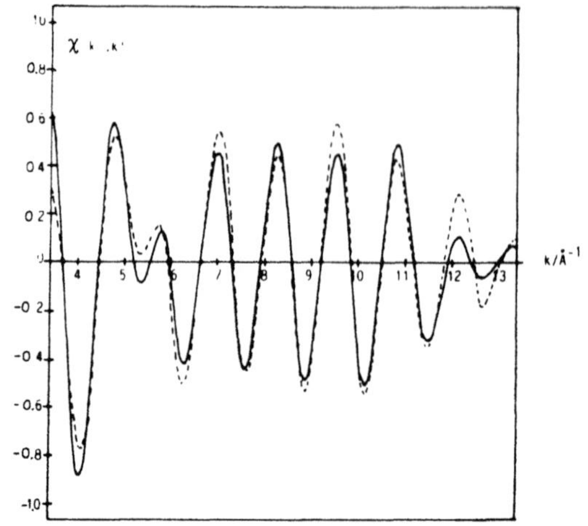

Figure 9. Rhodium K shell EXAFS. A,B) Catalysts reduced in hydrogen at 493 K; solid lines, experiment; dashed lines calculated spectra. A) Catalyst prepared from rhodium chloride; B) catalyst prepared from rhodium nitrate. C,D) Catalysts prepared respectively as in A,B, but after exposure to CO at room temperature. (*From Reference 37, with permission.*)

much greater stability in carbon monoxide that those prepared from the chloride. As shown in Fig. 10d, the rhodium EXAFS spectrum of this chlorine free catalyst is almost unaltered on exposure to CO at room temperature [47].

Catalyst characterisation should not be regarded as an end in itself, but rather as an aid to our understanding of reactivity. It is therefore important to ask if the changes in catalyst structure observed in CO influence catalytic reactivity, and if the chlorine from the catalyst precursor influences reactivity. Rhodium catalysts are of particular interest as they are uniquely able to form C_2 oxygenates, such as ethanol and acetaldehyde from synthesis gas, (a mixture of CO and hydrogen). The selectivity to these desired products is not high, rarely exceeding 50%, with hydrocarbons, and in particular methane forming the balance. Our EXAFS studies show that the disruptive effect of carbon monoxide on rhodium particles is reversed by the presence of synthesis gas. Metallic particles are reformed, in a process whose mechanism is not understood at all! The catalytic reactivity in synthesis gas is thus that of metal particles and not of metal carbonyl species. Chloride from the rhodium precursor does influence the selectivity of the reaction, enhancing the yield of oxygenates. The effect is significant but not dramatic, with the chlorine containing catalysts having a selectivity of 48% to oxygenates (including methanol), compared to 36% of oxygenates for catalysts prepared from the nitrate [48].

Attention is also drawn to the interesting 'ship in a bottle' phenomenon reported by Sachtler et al., where *growth* of palladium particles in zeolites is induced by CO [49-51]. Palladium was exchanged into sodium-Y zeolite from an 0.01 M solution of $Pd(NH_3)_4(NO_3)_2$ to give 5% loading. Palladium EXAFS data were analysed after reduction in hydrogen at 623 K and the Pd-Pd coordination found to be 4 ($\pm$ 10%), indicative of particles containing 7-8 atoms. On exposure to CO at room temperature the palladium particles grew to about double the size, with a Pd-Pd coordination number of 6 ($\pm$ 10%). It is argued that $Pd_{13}(CO)_x$ clusters are formed in the supercages of the Y zeolite, as pictured schematically in Fig. 10. The value of x is not known, but the authors suggest an analogy with the anionic cluster family $[Rh_{13}(CO)_{24}H_{5-n}]^{n-}$. The formation of the cluster is referred to as a 'ship in a bottle' synthesis since no conventional palladium carbonyl clusters are known, and the zeolite matrix must therefore play a stabilising role.

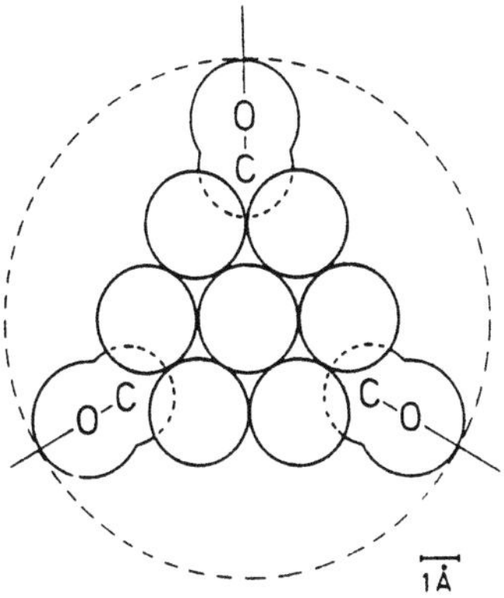

Figure 10. Proposed location of a $Pd_{13}(CO)_n$ cluster inside the supercage of zeolite Y. (*From Reference 49, with permission.*)

5. The Future

The application of synchrotron radiation in catalysis, particularly for X-ray absorption, is now established as a routine technique in the repertoire of the catalytic chemist. EXAFS will continue to be applied to a variety of structural problems and, of the topics discussed above, studies of *reactivity, chemisorption* and *alloy catalysts* offer the most potential for the future. The following points will help the method to be exploited to best advantage.

1) Experimental developments Data interpretation in EXAFS is usually limited by the available signal to noise ratio, so advances such as improved, brighter synchrotron sources, focusing mirrors and high sensitivity solid state detectors are very important. The availability of more intense sources will allow time resolved methods, and techniques such as total electron yield detection to make important contributions.
2) Better *in situ* capabilities. It is important to be able operate above atmospheric pressure and to make parallel measurements of catalyst activity and selectivity.
3) Greater synchrotron access. Although synchrotron time is scarce and expensive, it is clear that substantial allocations of beam time are necessary for successful *in situ* catalytic studies. Increased access will allow new challenges, for example catalyst ageing, to be tackled.
4) **CREDIBILITY** The full usefulness of the EXAFS method will not be realised until statistical assessments of experimental error and 'goodness of fit' become routine practise. Such analysis is relatively simple providing primary data are used [22]. Fourier transformation methods, which are unfortunately very widely employed, unfortunately destroy the statistical independence of data points. The straightforward application of statistical methods is thus very difficult.

6. References

1) The earliest theoretical studies appear to be due to A. Liénard, L'Eclairage Electrique, 16 (1898) 5.
2) H. Winick and S. Doniach, in H. Winick and S. Doniach (Eds.), *Synchrotron Radiation Research*, Plenum New York, 1980, pp 1.
3) *Synchrotron Radiation News*, published six times per year by Gordon and Breach Science Publishers, New York.
4) C.J. Hirschmugl, G.P. Williams, F.M. Hoffmann and Y.J. Chabal, Phys. Rev. Lett., 65 (1990) 480.
5) see e.g. J.C. Bertolini and B. Tardy, Surface Sci., 102 (1981) 131.
6) see e.g. G. Bergeret, T.M. Tri and P. Gallezot, J. Phys. Chem., 87 (1983) 1160.
7) For a full review see D.C. Koningsberger and R. Prins, (Eds.), *X-ray Absorption, Principles, Applications and Techniques*, J. Wiley, Chichester, 1988.
8) A complete description of the theory of EXAFS is given by P.A. Lee and J.B. Pendry, Phys. Rev., B11 (1975) 2795.
9) For a discussion specifically related to catalysis see R.W. Joyner, in H.H. Brongersma and R.A. van Santen, (Eds.), *Fundamental Aspects of Heterogeneous Catalysis Studied by Particle Beams, Plenum Press,* New York, 1991 pp 191.
10) R.W. Joyner, P. Johnston, G. Tuleouva and E.S. Shpiro, J. Catal., *in press* April 1993.
11) R.W. Joyner, Catal. Today, 9 (1991) 161.
12) G.N. Greaves, P.R. Moore, R.W. Joyner, and M.A. Morris, Vacuum, 38 (1988) 929.
13) M.M. Harding, personal communication.
14) M.Z. Papiz, S.J. Andrews, A.M. Damas, M.M. Harding and R.M. Highcock, Acta. Cryst., C46 (1990) 172.
15) G.M.T. Cheetham, M.M. Harding, P.J. Rizkallah, V. Kaucic and N. Rajic, Acta. Cryst., C47 (1991) 1361.
16) The powder pattern of the zeolite discussed in [14] was analysed by R.M. Highcock, G.W. Smith and D. Wood, Acta. Cryst., C41 (1985) 1391.
17) H.M. Rietveld, J. Appl. Crystallog., 2 (1969) 65.
18) E. Dooryhee, G. N. Greaves, A.T. Steel, R.P. Townsend, S.W. Carr, J.M. Thomas and C.R.A. Catlow, Faraday Disc. Chem. Soc., 89, (1990) 119.
19) E.S. Shpiro, R.W. Joyner, K.M. Minachev and P.D.A. Pudney, J. Catal., 127 (1991) 366.
20) see e.g. J.J. Burton, Catal. Rev., 9 (1974) 209. and S.N. Khanna, J.P. Bucher, J. Buttet and F. Cyrot-Lackmann, Surf. Sci., 127 (1983) 165.
21) P. Johnston, R.W. Joyner, and P.D.A. Pudney, J. Phys. Condensed Matter, 1, (1989) SB171.
22) R.W. Joyner, K.J. Martin and P.Meehan, J. Phys. C., Solid State Phys., 20 (1987) 4005.
23) M.N. Vargaftik, I.Y. Moiseev, D.I. Kochubey and K.I. Zamaraev Faraday Disc. Chem. Soc., 92 (1991) 13.
24) R.W. Joyner, in J.M. Thomas and R.M. Lambert, (Eds.), *Characterisation of Catalysts*, J. Wiley, Chichester, (1980) pp 237.
25) R.B. Greegor and F. W. Lytle, J. Catal., 63 (1980) 476.
26) L.B. Hansen, P. Stoltze, J.K Norskov, B.S. Clausen and W. Niemann, J. Phys. Chem., 64 (1990) 3155.
27) J.W. Couves, J.M. Thomas, D. Waller, R.H. Jones, A.J. Dent, G.E. Derbyshire and G.N. Greaves, Nature, 354 (1991) 465.

28) D. Waller, D. Stirling, F.S. Stone and M.S. Spencer, Faraday Disc. Chem. Soc., 87 (1989) 107.
29) R.W. Joyner, F. King, M.A. Thomas and G. Roberts, Catal. Today, 10 (1991) 417.
30) F.B.M. Van Zon, G. Visser and D.C. Koningsberger, in M.J. Phillips and M. Ternan, (Eds.) *Proc. 9th Intern. Congr. on Catlysis*, Chemical Inst. of Canada, Ottawa, 1988, pp 1386.
31) F.W. Lytle, Comments during plenary lecture at 9th Intern. Congr. on Catlysis, Calgary, July 1988; also this author, unpublished results.
32) F.W. Lytle, in M.J. Phillips and M. Ternan, (Eds.) *Proc. 9th Intern. Congr. on Catlysis,* Chemical Inst. of Canada, Ottawa, 1988, Volume 5, pp 54.
33) D.C. Koningsberger, in discussion of paper by M. Vaarkamp, J.T. Miller, F.S. Modica, G.S. Lane and D.C. Koningsberger, Preprints, 10th International Congress on Catlysis, Institute of Isotopes, Budapest, 1992, Paper O51.
34) J.H. Sinfelt, *Bimetallic Catalysts: Discoveries, Concepts and Applications*, J. Wiley, New York, 1983.
35) J.H. Sinfelt, G.H. Via and F.W. Lytle, J. Chem. Phys., 72 (1980) 4832.
36) G.H. Via, K.F. Drake Jr., G. Meitzner, F.W. Lytle and J.H. Sinfelt, Catal. Lett., 5 (1990) 25.
37) E,S. Shpiro, R.W. Joyner, P. Johnston and G. Tuleouva, J. Catal., *in press* April 1993.
38) J.H. Sinfelt, Catal. Rev., 3 (1969) 175.
39) see e.g. Y. Gauthier, Y. Joly, R. Baudoing and J. Rundgren, Phys. Rev., B35 (1987) 6216.
40) see e.g. H. Topsoe in J.P. Bonelle, B. Delmon and E.G. Derouane (Eds.), *Surface properties and catalysis by non-metals*, Reidel, Dordrecht, 1983; also B.S. Clausen, B. Lengeler and B.S. Rasmussen, J. Phys. Chem., 89 (1985) 2319; and B.S. Clausen, B. Lengeler and H. Topsoe, Polyhedron, 5 (1976) 199.
41) see e.g. G.M. Lamble, R.S. Brooks, D.A. King and D. Norman, Phys. Rev. Lett., 61 (1988) 1112.
42) see e.g. J. Stohr, R. Jaeger and S. Brennan, Surface Sci., 117 (1982) 503.
43) see e.g. D.W. Moon, S.L. Bernasek, J.P. Lu, J.L. Gland and D.J. Dwyer, Surface Sci., 184 (1987) 90.
44) G.D. Moggridge, T. Rayment, R.M. Ormerod, M.A. Morris and R.M. Lambert Nature, 358, (1992) 658.
45) H.F.T. Van't Blik, J.B.A.D. van Zon, T. Huizinga, J.C. Vic, D.C. Koningsberger and R. Prins, J. Phys. Chem., 87, (1983) 2264.
46) P. Johnston, R.W. Joyner, P.D.A. Pudney, E.S. Shpiro and B.P. Williams, Faraday. Disc. Chem. Soc., 89 (1990) 91.
47) P. Johnston and R.W. Joyner, to appear as a *Communication* in JCS Faraday Trans., 89 (1993) issue 5.
48) P. Johnston and R.W. Joyner, in T.J. Dines, C.H. Rochester and J. Thompson, (Eds.), *Catalysis andSurface Characterisation*, Royal Soc. of Chem., London, (1992) pp 51.
49) L.-L. Sheu, H. Knozinger and W.M.H. Sachtler, Catal.Lett., 2 (1989) 129.
50) Z. Zhang, H. Chen. L.-L. Sheu and W.M.H. Sachtler, J. Catal., 127 (1991) 213.
51) W.M.H. Sachtler, F.A.P. Cavalcanti and Z. Zhang, Catal. Lett., 9 (1991) 261.

INFRARED SPECTROSCOPY AS A PROBE OF SURFACE ACIDITY

H. KNÖZINGER
Institut für Physikalische Chemie
Universität München
8000 München 2
Germany

ABSTRACT. The application of infrared spectroscopy for the study of surface acidity requires suitable probe molecules. A classification of such probe molecules is given and their suitablility for discrimination between protonic (Bronsted) and non-protonic (Lewis) acidity, and for the determination of surface acidity in terms of acid strength and site density is discussed. Selected examples are described with special emphasis being payed to acidic properties of simple binary oxides, supported oxides and zeolites.

1. INTRODUCTION

Many technologically important reactions such as catalytic cracking and isomerization are acid catalyzed, the catalysts preferably being solid acids which allow the application of fixed or fluidized bed reactors [1,2]. Easy separation of products and reactants is possible when batch reactors are to be used and corrosion problems are minimized. A detailed understanding of solid acid-catalyzed reactions requires knowledge on the surface acidity in terms of quality (protonic vs non-protonic), acid strength and acid site density. Attempts to characterize the acid properties of solid catalysts have therefore been made very frequently during the past decades and a number of review articles and books on this subject have appeared [3-8]. Acidity concepts which apply to solution chemistry have been introduced into surface chemistry of solids, although "acid-base equilibria" in solution and on surfaces can frequently not be compared. This is partially due to the fact that solvation phenomena, which play an important role in acid-base equilibira in solution, will not occur on surfaces when basic probes (or reactants) are adsorbed from the gas phase. Moreover, most of the classical techniques including titration methods, determination of adsorption isotherms and isobars, and calorimetric measurements do not permit a discriminiation between acid sites of different quality, i.e. protonic (Brönsted) and non-protonic (Lewis) acid sites. Typical solid acid catalysts are binary or ternary amorphous or crystalline oxides which frequently expose both hydroxyl groups (protonic sites) and coordinatively unsaturated (cus) cations (non-protonic sites) on their surfaces simultaneously. The relative densities of these two types of acid centers depends on the state of hydroxylation or hydration of the particular material.

For a discrimination between these two types of centers, the nature of the interactions between basic probe molecules and oxide surfaces must be analyzed at a molecular level. Infrared spectroscopy has a high potential for investigations of intermolecular interactions and it has

R. W. Joyner and R. A. van Santen (eds.), Elementary Reaction Steps in Heterogeneous Catalysis, 267–285.

proved to be by far the most powerful method to study acid-base interactions on solid surfaces, namely oxide surfaces [9-13]. Infrared spectroscopy permits discriminations to be made between protonic and non-protonic acid sites based on the effects induced by the acid-base interaction in the spectral features of the adsorbed basic probe molecule and of surface acid groups, such as O-H oscillators. It also provides the principle potential of determining the density of acid centers of either quality by measuring integrated intensities of characteristic vibrational modes and it permits an estimate of quantities related to the interaction energy (and hence, "acid strength") between a surface acid site and the basic probe molecule. Hence, infrared spectroscopy should allow one to obtain information on the intensive and extensive properties of solid acid catalysts provided appropriate basic probes are being used.

In this article, principles for the selection of suitable basic probe molecules will be discussed and selected examples will be described for their application in studies of acidic properties of binary oxides, supported oxides and amorphous and crystalline (zeolitic) ternary oxides.

2. CRITERIA FOR THE SELECTION OF PROBE MOLECULES

It is useful to formulate properties, a molecule should possess when it is used to probe surface acid centers of a given catalytic material by means of infrared spectroscopy. The characterization of acid properties requires qualitative information on the nature of the acid sites, i.e. protonic (Brönsted) or aprotonic (Lewis) sites and quantitative information on the acid strength and on the site densities on the material's surface. The following criteria are considered to be important for the selection of one or more suitable probe molecules for a given problem and they may be used as a guideline although occasionally a particular probe may not fulfill all criteria simultaneously:

Criterion (i): The detectable spectral response induced by an acid-base interaction between the basic probe and a surface acid site must permit an analysis of the intermolecular interaction. In particular, it must discriminate between protonic and aprotonic sites. This analysis may be based on changes of intramolecular vibrational characteristics of the probe molecule and/or perturbations of the surface acid groups, e.g. O-H oscillators.

Criterion (ii): Quantities, such as e.g. frequency shifts, must be measurable with sufficient accuracy, and it must be possible to correlate these quantities with the strength of interaction between a probe molecule of given base strength and an acid site; the strength (energy) of interaction may be considered as a measure of the acid strength of the surface center. Even though the acid strength of centers on solid surfaces cannot be expressed in terms of thermodynamic quantities such as pK_a-values by this approach, one can at least establish a relative acidity scale for a series of materials.

Criterion (iii): Extinction coefficients for characteristic vibrational modes should be known so as to determine the density of surface centers and perhaps the relative abundance of centers of different quality or acid strength.

Criterion (iv): The extinction coefficients for characteristic vibrational modes should be high so that high detection sensitivity can be achieved even with materials of low surface areas.

Criterion (v): The probe molecule should have the appropriate base strength to induce an optimal acid-base interaction. Probe molecules can be classified with respect to the strength of interaction with a given acid site on the basis of the principle of hard and soft acids and bases (HSAB) [15]. This approach was first proposed by Burwell [16,17] as a basis for a qualitative discussion of adsorption. In general, "hard" indicates low polarizability and small size, while "soft" stands for high polarizability and large size. Hardness will increase, and softness correspondingly decrease, with increasing positive oxidation state and degree of coordinative unsaturation, and with decreasing size of an ion on a surface. Thus, the most typical members of the group of hard acid centers on solid (oxide) surfaces are the proton (Bronsted site) and small cus cations in high oxidation states (aprotonic Lewis sites). As regards potential probe molecules, ammonia NH_3 is to be classified as a hard base while carbon monoxide CO is a soft base.

The HSAB concept can be formulated as follows: hard acids prefer to associate with hard bases and soft acids prefer to bind to soft bases. This rule can be successfully utilized for a prediction of the relative strength of adsorption (acid-base) interaction. Thus, NH_3 as a hard base is known to adsorb strongly on oxide surfaces, whereas the soft base CO interacts only weakly with the hard acid centers on oxide surfaces.

Criterion (vi): The probe molecule should provide high specificity so as to discriminate between sites of the same quality having (perhaps only small) differences in acid strength. Hard bases such as NH_3 will typically allow to determine the quality of acid centers but will not necessarily discriminate between centers of different strengths. In contrast, soft bases such as CO due to their low interaction energies will respond more sensitively to small differences in acid strength (hardness) of the surface center. As a consequence, it may be desirable to use a series of bases of varying hardness for a full characterization of the acid properties of a catalytic material.

Criterion (vii): The molecular size of the basic probe should be as small as possible for several reasons. First, the acid center may be located in narrow pores or in cavities which are accessible only through channels or windows of narrow diameters (zeolites). Second, the acid site, particularly aprotonic Lewis acid sites, may be exposed in an oxygen vacancy and it may be sterically shielded by neighboring oxide ions if the probe (or the basic functional group of the probe) is too large. Third, a large probe molecule may adsorb on one acid center and block another center located in close proximity.

However, Benesi [18] and Jacobs and Heylen [19] have suggested that the steric hindrance in 2,6-disubstituted pyridines may be utilized to specifically detect protonic sites in the presence of aprotonic sites. Spectroscopic evidence indicated (vide infra) that this approach must be considered with some caution [20].

Criterion (viii): The reactivity of the probe molecule at the applied temperatures should be low so as to avoid any surface reactions other than acid-base interactions.

The following classes of probe molecules have been preferentially used in conjunction with infrared spectroscopy for the study of acid properties of solid surfaces (the probes are listed according to decreasing hardness): Aliphatic amines, ammonia, pyridines, nitriles, ketones, ethers, benzene, carbon monoxide. Within each class of probe molecules (except NH_3 and CO), the individual molecular properties can be modified by substitution in a controlled fashion. A brief evaluation of the properties of each of these groups of basic probe molecules will follow based on the criteria (i)-(vii).

3. PROBE MOLECULES AND THEIR PROPERTIES

Ammonia: NH_3 is a hard Lewis base and it is small in size. The various forms of expected chemisorbed NH_3 species, namely coordinated and protonated NH_3, can be detected and discriminated by their characteristic normal vibrations. The protonated form NH_4^+ gives rise to bands near 1450 and 3130 cm^{-1}, while NH_3 coordinated to aprotonic sites would show bands near 1250, 1630 and 3330 cm^{-1}. However, dissociative adsorption of NH_3 to form - NH_2 groups or NH_2^- ions has also been observed, e.g. on γ-Al_2O_3 when pretreated at high temperatures [21]. Due to its strong basicity, ammonia can be strongly bonded on a wide variety of sites and therefore it is not a very specific probe molecule. It may thus not always provide ideal properties for studies of surface acidity.

Amines (other than ammonia and pyridines):
Although a variety amines, particularly trimethylamine and n-butylamine have been widely used for surface acidity determinations [4-6], comparably few spetroscopic data of adsorbed amines are available. As with ammonia, coordinatively adsorbed amines held by aprotonic (cus) sites have been detected ($-NH_2$ bending mode of n-butylamine near 1605 cm^{-1}) on pure oxides [22-24], and protonated species (symmetric and antisymmetric deformation modes of $-NH_3^+$ near 1590 and 1510 cm^{-1}, respectively) were additionally found on silica-alumina and zeolites [23]. However, protonated species were also detected on n-butylamine adsorption on transition aluminas [23,24] due to the high basicity of these aliphatic amines. In addition, there is evidence for dissociative adsorption of n-butylamine [23] and chemical transformations at higher temperatures [25,26].

The interaction of the aliphatic amines with oxide surfaces is certainly still less specific than that of NH_3 because of their higher basicity and their greater molecular size. The influence of steric effects on amine adsorption has been discussed [26,27]. Medema et al. [27] concluded that the adsorbed amount of an amine on γ-Al_2O_3 primarily depended on its molecular cross-sectional area and not on its basicity.

In conclusion, amines and aliphatic amines in particular, cannot be considered as ideal probes for the characterization of surface acidity.

Pyridine and substituted pyridines: Pyridine, although less basic than ammonia, is still a fairly hard base. Its molecular size may give rise to steric hindrance in intermolecular interactions. The pyridine molecule can undergo coordination to cus aprotonic sites, it can be protonated to form the pyridinium ion PyH^+ on acidic OH groups and it can undergo H-bonding with less acidic groups. The infrared spectra of pyridine coordination compounds [28] are clearly distinct from those of PyH^+ [29] and of H-bonded pyridine, so that the corresponding surface species can be quite easily distinguished. The ring vibration modes 19b and 8a - according to the nomenclature introduced by Kline and Turkevich [30] - respond most sensitively to the nature of intermolecular interactions via the nitrogen lone pair electrons. These two modes are observed at 1440 to 1447 and at 1580 to 1600 cm^{-1}, respectively, for H-bonded pyridine, at 1535 to 1550 and around 1640 cm^{-1} for PyH^+, and at 1447 to 1464 and 1600 to 1635 cm^{-1} for coordination compounds. In addition, H-bonded species lead to characteristic shifts of the O-H stretching modes of surface hydroxyl groups to lower frequencies, the shift $\Delta\nu_{OH}$ being a measure of the H-bond energy.

The 8a mode of coordination compounds shifts to increasingly higher frequencies as the coordination bond energy (hardness of the coordination site) increases. It has also been suggested [31,32] that this vibrational mode of coordinated pyridine would be sensitive to the coordination number of the metal center, and a possible discrimination of Al^{3+} in octahedral and tetrahedral sites on alumina surfaces was inferred.

Alkyl substitution in 4-position increases the basicity of pyridines via inductive effects. The same is true for substitution in 2- and 6-position; in this latter case, however, the nitrogen lone pair is sterically shielded. The influence of these effects on the adsorption interaction with alumina surfaces has first been investigated by Knözinger and Stolz [20]. Benesi [18] has suggested that 2,6-dimethyl-substituted pyridine may be used as a specific probe for protonic sites. However, the work of Knözinger and Stolz [20] had previously shown that

2,4,6 - trimethylpyridine can still undergo coordination to Al^{3+} sites on alumina surfaces, although the interaction is sterically hindered and hence, weaker than with unsubstituted pyridine. Following the paper of Benesi [18], several workers have used substituted pyridines as probe molecules for acid sites on various oxide materials including zeolites [19,24,33-35]. 2,6-di-tert-butylpyridine is sterically so strongly hindered that coordination onto Lewis acid sites is definitely impossible [36]. It has been inferred from infrared studies that protonation of this probe molecule by protonic sites should be possible [37]. However, 2,6-di-tert-butylpyridine does not form N....H-O hydrogen bonds with silanol groups on silica surfaces but rather interacts with silanol groups via the π-electron system [38] as evidenced by a frequency shift of the O-H stretching vibration of 100 cm^{-1} which is typical for OH...π interactions. The same type of interaction was also observed on alumina [36], so that protonation turns out to be unlikely.

In conclusion, pyridine and substituted pyridines, though still being rather hard bases, provide interesting properties for studies of surface acidity. Therefore, some examples of their application will be discussed in more detail below.

<u>Nitriles:</u> Nitriles are soft bases and sterically less hindered than pyridines. They may thus be expected to be almost ideal specific probe molecules. The C-N stretching mode is sensitive to intermolecular interactions and typically shifts by 35 to 60 cm^{-1} to higher frequencies when coordination bonds are formed to aprotonic sites via the nitrogen lone pair electrons, while H-bonding leads to a shift of only 10-15 cm^{-1}. H-bonding would also induce a characteristic shift of the O-H stretching mode of surface hydroxyl groups.

Acetonitrile, CH_3CN, is a small molecule and has been used frequently as a probe molecule for surface acid sites on various materials [24,39-44]. However, the assignment of the C-N stretching mode of CH_3CN (ν_2 mode) in coordination compounds and H-bonded species may be rather confused due to the occurrence of Fermi resonance between the ν_2 mode and the ($\nu_3+\nu_4$) combination mode. This problem has been dealt with by Knözinger and Krietenbrink [45]. An easy solution of this problem is the use of deuterated acetonitrile, CD_3CN.

Unfortunately, acetonitrile shows a high reactivity relative to various oxide surfaces, namely those exposing nucleophilic OH groups [45]. Its applicability as a probe molecule must therefore be carefully tested for each individual material. These surface reactions can be suppressed by proper substitution of the methyl group for stronger electron-releasing groups. Thus, t-butylnitrile was shown [45] to adsorb on alumina only through coordination bonds onto cus Al^{3+} besides weak H-bonding with no chemical transformation occurring below 500 K. Successful applications of t-butylnitrile were reported by Scokart and Rouxhet [43] and Lavalley and Caillot [46]. Odenbrand et al. [47] in a study of silica-titania mixed oxides, also used t-butylnitrile as a probe molecule.

<u>Ketones:</u> Ketones are soft bases and their coordination to aprotonic sites and their H-bond acceptor strength are weak. The carbonyl stretching vibration responds sensitively to intermolecular interactions. Shifts to lower frequency of up to 150 cm^{-1} are induced when acetone is coordinated to cationic centers while H-bonding leads to shifts of 20 to 40 cm^{-1}. Unfortunately, acetone is rather reactive when interacting with oxide surfaces [9] and is therefore not an ideal probe molecule. However, as in the case of nitriles substitution at the carbonyl group by electron-releasing substituents leads to improved stability as shown by Schulz and Knözinger [48] for the adsorption of di-isopropyl- and di-tert-butyl ketone on γ-Al_2O_3.

<u>Ethers:</u> Lavalley and Caillot [46] have suggested the use of deuterated

dimethylether CD_3OCD_2H as a probe molecule. The CH stretching vibrations in dimethylether are known to be sensitive to the ether oxygen environment and to shift toward higher frequencies when the lone pair electrons become engaged in coordination bonds. The use of the pentadeuterated compound is advantageous since it removes complications due to Fermi resonance in the undeuterated compound. C-H stretching frequency shifts to higher values of 37 cm^{-1} and 70 cm^{-1} have been reported when CD_3OCD_2H was adsorbed on ZnO [46] and Al_2O_3 [46,49], respectively, while H-bonding on SiO_2 led to a shift of 28 cm^{-1} [46]. Thus, this soft base may find further application as a probe molecule for surface acid sites, although the steric situation may not be optimal.

Benzene: Benzene is an extremely soft base and may thus be expected to act as a highly specific probe. It has been applied for studies of the acidic or H-bond donor properties of surface OH groups via O-H...π interactions. Although this interaction will also perturb the internal normal modes of the benzene molecule, the analysis of the shift of O-H stretching vibrations induced by the O-H...π interaction was particularly emphasized [11,50] and led to a ranking of acidities of various surface OH groups on modified silica surfaces [51] as follows:

$$BOH < SiOH < GeOH < POH.$$

Amorphous silica-aluminas only show one O-H stretching frequency at 3750 cm^{-1} characteristic of silanol groups although their catalytic properties indicate the existence of protonic sites having much higher acid strength than SiOH groups. The adsorption of benzene gave two bands of perturbed O-H groups being shifted relative to the band of unperturbed groups at 3750 cm^{-1} by 110 cm^{-1} and 310 cm^{-1}, respectively [41]. The former shift is characteristic of SiOH groups, the latter for the expected stronger acid site, probably an OH group bridging an Al^{3+} and a Si^{4+} center: Si(OH)Al.

Thus, benzene may in fact be considered as a useful, highly specific probe molecule for the detection of protonic sites, at least for speciality applications, although its molecular size may not always be optimal.

Carbon monoxide: A molecule that combines the softness of benzene with small molecular size is carbon monoxide. The carbonyl stretching frequency responds very sensitively to coordination onto cationic sites leading to positive frequency shifts of up to 80-90 cm^{-1} relative to the gas phase frequency when coordinated to the hardest cation sites and to even negative shifts when coordinated to soft transition metal ions in low oxidation states possessing finite d-electron occupancy. The carbonyl stretching mode is also sensitive relative to H-bonding to protonic sites. A high extinction coefficient of this vibrational mode provides good spectral sensitivity.

As a consequence of the softness of CO, the heats of adsorption are typically low and low temperature measurements are required, particularly when site saturation for quantitative determination of site densities is necessary. Under these low-temperature conditions the CO-probe is certainly entirely unreactive.

The properties of CO permit detailed studies of aprotonic [10,12,52,53] and protonic [52,54] sites on oxide materials. CO must be expected to probe surface acid sites very specifically and fulfills practically all criteria mentioned above. It is thus to be considered as an almost ideal probe molecule. Several examples for the application of CO in probing acid sites on simple binary and complex multicomponent oxide surfaces and in zeolites will therefore be discussed in some detail below.

4. CHARACTERIZATION OF APROTONIC SITES

Aprotonic Lewis acid sites are created on oxide surfaces by dehydration and dehydroxylation by which processes cus cations become exposed at the surface. These centers are to be considered als relatively hard electron pair acceptor sites, their hardness being largely determined by their charge to radius ratio and by their coordination. As instructive examples of the characterization of aprotonic sites, the chemisorption of pyridine on alumina surfaces and the coordination of carbon monoxide on various oxide surfaces will be discussed below.

4.1 Pyridine and Substituted Pyridines

Partially dehydroxylated surfaces of transition aluminas expose Al^{3+} ions in various coordination environments [11,55]. Pyridine coordinates to these sites as shown by infrared spectroscopy [9-13,56,57]. An "outer" complex with a characteristic 8a mode at, 1616 cm^{-1} was formed around room temperature, which was then transformed into an "inner" complex via an activated process upon heating to > 520 K, the latter being characterized by an 8a mode at 1624 cm^{-1} [56,57]. The activation barrier between the two forms of coordinated pyridine was attributed to a steric restriction of the approach of the Al^{3+} cation to the pyridine nitrogen. In contrast, Morterra et al. [31,32] did not find any evidence for an activated chemisorption of pyridine on alumina. These authors reported the simultaneous formation at room temperature of three chemisorbed pyridine species of different bond energy, which were characterized by 8a modes at 1598-1600 cm^{-1}, 1613-1617 cm^{-1}, and 1623-1627 cm^{-1}, respectively. Only the low-frequency species was observed on α-Al_2O_3. It was therefore assigned to pyridine bonded to a doubly uncoordinated octahedral Al^{3+} cation or to octahedral Al^{3+} cations being exposed in a bridging vacancy. The two high-frequency species were assigned, in order of increasing polarizing power of the adsorption site, to coordinated pyridine bridging an octahedral and a tetrahedral Al^{3+} cation, and to pyridine bonded to a tetrahedral Al^{3+} cation.

The studies of Knözinger and Stolz [20] indicated that 2,4,6-trimethylpyridine can still coordinate to cus Al^{3+} sites despite the steric crowding around the nitrogen atom. However, the steric restrictions made a close approach between Al^{3+} and N-atom impossible so that the coordination interaction of this molecule was significantly weaker than that of unsubstituted pyridine on the same surface. In fact, preadsorbed 2,4,6-trimethylpyridine was almost quantitatively displaced by the weaker base pyridine [20]. These observations practically eliminate 2,6-dimethyl substituted pyridines as selective probes for the detection of protonic sites as originally suggested by Benesi [18]. As already mentioned above, the steric crowding at the nitrogen atom in 2,6-di-tert-butylpyridine renders coordination to cus Al^{3+} impossible [36], but also H-bonding and protonation do not occur [36,38,50].

The results of these studies on aluminas demonstrated the potential of pyridine and 4-substituted pyridines for the characterization of aprotonic acid centers and they indicated the possible influence of steric effects on the coordination interaction. Analogous phenomena are expected to occur on other oxide surfaces exposing cus cation sites.

4.2 Carbon Monoxide

The frequency shifts of the C-O stretching mode induced by coordination of the CO molecule to metal centers can be interpreted in terms of the bonding mechanism: The 5σ orbital of CO forms a donor bond to the metal with prevalent σ character. Considerable contributions to the stabilization of the metal-carbon bond come from the participation of metal d-orbitals and the antibonding π^* orbitals of the ligand (π-back donation). It is the increased electron density in the π^* orbital due to this back donation which leads to a decrease in the C-O force constant and stretching frequency, this effect overcompensating for an increase in the force constant by formation of the σ-bond. When, however, metal centers with zero or negligible d-electron densities (cations) are involved, the σ-bond becomes the dominating bond. Comparably weak bond energies and carbonyl stretching frequencies higher than the gas phase frequency are the consequence.

Adsorption of CO on oxide surfaces (preferably at low temperatures) typically shows this behaviour with the carbonyl frequencies varying between the gas phase frequency and values as high as 2240 cm^{-1}, depending on the type of oxide and its pretreatment. It can thus be anticipated that CO is in fact a useful probe for the nature of the coordination site (acid site) and the strength of the coordination bond (acid strength). Among the systems studied are the binary oxides Al_2O_3 [52,54,58-62], MgO [61,62], TiO_2 [59,61-65], Cr_2O_3 [53,66,67] and more complex multicomponent systems such as SiO_2-Al_2O_3 [52,61], vanadium oxide supported on TiO_2 [68], tungsten oxide supported on TiO_2 [69] and silica-titania mixed oxides [70].

The potential of CO for discrimination of adsorption sites in different coordinations was nicely demonstrated for transition aluminas, which expose cus Al^{3+} in octahedral and tetrahedral sites. A carbonyl band observed near 2190 cm^{-1} was attributed to CO coordinated to an octahedral Al^{3+} center [52,53] while carbonyl bands beyond 2200 cm^{-1} were related to CO coordinated to tetrahedral Al^{3+} sites [52,53,58,71]. Only the high frequency species was observed on SiO_2-Al_2O_3 [52,53,61} which exclusively exposes tetrahedral Al^{3+} cations.

Studies of the multicomponent systems, e. g. tungsten oxide on TiO_2, are particularly interesting, since here the infrared spectra of chemisorbed CO permit the differentiation between different elements being exposed, e.g. W^{6+} and Ti^{4+} [69]. The relative acid strength (hardness) of the two sites can be qualitatively characterized by the temperature and pressure dependence of the carbonyl band intensities, and the relative abundance of the two types of sites is correlated with their intensity ratios at saturation coverage.

It has also been shown by theoretical considerations [72] that the carbonyl stretching frequency should monotonously increase with the electric field strength at the cation (d^0 electron configuration) center. Several examples for such correlations have in fact been reported in the literature [52,53,61,73-75]. Zaki and Knözinger [52,53] have pointed out that oxidation states and coordination numbers of cation centers can be estimated from experimental carbonyl stretching frequencies by means of a universal frequency/field strength correlation, provided the carbonyl stretching frequencies are free of shifts due to coverage-dependent dynamic and static lateral interactions. This can most easily be achieved by using frequencies which are determined by extrapolation to zero-coverage [52,53].

4.3 Quantitation

Paukshtis and Yurchenko [12] have reviewed the attempts which have been made to correlate characteristic frequencies or frequency shifts of normal modes of chemisorbed probe molecules with isosteric heats of adsorption as a measure of the acid strength of the coordination site. They proposed linear correlations between isosteric heats of adsorption of CO and the carbonyl stretching frequency shift relative to the gas phase frequency, and between isosteric heats of adsorption of pyridine and the frequency of the 8a vibrational mode. Morterra et al. [76] in contrast reported a non-linear, parabolic correlation between carbonyl frequency shifts and enthalpies of adsorption of CO. These authors strongly suggested that such correlations are real although there might be no unique correlation curve but rather individual correlations for families of similar and comparable adsorbents.

A quantitative determination of aprotonic site densities becomes possible once extinction coefficients of characteristic absorption bands of the chemisorbed probe molecule are known. The determination of such extinction coefficients requires simultaneous measurement of adsorbed amounts and infrared spectra. Relatively few investigations of this kind have been reported so far. Details have been summarized by Paukshtis and Yurchenko [12] and by Boehm and Knözinger [11]. In several studies [33,77,78] the relative numbers of aprotonic and protonic acid sites have been determined by pyridine adsorption, which requires knowledge of the ratio of extinction coefficients of e.g. the 19b mode near 1450 cm^{-1} for the coordinated pyridine species and near 1545 cm^{-1} for the pyridinium ion. This ratio $\epsilon(1450)/\epsilon(1545)$ turned out to be close to unity. Similarly, the ratio of the corresponding band of coordinated (1579 cm^{-1}) and protonated (1642 cm^{-1}) 2,6-dimethylpyridine was found to be close to unity. Morterra et al. [31] by careful deconvolution of the bands of chemisorbed pyridine on η-Al_2O_3 in the region of the 8a-mode, succeeded in determining the absolute extinction coefficients for the various surface species, giving rise to bands at 1578, 1592, 1600, 1616 and 1626 cm^{-1} (vide supra). These authors emphasize, that the extinction coefficient of any adsorbed species is not transferable from one system to another, even if the chemical and spectroscopic characteristics are similar.

5. CHARACTERIZATION OF PROTONIC SITES

Binary and multicomponent oxides typically contain surface hydroxyl groups [11] and zeolites may contain protons as charge-compensating ions [79,80]. The O-H stretching frequency of unperturbed hydroxyl groups depends on the electronegativity of the coordination center of the group, the number of cationic coordination centers it is bonded to, the oxygen coordination number of these centers and the overall bond character of the particular material. The proton donor strength of a hydroxyl group should increase with increasing polarizing power and number of coordination centers and, hence, with decreasing O-H stretching frequency. However, the absolute frequency cannot be taken as a direct measure of the proton donor strength or acid strength of OH groups when comparing different types of materials, although the frequencies of different OH configurations on one oxide material may with some caution be taken as a relative measure of an acid strength sequence, as e.g. on aluminas [11,55].

The proton donor or acid strength of an OH group determines the ability of the group to either protonate an adsorbed basic probe molecule to form an ion pair:

$$OH + B \rightarrow O^- \ldots H^+B$$

or to undergo H- bonding with softer probe molecules:

$$OH + B \rightarrow O\text{-}H \ldots B$$

Whether or not proton transfer occurs, depends on the proton donor strength of the OH group and on the hardness (basicity) of the basic acceptor molecule. Paukshtis and Yurchenko [12] have strongly emphasized the proton affinity PA as a characteristic molecular property that controls proton transfer. The proton affinity of a base is defined as the enthalpy of the gas phase reaction $H^+B \rightarrow H^+ + B$. The respective intermolecular interaction will lead to a characteristic double minimum potential well at fixed O .... B distance and the shape of the potential separating the two minima (anharmonicity) will determine the probability of proton transfer and the life time of the protonated species will depend on the relative depths of minima at the O atom and the B molecule, respectively. Kazansky [81] has first pointed out these aspects and suggested to experimentally determine the anharmonicities and dissociation energies of O-H oscillators interacting with probe molecules [81].

For a given material, the formation of protonated species is obviously most likely with the hardest probe molecules such as aliphatic amines, ammonia and pyridine, which can be easily detected by means of the characteristic normal modes of their protonated forms (vide supra). Thus, silica-aluminas and H-zeolites will typically form NH_4^+ and PyH^+ species, while only NH_4^+ can be detected on aluminas and neither of these protonated species is formed on silica [11]. Experimental results of this kind provide information on the ability of surface OH groups to stabilize a protonated species of the applied base molecule for a period of time which must at least correspond to one vibrational period. Softer probe molecules (or reactants in catalytic reactions) may be protonated for shorter periods of time and will thus escape detection by infrared spectroscopy. Nevertheless, such species may be relevant in proton catalyzed reactions.

In certain cases, however, the individual protonated species may escape detection via their molecular normal modes. Instead a diffuse continuous absorption over the entire spectral range may be observed [38]. This phenomenon has been explained by the formation of extremely highly polarizable H-bonds which are presumably produced in dimeric species of the type

$$H_3N....H^+....NH_3 \quad \text{or} \quad Py....H^+...Py.$$

These bridges are characterized by a double minimum potential well which can be easily deformed due to the anisotropic and inhomogeneous environment of the surface species and thus leads to a quasi-continuum of vibrational states [38]. Moreover, the proton can fluctuate between the two equilibrium positions within the H-bond fairly rapidly.

Another approach toward the analysis of acid properties of surface OH groups uses their H-bonding ability. As indicated above, H-bonding will induce characteristic shifts of internal vibrations of the probe, but more importantly, typically rather significant shifts occur in the O-H stretching region $\Delta\nu_{OH}$ accompanied by the characteristic broadening of the O-H band and increase in integrated band intensity [50]. A few examples will be given below.

5.1 Hydrogen-bonding

When pyridine and substituted pyridines were adsorbed on silica (outgassed at 973 K), frequency shifts $\Delta\nu_{OH}$ between 515 and 855 cm^{-1} of the O-H stretching band were observed depending on the H-bond acceptor strength of the base. An example of hydroxyl infrared spectra obtained after adsorption of pyridine and 2,6-di-tert.-butyl pyridine is shown in Fig. 1 [38]. The band of unperturbed silanol groups at 3750 cm^{-1} clearly decreases in intensity on adsorption, while an intense and broad band develops at 3010 cm^{-1} on adsorption of pyridine. In contrast, the 2,6-di-tert.-butyl substituted pyridine is sterically crowded around the nitrogen, O-H...N bonding is not possible and an O-H...π interaction is thus produced as indicated by a small frequency shift of only 100 cm^{-1} relative to the position of the unperturbed O-H stretching mode (vide supra). The H-bond acceptor strength can be expressed by Hammett's

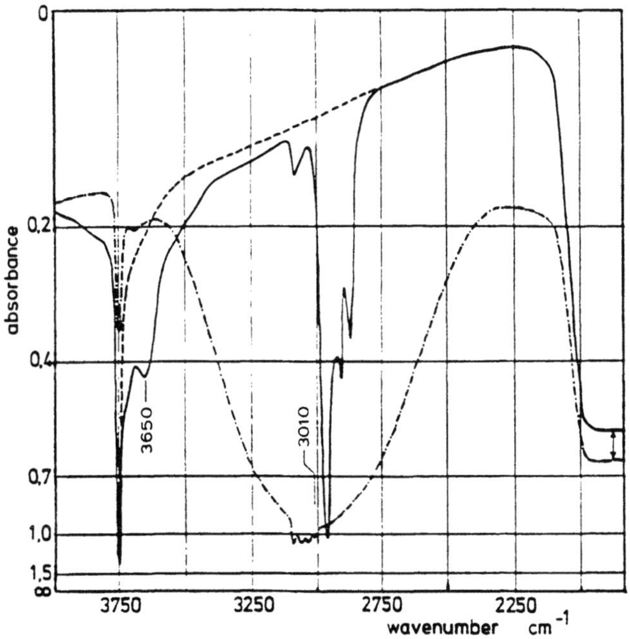

Fig. 1.IR spectra of silica after adsorption of pyridine (dash-dotted line) and 2.6-di-tert. butylpyridine (solid line). The background spectrum of silica prior to adsorption is represented by the dashed line.

polar substituent parameters and Fig. 2 shows a linear correlation between the frequency shift and this parameter for a large number of substituted pyridines. The points for para- and meta-substituted pyridines fall on a straight line which is also fitted by those adsorbates which carry one or two methyl or ethyl substituents in ortho-position. Even the point for 2,6-di-isopropyl-pyridine falls on the straight line within the limits of accuracy . This result seems to indicate that the so-called "ortho-effects" do not influence the H bonding of ortho-substituted pyridines onto surface silanol groups. Steric restriction should therefore not remarkably influence the H-bond strength for orthosubstituents with volumes smaller than that of the isopropyl group.

There are deviations apparent for three other ortho-substituted adsorbates in Fig. 2, particularly 2-methoxy-, 2-chloro- and 2-fluoro-pyridine. Various reasons may account for these deviations, which most probably work together. All these substituents possess p-electron pairs and the substituent effect therefore may not be purely polar but also mesomeric in nature. Furthermore, steric restriction could contribute more or less, at least in the case of the methoxy group and of chlorine. However, a third observation may probably account predominantly for the deviations of these compounds. Besides the broad and intense absorption band of the silanol groups interacting with the nitrogen lone-pair electrons, weaker and less displaced absorptions are observed in the infrared spectra on adsorption of these three pyridines, indicating dual interaction with silanol groups. These interactions will mutually affect each other.

These results show that with the substituted pyridines a variety of probes of varying base strength is at hand which may successfully be

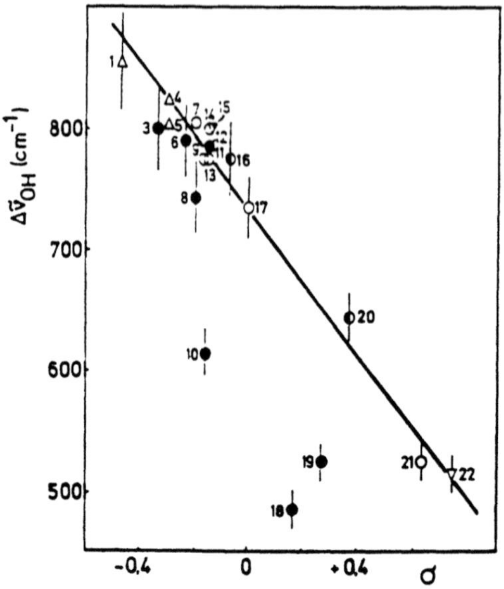

Fig. 2. Correlation of $\Delta\nu_{OH}$ with Hammett's polar substituent parameter: o para-, o meta-, ● ortho-substitution; Δ 2,6-di-substitution; ∇ other di-substituted pyridines.
(1) 2,4,6-trimethyl; (3) 2,4-dimethyl; (4) 2,6-diisopropyl; (5) 2,6-dimethyl; (6) 2,5-dimethyl; (7) 4-tert. butyl; (8) 2-tert. butyl; (9) 4-methyl; (10) 2-methoxy; (11) 2-methyl; (12) 4-ethyl; (13) 2-ethyl; (14) 4-isopropyl; (15) 3,5-dimethyl; (16) 3-ethyl; (17) pyridine; (18) 2-fluoro; (19) 2-chloro; (20) 3-chloro; (21) 4-cyano; (22) 3,5-dichloro.

applied for characterization of the acid properties of surface OH groups. However, some of these probe molecules may undergo surface chemical reactions on surfaces other than silica.

Several other correlations between frequency shifts $\Delta\nu_{OH}$ of silanol groups with various properties of basic probe molecules have been reported and were reviewed by Paukshtis and Yurchenko [12] and by Boehm and Knözinger [11].

Carbon monoxide does undergo H-bonding interactions with surface hydroxyl groups [52] and - as discussed above - is expected to be a very specific probe at low temperatures. CO adsorption on silica has been studied extensively by Ghiotti et al. [82] and by Beebe et al. [83]. An H-bonded complex of the type

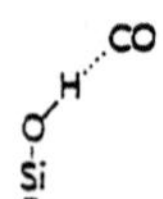

is undoubtedly formed at 77 K as indicated by the decrease in intensity of the band of unperturbed OH groups and the appearance of an intense band shifted to lower frequency by 93 cm^{-1} [83]. Simultaneously, the CO stretching mode was observed at 2158 cm^{-1}, i.e. shifted by 15 cm^{-1} to higher frequency as compared to the gas phase frequency of CO.

Aluminas are amphoteric and due to their structural characteristics

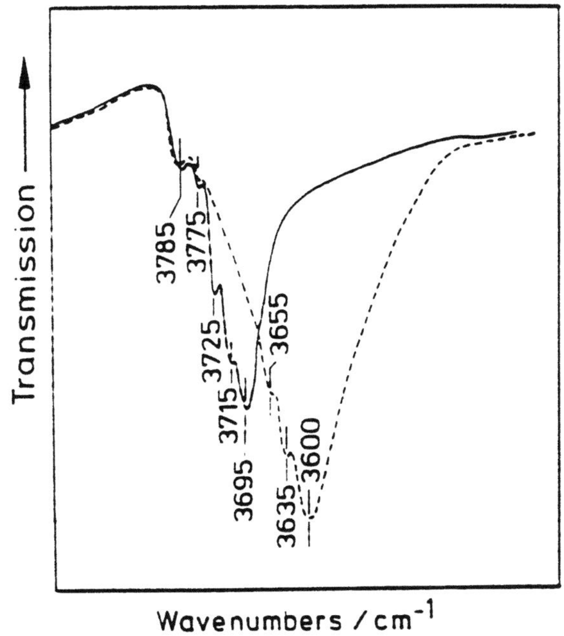

Fig. 3. Hydroxyl IR spectra of γ-Al_2O_3 before (solid) and after (dotted) adsorption of CO at 77 K.

with Al^{3+} cations in octahedral and tetrahedral coordination, their surfaces contain five different OH configurations which give rise to five individual O-H stretching bands [11,55]. Fig. 3 shows a low temperature hydroxyl spectrum of γ-Al_2O_3 (calcination at 773 K) in which these five bands can be clearly recognized [54]. The bands are attributed to type I, II, III hydroxyls with two forms of type I (a and b) and type II (a and b) OH's due to the presence of octahedral and tetrahedral Al^{3+} [11,55]. Simple estimations based on Pauling's electrostatic valence rule [55] suggested increasingly positive charge on the OH groups with decreasing O-H stretching frequency. In fact, Fig. 3 shows that after admission of CO the two bands at 3785 and 3775 cm^{-1} remain almost entirely unaffected, whereas the three bands at lower frequency, namely at 3725, 3715 and 3695 cm^{-1}, are completely eroded. Instead a partially resolved triplet appears at lower frequencies, with band positions at 3655, 3635 and 3600 cm^{-1}. The relative intensities of individual bands in the two triplets permits a correlation to be made between the unperturbed and associated OH groups. This correlation and the corresponding frequency shifts $\Delta\nu_{OH}$ are summarized in Table 1.

The results indicate -in agreement with the sequence of acid strengths predicted on the basis of the estimated net charges of the free hydroxyls - that the low frequency bands are representing the stronger H-bond donor groups. Moreover, the weak interaction with CO permits the distinction between AlOH...CO surface complexes of only slightly different H-bond energies. The increasing frequency shift for the individual associated complexes as the O-H stretching frequency of the unperturbed groups decreases, provides strong support for the increasing

TABLE 1: Hydroxyl band positions (cm^{-1}) of free and associated OH groups on γ-Al_2O_3 before and after CO adsorption at 77 K

Isolated OH's Type I	Type II	Type III	Associated OH's	$\Delta\nu_{OH}$
3785 (b)			3785	
3775 (a)			3775	
	3725 (b)		3655	70
	3715 (a)		3635	80
		3695	3600	95

H-bond donor capacity with increasing coordination number of the OH group. The $\Delta\nu_{OH}$-value 95 cm^{-1} for the low-frequency type III hydroxyl groups is almost identical with the frequency shift for silanol groups and, hence, suggests that the corresponding OH groups should possess an H-bond donor strength (acid strength) comparable to that of SiOH groups.

It is interesting to note that the two high-frequency (type Ia and Ib) hydroxyl bands remain unperturbed by CO. This behaviour resembles that of the OH groups of MgO and CeO_2 (vide supra) and is clearly indicating their strongly basic nature [52,54].

Such detailed information of the individual hydroxyl groups coexisting on oxide surfaces can hardly ever be obtained when stronger bases are being used as probe molecules.

Silica-aluminas are usually considered to develop stronger protonic acidity than pure silicas [5]. One would therefore expect a larger value for $\Delta\nu_{OH}$. However, low temperature infrared spectra of CO adsorbed on a SiO_2-Al_2O_3 ("High alumina" from Akzo Research) gave a broad band characteristic of H-bonding with a maximum at 3658 cm^{-1}, shifted relative to the position of unperturbed OH groups at 3738 cm^{-1} by only 70 cm^{-1} [54]. In agreement with earlier reports [84], bands characteristic of free OH groups associated with Al^{3+} were not detected. The small frequency shift seems to be in contrast with expectation. A closer comparison of the hydroxyl spectra of SiO_2-Al_2O_3 [54] and of SiO_2 [82,83] after CO adsorption shows, however, significant differences. The band typical for associated OH groups on SiO_2 [82,83] is relatively symmetric and the absorbance has declined to the baseline at wavenumbers below 3550 cm^{-1}. In contrast, the corresponding band for SiO_2-Al_2O_3 [54] shows some structure on its high-frequency flank and a finite absorption extends to wavenumbers even below 3400 cm^{-1}. This suggests that the OH groups on SiO_2-Al_2O_3 are heterogeneous in nature although individual O-H stretching bands cannot be resolved in the absence of the H-bonded adsorbate. The broad absorption extending from 3600 cm^{-1} towards lower frequencies particularly suggests the existence of OH groups with high H-bond donor strength. These groups might well be considered tentatively as those developing protonic acidity.

An interesting case is CO adsorption on zeolites. The acidic OH groups in zeolites are believed to bridge an Al^{3+} and a Si^{4+} center: Si(OH)Al [79,80]. H- bonding occurs when CO is adsorbed at 77 K and leads to $\Delta\nu_{OH}$ shifts of 250 to 340 cm^{-1}, depending on the type of zeolite and its composition. Simultaneously, the CO stretching mode shifts to higher frequency relative to the gas phase frequency by as much as 30 to 35 cm^{-1}. Table 2 summarizes reported data on CO adsorption on HNa-Y, H-ZSM-5 and

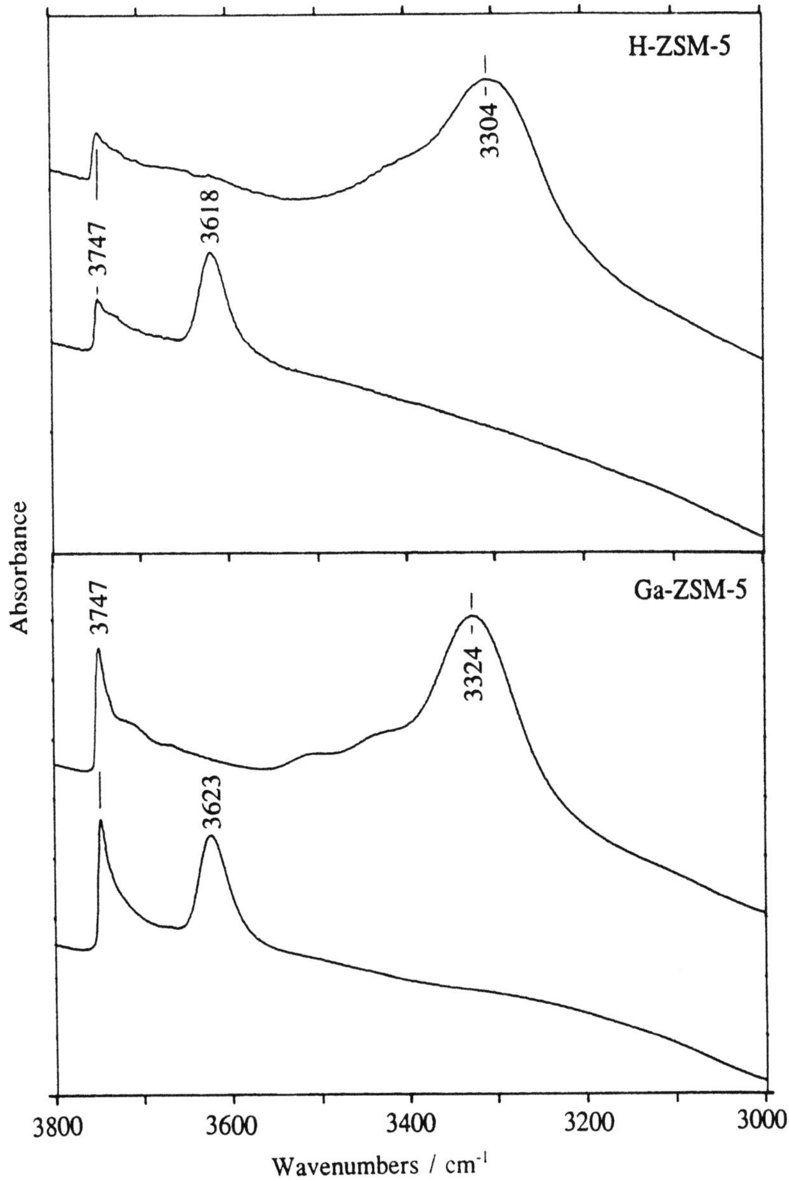

Fig. 4. Hydroxyl IR spectra H-ZSM-5 (SiO_2/Al_2O_3 ratio 26.8) and Ga-ZSM-5 (SiO_2/Ga_2O_3 ratio 26.8) at 80 K prior to (lower spectra) and after (upper spectra) CO adsorption. The bands at 3747 cm^{-1} are due to silanol groups, the bands near 3620 cm^{-1} are characteristic for the unperturbed acidic OH groups.

Table 2: O-H stretching frequencies of zeolites prior to and after CO adsorption at 77 K and the corresponding frequency shifts $\Delta\nu_{OH}$

Zeolite	SiO_2/Al_2O_3 ratio	ν_{OH}/cm^{-1} free OH	ν_{OH}/cm^{-1} OH...CO	$\Delta\nu_{OH}/cm^{-1}$	Ref.
H-ZSM-5	24	3620	3305	315	84
	13.6	3617	3305	312	85
	26.8	3619	3290	339	86
	41.5	3617	3295	322	86
H-Ga-ZSM-5	-	3622	3330	282	86
$H_{70}Na_{30}$-Y		3652	3377	275	84
$H_{95}Na_5$-X		3643	3347	296	85
$H_{70}Na_{30}$-Y		3647	3369	278	85
$H_{20}Na_{80}$-Y		3650	3385	265	85
$H_{90}Na_{10}$-Y		3640	3355	285	87

H-Ga-ZSM-5. The spectra in the O-H stretching regime of H-ZSM-5 and H-Ga-ZSM-5 are compared in Fig. 4, which -together with the data in Table 2 - clearly demonstrates that the $\Delta\nu_{OH}$-value measured upon CO adsorption for the gallium-form of ZSM-5 is significantly lower than that obtained for the aluminium-form consistent with the known relative acid strengths of these materials.

The examples presented clearly demonstrate the potential of CO as a highly specific probe molecule for the characterization of protonic acidity on surfaces, which permits a ranking of acid strength to be established on the basis of $\Delta\nu_{OH}$ shifts.

5.2 Quantitation

Most correlations reported in the literature were established for frequency shifts $\Delta\nu_{OH}$ of OH groups on the surface of a given oxide with molecular properties of a wide variety of probe molecules as briefly mentioned above. Paukshtis and Yurchenko [12] first reported a nonlinear correlation between $\Delta\nu_{OH}$ of silanols with the proton affinity PA of bases covering a wide range of PA-values between 200 and 900 kJ mole^{-1}. Attempts to correlate $\Delta\nu_{OH}$-values measured on adsorption of a given probe molecule on a family of solids with a characteristic property of the latter are much less frequent. Lercher and Noller [88] reported on a successful correlation of $\Delta\nu_{OH}$ of silica, alumina and magnesia mixed oxides induced by acetone adsorption with Sanderson's intermediate electronegativity [89]. A linear correlation of $\Delta\nu_{OH}$-values produced by benzene adsorption on various zeolites with Sanderson electronegativity was also found [79].

The determination of the concentration of protonic centers on surfaces from spectroscopic data requires knowledge of the extinction coefficients, as already discussed in section 4.3.

Acknowledgments

The work done in the authors laboratory was financially supported by the Deutsche Forschungsgemeinschaft (partly in the framework of the Sonderforschungsbereich SFB 338) and the Fonds der Chemischen Industrie.

REFERENCES

1. Gates B. C., Katzer J. R. and Schuit G. C. A. (1979) "Chemistry of Catalytic Processes", Mc Graw, New York.
2. Thomas C. L. (1970) "Catalytic Processes and Proven Catalysts", Academic Press, New York and London.
3. Benesi H. A. and Winquist B. H. C. (1978) Adv. Catal. 27: 98.
4. Forni L. (1974) Catal Rev. 8: 65.
5. Tanabe K. (1970) "Solid Acids and Bases", Kodansha, Tokyo, and Academic Press, New York and London.
6. Tanabe K. (1981), in: "Catalysis-Science and Technology", J. R. Anderson and M. Boudard, eds., Springer, Berlin, Heidelberg and New York, Vol. 2, p. 231.
7. Malinowski S. and Marczewski M. (1989) in: "Catalysis", The Royal Society of Chemistry, Cambridge, Vol. 8, p. 107.
8. Tanabe K., Hattori H., Yamaguchi T. and Tanaka T., eds. (1989), "Acid-Base Catalysis", Kodansha, Tokyo, and Verlag Chemie, Weinheim.
9. Knözinger H. (1976) Adv. Catal. 25:184.
10. Zecchina A., Garrone E. and Guglielminotti E. (1983) in: "Catalysis", The Royal Society of Chemistry, London, Vol. 6, p. 90.
11. Boehm H.-P. and Knözinger H. (1983) in: "Catalysis-Science and Technology", J. R. Anderson and M. Boudart, eds., Springer, Berlin, Heidelberg and New York, Vol. 4, p. 39.
12. Paukshtis E. A. and Yurchenko E. N. (1983) Russ. Chem. Rev. 52:42.
13. Peri J. B. (1984) in: "Catalysis-Science and Technology", J. R. Anderson and M. Boudart, eds., Springer, Berlin, Heidelberg and New York, Vol. 5, p. 171.
14. Kung M. C. and Kung H. H. (1985) Catal. Rev. 27:425.
15. Pearson R. G. (1963) J. Amer. Chem. Soc. 85:3533; and (1967) Chem. Brit. 3:103.
16. Burwell, Jr., R. L. (1973) in: "Catalysis-Progress in Research", F. Basolo and R. L. Burwell, Jr., eds., Plenum, New York, p. 51.
17. Burwell, Jr., R. L., Read J. F., Taylor K. C. and Haller G. L. (1969) Z. Phys. Chem. (Frankfurt am Main) 64:.18.
18. Benesi H. A. (1973) J. Catal. 28:176.
19. Jacobs P. A. and Heylen C. F. (1974) J. Catal. 34:267.
20. Knözinger H. and Stolz H. (1971) Ber. Bunsenges. Phys. Chem. 75:1055.
21. Peri J. B. (1965) J. Phys. Chem. 69:211.
22. Primet M., Pichat P. and Mathieu M.V. (1971) J. Phys. Chem. 75:1221.
23. Morimoto T., Imai J. and Nagao M. (1974) J. Phys. Chem. 78:704.
24. Healy M. H., Wieserman L. F., Arnett E. M. and Wefers K. (1989) Langmuir 5:114.
25. Tanaka M. and Ogasawara S. (1972) J. Catal. 25:111.
26. Koubek J., Volf J. and Pasek J. (1975) J. Catal. 38:385.
27. Medema J., van Bokhoven J. G. M. and Kuiper A. E. T. (1972) J. Catal. 25:238.
28. Nakamoto K. (1986) "Infrared Spectra of Inorganic and Coordination Compounds", Wiley-Interscience, New York.

29. Cook D, (1961) Can. J. Chem. 39:2009.
30. Kline C. H. and Turkevich J. (1944) J. Chem. Phys. 12:300.
31. Morterra C., Chiorino A., Ghiotti G. and Garrone E. (1979) J. Chem. Soc., Faraday Trans. I 75:271.
32. Morterra C. and G. Cerrato, (1990) Langmuir 6:1810.
33. Matulewicz E. R. A., Kerkhof F. P. J. M., Monlijn J. A. and Reitsma H. J. (1980) J. Colloid Interf. Sci., 77:110.
34. Miyata H. and Moffat J. (1980) J. Catal. 62:357.
35. Corma A., Rodellas C. and Fornes V. (1984) J. Catal. 88:374.
36. Knözinger H., Krietenbrink H. and Ratnasamy P. (1977) J. Catal. 48:436.
37. Dewing J., Monks G. T. and Youll B. (1976) J. Catal. 44:226.
38. Knözinger H. (1974) Surface Sci. 41:339.
39. Angell C. L. and Howell M. V. (1969) J. Phys. Chem. 73:2551.
40. Karge H. (1973) Surface Sci. 40:157.
41. Sempels R. E. and Rouxhet P. G. (1976) J. Colloid Interf. Sci. 55:263.
42. Scokart P. O., Declerck F. D., Sempels R. E. and Rouxhet P. G. (1977) J. Chem. Soc., Faraday Trans. I, 73:359.
43. Scokart P. O. and Rouxhet P. G. (1982) J. Colloid Interf. Sci. 86:96.
44. Busca G. (1986) Langmuir 2:577.
45. Knözinger H. and Krietenbrink H. (1975) J. Chem. Soc., Faraday Trans. I.,71:2421.
46. Lavalley J.-C. and Caillod J. (1980) J. Chim. Phys., 77:373.
47. Odenbrand C. U., Brandin J. G. M. and Busca G. (1992) J. Catal., 135:505.
48. Schulz W. and Knözinger H. (1976) J. Phys. Chem., 80:1502.
49. Caillod J. and Lavalley J.-C. (1980) J. Chim. Phys., 77:379.
50. Knözinger H. (1976) in:"The Hydrogen Bond", P. Schuster, G. Zundel and C. Sandorfy, eds., North Holland, Amsterdam, Vol. 3, p. 1263.
51. Fink P., Pohle W. and Köhler A., (1972) Z. Chem., 12:117.
52. Knözinger H. (1989) in: "Acid-Base Catalysis", K. Tanabe, H. Hattori, T. Yamaguchi and T. Tanaka, eds., Kodansha, Tokyo, and Verlag Chemie, Weinheim, p. 147.
53. Zaki M. I. and Knözinger H. (1989) J. Catal., 119:311.
54. Zaki M. I. and Knözinger H. (1987) Mater. Chem. Phys., 17:201.
55. Knözinger H. and Ratnasamy P. (1978) Catal. Rev., 17:31.
56. Knözinger H. and Stolz H. (1971) Fortschr. Kolloid. Polymer, 55:16.
57. Stolz H. and Knözinger H. (1971) Kolloid-Z., 243:71.
58. Della Gatta G., Fubini B., Ghiotti G. and Morterra C. (1976) J. Catal., 43:90.
59. Tsyganenko A. A., Denisenko L. A., Zwerev S. M. and Filimonov V. N. (1985) J. Catal., 94:10.
60. Zecchina A., Escalona Platero E. and Otero Areán C. (1987) J. Catal., 107:244.
61. Zaki M. I. and Knözinger H. (1987) Spectrochim. Acta, 43 A:1455.
62. Soltanov R. I., Paukshtis E. A. and Yurchenko E. N. (1982) Kinet. Catal. (Engl.), 23:135.
63. Tanaka K. and White J. M. (1982) J. Phys. Chem., 86:4708.
64. Busca G., Saussey H., Saur O., Lavelley J.-C. and Lorenzelli V. (1985) Appl. Catal., 14:245.
65. Morterra C. (1988) J. Chem. Soc., Faraday Trans I, 84:1617.
66. Zecchina A., Coluccia S., Guglielminotti E. and Ghiotti G. (1971) J. Phys. Chem., 75:2774.
67. Scarano D., Zecchina A. and Reller A. (1988) Surface Sci., 198:11.
68. Busca G. (1986) Langmuir, 2:577.
69. Hilbrig F., Schmelz H. and Knözinger H. (1992) Proc. 10. Intern. Congr. Catal., Budapest, in print.

70. Odenbrand C. U. I., Andersson S. L. T., Andersson L. A. H., Brandin J. G. M. and Busca G. (1990) J. Catal., 125:541.
71. Morterra C., Coluccia S., Garrone E. and Ghiotti G. (1979) J. Chem. Soc., Faraday Trans I, 75:289.
72. Hush N. S. and Williams M. L. (1974) J. Molec. Spectrosc., 50:349.
73. Angell C. L. and Schaffer P. C. (1966) J. Phys. Chem., 70:1413.
74. Harrison P. G. and Thornton E. W. (1978) J. Chem. Soc., Faraday Trans. I, 74:2703; and (1979) J. Chem. Soc., Faraday Trans. I, 75:1487.
75. Larsson R., Lykvist R. and Rebenstorf B. (1982) Z. Phys. Chem. (Leipzig) 263:1089.
76. Morterra C., Garrone E., Bolis V. and Fubini B. (1987) Spectrochim. Acta 43A:1577.
77. Hughes T. R. and White H. M. (1967) J. Phys. Chem., 71:2192.
78. Bourne K. H., Cannings F. R. and Pitkethly R. C. (1970) J. Phys. Chem., 74:2197.
79. Jacobs P. A. (1982) Catal Rev.-Sci. Eng., 24:415.
80. Rabo J. A. and Gajda G. J. (1989/90) Catal. Rev.-Sci. Eng., 31:385.
81. Kazansky V. B. (1977) Proc. Intern. Congr. Catal., London 1976, Vol. 1, p. 50.
82. Ghiotti G., Garrone E., Morterra C. and Boccuzzi F. (1979) J. Phys. Chem., 83:2863.
83. Beebe T. P., Gelin P. and Yates J. T. (1984) Surface Sci., 148:526.
84. Kustov L. M., Kazansky V. B., Beran S., Kubelková L. and Jiru P. (1987) J. Phys. Chem., 91:5247.
85. Kubelkova L., Beran S. and Lercher J. A. (1989) Zeolites, 9:539.
86. Mirsojev I., Ernst S., Weitkamp J. and Knözinger H., unpublished.
87. Echoufi N. and Gélin P. (1992) J. Chem. Soc. Faraday Trans., 88:1067.
88. Lercher J. A. and Noller H. (1982) J. Catal., 77:152.
89. Sanderson R. T. (1972) "Chemical Bonds and Bond Energy", Academic Press, New York.

THE EFFECTS OF COADSORBATES ON THE BEHAVIOUR OF SURFACE SPECIES AND SITES IN CATALYSIS BY MEANS OF EXAFS AND FTIR

Y. IWASAWA
Department of Chemistry, Faculty of Science, The University of Tokyo
Hongo, Bunkyo-ku
Tokyo 113
Japan

ABSTRACT. The discontinuous behaviour of reaction intermediates and active sites in ethanol reaction on Nb catalysts and in water-gas shift reaction on MgO, ZnO and Rh/CeO_2 by means of EXAFS and FTIR is discussed in relation with the effect of coadsorbates as an important aspect of catalysis, where new reaction paths are opened by coadsorbates or adjacent active sites.

1. Introduction

The nature of the interaction between neighboring adsorbates and its role in surface chemical processes must be clarified, since most catalytic reactions proceed on surfaces covered with different kinds of adsorbates. Reaction intermediates, in general, coexist with reactants and products adsorbed on catalyst surfaces under reaction conditions, rather than occurring in isolated distribution without mutual interaction, and the intermediates are thought to be influenced by the coadsorbed species. In several systems, the activation of the intermediate by coadsorbate is suggested[1-4]. Thus it is important to know the behaviour of intermediates in the presence of coadsorbates in order to understand essential factors for the genesis of solid catalysis.

The coadsorption of different molecules on metal single crystals has also been conducted. The successful works are the observation of formate formation from CO+OH on Rh(100)[5] and the enhancement of NH_3 dissociation by coadsorbed CO on Ru(001)[6]. It was proposed that the reactant coverages high enough to force CO and OH into adjacent sites were required for the reaction to proceed[5]. The coadsorption of NH_3 and CO on Ru(001) leads to the formation of mixed ordered layers with a (2x2)

R. W. Joyner and R. A. van Santen (eds.), Elementary Reaction Steps in Heterogeneous Catalysis, 287–304.

structure. The dissociation of NH_3 decreased initially with increasing CO coverage because of its usual poisoning effect, passed a minimum at θ_{co}=0.3, then surprisingly increased with a further increase of CO coverage, and eventually reached a saturation value above θ_{co}=0.5, where NH_3 dissociation is enhanced by a factor of ca.6 by the coadsorbed CO. HREELS spectra indicated that the C_{3v} molecular axis of NH_3 is tilted in the coadsorbed layers. The tilting of NH_3 due to the interaction with CO facilitates electron donation from Ru 4d to LUMO of NH_3, leading to the N-H bond dissociation[6]. The behaviour of NH_3 dissociation is non-linear as a function of coadsorbed CO coverage. These results indicate that local adsorbate-adsorbate interaction plays an essential role in surface chemical reaction.

Besides the effect of coadsorbates, the behaviour of active site may also be influenced by the adjacent active site in a non-linear manner. In order to study this subject, molecular-level design of catalyst is required. Although the efforts on design of excellent catalysts have acutely been difficult challenges, recently molecular-level catalyst preparation has become reallistic on the basis of modern physical techniques and accumulated knowledge of oxide surfaces[7]. Active molecular solid catalysts provide an effective approach for reaction mechanism in a molecular scale. This kind of approach also provides new implication on catalyst development.

In this article the discontinuous behaviour of reaction intermediates and active sites in catalysis are reviewed. Particularly, discontinuous chemistry of coadsorbate effects in water-gas shift reaction on MgO[8], ZnO[9] and Rh/CeO_2[10], and in ethanol reaction on Nb-monomer[1] and -dimer[11] catalysts are discussed.

2. Experimental

2.1. WATER-GAS SHIFT REACTION

MgO powder was obtained by the thermal decomposition of $Mg(OH)_2$, which was made by the precipitation of $Mg(NO_3)_2$(99%) with NH_3(aq). $Mg(OH)_2$ powder was heated under vacuum at a rate of 3.3 K min^{-1} to 673 K and kept at this temperature for 3 h. Zinc oxide powder used in this work was Kadox-15 from New Jersey Zinc Co. CeO_2 was obtained by calcination of $Ce(OH)_3$ aat 673 K for 4 h. $Ce(OH)_3$ was obtained by precipitation of $Ce(NO_3)_3$(Soekawa Co., 99%) with NH_3(aq). Rh-doped CeO_2 catalysts were prepared by a usual impregnation method using aqueous solution of $Rh(NO_3)_3$. The loading of Rh was 0.2 wt%.

IR spectra were recorded on a JEOL JIR-10 Fourier-transform IR spectrometer with a double beam using a liquid nitrogen-cooled MCT detector. The measurements were performed at room temperature or at reaction temperatures. TPD and kinetic studies were conducted in a closed circulating system. Desorption species or reaction products were analyzed by MS spectrometer and gas chromatograph.

2.2. Nb-MONOMER and Nb-DIMER

The SiO_2-attached Nb monomers were prepared by the reaction between $Nb(\eta^3\text{-}C_3H_5)_4$ and surface OH groups of SiO_2(pretreated at 673 K for 2 h in situ before use as support), followed by treatments with H_2 and O_2 as illustrated in Scheme 1. The surface species(A-C) were characterized by EXAFS, FTIR, Raman, XPS, ESR, UV, and TPD[1]. The Nb monomer catalyst(C) was analyzed in detail by EXAFS analysis. The bond distances determined are shon in Scheme 1. No Nb-Nb bond was observed. Nb-Si(surface) distance(0.326 nm) was found besides Nb=O(0.168 nm) and Nb-O(0.193 nm), giving an evidence that the Nb atom is chemically bound to the SiO_2 surface.

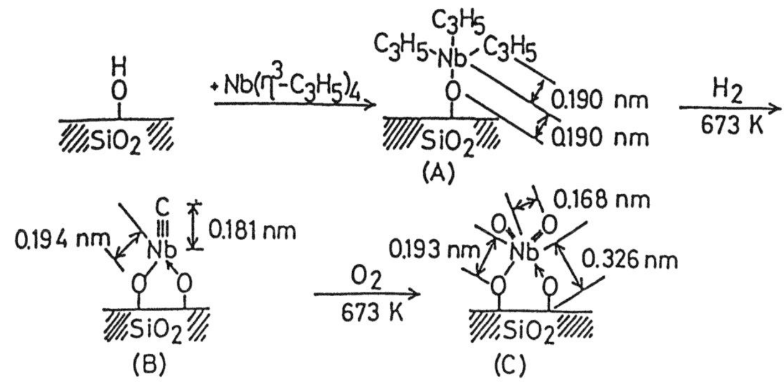

Scheme 1. Preparation steps and structures for Nb monomers on SiO_2.

The dimeric Nb complex$[Nb(\eta^5\text{-}C_5H_5)H\text{-}\mu\text{-}(\eta^5,\eta^1\text{-}C_5H_4)]_2$ was synthesized from the reaction of $NbCp_2Cl_2$ with NaH in water-free THF[12]. The dimeric Nb complex was interacted with SiO_2(Aerosil 200) at 313 K under high purity Ar(99.9999%), followed by treatments with H_2 at 823 K and O_2 at 773 K, to obtain the Nb dimer catalyst(E) as shown in Scheme 2[11]. The catalyst was characterized by EXAFS, XANES, FTIR, UV, XPD, and TPD.

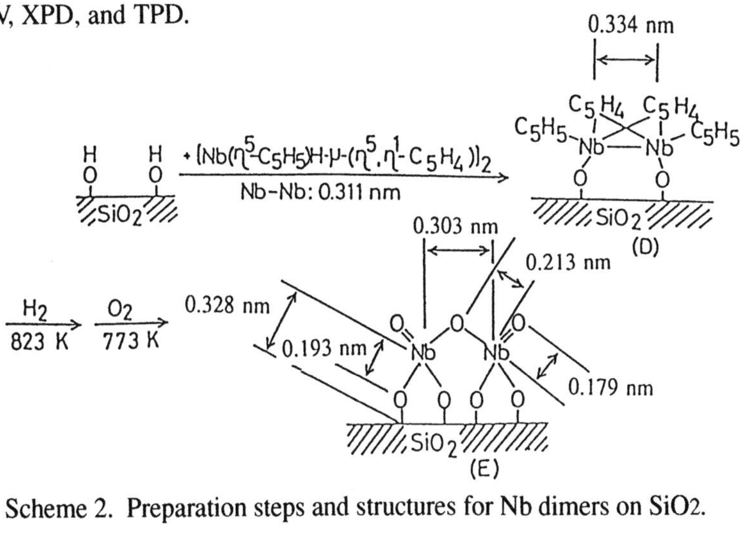

Scheme 2. Preparation steps and structures for Nb dimers on SiO_2.

2.3. EXAFS ANALYSIS

EXAFS measurements for Nb samples were carried out at 80 K at BL 10 B and 6B of Photon Factory in National Laboratory for High Energy Physics. The Fourier-filtered data were analyzed by means of the least-sqaure curve-fitting method by using the theoretical EXAFS eqn[13,14].

$$\chi(k)=\Sigma N_jF_j(k)\exp(-2\sigma_j^2k^2)\sin(2kr_j+\phi_j(k))/kr_j^2 \ ,$$

where N_j, r_j and σ_j represent the coordination number, bond length and Debye-Waller factor, respectively. The curve-fitting analysis for the Nb K-edge EXAFS data was performed by using empirical parameters extracted from Nb foil, $LaNbO_4$, $NbCp_2Cl_2$, and $NbSi_2$ for Nb-Nb, Nb-O, Nb-C, and Nb-Si bonds, respectively.

3. Results and Discussion

3.1. WATER-GAS SHIFT REACTION(WGSR)

3.1.1. MgO

The WGSR system has advantages for studying the behaviour of reaction intermediates as follows: (1) As the local structures of the surface OH groups are estimated by the frequency of OH stretching vibrations(ν(OH)), we can estimate the structures of sites on which the formate intermediates exist. The structure of the formate can be determined by comparing the value of $\Delta\nu=\nu_{as}(OCO)-\nu_s(OCO)$ with those of known metal complexes[15]. Thus the surface formates which have been demonstrated to the intermediates of WGSR([16,17], methanol decomposition [4], and formic acid decomposition[18] can be well characterized, at least in the absence of coadsorbates.

The local structures of OH groups on MgO have been characterized in detail by FTIR as shown in Figure 1[19]. Only terminal OH groups reacted with CO to form surface formates.

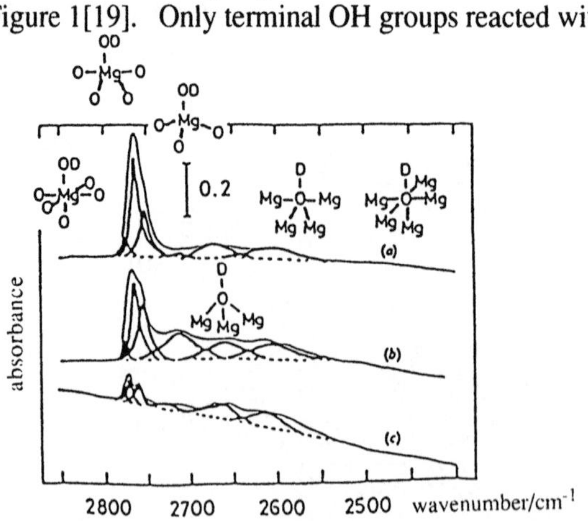

Figure 1. Typical IR spectra of OD groups on three different MgO.

The structure of the formates can be determined by the value of $\Delta\nu=\nu_{as}(OCO)-\nu_s(OCO)$ [20]. When $\Delta\nu$ is larger or smaller than that of a free ion, the formate is considered to be unidentate or bidentate , respectviely. When it is as large as that of a free ion, it is assigned to the bridge-type structure. The unidentate formate is the first surface formate produced at low temperatures, followed by bidentate and bridged formates with increasing temperature. The unidentate formate was preferablely converted to the bridged formate in the presence of water vapor. The OH groups on an ideal MgO(001) flat surface are known to be unstable and readilly removed by evacuation at 673 K[21], the temperature at which the samples were usually pretreated. Therefore, most of the OH groups remaining after pretreatment of MgO may be located on coordinatively unsaturated sites such as step and kink sites. Neighbouring coordinatively unsaturated Mg atoms are necessary to form bridge-type formates which are the main species under catalytic reaction conditions.

Table 1 shows the rate constants for the decomposition of surface formates with and without coadsorbed water. It should be noted that k_+ for the decomposition of H_2+CO_2 was zero in the absence of water, whereas in the presence of water, the formation of H_2 and CO_2 was observed; k_- for the backward decomposition to OH and CO became 10 times smaller than that without water. As there are a large number of OH groups on MgO surface, the selectivity of the formate

Table 1. Rate constants for the decomposition of surface formates as a function of P_{H_2O}.

P_{H_2O}/Pa	$k_+/10^{-4}s^{-1}$	$k_-/10^{-4}s^{-1}$
0	0	13
67	1.1	1.9
670	1.4	0.5

at 600 K.

decomposition to CO_2+H_2 or CO+OH might be thermodynamically affected. However, the decomposition to CO_2+H_2 was never observed as shown in Table 1, 100% of the surface formate being decomposed to CO and OH. The introduction of water vapor increased the total quantity of OH groups+adsorbed water only two-three times at reaction temperatures. The decomposition is kinetically controlled, not controlled by thermodynamical equilibrium. Thus these results show that *adsorbed water molecules create a reaction channel for the dehydrogenation of formates to form CO_2 and H_2.* Surface formates were stabilized by the presence of water as revealed by an upshift in the TPD peak. The rate constant k- is more than 10 times smaller than that in the absence of water. Water molecules are also stabilized by surface formates. Without surface formates, a desorption peak appears at 490 K in the TPD spectra. On the other hand, a desorption peak appears at 520 K when they coadsorb with formates. From the above results, water is suggested to adsorb at coordinatively unsaturated sites. This is also supported by the IR peak shifts. Table 2 shows that the absorption bands and decomposition rate constants fo formate in the presence of water, methanol, pyridine, and THF. ν_{as}(OCO) and ν_s(OCO) were observed to shift in the presence of these electron-donating compounds. The shift of ν(CD) was much less. These results indicate that water molecules behave as electron donors toward coordinatively unsaturated Mg ions. On the other hand, pyridine and THF did not change the values of k_+ and k_-; that is, these electron donors did not promote the forward reaction to form CO_2 and H_2. However, effects on rate constants and peak shifts similar to those of water were observed with methanol, which is a weak electron donor, but can hydrogen-bond with O anions on the MgO surface. As a common feature of water and methanol in adsorption, a coadsorption strcuture involved in WGSR is shown in Scheme 3, where the water molecules donates

Table 2. IR bands(cm^{-1}) and decomposition rate constants($10^{-4}s^{-1}$) of bridge formate at 600 K in the presence of electron donors.

Molecules	ν(CD)	ν_{as}(OCO)	ν_s(OCO)	k_+	k_-
None	2120	1615	1321	0	13
Water	2130	1593	1333	1.4	0.5
Methanol	2125	1600	1334	0.5	7
Pyridine	2121	1595	1334	0	12
THF	2120	1610	1330	0	13

electrons to the Mg ion by the lone-pair electron of the oxygen atom and withdraws electrons form the lattice oxygen atom with the hydrogen atom. The energy level of the 3s orbital of Mg in Scheme 3 is increased by the coordination of water-hydrogen to the lattice oxygen(bridge OH), making it easier for a chemical bond to form between Mg and O(bridge formate) in Scheme 3 and resulting in the decrease in k_- and the promotion of the forward decomposition of formates.

Scheme 3 shows the backward decomposition of the bridge formate and the catalytic cycle for WGSR. Water dissociates preferably at edge or corner sites on the MgO surface to produce on-top and bridge OH groups. CO reacts with the on-top OH groups to produce formates as proved by IR. The formates decompose into CO and OH in the absence of water. However, in the presence of water, 70% of the formates decompose to CO_2 and H_2. The reaction rate increased linearly with the amount of adsorbed water. Water molecules are suggested to interact more strongly with Mg sites coordinated by formates than with other Mg sites without formate ligands. The electron-donating and -withdrawing property of water molecules are essential for the forward decompo-sition of the formates to produce CO_2 and H_2. The interaction of water also prevents the backward decomposition of formates to CO and OH.

3.1.2. ZnO

On ZnO surface, the rate of the formate docomposition was promoted by a factor of more than 10 by the presence of a second water molecule. The activation energy of the decomposition of the formates decreases in the presence of water; 155 kJ mol-1 under vacuum and 109 kJ mol-1 with ambient water.

TPD spectra of the surface formates formed by the reaction between surface OD groups and CO at 493 K were taken at a heating rate of 3 K min^{-1} in Figure 2(left). Desorption peaks for CO and D2 appeared at 550 K. The ratio of CO to D2 was about 7:3. The desorption of CO2 gave no definite peak, showing a gradual increase with temperature. Figure 2(right) shows TPD spectra of the surface formates coadsorbed with water. Water vapor was introduced at 303 K. Desorption of D2 and CO was observed at 550 K, and that of CO2 and water appeared at 660 and 720 K, respectively. In contrast to the case of the absence of water, the amount of desorbed CO was very small, where the equal amount of D2 and CO2 were preferentially formed.

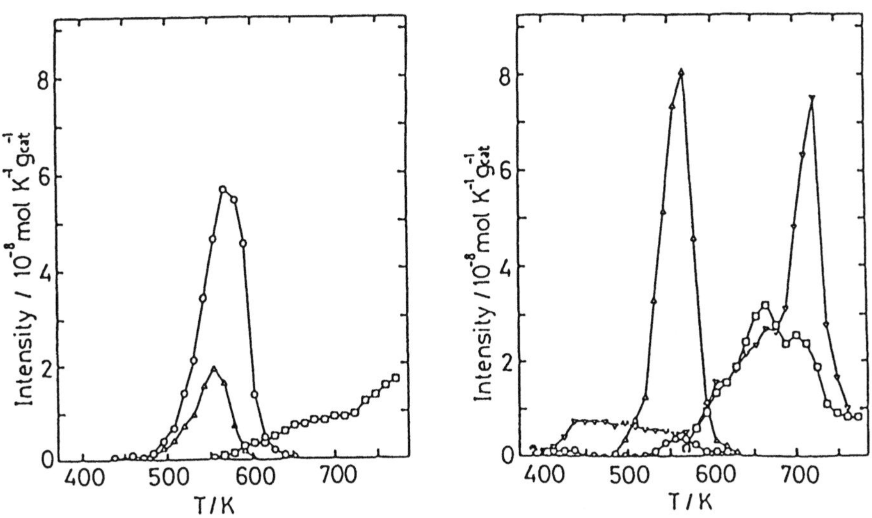

Figure 2. TPD spectra of surface formate(DCOO) produced by CO and surface OD (left) and TPD spectra of surface formate(DCOO) and D2O coadsorbed on ZnO(right); (O) CO, (□) CO_2, (△) D_2, (▽) D_2O.

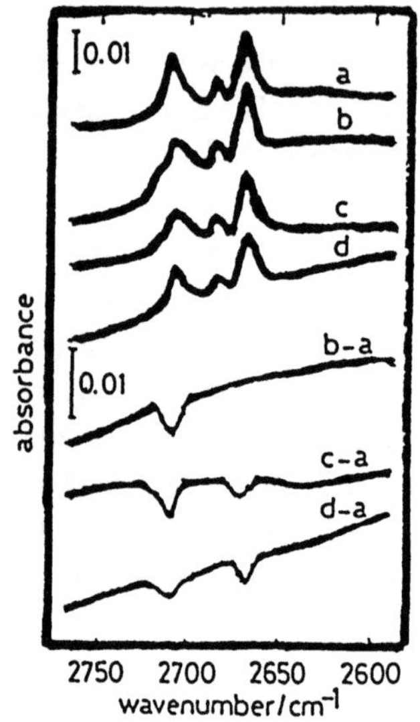

Figure 3. ν(OD) stretching bands of OD groups on ZnO(D); (a) after evacuation at 673 K; (b) after exposing sample (a) to CO at 473 K (P_{CO} = 1.33 × 10^3 Pa); (c) after evacuating sample (b) at 573 K, (d) after evacuating sample (c) at 673 K.

The local structure of surface OH groups on ZnO has been investigated by means of IR [22-25]. Taking account of the crystal structure of ZnO, the absorption bands at 2715, 2706, 2682, and 2669 cm^{-1} in Figure 3 are attributable to on-top OD groups on three-coordinated Zn ion, on-top OD groups on two-coordinated Zn ions, bridge OD groups and three-fold OD groups, respectively.

In Figure 3 the peak at 2706 cm^{-1} decreased when CO was introduced at 473 K and at the same time surface formates were produced, suggesting that OD groups on two-coordinated Zn ions react with CO to produce surface formates. When the sample with the formates was evacuated at 573 K at which temperature the formates are decomposed, the bands at 2682 and 2669 cm^{-1} decreased as shown in Figure 3. This implies that the deuterium atoms of bridge and threefold hollow OD groups reacted with the D atoms of formate(DCOO) ions to form D2.

The forward rate of the formate decomposition increased more than 10 times by the presence of ambient H2O, accompanied with the decrease in the activation energy from 155 kJ mol^{-1} to 109 kJ mol^{-1} as shown in Table 3. This fact demonstrates that water activates the surface formate to promote the forward decomposition for the formation of H2+CO2(ad). Electron donors also increased the decomposition rate. The driving force for the forward decomposition of formates is suggested to be electron donation to the Zn ion on which the bidentate formate adsorbs. Again, the peak shift for the bidentate formate by water addition suggests a certain interaction between them.

Table 3. Rate constants for the formation of hydrogen or carbon dioxide in various combinations of labeled water and bidentate.

Combination	$k_+/10^{-4}s^{-1}$	Activation Energy/kJ mol^{-1}
$DCOO^-$ alone	0.34	155
$DCOO^-$ + D_2O	5.3	109
$DCOO^-$ + H_2O	5.4	
$HCOO^-$ + D_2O	7.0	
$HCOO^-$ + H_2O	7.9	
Carbonate + D_2O	1.0	
Carbonate + H_2O	1.0	

at 530 K; water: 400 Pa.

An isotope effect in the decomposition rate was observed with the hydrogen atom of formate as shown in Table 3. Almost no isotope effect was observed with the hydrogen atoms of surface hydroxyls. The decomposition rate of HCOO was 1.4 times larger than that of DCOO. Thus the rate-determining step is suggested to be the dissociation of the CH bond of formate.

Scheme 4 shows the reaction mechanism of catalytic WGSR on ZnO surface. The first water molecule dissociates at mainly Zn2c-O3c pair to form OH groups. On-top OH groups on Zn react with CO to produce bidentate formates through unidentate formates. Seventy percent of the formates reversibly decomposed to the original OH and CO under

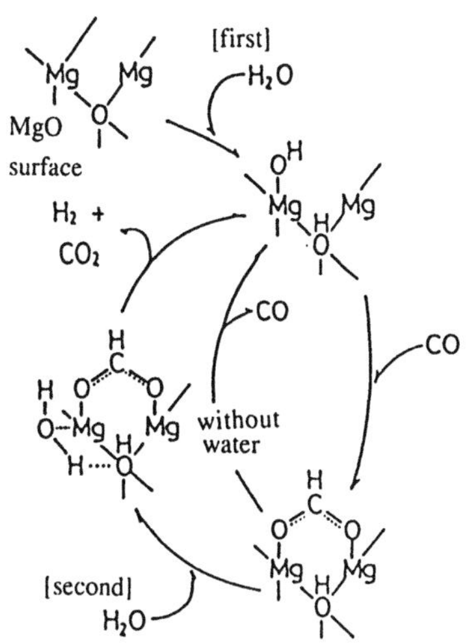

Scheme 3. WGSR mechanism on MgO.

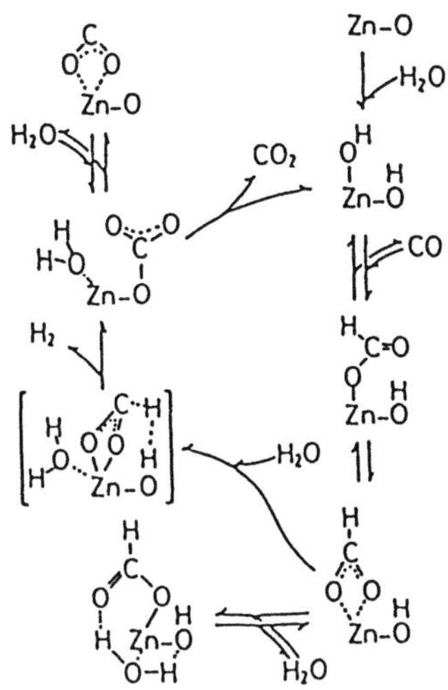

Scheme 4. WGSR mechanism on ZnO.

vacuum and only 30% of them reacted with the hollow-site OH groups to produce H_2 and CO_2(ad). When the second water molecule adsorbed on the Zn atom, almost 100% of the bidentate formates were able to decompose to H_2 and CO_2(ad)(unidentate carbonate) as illustrated in Scheme 4. The rate was promoted by a factor of more than 10 as compared with that in the absence of the second water molecule. An isotope effect was observed with the hydrogen atom of the formate and not observed with the hydrogen atom of the OH and introduced H_2O, indicating that the rate-determining step in the decomposition of formate or equally in the formation of H_2 is the dissociation of the C-H bond of the formate. The CO_2(ad) produced by the decomposition of the bidentate formate in the presence of the second water molecule is the unidentate carbonate as proved by IR.

When undissociative electron donors such as pyridine and THF were admitted to the carboxylate surface, the carboxylate was converted to the unidentate carbonate at room temperature, but no desorption of CO_2 was observed. When a dissociative electron donor such as water was introduced, the carboxylate was converted to the unidentate carbonate and the desorption of CO_2 by the decomposition of the carbonate was enhanced. Thus the decomposition of the unidentate carbonate to form CO_2 was also markedly promoted by the second water molecule adsorbed on the Zn atom. The accompanying dissociation of the adsorbed water to Zn-OH and O-H may assist the CO_2 desorption from the carbonate. In the steady-state reaction the rates of the two reaction steps(formate--->carbonate+H_2 and carbonate--->CO_2) are balancing in the presence of adsorbed water.

3.1.3. Rh/CeO_2

CeO_2 was contrasted to MgO and ZnO; the surface formate on CeO_2 was stabilized by water coexistence, rather than the promotion of WGSR, where the selectivity to H_2 and CO_2 only increased as a result of the more suppression of the backward decomposition of formate by water coexistence. No reactant-promoted effect on CeO_2 may be due to the intrinsic nature of CeO_2 or due to the stable structure of CeO_2 surface. We have tried to modify the properties of CeO_2 surface by doping Rh to CeO_2. Rh catalysts with CeO_2 have been used automobile exhaust catalysts[26], in which systems WGSR also proceeds.

The decomposition of the surface bidentate formate on Rh/CeO_2 was markedly affected by the presence of water. Sixty three percents of the formate decomposed to CO+OH (backward) and 37% of them decomposed to H_2+CO_2(ad)(forward) in vacuum. When 0.67 kPa of water vapor coexisted, almost 100% of the surface formates decomposed toward the forward direction and the backward decomposition was not observed as shown in Table 4. Besides the enhancement of the decomposition selectivity, the rate constant of decomposition increased in the coexistence of water; the value of k_+ became 100 times as large as that without water(Table 4). The activation energy of the forward decomposition decreased from 56 kJ mol^{-1}(under vacuum) to 33 kJ mol^{-1}(with water).

Table 4. Rate constants for the decomposition of bidentate formates under vacuum and in the presence of various coadsorbates.

Catalysts	Temp./K	Coadsorbed Species	$k_+ + k_-/s^{-1}$	k_+/k_-	$E_a(k_+)$	$E_a(k_-)$
CeO_2	573 K	None	3.15×10^{-3}	30/70	207	94
		H_2O	1.75×10^{-4}	50/50	264	193
		CH_3OH	6.09×10^{-4}			
		NH_3	8.81×10^{-4}			
		Pyridine	1.60×10^{-3}			
Rh/CeO_2	443 K	None	1.1×10^{-5}	35/65	56	122
		H_2O	1.12×10^{-3}	100/0	33	
		CH_3OH	1.58×10^{-3}			
		Pyridine	5.8×10^{-4}			
		THF	2.0×10^{-4}			

Coadsorbed molecules of 670 Pa were introduced.

These results demonstrate that the change of the selectivity(forward/backward) is not caused by an equilibrium shift but ascribed to the change of reaction path/reaction mechanism.

This promoting effect was also observed when methanol, THF, pyridine were introduced. However, the promotion by H_2O and CH_3OH was larger than that by pyridine and THF. The former molecules can interact with Ce-O pairs at the surface in an electron donor-acceptor pair mode, but the latter molecule interact with Ce ions alone. Electron transfer from the oxygen atom of Ce-O pair to the OH hydrogen of H_2O or CH_3OH through hydrogen bonding decreases the diameter of surface oxygen ion and the electron donation from the oxygen lone pair electrons to Ce ion becomes more strongly.

The iostope effects were observed with the hydrogen atom of formate and not observed with the hydrogen atom of water molecule. The result is similar to that observed on ZnO, where the rate determining step of the formate decomposition is suggested to be dissociation of the CH bond of the bidentate formate. In contrast, the isotope effects on CeO_2 were observed with the hydrogen atoms of both formate and water, where the activation of both the C-H bond of the formate and the O-H bond of the adjacent bridge OH groups in the transition state is necessary. The local structure and arrangement of the Ce-O reaction sites appear to be different between Rh/CeO_2 and CeO_2 surfaces, which must be caused by the difference of the degree of surface reduction.

The frequencies of the IR peaks of adsorbed species such as formate, carbonate and hydroxyl groups on Rh/CeO_2 were almost the same as those on CeO_2. It means that the electronic states of Ce-O pairs on CeO_2 and Rh/CeO_2 are not different. By addition of a small amount of Rh(0.2 wt%), the rate of WGSR increased tremendously. The mechanism for WGSR is essentially the same as that for ZnO.

3.2. ETHANOL REACTION ON Nb MONMOER/SiO2

Niobium has been considered to be insufficient element for catalysis although niobium oxide has been studied largely in view of SMSI phenomenon[27] or acidity[28]. In fact, Nb_2O_5 was almost inactive and non-selective for ethanol conversion[1]. Also the usual impregnated Nb catalysts did not behave as good catalysts as shown in Table 5[1].

Table 5. Typical results for activities and selectivities of three different Nb catalysts in the dehydrogenation and dehydration of ethanol at 523 K,

Catalyst	Initial rate*		Selectivity/ %		
	(i)	(ii)	Ethene	Diethyl Ether	Acetaldehyde
Nb-Monomer					
2.0 wt%	1.20	0.05	2.8	1.1	96.1
Activation Energy/kJ mol^{-1}			---	---	65
Nb-Dimer					
0.37 wt%	0.01	0.17	32.6	63.9	3.5
1.0 wt%	0.01	0.19	34.2	62.0	3.8
Activation Energy/kJ mol^{-1}			119	79	68
Impreg.Cat.					
0.45 wt%	0.05	0.12	26.0	45.3	28.6
Activation Energy/KJ mol^{-1}			129	85	70

*EtOH: 3.3kPa, mmol min^{-1} g-Nb^{-1}; (i)dehydrogenation, (ii)dehydration.

The SiO_2-attached Nb catalyst(C) exhibited a high activity and selectivity for the dehydrogenation of ethanol[1]. Acetaldehyde and hydrogen were stoichiometrically produced during the dehydrogenation(Figure 4). When the ambient ethanol was evacuated in the course of the dehydrogenation at 523 K, the reaction completely stopped as shown in Figure 4. The dehydrogenation began again by introducing ethanol vapor. Ethanol dissociatively adsorbed on Nb to form OH and C_2H_5O groups as proved by IR(Figure 4). These species still remained on the surface after evacuation for 10 min at 523 K; the typical peaks for the adsorbed species are shown in Figure 4. The intensity of the peaks during the period when the reaction did not proceed was almost the same as that during the steady-state reaction. The XANES 1s--->4d peak intensity for the catalyst saturated with OH and C_2H_5O groups(F) was half of that for the catalyst(C), suggesting that one of two Nb=O bonds reacted with one ethanol molecule. These are in good agreement with the volumetric measurements of the amount of ethanol adsorbed.

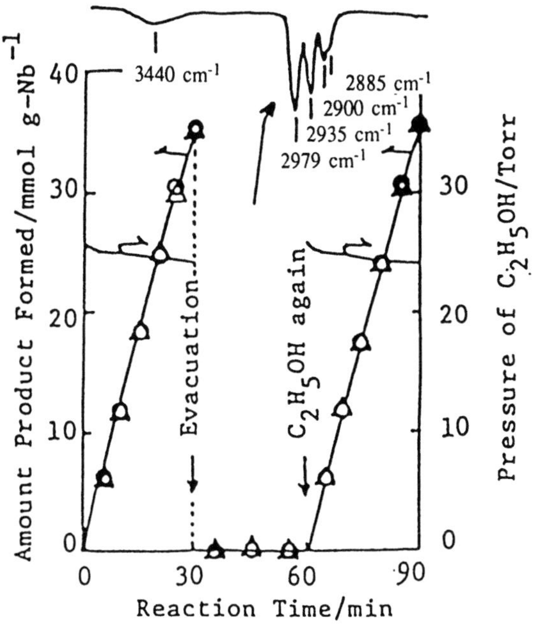

Figure 4. The dehydrogenation of ethanol on Nb-monomer/SiO2; O: H_2, Δ: CH_3CHO.

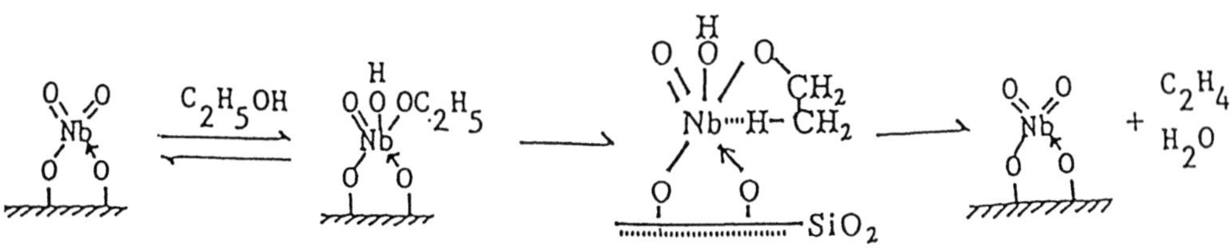

(C) (F) (G) (C)

The adsorbed species(F) under vacuum were too stable and it did not react further. By heating the (F) to much higher temperatures than 600 K, the species(F) was dehydrated to ethene and water as shown in Figure 5.

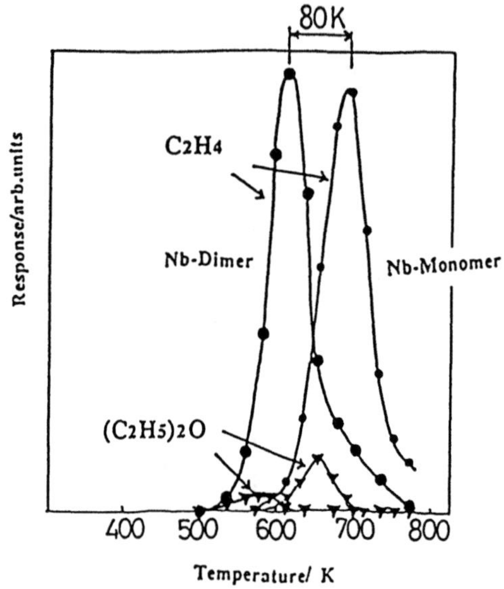

Figure 5. TPD spectra from adsorbed ethanol on Nb monomer/SiO_2 and Nb dimer/SiO_2.

On the other hand, when the (F) was exposed to ethanol, the dehydrogenation of ethanol was found to proceed at much lower temperatures like 423 K, with ca.100% selectivity. Thus the ambient ethanol is necessary for dehydrogenation.

When various electron donors were admitted onto the species(F), H_2 and acetaldehyde formations from OH/C_2H_5O started as shown in Figure 6. The abstraction of β-hydrogen of C_2H_5O group was assisted by electron donating reagents.

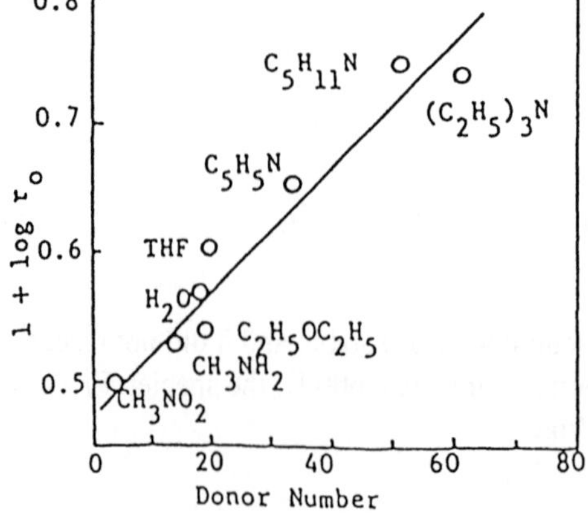

Figure 6. The correlation between initial rates of H_2(CH_3CHO) formation from (F) and donor numbers of nucleophillic reagents.

The catalytic dehydrogenation is expressed as follows.

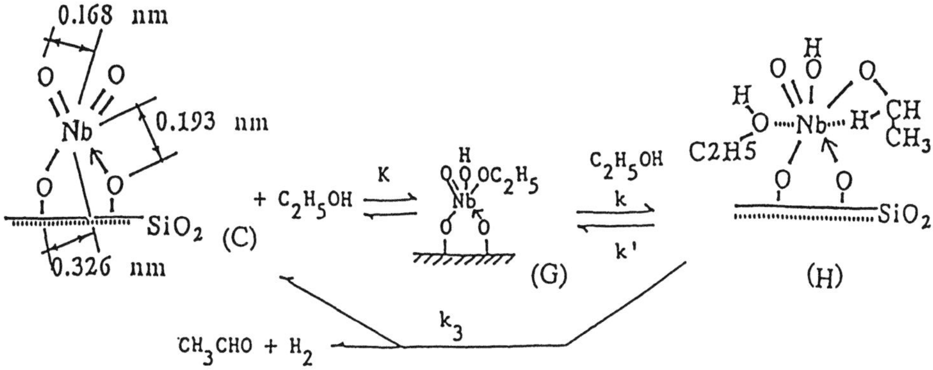

From kinetics, the rate constant k3 was determined to be 1.5 mmol min^{-1} $g\text{-}Nb^{-1}$. The equilibrium constant K'=k/k' was obtained to be 0.12 $Torr^{-1}$. The value of K' was smaller than 0.2 $Torr^{-1}$ for the equilibrium constant K obtained form the IR studies. The first ethanol adsorbed on Nb monomer is dehydrogenated to form H2 and acetaldehyde by assist of the second ethanol molecule, not by direct collision from ambient ethanol.

Even if an adsorbed species at surface was too stable to react under vacuum, the catalytic reaction was possible to proceed through the same species in the presence of ambient gas. *The present results show the switchover of the reaction path from dehydration(γ-hydrogen abstraction) to dehydrogenation(β-hydrogen abstraction) by the second ethanol molecule.*

3.3. ETHANOL REACTION ON Nb-DIMER/SiO2

The aim of this study is to find the effect of the second Nb atom instead of the second ethanol molecule(cf.3.2) and to understand the cooperative catalysis of two adjacent Nb atoms, in other words, whether the dimer is merely the double of the monomer, or not [11].

The EXAFS analysis for the dimer structure(E) in Scheme 2 is shown in Table 6. The observation of Nb-Nb bond at 0.303 nm with C.N. of 0.9 demonstrates the retention of a dimeric structure on SiO2. The value of 0.303 nm is shorter than 0.311 nm for the dinuclear Nb complex having a direct interaction between two Nb atoms. The Nb-Si length was determined as 0.328 nm with C.N. of 2.3. The distance is similar to 0.326 nm for the Nb monomer. There are two different Nb-O bonds at 0.193 nm and 0.213 nm. The former distance is the same as Nb-O (surface) distance for the monomer(C), and the C.N. of 1.9 is the onset of Nb-Si. Thus the 0.193 nm bond is assigned to the bonding of Nb with two surface oxygen atoms as shown in Scheme 2. The bond of 0.179 nm is straightforwardly assigned to Nb=O bond. The remaining Nb-O bond at 0.213 nm with C.N. of 1.1 is due to the oxygen-bridged structure(E) with the Nb-Nb separation of 0.303 nm. The similar bond distances(0.209-0.211 nm) have been reported for Nb-O(bridged) bonds for Nb-oxide monolayers on SiO_2[29]. The XANES spectrum showed a distinct 1s--->4d transition, which implies a tetrahedral/disstorted tetrahedral symmetry.

Table 6. Best-fit result for the EXAFS data of the Nb dimer on SiO_2(E).

Bond	C.N.	r/nm[b]	DW/nm[c]
Nb=O	0.6 ± 0.2	0.179	0.008
Nb-O	1.9 ± 0.2	0.193	0.005
Nb-O	1.1 ± 0.1	0.213	0.005
Nb-Nb	0.9 ± 0.1	0.303	0.006
Nb-Si	2.3 ± 0.2	0.328	0.004

a: 1.0 WT%-Nb, b: distance(±0.003), c: Debye-Waller factor,
Filtering range: 0.13-0.35 nm(2) and 0.10-0.35 nm(4).

Ethanol(3.3 kPa) was admitted onto (E) to form $Nb(OH)(OC_2H_5)$ at 373 K, followed by evacuation for 30 min, exhibiting the IR peaks at 3471 cm^{-1}(ν(OH), and 2979, 2937, 2900, and 2883 cm^{-1}(ν(CH)) similar to those for the Nb monomer. Then, the TD spectrum was taken in Figure 5. Ethene was desorbed. The peak temperature(610 K) for ethene was lower by 80K than 690 K for the Nb monomer, suggesting the promotion of ethene formation with the dimer. The characteristic feature of the EXAFS analysis for the sample saturated with ethanol at 523 K is the disappearance of the Nb=O bond. Instead, the C.N. of the Nb-O bond at 0.192 nm increased to 3.9 by 2 from 1.9 for the catalyst(E), which agrees with the formation of $Nb(OH)(OC_2H_5O)$ by the dissociative adsorption of ethanol. The other bonds remained unchanged, demonstrating no change in the oxygen-bridged dimeric structure.

It was found that the Nb dimer catalyst always showed a high selectivity for dehydration of ethanol irrespective of the presence or absence of the second ethanol in Table 5. This is contrasted to the catalysis of the monomer. The dehydrogenation was remarkably suppressed and the dehydration was promoted on the dimer. The results imply that the Nb dimer on SiO_2 has an acidic character. *It is to be noted that the change of the number of Nb atoms in active sites form one to two metal atoms gave rise to the complete reverse of basicity/acidity in catalytic properties*[11].

The rate of ethene formation was accelerated 1.7 times by replacing C_2H_5OH by C_2D_5OD. The substitution of the OH hydrogen with deuterium gave almost no effect on the rate. The inverse isotope effect can be explained by a late transition-state theory. By taking into account the vibrational partition functions, the ratio of the rate constants(k_D/k_H) is expressed by the following eqn.

$$k_D/k_H = \Pi \exp[\mathbf{h}(\nu_{Hi}^{\ddagger} - \nu_{Di}^{\ddagger} - \nu_{Hi} + \nu_{Di})2\mathbf{k}T]$$

We adopted ν(OH) of H_2O and ν(CH) of methylene group as the vibrational modes of the transition state(K), while the observed values were used for the vibrational modes of $Nb(OH)(OC_2H_5)$(J). The calculated inverse isotope effect was 1.6, which is in good

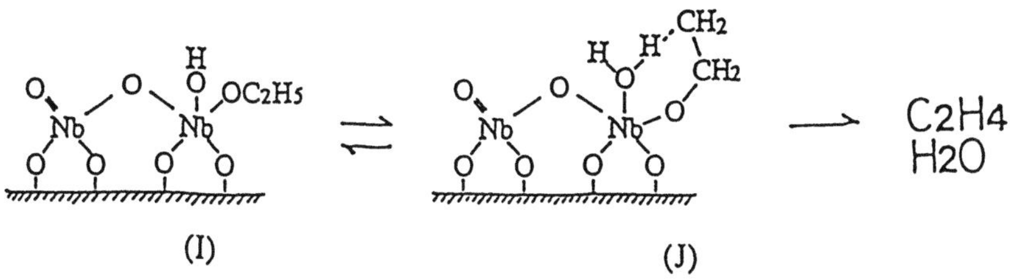

agreement with the experimental value.

The adjacent Nb atom in the dimer(E) may change the local conformation of Nb sites. In the dimer structure the access of a second ethanol molecule to the Nb atom with a first ethanol molecule in a preferable conformation is difficult unlike to the case of the monomer(C). In other words, the assisted-dehydrogenation[1] is not possible with the dimer. The charge population on Nb in the dimer reduced to 92% of that of the monomer. This trend leads to an increase in acidic character of the Nb site.

4. Conclusions

In this article the reactant-promoted reaction mechanisms opened by the activation of reaction intermediates by coadsorbed, second reactant-molecules were described. The first reactant-molecule is converted to an intermediate and the intermediate is converted to the products by the assist of the second reactant-molecule due to intermediate-reactant interaction. In this case the intermediate is not necessarily reactive. Unreactive species can be reactive intermediate by this mechanism, which may provide new implication on a way of catalyst preparation. Besides the promotion of reaction rate, the reaction path(selectivity) is possible to be changed by intermediate-coadsorbate interaction. The first and second reactant-molecules play different roles and behave in a discontinuous (non-linear) manner.

5. References

1. Nishimura, M., Asakura, K., and Iwasawa, Y., Proc.9th Internat. Congr. Catal. (Phillips, M.J. and Terna, M. eds), 4, 1842(1988).
2. Cant, N.W. and Bell, A.T., J.Catal., 73, 257(1982).
3. Ueno, A., Onishi, T., and Tamaru, K., Trans.Faraday Soc., 67, 3585(1971).

4. Yamashita, K., Naito, S., and Tamaru, K., J.Catal., 94, 353(1985).
5. Gurney, B.A. and Ho, W., J.Chem.Phys., 87, 1376(1987).
6. Sasaki, T., Aruga, T., Kuroda, H., and Iwasawa, Y., Surf.Sci., 240, 223(1990).
7. Iwasawa, Y., Adv.Catal.(Eley, D.D., Pines, H., and Weisz, P.B. eds), 35, 187 (1987).
8. Shido, T., Asakura, K., and Iwasawa, Y., J.Catal., 122, 55(1990).
9. Shido, T. and Iwasawa, Y., J.Catal., 129, 3443(1991).
10. Shido, T. and Iwasawa, Y., J.Catal., in press.
11. Ichikuni, N. and Iwasawa, Y., Proc. 10th Internat. Congr. Catal.(1992), in press.
12. Lemenovskii, D.A., Urazowski, I.F., Nivant'ev, I.E., and Pervalova, E.G., J.Organomet.Chem., 292, 217(1985).
13. Teo, B.K., "EXAFS:Basic Principles and Data Analysis", Inorganic Concepts, Springer, vol.4 (1986).
14. Asakura, K. and Iwasawa, K., J.Chem.Soc., Faraday Trans.l, 84, 2445(1988).
15. Nakamoto, K., "Infrared and Raman Spectra of Inorganic and Coordination Compounds" 3rd ed., Wiley(1987).
16. Sholten, J.J.F., Mars, P., Menon, P.G., and Hardeveld, R.Van., Proc. 3rd Internat. Congr. Catal. Amsterdam, 2, 881(1965).
17. Ueno, A., Onishi, T., and Tamaru, K., Trans.Faraday Soc., 66, 756(1970).
18. Noto, Y., Fukuda, K., Onishi, T., and Tamaru, K., Trans.Faraday Soc., 63, 3081(1967).
19. Shido, T., Asakura, K., and Iwasawa, Y., J.Chem.Soc., Faraday Trans.l, 85, 441(1989).
20. ref.15.
21. Onishi, H., Aruga, T., Egawa, C., and Iwasawa, Y., Surf.Sci., 191, 479(1987).
22. Atherton, K., Newbold, G., and Hockey, J.A., Discuss.Faraday Soc., 52, 33 (1971).
23. Dent, A.L. and Kokes, R.J., J.Phys.Chem., 63, 3781(1969).
24. Tsyganenko, A.A. and Fillimonov, V.N., J.Mol.Struct., 19, 379(1973).
25. Mattman, G., Oswald, H.R., and Schweizer, F., Helv.Chim.Acta, 55, 1249(1972).
26. Wei, J., Adv.Catal.(Eley, D.D., Pines, H., and Weisz, P.B. eds), 24, 57(1975).
27. Baker, R.T., Tauster, S.J., and Dumesic, J.A. eds., "Strong Metal-Support Interactions", ACS Symp.Ser. 298, American Chemical Society(1986).
28. Iizuka, T., Ogasawara, S., and Tanabe, K., Bull.Chem.Soc.Jpn., 56, 2927(1983).
29. Shirai, M., Ichikuni, N., Asakura, K., and Iwasawa, Y., Catal.Today, 8, 57(1990).

HOW CAN SCANNING TUNNELING MICROSCOPY, COUPLED WITH THEORY, HELP US UNDESTAND SOME ELEMENTARY STEPS IN CATALYSIS?

P. SAUTET[1],
Institut de Recherche sur la Catalyse, 2 Avenue A. Einstein, 69626 Villeurbanne cedex, France
Laboratoire de Chimie theorique, Ecole Normale Superieure de Lyon, 69364 Lyon Cedex 07, France

J.C. DUNPHY and M.B. SALMERON
Materials Science Division, Lawrence Berkeley Laboratory, Berkeley, CA 94720, USA

ABSTRACT. Scanning Tunneling Microscopy, complemented by a theoretical analysis of the images, can give important informations on static steps of an heterogeneous reaction mechanism, such as the adsorption of atoms and molecule on a metallic surface. The examples of the adsorption of S on Re(0001) and of benzene on Rh(111) are detailed after a presentation of the theoretical approach. The STM image is shown to be strongly dependent on the geometry and symmetry of the adsorption sites and to give important informations on the electronic structure of the adsorbate.

1. Introduction

The scanning Tunneling Microscope has recently emerged as a major tool for the study of conducting surfaces with an atomic resolution[2]. It has been applied in surface science to solve several surface structure problems[3]. It has also been used for the study the adsorption of atoms and in some cases of molecules on metallic surfaces[4]. However the interpretation of such images is usually difficult for several reasons. First, the metallic substrate is generally not resolved on an image with adsorbates, and this even if atomic resolution images can be obtained for the bare surface[5]. This is due to the fact that a high tunneling current is necessary for obtaining the atomic resolution on metals, the tip being close to the surface. On the contrary, the study of adsorbates requires a low current, to avoid displacement of the adsorbate by the tunneling tip. Moreover the STM image is very far from a van der Waals representation of the surface geometry since it is a probe of the surface wavefunction at the Fermi level electron energy. As a consequence, the analysis of the image is not strait forward since the adatoms can appear as a bump, as a hole, or not at all in the case of an adsorbed molecule. Therefore if the symmetry and period of the adlayer can be easily addressed with an STM image, the adsorption site and geometry can not be usually directly obtained.

R. W. Joyner and R. A. van Santen (eds.), Elementary Reaction Steps in Heterogeneous Catalysis, 305–311.

In order to go beyond the visual analysis of the image, a theoretical method has been developed to calculate STM images[6]. In a first step this new theoretical technique will be introduced. We will show then how, by comparison between experimental and theoretical images, the adsorption site and the adsorption geometry of the adatom can be obtained from STM data. This analysis will be detailed on the example of the (2x2) adsorption of sulfur on a Re(0001) surface. On the second case of benzene on Rh(111) the relation between the STM image and the symmetry of the adsorption site will be detailed. The relation with the molecular activation upon adsorption will be underlined.

2. Method of calculation

In order to simulate an STM image one needs to be able to calculate the tunneling current between a tip and a surface as a function of their x,y,z relative position. The actual trend is to use the Tersoff-Hamman local-density-of-states (LDOS) approximation[7]. Our approach does not rely on this approximation and its main characteristic is to consider the electronic tunneling between tip and surface as a scattering phenomenom[5]. The probability for a conduction electron of the surface to be transmitted through or reflected by the tunnel gap is directly calculated and expressed in a scattering matrix. The tunneling current is then calculated from the scattering matrix with the Landauer formula[8], valid at low polarization voltage. Since the current can be obtained with this method for any xyz position of the tip, the simulation of the STM image is then direct.

This scattering matrix through the tunnel gap, for the conduction electrons of the surface and tip, is calculated via a propagative technique. The calculated system includes simultaneously the surface, the adsorbate layer and the tip and it is infinite in the z direction perpendicular to the surface. No slab approximation is used to described the tunneling interface. The system is finite in the xy direction with adequate cyclic boundary conditions. Each atom is associated with an orbital basis set and the quantum chemistry inputs for the recursive calculation of the scattering matrix are the Hamiltonian and overlap matrix elements on this basis set. While any calculation method of these matrix element could be used, the Extended Hückel approximation is applied for the present calculation for simplicity.

3. STM images and the determination of the adsorption site

Adsorption is a very important first step of most heterogeneous catalysis mechanisms. The site of the adsorption and the coordination of the adsorbate on the surface generally determines the activation and the further reactivity and selectivity of this adsorbate compare to its gaz phase behavior. It is therefore most important to decide if the scanning tunneling microscope can be used as a technique of direct visualization of the adsorption site. As already mentioned, the difficulty for this determination is that the first metal layer is not seen on the image and that only the signal of the adsorbate is present. However this signal of the adsorbate is an indirect signature of its electronic structure and interaction with the substrate, and therefore should contain the information on the adsorption site.

Theoretical STM images have been calculated for the 2x2 adsorption of sulfur on the (0001) surface of Rhenium for the hollow, bridge and top sites, as a test of the dependance of the STM image on the adsorption site[9]. Three shapes of images have been found experimentally: The first image shows an array of round bumps in a 2x2 arrangement. In the second image the bump has a marked triangular shape, while in the third type the bright area extends between the sulfur atoms to built a honeycomb lattice. These three types of images are shown in Figure 1. It has been shown previously that this change in the image shape results from a modification of the atomic structure of the tip end[9,10]. From the analysis of images with point defects, it was also shown that the bump is located on the sulfur atom in the case of the ball and triangle images, and that similarly the sulfur atom is on the bright honeycomb lattice in this third case.

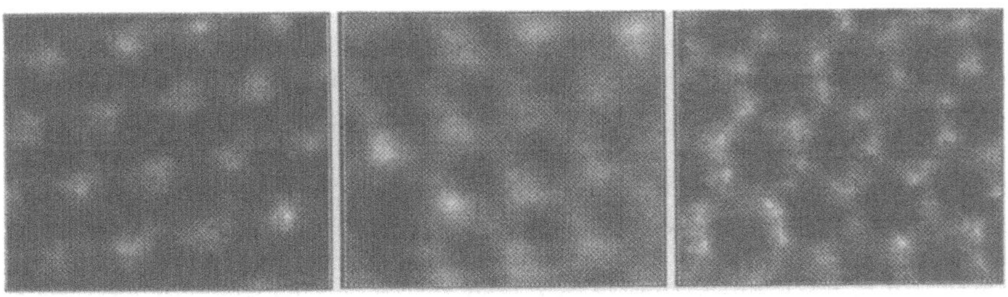

a) b) c)

Figure 1: The three types of STM images obtained for S(2x2) on Re(0001); a) ball, b) triangle, c) honeycomb.

Figure 2 shows the calculated images for a S atom on a top,bridge and hollow site on the surface. In each case the bond distance is fixed to the sum of the Re and S covalent radii. For simplicity only one type of tip structure is considered here, a tip ending with a Pt metal atom, but the result can be extended to the other types of tips considered in [9].

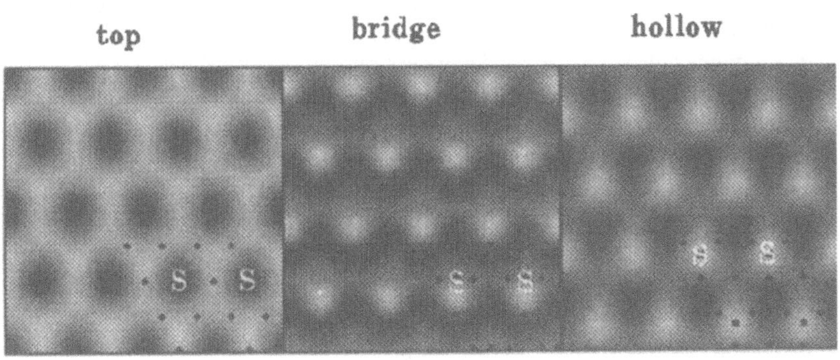

Figure 2: Influence of the adsorption site on the calculated image of S(2x2) on Re(0001).

The shape of the image is found to be strongly dependent on the adsorption site. In the case of the top adsorption, the resulting image is a honeycomb lattice but the sulfur atom is located in the dark spot of the image, in contradiction with the experimental observation. The bridge conformation shows a marked two-fold image, that has a very small corrugation, and is not seen experimentally. Therefore only the 3-fold hollow site gives a theoretical image that compares well with one of the experimental images, in which the S atoms appear as overlapping triangles (fig 1b). It was shown that for this same hollow adsorption geometry the ball image (fig 1a) is obtained theoretically with a tip ending with a S or C atom, characterized by a small electronic radius of the probing atom, while the honeycomb image (fig 1c) is obtained with a "large" tip, ending with a microfacet of two or three metal atoms.There are two types of hollow sites on a (111) face of an fcc metal. However the fcc hollow and the hcp hollow sites give a similar calculated image and their comparison would require a very precise quantitative analysis.

The appearance of the image is therefore controlled by the electronic size of the tip. Large tips can probe simultaneously three S atoms on the surface, when located in the middle of these three atoms. The electronic interferences caused by the multiple tunneling channels in this position is the origin of the change in the STM image, starting from the "canonical" ball image. The increase of the current in one inter-sulfur area on the surface changes the image to a triangle shape for a tip ending with one metal atom and ultimately to a honeycomb shape for a tip ending with three metals atoms. These electronic interferences appear as a fundamental process for the imaging of adsorbates and as a key for the precise interpretation of the images. They induce large differences between the top, bridge and hollow adsorption images, which allow the determination of the favored site.

The "gross" geometry being determined, the following step is to analyze how the STM image can be sensitive to the "fine" geometry that is the precise position of the S atom in the hollow site.

4. STM images and the determination of the adsorption geometry

In this simple case of an atomic adsorption, the position of the adsorbate in the 3 fold symmetric site is only controlled by the height of the atom, compared to the metal top layer. The relaxation of the substrate is not considered and the surface geometry is therefore frozen in a bulk termination structure. The calculated STM image for the hollow site as a function of the sulfur height is given in Figure 3.

The image is clearly strongly dependent on this structural parameter. Short bonding distances give a ball type image, with the bump located on the sulfur atom. Increasing the distance turns the bump shape into a triangle, by augmenting the current in one of the inter-sulfur position[10]. A honeycomb image is obtained when this inter-sulfur current equals the on-sulfur one. At higher distances this inter-sulfur current dominates and an image of balls is obtained again, but with the bumps located between the adatoms. This change in the image with height comes from the fact that the interference effects is not only controlled by the tip shape but also by the adsorption geometry. In this example, the interferences have been shown to be small for a short sulfur-surface distance and to increase with this distance, causing marked changes in the image aspect.

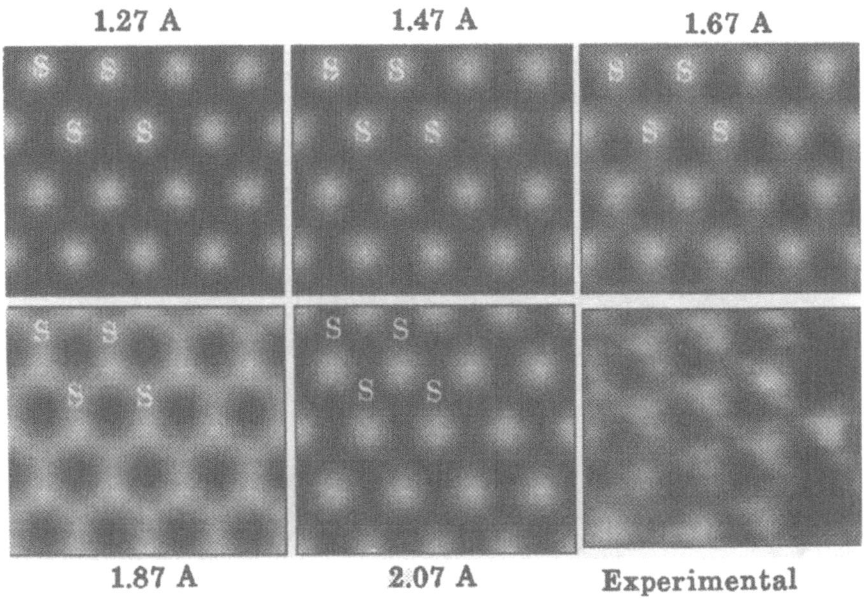

Figure 3: Calculated images of S(2x2) on Re(0001) as a function of the height (Å) of the S atom with respect to the surface metal layer.

The main problem is this dual dependence of the image on the tip and on the surface structure. In order to remove any ambiguity in the analysis of the image, it would be necessary to get more information on the system, for example information on the precise tip structure or a series of images with different tunneling gap conditions that could only be associated with a single couple of surface and tip structures. The case of a molecular adsorbate introduces a new difficulty: the atoms of the adsorbate are no longer seen individually, but the molecule is seen as a single entity with some internal details. This will be illustrated in the following for the case of benzene on Rhodium(111).

5. Images of adsorbed molecules: Benzene on Rh(111)

The calculated image of benzene on Rh(111) is shown in Figure 4. The 3-fold hollow geometry obtained by LEED has been chosen for this calculation and the tip is Rh ending with one Rh atom. The molecule appears on the image as three lobes, in good agreement with the experimental image, and in contrast with its C_6H_6 formula[11].

It can be noticed that these three lobes are not located on carbon atoms of the molecule, but in between the atoms, near three of the C-C bonds. The bumps are located near the bonds that are located between the metal atoms. The bonds seen in the image are therefore not the one that participate in the interaction of the molecule with the surface. More precisely it can be shown that the major contribution in the tunneling current comes from the antibonding e* orbital which is the LUMO of the molecule. This orbital looses its C_6 symmetry when the molecule interacts with the C_3 hollow site: it is polarized by mixing with the occupied e orbital through the metal atoms. The C_3 and not C_6 appearance of the molecule results from this orbital mixing. The small geometric Kekule distortion of the molecule upon adsorption does not affect the STM image significantly, at least much less than the electronic orbital distortion due to the interaction.

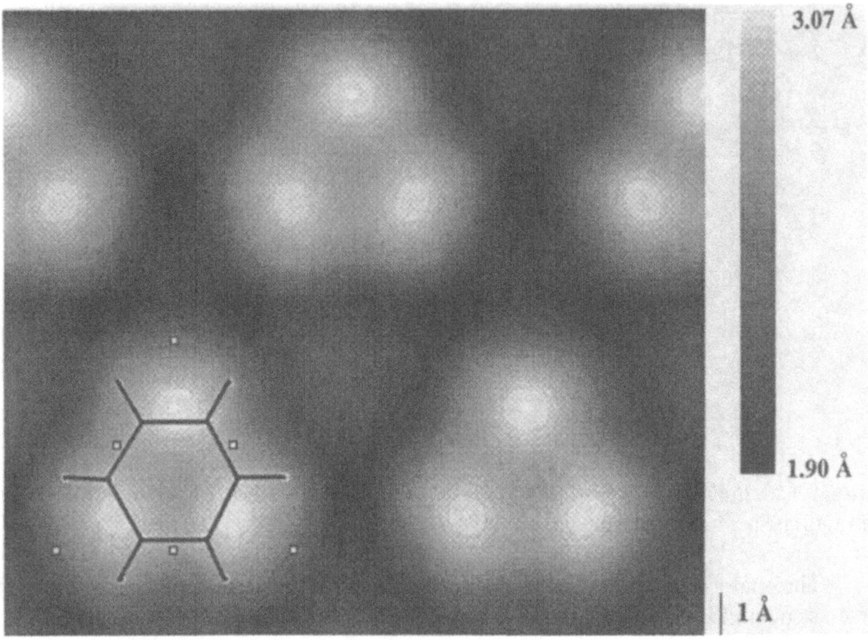

Figure 4: Calculated image of benzene on Rh(111). A stick model of one of the benzene molecules is given together with the neighboring Rh atoms.

The STM image therefore shows the anti-bonds of lowest energy of the molecule and how these are localized due to the interaction with the surface. This low lying vacant orbital is very important for the activation and the reactivity of the benzene molecule and the STM image gives a direct view of this orbital. As a consequence, the information given by a STM image must not be seen only as a limited geometrical and structural information, but also as a unique view of the electronic structure of the molecule for energies close to the Fermi level and of its potential reactivity[12].

6. Conclusion

The specific information brought by the STM in the field of adsorbate structure and eventually reactivity has been discussed in the case of S on Re(0001) and benzene on Rh(111). The key importance of a theoretical analysis of the images for this purpose has been underlined. These theoretical calculations help understand the image and extend by a large amount the information obtained from the image. For the case of S on Re, the adsorption site and geometry have been analyzed. The example of benzene on Rh(111) shows how the STM image of the molecule is affected by the interaction with the surface and by the symmetry of the adsorption site. Most important, it shows that the STM image is not only a structural information on the adsorbed molecule but mainly a view of its electronic structure near the surface Fermi level. These molecular levels being those involved for the molecule surface reaction, the STM appears therefore as a powerful and unique tool for the analysis of the activation and potential reactivity of an adsorbed molecule.

The problem of the surface reconstruction upon adsorption has also been addressed with the STM in the case of high coverage adsorption of Sulfur on a Mo(100) surface[13]. The S adsorption is shown to cause a merging together of the steps on the surface, the step-step interaction being changed from repulsive for the bare surface to attractive for the adsorbate covered surface. This important movement of atoms results in the creation of new facets on the surface. The main problem of the STM for Catalysis is its poor actual time resolution, that exclude for now the study of quickly moving or reacting adsorbates. The study of slowly moving adsorbates, with a residence time on a site of ~ 0.1 s, has however been shown to be possible[14].

Complemented by a theoretical approach, the STM therefore appears as an original and important tool for surface science and catalysis. It creates an unique link between the reality of the experiment and the conceptual wavefunctions and electronic structures of the theoreticians.

REFERENCES

1) On Leave from Materials Science Division, Lawrence Berkeley Laboratory, Berkeley, CA 94720, USA
2) G. Binnig, H. Rohrer, Ch. Gerber and E. Weibel, Phys. Rev. Lett. 49 (1982) 57.
3) D.F. Ogletree and M.B. Salmeron, Progr. Sol. St. Chem. 20 (1990) 235.
4) R.Q. Hwang, D.M. Zeglinsky, D.F. Ogletree, A. Lopez vazquez-de-Parga, G.A. Somorjai and M.B. Salmeron, Phys. Rev. B 44 (1991) 1914.
5) J.A. Stroscio and D.M. Eigler, Science 254 (1991) 1319.
6) P. Sautet and C. Joachim, Ultramicroscopy 42-44 (1992) 115.
7) J. Tersoff and D.R. Hamann, Phys. Rev. B 31 (1985) 805.
8) R. Landauer, Phil. Mag. 21 (1970) 863.
D. Fisher and P.A. Lee, Phys. Rev. B 23 (1981) 6851
9) J.C. Dunphy, D.F. Ogletree, M.B. Salmeron, P. Sautet, M.-L. Bocquet and C. Joachim, Ultramicroscopy 42-44 (1992) 490.
10) P. Sautet, J.C. Dunphy and M.B. Salmeron, to be published.
11) P. Sautet and C. Joachim, Chem. Phys. Lett. 185 (1991) 23.
12) P. Sautet and C. Joachim, Surf. Sci. 271 (1992) 387.
13) J.C. Dunphy, C. Knight, P. Sautet, D.F. Ogletree, G.A. Somorjai and M.B. Salmeron, Surf. Sci. in press (1993).
14) J.C. Dunphy, P. Sautet, D.F. Ogletree, O. Dabbousi and M.B. Salmeron, Phys. Rev. B. in press (1993).

Session Three, *In-situ Methods*

Rapporteur V. PONEC
Gorlaeus Laboratory
Leiden University
PO Box 9502
2300 RA Leiden
The Netherlands

The main topics of the presentations and the discussions in this session were synchrotron radiation techniques, in particular X-ray absorption spectroscopy, (EXAFS and NEXAFS), infra-red spectroscopy and scanning tunnelling microscopy. Transient methods are discussed in session 5 by Frennet and UV/visible spectroscopy is mentioned briefly by Rooney, (session 1). Some other *in-situ* techniques, like electron microscopy on catalytic reactors, positron emission tomography and the wider use of radio-active isotopes in studies on the mechanism, were therefore outside the scope of the workshop. However, the choice of the techniques to be discussed was not just incidental and the discussion confirmed the merits and a good future for the above mentioned techniques in catalysis.

The analysis of EXAFS data is slowly reaching a mature state. With the wide use of model compounds for the calibration of the necessary parameters, the reliability of this technique in the determination of the main structural parameters, the distance and sometimes also the composition of the catalysts is rather high. However, it does not mean that the problems of the technique itself have been solved. The worries expressed in discussion by this Rapporteur concerned in the first place the number of coordinated atoms. Estimates by many authors sets the error at a level of ± 10% (maximum average number of nearest neighbours considered is 12), but examples are known (Joyner) when this number is surprisingly high, eg. for 'O' in 'Metal-C≡O' species due to multiple scattering. On the other hand it is surprising how often a very low number N is found, ie. very small particles (few atoms) claimed to be found by EXAFS analysis. Renouprez mentioned that by making the analysis simultaneously for more than one shell, including the first, nearest neighbour shell, the values of N can be made more reliable. Joyner recalled that N and the Debye-Weller factor both enter the equation used for EXAFS analysis and the process of optimising the parameters is often to trade one of them against the other. High resolution electron microscopy (HREM) confirms the particle size derived from the values of N found by EXAFS analysis in the cases which he discussed. Joyner also mentioned, with regret, that the average thermal deviation of the lattice distances (the Debye-Waller factor, $\overline{u}^2$) has been up to now been largely unused. Ponec pointed out the following interesting results from the literature: (i) reactions which are known to be sensitive for the ensemble size are often also sensitive to the particle size (eg. hydrogenolysis of hydrocarbons, dissociation of CO); (ii) small particles usually show (in EXAFS) higher values of ($\overline{u}^2$) than the large ones and calculations by molecular dynamics show that at the usual reaction temperatures (400-800 K) the disorder in the outermost layer may be so large that availability of the larger ensembles required by dissociations mentioned above (C-O, C-C) could be appreciably suppressed. If this idea is correct, one would expect a larger particle size effect with, for example, Pt than with Rh. It could be worthwhile to check this idea, by analysis of the Debye-Waller values, performed in parallel to catalytic measurements.

R. W. Joyner and R. A. van Santen (eds.), Elementary Reaction Steps in Heterogeneous Catalysis, 313–315.

Knözinger raised a question about the applicability of EXAFS, asking specifically where the limits are of the pressure and temperature which can be used in *in-situ* EXAFS measurements. Iwasawa answered that according to his experience the limit is somewhere about 30 bar total pressure and about 600 K. Above 600 K, the particles behave (as molecular dynamics also predicted, see above) like in a state of percolation. Sautet remarked that perhaps for special cases (Joyner recalled here the flat surfaces studied by Hayden) a combination of Scanning Tunnelling Microscopy and of EXAFS would be very useful.

Infra-red spectroscopy is the technique which has been used for the longest time of all three methods discussed. There is no doubt about its value for identification of various adsorbed species on the surface of catalysts and for the characterisation of surfaces (acid-base or redox properties), the matter of concern is the extent to which it provides a reliable identification of reactive intermediates. This point was stressed by the discussion (Basset, Iwasawa) concerning the role of formates, 'how are they formed and are they the reactive intermediates of methanol-synthesis and water-gas shift reaction?' The opinion of the audience was divided on this point, some denied any role of formates in the WGS reaction, others suggested that formates are probably not formed directly, but by a conversion of a through carbon adsorbed carboxylic acid. However, the arguments in favour of formates as intermediates, directly formed from OH groups and CO, were equally strong.

Bassett suggested the characterisation of the acid-base properties of oxide surfaces by monitoring the reactions of carbonyls, since he believed that this interaction is well understood. However, not everyone agreed with Bassett's idea that CO is 'inserted' for example into Al-O-H, to produce a monodentate carboxylate group. Iwasawa preferred a mechanism by which adsorbed CO is hydrogenated in a bidentate formate group. He mentioned that production of atomic hydrogen on basic oxides leads immediately to the formation of formates from adsorbed CO, but formation of acidic OH from H_{ads} has never been seen.

A very important problem, accepted by almost everybody, is the necessity to have a reliable method for characterising *basic* centres on catalyst surfaces, at least as effectively as can be done for acidic sites. Knozinger, Rooney, Bassett among others supported the use of fluorohydrocarbons for this purpose.

Hayden, Rooney and Pacchioni all mentioned that making and using the analogy between the reaction of mononuclear organometallic complexes and those on large clusters or extended surfaces, is often very useful (see contributions by Bassett and Rooney), but can be also very misleading. An extended metal or oxide surface presents to the molecules in the fluid phase, binding which is of a multi-site character. This is a situation which cannot be easily created in soluble mononuclear coordination compounds. Also the electronic structure of a single transition metal atom or ion can differ substantially from that of an atom in a metal.

Scanning tunnelling microscopy, the youngest method discussed, distinguishes itself by the greatest promise and important uncertainties at the same time. Since the promise is so attractive and the difficulties do not seem to be of a fundamental kind, the method will spread widely in the near future. However, the excellent review presented to the workshop indicated that the situation might be similar to that with high resolution electron microscopy, some years ago. It is irresistibly tempting to see the STM image as a direct picture of the surface. However, the theoretical analysis of the 'S on Mo' system, presented by Sautet, revealed that the STM picture is dependent on the (i) structure of the surface and its electronic properties (ii) the elevation of the tip above the surface and (iii) the structure of the very top of the tip (one or more atoms). Discussion between Sautet, van Geem and Wells revealed that it is very difficult to be sure at this stage, without a parallel theoretical analysis, what the images seen by STM really are.

Joyner summarised the discussion on some points which were common to all problems discussed in relation to the various techniques mentioned above. The availability of large ensembles of atoms or of extended and stable flat surfaces is probably an important aspect of highly active metal surfaces in reactions where the dissociation of a strong bond is required. This is true for dissociation of C-C, C-O, C-N, N-N bonds, reactions which either do not take place with coordinated complexes of the metal element under study, or which operate by a substantially different mechanism on metals and on coordination compound (of the same element). If one wishes to draw analogies between the behaviour of (modified) metal surfaces and coordination (soluble) compounds one must look more in the direction of, for example, hydrogenation, CO insertions, etc.

Finally, a very personal word reflecting the opinion of the Rapportuer. It is certainly a good thing that it is required by the NATO committee to document and to report on the discussion and to arrive at certain conclusions. However, it would be a serious error to think that the most efficient approach for funding bodies or the most rewarding directions of future research can be identified by majority vote. For example the birth of relativity theory and the uncertainty principle were not accompanied by the applause of the majority in the scientific community of their time. The best new ideas are always born as the opinion of the minority, and often of only a single person.

QUANTUM CHEMISTRY OF CHEMISORPTION AT METAL AND OXIDE SURFACES: A CLUSTER MODEL APPROACH

G. PACCHIONI [a] and P. S. BAGUS [b]
[a] *Dipartimento di Chimica Inorganica e Metallorganica, Universita' di Milano, via Venezian 21, 20133 Milano, Italy*
[b] *IBM Research, Almaden Research Center, 650 Harry Road, San Jose, California 95120-6099, USA*

ABSTRACT. Coadsorption of K and CO at the Cu(100) surface and adsorption and reaction of CO and CO_2 at surface and defect sites of MgO are reviewed. The cluster technique is used to determine optimal adsorption geometries, the physical properties of the adsorbed species, in particular the vibrational frequencies, and the nature of the surface chemical bond. It is shown that the large red-shift of the CO ω_e when K is coadsorbed on a metal surface can be explained only with very short K-CO distances. Chemisorption at oxide surfaces is also discussed and the important role of the surface defects in determining the reactivity of the MgO surface is shown.

1. INTRODUCTION

The theoretical study of chemisorption of atoms and molecules on solid surfaces is intimately connected with both molecular (or atomic) and solid state aspects of the problem. For this reason, many theoretical methods have been used, based on previous experience in molecular chemistry and solid state physics, to investigate chemisorption at clusters and surfaces. Applications of quantum mechanical methods to chemisorption problems make use, in general, of the cluster model technique where the substrate, a metal, semiconductor or ionic surface, is represented by a finite number of atoms [1]. Implicitly, this procedure relies on the fact that an adsorbate induces a rather local perturbation on the substrate. Embedding techniques are available to adapt the cluster model to the specific boundary conditions of chemisorption. They range from special localization procedures [2] to sophisticated Green's function techniques [3]. On the other hand, there is enough practical experience nowadays that useful contributions to selected problems can be obtained from a direct cluster approach. The electronic wavefunctions for these cluster models have special advantages for understanding the local characteristics of the surface chemical bond. The cluster wavefunctions, obtained with molecular orbital theory, make it possible to relate chemical concepts developed to describe chemical

R. W. Joyner and R. A. van Santen (eds.), Elementary Reaction Steps in Heterogeneous Catalysis, 317–340.

bonds in molecules to the very closely related chemical bonding at the surface of a material.

The most widely studied subject within the cluster model approach is atomic and molecular adsorption at metal surfaces. Some examples of adsorption at semiconductor surfaces have also been reported [1]. Only recently, however, this computational technique has been applied to the study of ionic crystals and in particular to the adsorption and reaction of molecules at the surface of ionic, or partially ionic, oxides.

In this paper we present two examples of application of the cluster model technique. First, we consider a very important and new problem in this area, i.e. the description of coadsorbed atoms and molecules on a cluster. Until now, due to computational difficulties, it has been difficult to treat clusters large enough to accomodate more than a single adsorbed molecule and just a few examples have been reported. The problem investgated is the coadsorption of a CO molecule in the presence of K atoms, a topic of great technical importance in catalysis with promoters. In the second part of the paper we present some recent results on the adsorption and reaction of CO and CO_2 molecules on the MgO surface and we show the important role of surface defects, like steps and corners, in determining the reactivity of a real, polycrystalline oxide.

2. COADSORPTION OF K AND CO ON THE Cu(100) SURFACE

Because of their fundamental and applied interest, coadsorption of CO and alkali metal atoms on metal surfaces has been intensively studied [4]. The alkali atoms influence the electronic structure of the CO molecule and lead to strong changes in its chamisorption properties. One of the most dramatic changes induced by the coadsorbed alkali atoms is the decrease in the C-O vibrational frequency, ω_e, which in some case can be as low as 1200-1400 cm^{-1}. On an alkali-free surface the normal values are in the range 1900-2100 cm^{-1} [4]; for comparison, the free CO ω_e is 2143 cm^{-1} [5].

Several mechanisms have been suggested to explain the observed decrease in ω_e and the enhanced catalityc activity of CO coadsorbed on transition metal surfaces in the presence of promoters, in particular alkali metal atoms [4]. These include (a) a "through-metal" effect, based on the electron donation from the alkali atom to the surface with consequent increase of the capability of the metal to backdonate charge into the $2\pi^*$ molecular orbital (MO) of CO, (b) the rehybridization to sp^2 of the metal-CO bond, with reduction of the triple bond character of the C-O bonding, (c) a direct interaction between the outermost s orbital of the alkali metal and the $2\pi^*$ MO of CO [6], (d) the electrostatic interaction between the electric field created by the positively charged alkali metal ions and the dipole moment of adsorbed CO [7,8], and (e) geometric effects [9]. It is likely that some of these effects work together.

In this work we have considered the coadsorption of two K atoms and a CO molecule adsorbed normal to the surface at a bridge site of a Cu_{32} cluster model of the Cu(100) surface. The cluster contains 20 Cu atoms in the first layer and 12 Cu atoms in the second, see Fig. 1. The Cu atoms have been fixed at their bulk position with a Cu-Cu distance of 4.80 bohr (2.54 Å). The K atoms have been placed at three different sites, Site A, B, and C, respectively. In Site A and C the K atoms are adsorbed on a bridge site, Site B corresponds to a four-hollow open site. The open positions are known to be preferred for adsorption of alkali atoms [10-12], although examples of on-top adsorption have been reported recently [13,14]. The CO molecule is adsorbed normal to the surface at a bridge site and lies in between the two K atoms. Cu_{32} is large enough to evaluate how "local" is the interaction between K and CO. Recent experiments [15] on Cu(100)/K/CO have shown a strong dependece of the CO ω_e on the K coverage. For K coverages up to 0.14 there is a line in the region around 2100 cm^{-1}. For K coverages of 0.12-0.21 there is a line at about 1960 cm^{-1}, and only at higher coverages several peaks in the range 1200-1600 cm^{-1} are observed [15]. This suggests that the distance between CO and the coadsorbed K atoms must be rather short.

Hartree-Fock, Self-Consistent-Field, SCF, calculations have been performed to determine the chemisorption properties of the Cu/K/CO system. To reduce the computational effort, Effective Core Potentials (ECP) have been employed to replace the atomic cores. The two Cu atoms nearest to CO have been treated with an 11-electron ECP [16] which explicitly includes only the 3d and the 4s electrons of Cu; the Gaussian type basis set used for these Cu atoms is [3s2p5d/2s1p2d] [16]; the remaining 30 environmental Cu atoms of the Cu_{32} cluster were treated with a 1-electron ECP and a [4s1p/2s1p] basis set [17]; in these atoms, therefore, the 3d electrons are not included and only the 4sp conduction

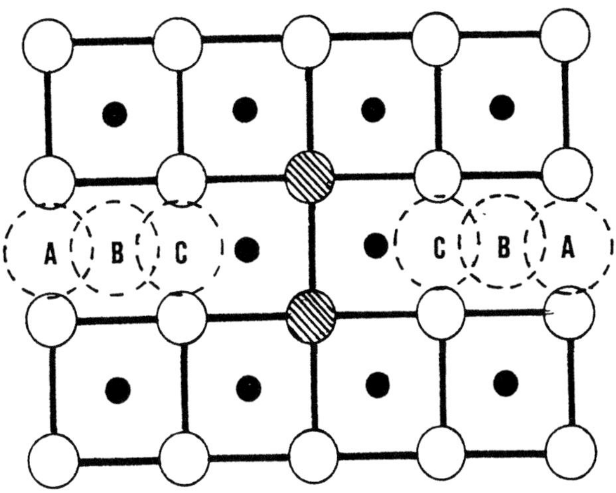

Figure 1 - Geometry of the $Cu_{32}/K_2/CO$ cluster model of the Cu(100)/K/CO system. The Cu_{32} cluster contains 20 atoms in the first layer and 12 atoms in the second. The CO molecule is adsorbed on a bridge site. Three sites have been considered for the adsorption of the two K atoms: Site A, Site B, and Site C, as indicated in the figure. The Cu atoms where CO is bridge bonded are shown with hash marks.

band electron is explicitly considered. The two K atoms were treated with a 9-electron ECP [16] including the 3s, 3p, and 4s K electrons; the basis set is [5s5p/4s,4p]. Finally, the CO molecule is described at the all electron level employing a [9s5p/4s3p] basis set [18].

2.1. Cu/K interaction

The nature of the interaction of alkali metal atoms with metal surfaces is a controversial subject. The classical picture, originally proposed by Langmuir [19], assumes that the adsorbed alkali atom donates one electron to the substrate and becomes a cation in the limit of zero coverage. This model is supported by a large body of experimental and theoretical data [20-24]. Recently, a theoretical model has been proposed which is consistent with the alkali atom being covalently bonded to the surface at every coverage [25,26]. This picture, however, is not supported by recent investigations [22-24]. The determination of the nature of the alkali/metal bonding, i.e. ionic or covalent, is not purely academic given the important consequences that these two kinds of interaction may have on the coadsorption with CO molecules.

One of the most convincing indications that the bonding of K to a Cu surface is, indeed, ionic, comes from the analysis of the dipole moment curve for the vertical displacement of the adsorbed K atom above the surface. For an ionic adsorbate, the $\mu(z)$ curve is linear and the slope, $d\mu/dz$, is large. For a covalent adsorbate, like CO, $d\mu/dz$ is small and the curve may deviate from being linear. In a series of papers we have clearly shown the usefulness of this measure for determining the character of a surface chemical bond [22,27,28]. The $\mu(z)$ curve can be expanded in a polynomial series as

$$\mu(z) = M_0 + M_1(z-z_e) + M_2(z-z_e)^2 + \ldots$$

where M_0 is the value of μ for $z=z_e$, $M_1 = d\mu/dz$ is the slope of the curve for $z=z_e$, and $M_2 = d^2\mu/dz^2$ gives a measure of the curvature. For a Cu_{25}/K cluster, we found [22] that $M_1 = 0.90$ and $M_2 = 0.08$, suggesting that the bonding is highly ionic; for comparison, the motion of a point charge of +1 on a Cu_{25}^- cluster gives $M_1 = 1.37$; M_1 differs from 1 because the polarization of the substrate also contributes to the value of the dipole moment. Other theoretical measures indicate that the charge associated with the adsorbed K atoms is close to +1 [22]. The analysis of the dipole moment curve is important because is connected to a measurable quantity. In fact, an ionic adsorbate, with a large dynamical dipole moment, $M_1=d\mu/dz$, will have an intense vibrational absorption; this latter, in fact, is proportional to the square of the dynamical dipole moment [22]. This is exactly the case for the adsorption of K atoms on a Cu surface [20]. Thus, we have largely based the analysis of the bonding of two coadsorbed K atoms on a Cu_{32} cluster on the shape and the slope of the dipole moment curve for their motion above the surface.

In Site A of Cu_{32}/K_2 the two K atoms are separated by 19.2 bohr; the covalent radius of K is about 3.7 bohr, hence the two atoms should be non-interacting in Site A. In Site B the K atoms are adsorbed in a four-fold hollow position and are separated by 14.4 bohr. In Site C, the K-K distance is only twice the nearest neighbor distance, 9.6 bohr. This distance is large compared to the sum of the K covalent radii, but it gives rise to a strong electrostatic repulsion between two full K^+ cations. As the coverage increases, the ionicity of the adsorbates is expected to decrease as the consequence of this Coulomb repulsion.

TABLE 1 - Chemisorption properties of two K atoms chemisorbed on three different sites of a Cu_{32} cluster model of the Cu(100) surface. See Fig. 1 for definition of the three sites. The results are for the valence configuration $9a_1^2 5a_2^2 7b_1^2 6b_2^2$. In the Cu_{32}/PCs system the two K atoms have been replaced by PCs of +0.5

	Site A r(K-K)=19.2 bohr		Site B r(K-K)=14.4 bohr		Site C r(K-K)=9.6 bohr	
	Cu_{32}/K_2	Cu_{32}/PCs	Cu_{32}/K_2	Cu_{32}/PCs	Cu_{32}/K_2	Cu_{32}/PCs
z_e, bohr	5.76	---	5.48	---	5.43	---
ω_e, cm^{-1}	122	---	127	---	127	---
$M_1=d\mu/dz$	1.13	1.34	1.33	1.33	1.21	1.34
$M_2=d^2\mu/dz^2$	-0.05	-0.03	0.04	0.00	-0.20	0.01
D_e/at, eV	0.61	---	0.74	---	0.65	---

Cu_{32} has a valence ground state configuration $8a_1^2 5a_2^2 7b_1^2 6b_2^2$; a second state, $8a_1^2 5a_2^2 6b_1^2 7b_2^2$ lies only 0.19 eV above the ground state. In Cu_{32}/K_2 there are two lowest close-lying states, $9a_1^2 5a_2^2 7b_1^2 6b_2^2$, the ground state, and $9a_1^2 5a_2^2 6b_1^2 7b_2^2$, 0.18 eV higher in energy. The addition of the two K atoms results in the filling of the Cu_{32} cluster HOMO, an a_1 orbital of Cu 4sp character highly delocalized over the Cu atoms.

The chemisorption properties of the Cu_{32}/K_2 cluster model are given in Table 1 for the three sites considered. The potential energy and the dipole moment curves for the vertical motion of the two K atoms are shown in Fig. 2 and Fig. 3. Of the three sites considered, Site B, corresponding to adsorption in the four-fold hollow site, is the most stable. In this site, each K atom is bound by 0.74 eV; adsorption at the bridge sites is less favorable by about 0.1 eV, see Table 1. The potential energy curves are rather flat, Fig. 2, as shown also by the low vibrational frequency, about 120-130 cm^{-1}. The equilibrium distances are in the range 5.5-5.75 bohr. For Site B, both the ground and the first excited configuration have been investigated, but the chemisorption properties are very similar for the two states. This indicates that the cluster is large enough to avoid strong dependence of the bonding properties on the electronic state chosen.

Figure 2 - Potential energy curves for the adsorption of two K atoms on Site A, Site B, and Site C of Cu_{32} (as marked in the figure), see also Fig. 1. The interaction energy is defined as $E(Cu_{32}) + 2E(K) - E(Cu_{32}/K_2)$.

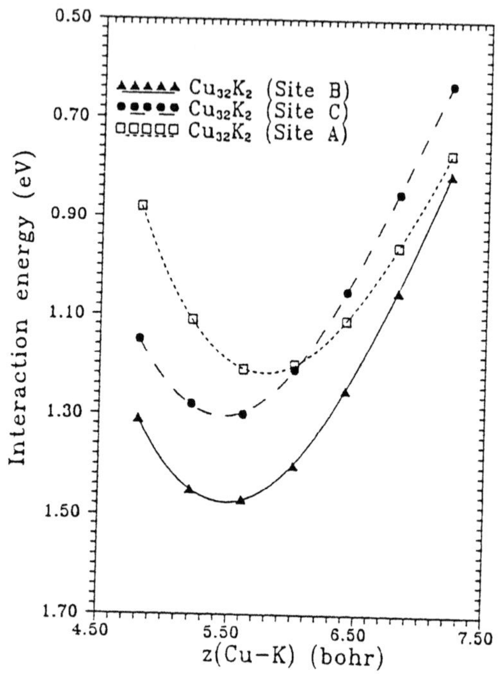

To establish the character of the bonding in Cu_{32}/K_2 we have determined and compared the slope and the curvature of the dipole moment curves with an "electrostatic" model where the two K atoms have been replaced by two point charges, PCs. For an ideal ionic bond, Cu_{32}^{2-}/K_2^{2+}, and in absence of any polarization, M_1=2 and M_2=0. The $\mu(z)$ curves for the three sites exhibit a large slope and are linear or almost linear for small displacements about z_e, see Fig. 3. The computed slopes, however, M_1=1.1-1.3, indicate that the net charge associated with two adsorbed K atoms is smaller than +1. We have determined the dipole moment curve for the vetical motion of two PCs=+0.5 above Cu_{32}, see Cu_{32}/PCs in Table 1 and Fig. 3. For Sites A and B, the two curves, Cu_{32}/K_2 and Cu_{32}/PCs, have a remarkably similar slope and curvature. A difference in the curvature of the $\mu(z)$ curves for the Cu_{32}/K_2 and the Cu_{32}/PCs clusters is found for Site C. This is due, in particular, to the decrease of $d\mu/dz$ as the distance is increased, see Fig. 3. In Site C, the two K atoms are separated by 9.6 bohr only. When the two K atoms are at the equilibrium position, the net charge on K is screened by the large polarization of the Cu conduction band electrons. As the two atoms are moved away from the surface, this screening effect is reduced and the Coulomb repulsion between the two cations increases. As a consequence, charge flows back from the surface to the adsorbed K atoms to reduce the Coulomb repulsion. The change in the nature of the adsorbate as the distance is increased is reflected in the change of the slope of the $\mu(z)$ curve.

Figure 3 - Dipole moment curves for the vertical motion of two K atoms above the Cu_{32} cluster. The curves are given for K atoms on three sites of Cu_{32}, Site A, Site B, and Site C, respectively, (as marked in the figure), see also Fig. 1. Curves are given also for the motion of two point charges, PC=+0.5 (Site B), and for the motion of the two K atoms adsorbed in Site C in the presence of a coadsorbed CO molecule, $Cu_{32}/K_2/CO$ cluster.

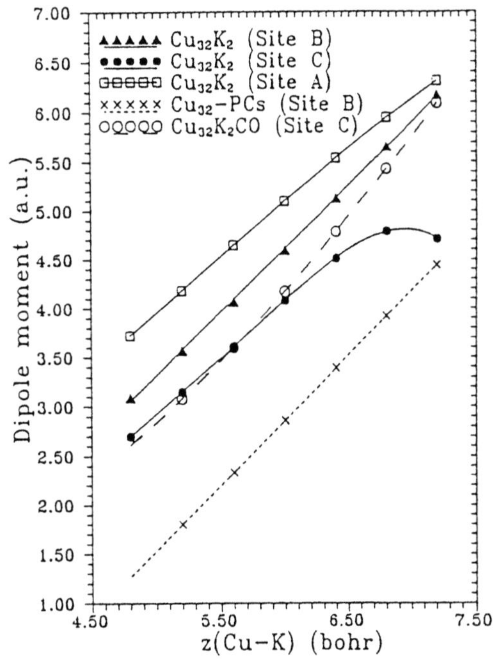

The present results indicate that the coadsorbed K atoms on the Cu_{32} cluster model of the Cu(100) surface are partially ionic on all the three sites considered; it is possible to estimate that the net charge on the K atom is close to about +0.5 e. This is less than the charge associated to a single K atom adsorbed on a Cu_{25} cluster, about +1 e, because of the Coulomb repulsion between the two adsorbed K ions.

2.2. Cu/CO interaction

On a clean Cu(100) surface CO is preferentially bound at on-top sites [29]. Bound states for CO on-top of Cu cluster models of the Cu(100) surface have been found [30]. On the other hand, there is no experimental evidence of bound states for CO bridge bonded to Cu(100), a result confirmed by Hartree-Fock calculations on a Cu_{24}/CO cluster [31]. Thus, it is not surprising that the potential energy curve for the adsorption of CO on the bridge site of the Cu_{32} cluster is purely repulsive. Nevertheless, the choice of considering this site for CO adsorption is justified by the fact that in the presence of coadsorbed K atoms CO is preferentially adsorbed at the bridge sites [32].

The occupied 4σ, 1π, and 5σ MOs of CO are much deeper in energy than the 4sp-band in Cu_{32}. For sake of clarity, we identify the lowest states of Cu_{32}/CO by giving only the valence Cu levels; the two lowest configurations are thus the $8a_1^2 5a_2^2 7b_1^2 6b_2^2$ state separated by 0.23 eV from the $8a_1^2 5a_2^2 6b_1^2 7b_2^2$ state. For z(Cu-CO)=3.60 bohr, the two states are unbound with respect to Cu_{32} and CO by 0.39 and 0.44 eV, respectively.

TABLE 2 - Chemisorption properties of CO perpendicularly adsorbed on the bridge site of the Cu_{32} cluster model of the Cu(100) surface. The CO molecule is 3.60 bohr above the cluster first layer.

	Cu_{32}/CO	$[Cu_{32}/CO]^{2-}$
valence configuration	$8a_1^2 5a_2^2 7b_1^2 6b_2^2$	$9a_1^2 5a_2^2 7b_1^2 6b_2^2$
r_e(C-O), bohr	2.125	2.130
ω_e(C-O), cm^{-1}	2227	2192
$\Delta\omega$(C-O),[a] cm^{-1}	-43	-78
$D_e(Cu_{32}$-CO), eV	-0.39	...
$M_1=d\mu$(C-O)/dr(C-O)	-1.86	-1.90
$P(2\pi^*)$[b]	0.18	0.19
ΔE,[c] eV	0.00	0.31

a) ω shift measured with respect to free CO $\omega_e=2270$ cm^{-1}. b) Occupation of the $2\pi^*$ MO of CO according to the Mulliken population analysis. c) Energy separation.

The C-O vibrational frequency in Cu_{32}/CO, computed by keeping fixed the distance between the surface and the CO center of mass, is reduced by 43 cm^{-1} with respect to free CO. This is a small red-shift, comparable to that observed experimentally for Cu(100)/CO [4]. The lowering is due to the classical back donation of charge from the surface into the $2\pi^*$ MO of CO [33]. The occupation of this empty orbital of CO is 0.18 e according to the Mulliken population analysis.

The addition of two extra electrons to the cluster may be viewed as a simple representation of an indirect "through-metal" interaction between K and CO; the charge donated from K to the substrate metal is available to reinforce the back bonding to CO. The addition of two extra electrons to Cu_{32}/CO results in the $9a_1^2 5a_2^2 7b_1^2 6b_2^2$ state; the total energy of $[Cu_{32}/CO]^{2-}$ is 0.31 eV higher than that of Cu_{32}/CO. The red-shift of the CO ω_e for $[Cu_{32}/CO]^{2-}$, 78 cm^{-1}, see Table 2, is small compared to the large shifts observed when K is coadsorbed at high coverage with CO [4]. The small additional shift toward lower vibrational frequencies is consistent with a modest increase of both the occupation of the $2\pi^*$ MO of CO and of the CO dynamic dipole moment, see Table 2.

2.3. Cu/K/CO interaction

To study the coadsorption of CO and K on the Cu(100) surface we have used two cluster models. In the $Cu_{32}/K_2/CO$ cluster all the atoms are represented with their valence electrons; the character of the K atoms in this "real" model is determined by the SCF wavefunction. Then, we used an "electrostatic" model where it is assumed that the two adsorbed K atoms donate an entire electron to the substrate which becomes therefore Cu_{32}^{2-}; in this model the two K ions are replaced by a +1 PCs. In both cases the CO molecule is at fixed distance from the surface, z_{CO}=3.60 bohr, while the K atoms, or the PCs, are placed at the equilibrium positions found for the three sites of the Cu_{32}/K_2 cluster, Table 1. In the "electrostatic" model, the only electronic interaction between CO and the rest of the system is the back donation from the Cu 4sp levels into the $2\pi^*$ MO of CO; no direct electronic K-CO interaction is possible. In response to the presence of the two PCs, the Cu_{32} cluster polarizes. In a perfect conductor this polarization results in the formation of an image charge below the cluster first layer. The combined presence of the positive charges above the surface and of the negative image charges below it generates a non-uniform electric field which interacts with the permanent dipole moment of the CO molecule normal to the surface [8]. This purely electrostatic interaction, sometimes referred to as a Stark effect, contributes significantly to the lowering of the CO ω_e [8]. The electrostatic model however suffers from some limitations. In fact, it is assumed that the the two K atoms donate an entire electron to the surface while we have shown that the charge associated to K in Cu_{32}/K_2 is about +0.5.

TABLE 3 - Chemisorption properties of two K atoms chemisorbed on two different sites of a Cu_{32} cluster model of the Cu(100) surface in the presence of a coadsorbed CO molecule. See Fig. 1 for definition of the three sites. The results are for the valence configuration $9a_1^2 5a_2^2 7b_1^2 6b_2^2$.

	$Cu_{32}/K_2/CO$	
	Site B	Site C
z_e(Cu-K), bohr	5.59	6.34
ω_e (Cu-K), cm^{-1}	122	109
$M_1=d\mu(Cu\text{-}K)/dz(Cu\text{-}K)$	1.35	1.48
$M_2=d^2\mu/dz^2$	0.02	0.12
D_e/at, eV	0.69	0.33

The presence of the CO molecule between the two K atoms provides an efficient way to screen the two ions and to reduce their Coulomb repulsion. In Site C the slope of the $\mu(z)$ curve for the vertical motion of K is considerably larger when the CO molecule is coadsorbed than for the Cu_{32}/K_2 cluster, Fig. 3. Also the curvatures are different. In particular, for z(Cu-K) larger than the equilibrium position, the $\mu(z)$ curve remains linear in $Cu_{32}/K_2/CO$ while it deviates from linearity in

Cu_{32}/K_2 (Site C), see Fig. 3. The role of the Cu 4sp electrons in screening the two K ions in Cu_{32}/K_2 is taken by the CO molecule in $Cu_{32}/K_2/CO$. In Cu_{32}/K_2, for Cu-K distances larger than z_e, electronic charge flows back from Cu to K to reduce the Coulomb repulsion between the two K ions and the $\mu(z)$ curve changes its curvature; on $Cu_{32}/K_2/CO$ the screening is provided more efficiently by the interstitial CO molecule and the K adsorbates keep their partial ionic character. This effect is not observed when the two K atoms are sufficiently distant, as in Site B; the $\mu(z)$ curves for Cu_{32}/K_2 and $Cu_{32}/K_2/CO$ are very similar, see Tables 1 and 3. Thus, we have some evidence of an increased ionicity of the two K atoms when the CO molecule is coadsorbed, although the net charge on K is far from being +1. A second limitation of the "electrostatic" model is that the Cu_{32} cluster has only two layers and probably does not provide a sufficiently delocalized source of electrons to allow the formation of an image charge. Nevertheless, the comparison of the "real" with the "electrostatic" model will give some information about the relative importance of the "through-metal", the "direct", and the "electrostatic" interactions of coadsorbed K and CO.

When the CO molecule is coadsorbed with the two K atoms in Site A there is a moderate additional shift of the CO ω_e toward lower frequencies with respect to Cu_{32}/CO, cf. Tables 2 and 4. $\Delta\omega \simeq -91\ cm^{-1}$ is only slightly larger than for $[Cu_{32}/CO]^{2-}$; the occupation of the $2\pi^*$ level, 0.21 e, is not significantly larger than for a single CO molecule with no coadsorbates, see Table 4. In Site A the two K atoms are about 10 bohr apart from CO, and their effect on the CO bond length and vibrational frequency is practically negligible. For the same reason, the "electrostatic" model for the Site A gives very similar values of the CO r_e and ω_e as the "real" model. Thus, it is possible to conclude that at this K-CO distance there is no appreciable influence of K on the CO chemisorption properties.

For Site B of $Cu_{32}/K_2/CO$ the CO ω shift, $-176\ cm^{-1}$, is considerably larger than for Cu_{32}/CO; the decrease of the CO ω_e and the corresponding elongation of the C-O bond distance, Table 4, are consistent with the augmented occupation of the $2\pi^*$ orbital. However, the ω shift is much smaller than that experimentally observed for Cu/K/CO [15]. Also in this case the "electrostatic" model gives very similar values of the CO chemisorption properties as the "real" model but it is possible that this similarity is fortuitous and arises from the sum of different cancelling effects. Thus, also in Site B, where the distance between K and CO, about 7 bohr, is larger than the sum of the covalent radii of K and CO, the effect of K on the CO stretching is modest.

A dramatic lowering of the CO vibrational frequency is found when CO is coadsorbed with the K atoms in Site C.In this case, the ω shift, $-631\ cm^{-1}$, is not too different from that observed experimentally. This shift is one order of magnitude larger than for Cu_{32}/CO and $[Cu_{32}/CO]^{2-}$ and is accompanied by a large increase of the population of the $2\pi^*$ level of CO, see Table 4. The shift found with the "electrostatic" model, $[Cu_{32}/PC/CO]^{2-}$, is also very large, $-475\ cm^{-1}$. This is due in part to the fact that the two PCs induce a considerable polarization of the Cu conduction band electrons; this causes a large overlap with the CO empty levels with consequent increase of the $2\pi^*$ occupation, see Table 4. The effect is similar to that obtained with the "real" $Cu_{32}/K_2/CO$ model but in this latter case the ω shift, the C-O bond length, and the $2\pi^*$ occupation are larger than with the PCs model, see Table 4. This suggests that some direct charge transfer could occur between the K atoms and CO. Another possibility, however, is that the different polarization of the metal substrate in response to the presence of a real K ion or of a point charge is responsible for the larger population of the CO $2\pi^*$ MO without assuming a direct K-CO charge transfer.

TABLE 4 - Chemisorption properties of CO perpendicularly adsorbed on the bridge site of the Cu_{32} cluster model of the Cu(100) surface in the presence of two coadsorbed K atoms or of two point charges, PCs. The CO molecule is 3.60 bohr above the cluster first layer. The results are for the valence configuration $9a_1^2 5a_2^2 7b_1^2 6b_2^2$.

	"Real" model $Cu_{32}/K_2/CO$			"Electrostatic" model $[Cu_{32}/PC/CO]^{2-}$		
	Site A	Site B	Site C	Site A	Site B	Site C
r_e(C-O), bohr	2.133	2.149	2.226	2.136	2.150	2.208
ω_e(C-O), cm^{-1}	2179	2094	1639	2166	2093	1795
$\Delta\omega$(C-O),[a] cm^{-1}	-91	-176	-631	-104	-177	-475
$D_e(Cu_{32}$-CO), eV	-0.40	-0.50	-1.18	...	...	...
$M_1=d\mu$(C-O)/dr(C-O)	-1.23	-1.27	-1.62	-1.91	-1.94	-1.75
$P(2\pi^*)$[b]	0.21	0.27	0.57	0.21	...	0.53
ΔE,[c] eV	0.15	0.00	0.85	0.65	0.58	0.00

a) ω shift measured with respect to free CO ω_e(SCF)=2270 cm^{-1}. b) Occupation of the $2\pi^*$ MO of CO according to the Mulliken population analysis. c) Energy separation for the three sites.

The fact that a direct charge transfer from K to CO can occur in principle can be shown by means of a very symple model, K_2/CO, where a CO molecule is placed in between two K atoms; the distance between K and CO is the same as in $Cu_{32}/K_2/CO$, Site C. This corresponds to a distance between K and the O atom of CO, about 5 bohr, which is slightly shorter of the sum of the covalent radii of K, 3.70 bohr, and O, 1.40 bohr. We have considered three states for K_2/CO: the lowest one is a triplet with the two unpaired electrons in the a_1, bonding, and b_1, antibonding, combinations of the K 4s orbitals. This configuration is unbound with respect to CO + 2K atoms by 0.53 eV, see Table 5. Two excited configurations, corresponding to a doubly occupied a_1 or b_1 MO, respectively, are about 2 eV higher in energy, see Table 5. For symmetry reasons, in order to overlap with the CO $2\pi^*$ MO, the K 4s electrons must be in a b_1 level. In this case, because of the short K-CO distance, the diffuse 4s orbitals of K can overlap with the CO empty levels with consequent charge transfer from K to CO. This results in very large ω shifts, of about 600 cm^{-1} for the $(b_1)^1(a_1)^1$ triplet state and of about 900 cm^{-1} for the $(b_1)^2$ state; as expected, no shift is found for the $(a_1)^2$ state, see Table 5.

These large shifts could be due to the K-CO charge transfer but also to the Pauli repulsion between the occupied K and CO orbitals. In this second hypothesis the CO molecule stretches to reduce the "steric" repulsion with the neighboring K atoms. To test this, we have computed the C-O stretching frequency by keeping fixed the K and CO charge distributions to their atomic and molecular characters. At this Frozen Orbital, FO, level, the CO ω_e is virtually the same as for gas-phase CO, see Table 5. The strong weakening of the C-O bonding in K_2/CO is therefore to be ascribed entirely to the increased occupation of the antibonding $2\pi^*$ MO of CO and not to the "compression" of the K and CO charge distributions. This result however, obtained with the simple K_2-CO model, is not sufficient to proof the occurence of a direct K-CO charge transfer on a surface. Work is in progress to further analyze this point.

TABLE 5 - Vibrational properties of CO in K_2/CO. The CO molecule lies between two K atoms placed at the same distance as in Site C of Cu_{32}/K_2/CO.

state	$(b_1)^1(a_1)^1$		$(b_1)^2$	$(a_1)^2$
	FO[a]	Full SCF	Full SCF	Full SCF
r_e(C-O), bohr	2.119	2.337	2.276	2.122
ω_e(C-O), cm^{-1}	2280	1639	1377	2258
$\Delta\omega$(C-O),[b] cm^{-1}	+10	-631	-893	-12
D_e(2K+CO), eV	-2.03	-0.53	-2.25	-2.45

a) Frozen Orbital, see text. b) ω shift measured with respect to free CO ω_e(SCF)=2270 cm^{-1}.

These results provide strong evidence that, in order to explain the large decrease of the CO vibrational frequency experimentally observed, the distance between the coadsorbed K atoms and CO must be rather short. We have shown that for distances larger than the sum of the covalent radii of the coadsorbed species the effect on the CO physical properties is small. Of course, at the short K-CO distances of Site C there is a considerable repulsion between the two charge distributions. In Cu_{32}/K_2, Site C is less stable than Site B, but nevertheless the K atoms are strongly bound by about 0.65 eV/atom. In $Cu_{32}/K_2/CO$, the repulsion with the coadsorbed CO molecule reduces the strength of the Cu-K bond; for Site B this reduction is negligible, from 0.74 eV to 0.69 eV, see Tables 1 and 3, indicating that there is virtually no repulsion between K and CO at these distances. For Site C, on the contrary, the interaction energy of a K atom with the substrate goes from 0.65 eV without CO to 0.33 eV with CO, see Tables 1 and 3. The repulsive interaction between K and CO is demonstrated also by the longer equilibrium bond distance of K from the Cu surface when CO is present, see Table 3. The CO molecule, which is unbound in Cu_{32}/CO by 0.39 eV, see Table 2, is unbound in $Cu_{32}/K_2/CO$, Site C, by 1.18 eV; however, while in Cu_{32}/CO the Cu_{32}-CO potential energy curve is purely repulsive, in the presence of the two K atoms there is local minimum for a Cu-CO height of about 3.4 bohr. In other words, the CO molecule is trapped on the cluster by the adjacent (fixed) K atoms.

To summarize, in order to reproduce the very large red-shift of the CO ω_e very short distances between CO and K must be considered, at the price of reducing the stability of the species chemisorbed on the surface. However, this is consistent with the observation that at low K coverage the ω shift is relatively small and that high K coverages are required in order to observe C-O vibrational frequencies of the order of 1200-1400 cm^{-1} [15].

3. CHEMISORPTION AND REACTIVITY OF CO AND CO_2 AT THE SURFACE OF MgO: ROLE OF SURFACE DEFECTS

The surface chemistry of metal oxides is of great importance in many catalytic processes and has been intensively investigated in the last three decades [34,35]. Several experimental techniques have been developed to characterize the adsorbed species at the surface of a metal oxide. One of the most widely used techniques is the infra-red, IR, specroscopy of the surface species [36]. This technique has the advantage of being non destructive and can be employed to study powders and polycrystalline samples, more closely related to the topography of an actual catalyst, under normal conditions (e.g. in air, in solution, etc.). Only recently it has been possible to study adsorption on well defined oxide single crystal surfaces under Ultra-High Vacuum, UHV, conditions [37]. These studies have shown that the chemistry of polycrystalline oxide surfaces is very rich and complex; on the other hand, regular, defect-free oxide surfaces are much less reactive and usually give rise to the formation of weakly physisorbed species. Thus, the surface defects play a very important role in determining the reactivity of the surface of an oxide material.

We have considered the interaction of two gas-phase molecules, CO and CO_2, at the surface of an ionic oxide like MgO. In particular, we have focussed the analysis on the interaction of the two molecules with defect sites like steps or corners, see Fig. 4, and compared the interaction at these sites with the bonding at regular surface sites.

We have performed all electron cluster model calculations on the interaction of CO and CO_2 with both cationic, Mg^{2+}, and anionic, O^{2-}, sites of the MgO(100) surface. We have considered regular, five-coordinated, surface sites, Mg_{5c}^{2+} and O_{5c}^{2-}, and defect, four-coordinated, step sites, Mg_{4c}^{2+} and O_{4c}^{2-}. Three-coordinated, corner, Mg_{3c}^{2+} sites have been considered only for CO adsorption. $[MgO_5]^{8-}$, $[MgO_4]^{6-}$, $[MgO_3]^{4-}$, see Fig. 4-5, are the clusters representing, respectively, a surface, five-coordinated, a step, four-coordinated, and a corner, three-coordinated, Mg^{2+} site. $[OMg_5]^{8+}$ and $[OMg_4]^{6+}$, Fig. 6, are the two clusters representing the surface and step O^{2-} chemisorption sites, respectively. In all these clusters only the atom where the interaction takes place and the first nearest neighbors have been included. The rest of the ionic crystal has been simulated by a very large set of point charges, PC=$\pm$2, placed at the lattice positions. This array of PCs is large enough to ensure that the Madelung potential in the chemisorption region is reproduced with reasonable accuracy [38]. The entire system, cluster + PCs, is neutral. Of course, in this model perfect ionicity is assumed for the MgO crystal.

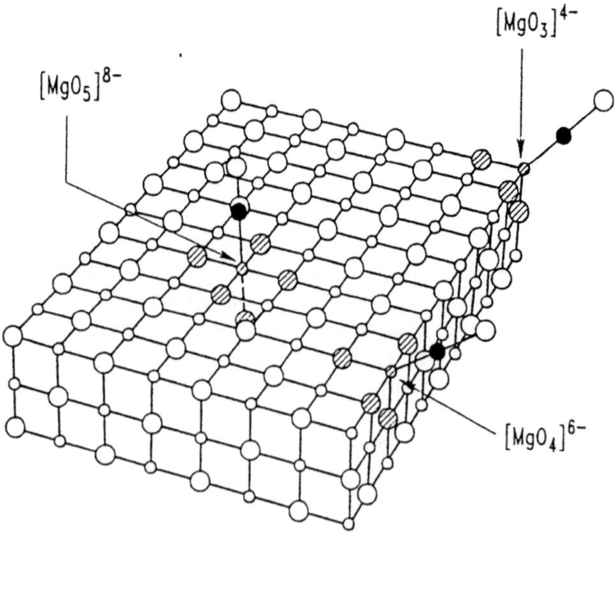

Figure 4 - Schematic representation of CO at surface, edge, and corner sites of MgO(100); the different equilibrium CO orientations are indicated. The atoms explicitly included in the cluster for the different sites are shown with hash marks.

Hartree-Fock SCF wavefunctions have been determined within the MO-LCAO approximation. The basis sets used are: Mg^{2+} [13s8p/6s3p], O^{2-} [11s7p/5s3p] [38]; the C and O atoms of CO were treated with a [9s5p/4s3p] basis set [18]; for CO_2 one d polarization function was added (α_d(C) =0.657, α_d(O)=0.74). For the special cases where the interaction of CO_2 occurs at the oxide anions, with possible formation of a surface carbonate, the same d polarization function used for the description of the oxygen atoms in CO_2 was added to the O^{2-} anion of the $[OMg_5]^{8+}$ and $[OMg_4]^{6+}$ clusters. Relaxation of the Mg^{2+} or O^{2-} chemisorption sites as well as the equilibrium geometry of the adsorbed CO_2 molecule were determined by performing a geometry optimization for the SCF wavefunctions by means of analytical gradient techniques as implemented in the HONDO8 [39] set of programs. Correlation effects have been introduced in few, selected, cases by performing single-double configuration interaction, SD-CI, calculations at the SCF optimal geometry. For further details see ref. 40,41.

3.1. Adsorption of CO at defect sites of MgO

The bonding of CO to the surface of an ionic oxide is dominated by electrostatic contributions [42-45]. Chemical bonding effects like charge donation from the 5σ MO of CO to the metal cations of the surface or back donation from the occupied d orbitals of transition metal oxides to the virtual $2\pi^*$ MO of CO are very small and do not significantly contribute to the stability of the chemisorbed CO molecule [42]. CO is pependicularly bound at carbon at the cationic sites of the oxide surface by the effect of the interaction between the electric field due to the cation and the permanent multipole moments, dipole, quadrupole, ocutpole, etc., of CO. This interaction, however, is rather weak, about a quarter of an eV in SCF [42], in agreement with recent experimental values [46]; correlated calculations give very similar values of the binding energy, see Table 6. Hence, CO is weakly physisorbed at the regular five-coordinated sites of the MgO surface at very low temperatures and easily desorbs as the temperature is increased.

At defect, low-coordinated, sites like steps and corners, however, we find that the bond distance between the Mg^{2+} cation and CO is shorter and the bonding is stronger with respect to five-coordinated surface sites, see Table 6. Also the vibrational frequencies of the CO molecules adsorbed at these sites, are significantly different than for a surface site. In particular, there is a larger shift of the CO ω_e toward higher frequencies for low-coordinated sites. These differences in the chemisorption properties of the three sites considered, see Fig. 4, indicate some changes in the mechanism of the bonding as one goes from regular to defect sites.

In order to identify these differences we have performed a decomposition of the interaction energy between CO and the various sites of MgO according to the Constrained Space Orbital Variation, CSOV [47,48]. The results are for unrelaxed substrates and for a CO geometry close to equilibrium. At each step of the CSOV, the interaction energy is defined as E_{int}(CSOV step n) = E(CO) + E(MgO) - E(MgO/CO; CSOV step n). The change in E_{int} between CSOV step n and the preceeding step n-1 represents the energetic importance of the new variational freedom allowed at step n [47,48], see Table 6.

The first step of the CSOV consists in simulating the ionic surface by an array of PCs. The Mg^{2+} cation and the O^{2-} anions are represented by

PC=+2 and PC=-2, respectively. At this first step, the charge density of the CO molecule is fixed as it is for the SCF wavefunction of the isolated molecule for the appropriate C-O bond distance. The value of E_{int} at this step provides a measure of the importance of the pure electrostatic interaction. At step 1, E_{int} is considerably larger for corner and edge sites than for a surface site, Table 6, because in the former cases the CO is adsorbed at a low-coordinated cation where the local electric field is stronger. In the second step of the CSOV we replace the few PCs defining the local chemisorption site with real Mg^{2+} and O^{2-} ions, but we do not allow the two charge densities, cluster and CO, to relax. The charge density for $[MgO_x]$ is fixed as that for the SCF wavefunction for the isolated cluster and the CO charge density is kept at that for the isolated molecule. However, at this step 2, the wavefunction formed from the superposition of the frozen charge densities of $[MgO_x]$ and CO, is normalized [47,48]. This step measures the extent of Pauli repulsion originating from the interpenetration of electronic charge at equilibrium positions, see Table 6. This energy contribution, which is purely repulsive, depends on the surface-CO distance but also on the number of O^{2-} ions on the surface. In $[MgO_5]^{8-}$/CO there are four spatially extended O^{2-} anions which prevent the CO molecule from approaching too close to the surface; in $[MgO_4]^{6-}$/CO there are only two O^{2-} ions on the first layer of the cluster, while in a corner site the O^{2-} ions are in the second layer, see Fig. 4. The Pauli repulsion is smallest for the surface and largest for the edge, Table 6, because it results from the balance of the two effects, i.e. the distance from the surface and the number of O^{2-} ions which are nearest neighbors of the CO molecule.

The substrate polarization, step 3, measured by allowing the MgO cluster orbitals to vary, is negligible for all three sites. Also the MgO to CO charge transfer, measured at CSOV step 4, is very small, Table 6. In CSOV step 5, the CO molecule is free to polarize but the $[MgO_x]$ charge density is frozen as determined at the previous CSOV step. The CO polarization can be viewed as arising from two origins. The first origin reduces the Pauli repulsion; the second origin is the electric field or the electrostatic potential at the bonding site due to the Mg^{2+} and O^{2-} ions. The CO polarization is considerably larger for the corner site than for the regular surface site; the edge site is intermediate. The dominant effect determining the size of this polarization is the electric field associated with the different CO adsorption sites. The charge donation from the CO molecule to the surface, see CSOV step 6 in Table 6, is small, and similar for the three sites. Differently from what is often believed, there is little or no σ-donation from CO to the ionic surface, even for a low-coordinated cation [38,42].

The different binding energies computed for CO on the three sites arise principally from the very different local electric field at the surface cation. For a regular, five-coordinated surface site this electric field is small, the initial electrostatic attraction is partially offset by the Pauli repulsion and the weak bonding is due to the electrostatic term plus the CO polarization; for a four- and a three-

TABLE 6 - Contributions, from SCF wavefunctions, from various bonding mechanisms to the interaction energy, E_{int}, in eV, for CO chemisorbed on cluster models of surface, edge and corner sites of the MgO surface. The analysis refers to r(C-O)=2.15 bohr. The total binding energy, D_e in eV, and the vibrational frequency, ω_e in cm^{-1} are given for SCF and CI wavefunctions.

cluster	MgO_5-CO	MgO_4-CO	MgO_3-CO
site	surface	edge	corner
z(Mg-C), in bohr	4.90	4.60	4.50
Electric field at Mg^{2+}, a.u.	0.019	0.033	0.062
Bonding contributions			
1. Electrostatic interaction	+0.24	+0.42	+0.48
2. Pauli repulsion	-0.21	-0.34	-0.28
3. MgO polarization	+0.02	+0.01	0.00
4. MgO ---> CO charge transfer	+0.06	+0.04	+0.04
5. CO polarization	+0.06	+0.12	+0.27
6. CO ---> MgO charge transfer	+0.06	+0.07	+0.07
total D_e (SCF)	+0.23	+0.33	+0.59
total D_e (SD-CI)	+0.32	+0.49	+0.83
$\omega_e/\Delta\omega$, cm^{-1} (SCF)[a]	2301/+31	2325/+55	2367/+97
$\omega_e/\Delta\omega$, cm^{-1} (SD-CI)[b]	2110/+43	2122/+55	2140/+73
$\omega_e/\Delta\omega$, cm^{-1} (experimental)[c]	2156/+13	2164/+21	2203/+60

a) $\Delta\omega$ is computed with respect to free CO ω_e(SCF) = 2270 cm^{-1}. b) $\Delta\omega$ is computed with respect to free CO ω_e(CI) = 2067 cm^{-1}. c) From ref. 49 and 50; $\Delta\omega$ is computed with respect to free CO ω_e(exp.) = 2143 cm^{-1}.

coordinated site, e.g., an edge or a corner site, the electric field becomes larger hence both the electrostatic attraction and the CO polarization increase. As a consequence, the chemisorption energy of a CO molecule on a corner site is almost three times larger than on a five-coordinated site; this is found for both the SCF and the correlated CI wavefunctions, see Table 6.

Experimentally it has been observed that the main line in the IR spectrum of CO on MgO microcrystals and on MgO powders, at 2156 cm^{-1}, is accompanied by a low-intensity feature at 2164 cm^{-1} [49,50] which has been attributed to CO molecules adsorbed on the edges of the crystallite [49,50]; the low intensity is consistent with the fact that the edge sites do not exceed the 5% of the total number of surface sites on the crystallite, the rest being regular, non-defect, surface sites. When CO is adsorbed on polycrystalline MgO, which has an high concentration of defect sites, a third band is observed at 2303 cm^{-1} [49]. This band has been assigned to CO adsorbed on low-coordinated Mg_{3c}^{2+} sites; an assignment which is fully supported by the present calculations, see Table 6. The SCF ω_e for an edge Mg_{4c}^{2+} site is 24 cm^{-1} larger than for the regular site, compared to the experimental assignement of the frequency shift being $\simeq$+10 cm^{-1}. For a corner, Mg_{3c}^{2+}, site, the SCF ω_e is 66 cm^{-1} larger than for the regular, Mg_{5c}^{2+}, site while the experimental ω_e is $\simeq$50 cm^{-1} larger than for the regular site. A similar trend is found when ω_e(C-O) is computed with CI wavefunctions [49,50].

The shift is basically due to two mechanisms: the interaction between the non uniform electric field at the surface and the CO dipole, and the "wall effect"; both mechanisms act to raise ω_e. For a coordinatively unsaturated corner site the field-dipole interaction is dominant with respect to the wall effect; for a five-coordinated surface site the field-dipole term is small and the shift is largely due to the wall effect. This is different from the interpretation which attributes the shift to the formation of a surface σ-complex where charge flows from the CO 5σ orbital to the substrate. These results clearly indicate that the defect sites of an ionic surface give rise to very different adsorbed species mainly because of the different electric field associated with a surface and a defect site. In the next paragraph we will show that this can have very important consequences for the reactivity of these sites.

3.2. Adsorption of CO_2 at defect sites of MgO

Differently from CO which bounds on metal as well as on oxide surfaces with the C atom down and the molecular axis parallel or slightly tilted from the surface normal, CO_2 can bound in several modes at the surface of an oxide. At least five different bonding modes have been found for CO_2 coordinated to metal atoms or in metal complexes: side-on, C-coordination, chelating bent, end-on bent, and end-on linear. All these bonding modes are possible candidates for the bonding of CO_2 with the cationic sites of an ionic crystal. In addition, CO_2 can bound with the oxide, O^{2-}, sites of an oxide surface and form surface carbonates.

Various orientations of the CO_2 molecule on the $[MgO_5]^{8-}$ and $[MgO_4]^{6-}$ cluster models of surface and step sites, respectively, of the MgO surface have been considered, see Fig. 5; however, only the end-on linear bonding mode is stable. All the other orientations considered, side-on and C-coordination in particular, are unfavorable and spontaneously dissociate into CO_2 and MgO fragments. The dissociation energy, D_e, of CO_2 adsorbed on a surface Mg^{2+} site is 0.18 eV. The bonding is slightly stronger, about one third of an eV, on a step Mg_{4c}^{2+} site, see Table 7. On a low-coordinated site like Mg_{4c}^{2+} the bonding is stronger for the same reasons discussed for the adsorption of CO: there are only two O^{2-} ions on the first layer of the cluster instead of four as for a surface Mg_{5c}^{2+} cation, Fig.s 4 and 5; the reduced repulsion with the bulky O^{2-} surface anions allows the CO_2 molecule to come closer to the surface and to interact more strongly with the cation; notice that the surface-CO_2 distance for a Mg_{4c}^{2+} step site is about 0.2 Å shorter than for a surface site, Table 7. Correlation effects, included through MP2 calculations, only moderately increase the stability of the surface complexes, Table 7. To summarize, CO_2 physisorbs at surface cations with its axis normal to the surface, see structures **1** and **3** of Fig. 5, forming a linear end-on surface complex; the CO_2 electronic structure is relatively unperturbed, as shown for instance by the very small changes in the C-O equilibrium bond distances with respect to the free CO_2 molecule, see Table 7.

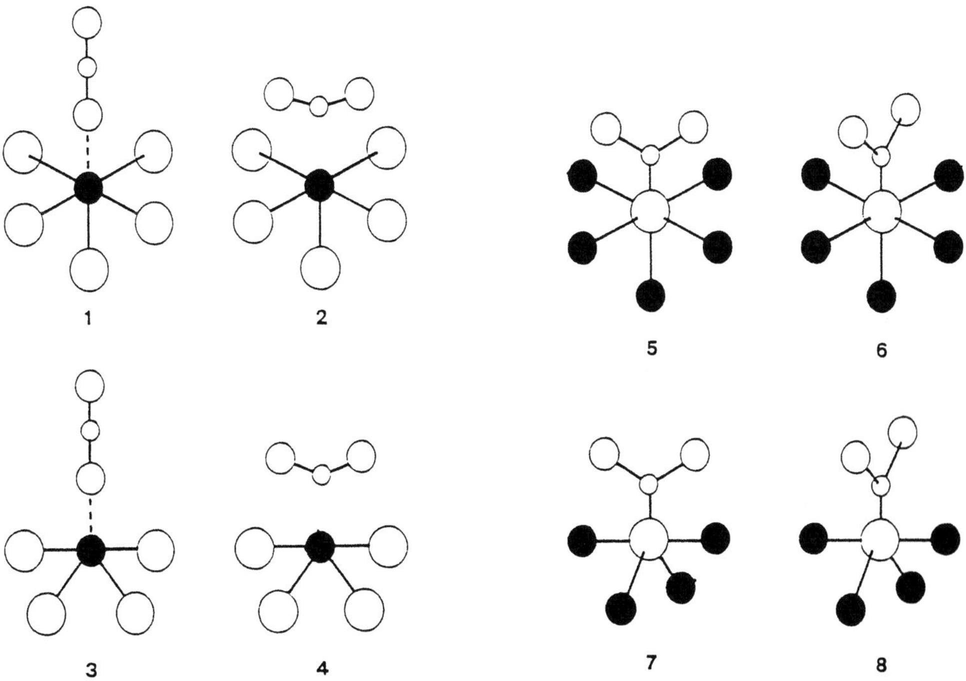

Figure 5 - Cluster models of CO_2 interacting with (a) a five-coordinated Mg_{5c}^{2+} surface site, 1 and 2, (b) a four-coordinated Mg_{4c}^{2+} step site, 3 and 4, (c) a five-coordinated O_{5c}^{2-} surface site, 5 and 6, and (d) a four-coordinated O_{4c}^{2-} step site, 7 and 8. The clusters are embedded in a large array of point charges (not shown) to reproduce the Madelung potential at the adsorption site. In 1 and 3 CO_2 is physisorbed; in 7 CO_2 is chemisorbed (surface carbonate); the other structures are unstable toward dissociation.

The interaction of CO_2 with the oxide, O^{2-}, sites of MgO can lead to the formation of a surface carbonate. This can be schematically described as a two-electron, two orbitals HOMO-LUMO interaction involving charge transfer from a filled sp_z hybrid orbital of O^{2-} to the empty $2\pi_u^*$ MO of linear CO_2. This charge transfer induces the bending of the CO_2 molecule. The same mechanism leads to the formation of C-coordinated CO_2 on transition metal surfaces [51], atoms [52], or complexes [53]. The formation of carbonates at the surface of MgO has been investigated with the help of the $[OMg_5]^{8+}/CO_2$ and $[OMg_4]^{6+}/CO_2$ cluster models, see Fig. 5.

At a five-coordinated O^{2-} surface site, the formation of a carbonate is an endothermic process; the surface carbonate, in fact, is 0.4-0.6 eV less stable than the separated units, $[OMg_5]^{8+}$ and CO_2, depending on the orientation of the molecule, see structures **5** and **6** and Table 8. At a distance of about 1.5 Å from the surface, we found a local minimum on the potential energy surface; a very small energy barrier separates this minimum from the dissociation limit, see Fig. 6. The small energy difference between structures **5** and **6** suggests that the electrostatic attraction between the surface Mg^{2+} cations and the oxygen atoms of CO_2, which become negatively charged because of the partial charge transfer from the surface, plays a non-negligible role in the stabilization of the surface carbonate. In fact, in **5** the shortest distance between a Mg^{2+} ion and the O-end of CO_2 is about 2.5 Å while it is 2 Å in **6**. This effect has

TABLE 7 - Optimal geometrical parameters and binding energy from SCF wavefunctions for CO_2 linearly bound at surface, five-coordinated, and step, four-coordinated, Mg^{2+} sites of the MgO(100) surface. See Fig. 5 for structure identification.

Site and structure			r_e[a], Å			D_e, eV
			Mg-O1	O1-C	C-O2	
Surface site	Mg_{5c}^{2+}-O1-C-O2	(1)	2.411	1.144	1.135	0.18 (0.12[b])
Step site	Mg_{4c}^{2+}-O1-C-O2	(3)	2.217	1.147	1.132	0.33 (0.29[b])

a) Free CO_2: r_e(CO)=1.141 Å. b) Corrected by the basis set superposition error.

a dramatic influence on the reactivity of the step O_{4c}^{2-} site, see Table 8. The surface carbonate **7** is bound by 1.60 eV. Structure **8**, on the contrary, is 0.5 eV less stable than the separated fragments. The two structures, **7** and **8**, differ only by the fact that the CO_2 molecule is rotated by 90°, see Fig. 5. In **7** the shortest distance between Mg^{2+} and the O-end of CO_2 is about 2 Å, while it is about 3 Å in **8**. Clearly, local electrostatic effects largely contribute to stabilize a given conformation. The different stability of the two forms is reflected also in the equilibrium geometry and in particular in the O-CO_2 distance which is substantially shorter for the bound complex **7**, see Table 8.

The completely different reactivity of a low-coordinated, defect, site with respect to a surface site is clear also from the analysis of potential energy curves for the $[OMg_n]^{m+}/CO_2$ interaction, see Fig. 6. For each, fixed, distance of CO_2 from a surface or a step oxide anion the C-O bond lengths and the O-C-O angle have been optimized. On a O_{4c}^{2-} step site the surface carbonate forms with no activation barrier; the interaction is of long-range nature and a weakly bound complex forms at relatively large distances, 2.5-3 Å, from the surface. As the CO_2 molecule approaches the O_{4c}^{2-} site, charge starts to flow into the $2\pi_u^*$ orbital and the molecule begins to bend. At the equilibrium distance, $r(O-CO_2)=1.40$ Å, the C-O distances are 1.22 Å and the internal angle is 134°, Table 8. At a surface O_{5c}^{2-} site the interaction is always

TABLE 8 - Optimal geometrical parameters and binding energy from SCF wavefunctions for CO_2 bound at surface, five-coordinated, and step, four-coordinated, O^{2-} sites of the MgO(100) surface with formation of surface carbonate. See Fig. 5 for structure identification.

Site and structure			r_e, Å O_{nc}-C	r_e, Å C-O	α_e O-C-O	D_e, eV
Surface site	$[O_{5c}-CO_2]^{2-}$	(**5**)	1.525	1.202	140°	-0.62
		(**6**)	1.506	1.206	138°	-0.40
Step site	$[O_{4c}-CO_2]^{2-}$	(**7**)	1.398	1.224	134°	1.60
		(**8**)	1.481	1.211	136°	-0.52

repulsive, Fig. 7; however, as the CO_2 molecule is pushed toward the surface oxide ion, charge transfer to the adsorbate occurs and the CO_2 molecule bends; the bending angle and the C-O bond lengths for a given O-CO_2 distance are not too different from those found for the O_{4c}^{2-} step site [41]. At a O-CO_2 distance of about 1.6-1.8 Å the curve exhibits a small maximum followed by a shallow local minimum as the distance is further reduced. The height of the barrier, however, is too small to speculate about its electronic origin and could be due to computational artifacts as well.

Thus, the same mechanism leads to the formation of a carbonate at a surface and at a defect site, although the energetics of the two processes is completely different. This is due to the distinct nature of five- and four-coordinated surface O^{2-} ions. The O^{2-} ion is unstable in the gas-phase, where it dissociates into O^- + e^-, but is stabilized in ionic crystals by the Madelung potential; the Madelung potential for a surface O_{5c}^{2-} site is considerably larger than for a step O_{4c}^{2-} site

Figure 6 - SCF potential energy curves for the interaction of CO_2 with a surface, five-coordinated, O_{5c}^{2-} ion, —●—, structure 5 of Fig. 6, and with a step, four-coordinated, O_{4c}^{2-} ion, —▲—, structure 7 of Fig. 6. For each fixed O-CO_2 distance, the C-O bond lengths and the O-C-O internal angle have been reoptimized.

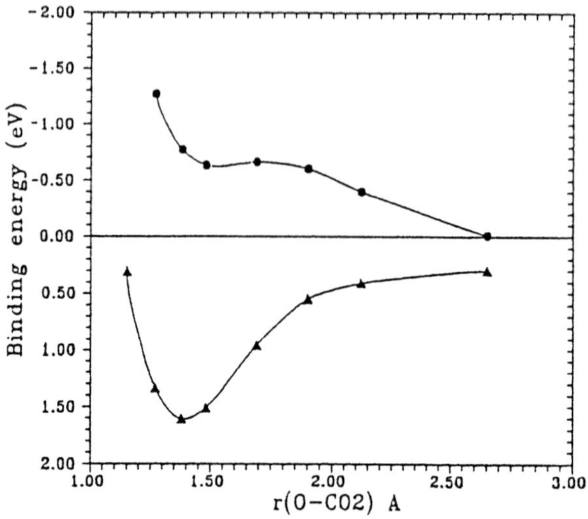

[54]. As a consequence, the electron cloud of a step O_{4c}^{2-} site is more spatially diffuse and can overlap with the CO_2 empty orbitals at much larger distances from the surface. The electrostatic potential at a step site is always larger than at a surface site; at about 5 Å from the surface, the electrostatic potential is virtually zero for a surface O^{2-} ion while is still large for a step site [41]. The interaction is of HOMO-LUMO type, and the position of the HOMO in the two sites is also very important: the charge transfer will occur at a lower energy cost if the "donor" orbital is at higher energy. Indeed, the a_1 HOMO of an O_{4c}^{2-} site is about 1.9 eV higher in energy than the corresponding level for a O_{5c}^{2-} site. This is roughly the difference in stability between the two surface carbonates **5** and **7**, see Table 8 and Fig. 6.

ACKNOWLEDGMENT. This work has been supported in part by a grant from Progetto Finalizzato Chimica Fine II of the Italian CNR. The support of NATO through the Collaborative Research Grant N. 900031 is gratefully acknowledged.

4. REFERENCES

[1] G. Pacchioni, P.S. Bagus, and F. Parmigiani (Eds.), *Cluster Models for Surface and Bulk Phenomena*, NATO ASI Series, Vol. 283, Plenum, New York 1992, and references therein.
[2] J.L. Whitten, in ref. 1, p. 375.
[3] C. Pisani, R. Orlando and R. Nada, in ref. 1, p. 515.
[4] H.P. Bonzel, Surf.Sci. Rept. 8 (1987) 43 and references therein.
[5] G. Herzberg, *Molecular Spectra and Molecular Structure*, Vol. 1, Van Nostrand, Princeton 1950.
[6] S. Katsuki, Y. Sakai, and E. Miyoshi, Surf. Sci. 220 (1989) 181; E.Miyoshi, Y. Sakai, and S. Katsuki, Surf. Sci. 242 (1991) 531.
[7] N. Rösch, A. Görling, P. Knappe, and J. Lauber, Vacuum 41 (1990) 150.
[8] G. Pacchioni and P.S. Bagus, Phys. Rev. B 40 (1989) 6003.
[9] P.E.M. Siegbahn, Surf. Sci 269/270 (1992) 276.
[10] M. Van Hove, S.Y. Tong, and N. Stoner, Surf. Sci. 54 (1976) 259.
[11] B.A. Hutchins, T.N. Rhodin, and J.E. Demuth, Surf. Sci. 54 (1976) 419.
[12] C.T. Campbell, Annu. Rev. Phys. Chem. 41 (1990) 775.
[13] D. Fischer, S. Chandavarkar, I.R. Collins, R.D. Diehl, P. Kaukasoina and M. Lindroos, Phys. Rev. Lett. 68 (1992) 2786.
[14] M. Kerkar, D. Fisher, D.P. Woodruff, R.G. Jones, R.D. Diehl, and B. Cowie, Phys. Rev. Lett. 68 (1992) 3204.
[15] L.H. Dubois, B.R. Zegarski and H.S. Luftman, J. Chem. Phys. 87 (1987) 1367.
[16] P.J. Hay and W.R. Wadt, J. Chem. Phys. 82 (1985) 299.
[17] P.S. Bagus, C.W. Bauschlicher, C.J. Nelin, B.C. Laskowski, and M. Seel, J. Chem. Phys. 81 (1984) 3594.
[18] F.B. van Duijeneveldt, IBM Res. Report N. RJ 945, 1971.
[19] L. Langmuir and K.H. Kingdon, Phys. Rev. 21 (1923) 381.
[20] S.A. Lindgren, C. Svensson, and L. Walldén, Phys. Rev. B 42 (1990) 1467.
[21] E.L. Garfunkel, X. Ding, G. Dong, S. Yang, X. Hou, and X. Wang, Surf. Sci. 164 (1985) 511.
[22] G. Pacchioni and P.S. Bagus, Surf. Sci. 269/270 (1992) 669.
[23] G.A. Benesh and D.A. King, Chem. Phys. Lett. 191 (1992) 315.
[24] M. Scheffler, C. Droste, A. Fleszar, F. Maca, G. Wachutka, and G. Barzel, Physica B 172 (1991) 143.
[25] H. Ishida, Phys. Rev. B 38 (1988) 8006; 42 (1990) 10899.
[26] B.N.J. Persson and H. Ishida, Phys. Rev. B 42 (1990) 3171.
[27] P.S. Bagus, G. Pacchioni, and M.R. Philpott, J. Chem. Phys. 90 (1989) 4287.
[28] G. Pacchioni, P.S. Bagus, C.J. Nelin, and M.R. Philpott, Int. J. Quant. Chem. 38 (1990) 675.
[29] B.A. Sexton, Chem. Phys. Lett. 63 (1979) 451.
[30] K. Hermann, P.S. Bagus and C.J. Nelin, Phys. Rev. B 35 (1987) 9467.
[31] K. Hermann and P.S. Bagus, Appl. Phys. A 44 (1987) 63.
[32] G. Pirug and H. P. Bonzel, Surf. Sci. 199 (1988) 371.
[33] G. Blyholder, J. Phys. Chem. 68 (1964) 2772.
[34] H.H. Kung, *Transition Metal Oxides: Surface Chemistry and Catalysis*, Elsevier, Amsterdam, 1989.

[35] V.F. Kiselev and O.V. Krilov, *Adsorption and Catalysis on Transition Metals and their Oxides*, Springer, Heidelberg, 1989.
[36] A.A. Davydov, *Infrared Spectroscopy of Adsorbed Species on the Surface of Transition Metal Oxides*, Wiley, New York 1990.
[37] H. Kuhlenbeck, G. Odörfer, R. Jaeger, G. Illing, M. Menges, Th. Mull, J.-H. Freund, M. Pölchen, V. Staemmler, S. Witzel, C. Scharfschwerdt, K. Wennemann, T. Liedtke and M. Neumann, Phys. Rev. B 43 (1991) 1969.
[38] G. Pacchioni, G. Cogliandro and P.S. Bagus, Int. J. Quant. Chem. 42 (1992) 1115.
[39] M. Dupuis, A. Farazdel, S.P. Karna, and S.A. Maluendes, in *Motecc: Modern Techniques in Computational Chemistry*, E. Clementi (Ed.) (Escom, Leiden 1991).
[40] G. Pacchioni, T. Minerva, and P.S. Bagus, Surf. Sci. 275 (1992) 450.
[41] G. Pacchioni, Surf. Sci., in press (1993)
[42] G. Pacchioni, G. Cogliandro, and P.S. Bagus, Surf. Sci. 255 (1991) 344.
[43] C. Pisani, R. Dovesi, R. Nada, and S. Tamiro, Surf. Sci. 216 (1989) 489.
[44] R. Dovesi, R. Orlando, F. Ricca, and C. Roetti, Surf. Sci. 187 (1987) 267.
[45] M. Polchen and V. Staemmler, J. Chem. Phys. 97 (1992) 2583.
[46] J.W. He, C.A. Estrada, J.S. Corneille, M.-C. Wu and D.W. Goodman, Surf. Sci. 261 (1992) 164.
[47] P.S. Bagus, K. Hermann and C.W. Bauschlicher, J. Chem. Phys. 81 (1984) 1966.
[48] P.S. Bagus, K. Hermann and C.W. Bauschlicher, J. Chem. Phys. 80 (1984) 4378.
[49] A. Zecchina, S. Coluccia, G. Spoto, D. Scarano and L. Marchese, J. Chem. Soc. Faraday Trans. 84 (1990) 703.
[50] L. Marchese, S. Coluccia, G. Martra, and A. Zecchina, Surf. Sci. 269/270 (1992) 135.
[51] H.J. Freund and R.P. Messmer, Surf. Sci. 172 (1986) 1.
[52] G.H. Jeung, Mol. Phys. 65 (1988) 669.
[53] S. Sakaki, K. Kitaura and K. Morokuma, Inorg. Chem. 21 (1982) 760.
[54] E.A. Colbourn, Surf. Sci. Rep. 15 (1992) 281.

REACTION DYNAMICS AT SURFACES

S. Holloway
Surface Science Research Centre
University of Liverpool
PO Box 147
Liverpool L69 3BX
UK

ABSTRACT. This work is a brief review of some of the more recent advances in our understanding of the reaction between small molecules and metal surfaces. Particular emphasis is placed upon the dissociation reaction and its dependence on molecular and surface properties.

1. Introduction

There is a direct mathematical link between reaction cross-sections and the familiar rate constants that appear in every entry-level textbook in physical chemistry [1]. By performing the relevant averages over the initial state distributions and summing over the final states, it is possible to pass from dynamics to kinetics. The most important question is whether or not this connection is simply a mathematical necessity or whether there is really anything to learn by studying state-to-state cross-sections or micro-canonical rate constants. I am not sure that I can answer this question in any convincing manner in this short article but what I shall attempt is to give a selection of recent interesting examples of surface dynamical phenomena and their rationalization in terms of microscopic models of the gas-surface interaction.

In studies of reactivity at surfaces, there is an implicit assumption that there is a bottleneck somewhere in some multi-dimensional space describing the nuclear motion, that has to be negotiated before products are observed. This is often written in the familiar language of transition state theory [2],

$$A + B \rightarrow A^* + B \rightarrow A_{ads}B$$

where the gas-phase species, A, needs to attain some particular configuration A* in position, geometry, internal state etc. before it will adsorb on the surface, B. An example of this might be the dissociation of a diatomic molecule or the abstraction of an atom from a polyatomic via an activated complex, both of which will be discussed in later sections. The transition state A*, typically has a very short lifetime, on the order of picoseconds, and therefore does not lend itself to detailed scrutiny offered by the vast range of surface science spectroscopic methods. Consequently one is frequently left to theorise (or even

R. W. Joyner and R. A. van Santen (eds.), Elementary Reaction Steps in Heterogeneous Catalysis, 341–358.

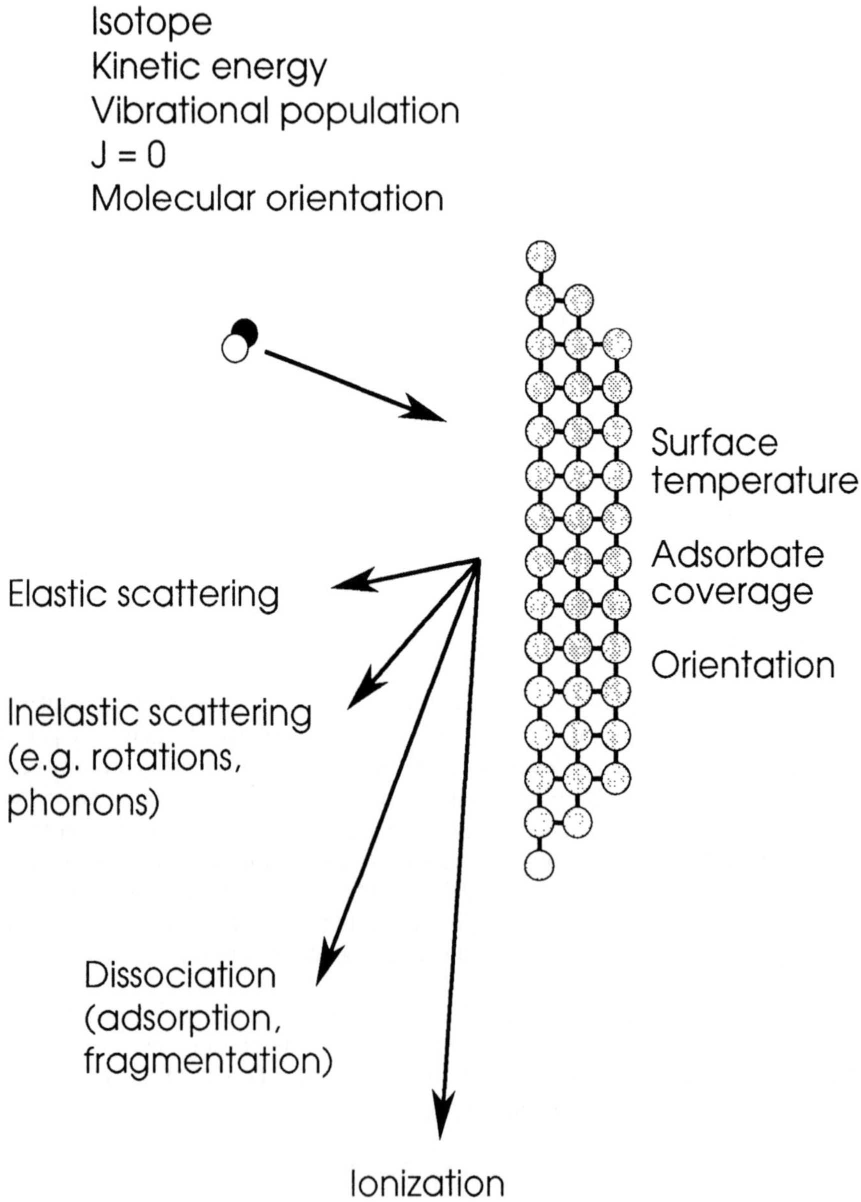

Figure 1: *Schematic figure showing some of the possible fates for a state-prepared molecule incident on a metal surface.*

philosophise) as to its presence. The interrogation of such bottlenecks by scattering methods opens up a new line of investigation for studying reactive processes. In addition, experimental dynamical studies can now offer new methods for investigating how energy can re-distribute itself within a molecule following collision with a surface. This is extremely important in the study of vibrationally or rotationally activated processes. Plainly put, if energy is selectively placed in a vibrational degree of freedom, for example, will their be a net change in the reaction cross-section for dissociation?

A typical dynamics experimental configuration is shown in **Fig. 1**. A mono-energetic beam of molecules (here referred to as species A_2) is produced from a supersonic source [3]. The expansion process used to generate the beam results in molecules having very low internal energies ($\leq$ 10 K), a FWHM of speed distributions typically $\geq$ 10% and variable translational energies in the range 0.05 to 10 eV. For studying state-to-state reaction cross sections, laser radiation may be employed to select the initial vibrational and rotational states and additionally probe the scattered molecules [4]. As in conventional studies, surface temperature, crystallographic orientation and adsorbate coverage may be controlled to provide a well defined 'target'. As indicated in **Fig. 1** the number of possible 'open channels' for the scattered particles is large. Broadly speaking the A_2 molecule can either be preserved intact or dissociate, with the products being trapped on the surface or scattered into the gas phase. Possible variations on this theme include inelastic scattering (vibrational, rotational excitation of the molecule; phonon, electron hole pair excitations in the surface), ionization, etc. etc.

2. Potential energy surfaces and the transition state approximation

In **Fig. 2** is shown a typical diagram which schematically shows the process of a particle desorbing from a surface. Three questions concerning this diagram need answering: *(i)* what is the black bump and what are its origins? *(ii)* what is a reaction coordinate? and *(iii)* why is the rate given by the Arrhenius expression?

The black bump represents a rudimentary potential energy surface (PES) which contains an activation barrier which separates reactants (e.g. A_2 in the gas phase) from products (e.g. A_2 or 2A adsorbed on a surface). **Fig. 3** shows a model PES for H_2 interacting with a Ni surface, which was first suggested by Lennard Jones in 1932, and has since become the standard vehicle for discussing the dissociative adsorption of molecules at metal surfaces [5]. In this PES, the activation barrier of magnitude P is shown at point K, D is the dissociation energy of molecular species AB, Q is the sum of chemisorption energies for species A and B and L is the equilibrium chemisorption location.

To obtain a 'real' PES describing a reactive encounter of a molecule with a metal surface is a formidable task, but the solution of the Schrödinger equation is facilitated by the large disparity in mass of the electrons and nuclei. As a molecule approaches a surface, suppose the coordinates of the electrons are given by $\mathbf{r} \equiv (\mathbf{r}_1, \mathbf{r}_2, \mathbf{r}_3, \ldots .\mathbf{r}_n)$, and the nuclei by $\mathbf{R} \equiv (\mathbf{R}_1,\mathbf{R}_2,\mathbf{R}_3, \ldots .\mathbf{R}_N)$. The total Hamiltonian may be expressed as,

$$\mathcal{H} = \mathcal{T}_R + \mathcal{H}_0 \tag{1}$$

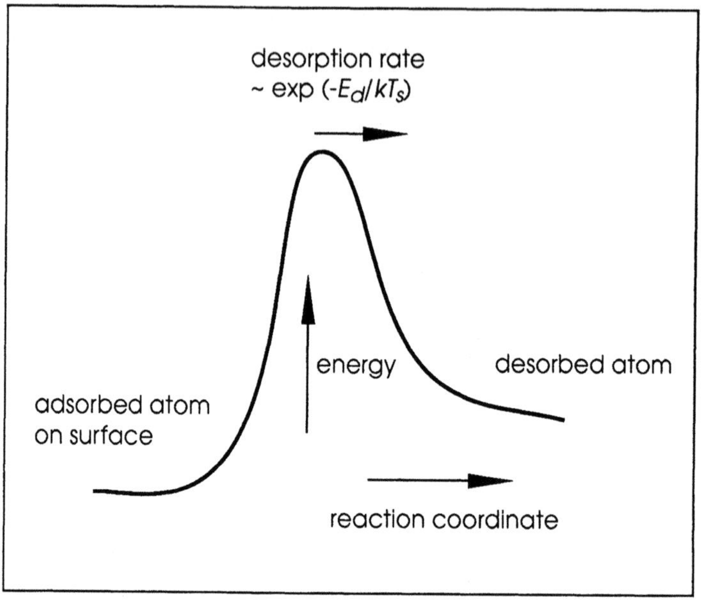

Figure 2: *A typical one-dimensional representation of the desorption of an atom from a surface.*

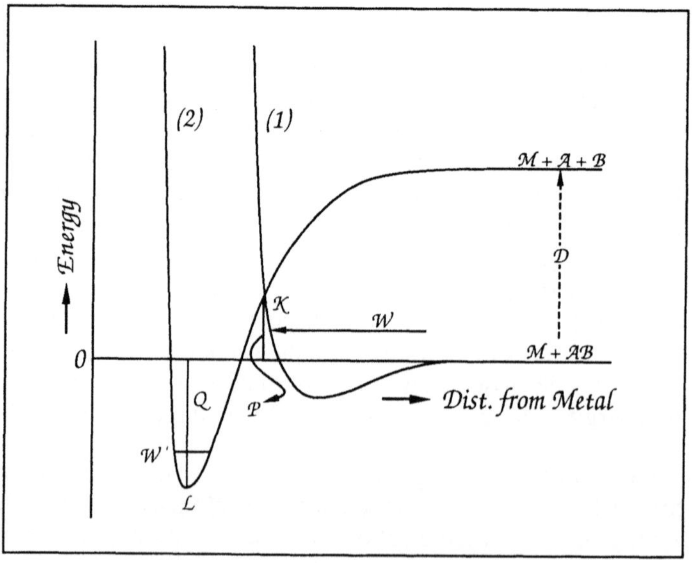

Figure 3: *The Lennard-Jones potential energy curves for the dissociative adsorption of a diatomic on a metal surface.*

where

$$\mathcal{T}_R = \sum_{M=1}^{N-1} -\left(\frac{1}{2\mu_M}\right)\nabla_M^2 \tag{2}$$

is the nuclear kinetic energy operator and

$$\mathcal{H}_0 = \sum_{i=1}^{n} -\frac{1}{2}\nabla_i^2 + \sum_{i=1}^{n-1}\sum_{j>i}^{n}\frac{1}{|r_i - r_j|} - \sum_{i=1}^{n}\sum_{M=1}^{N}\frac{z_M}{|r_i - R_M|}$$
$$+ \sum_{M=1}^{N-1}\sum_{M'>M}^{N}\frac{z_M z_{M'}}{|R_M - R_{M'}|} \tag{3}$$

is the electronic Hamiltonian for *fixed* nuclear positions. By selecting a suitable basis set for the electronic wave functions, $\phi_k(\mathbf{r};\mathbf{R})$, the total system wavefunction $\Psi(\mathbf{r},\mathbf{R})$ may be expressed as the linear combination

$$\Psi(\mathbf{r},\mathbf{R}) = \sum_{k} \phi_k((\mathbf{r};\mathbf{R})\ \chi_k(\mathbf{R})\ . \tag{4}$$

The wavefunction $\chi_k(\mathbf{R})$ describes the motion of the nuclei on the PES associated with the electronic state k. Substituting Eqs. 1-4 into the Schrödinger equation

$$[\mathcal{H} - \mathcal{E}]\Psi(\mathbf{r},\mathbf{R}) = 0 \tag{5}$$

one obtains the infinite set of coupled equations

$$[\mathcal{T}_R + U_{kk} - \mathcal{E}]\chi_k = -\sum_{k'\neq k}[\mathcal{T}'_{kk'} + U_{kk'}]\chi_{k'}\ , \tag{6}$$

where

$$U_{kk'} = <\phi_k \mid \mathcal{H}_0 \mid \phi_{k'}> \tag{7}$$

and

$$\mathcal{T}'_{kk'} = \sum_{M=1}^{N-1} -\left(\frac{1}{2\mu_M}\right) <\phi_k \mid \nabla_M \mid \phi_{k'}>\cdot\nabla_M\ . \tag{8}$$

It is the diagonal elements of $U_{kk'}$ which are the potential energy surfaces governing the nuclear motion in a particular electronic state, k. The off diagonal terms $\mathcal{T}'_{kk'}$ and $U_{kk'}$

give rise to transitions between the various electronic states and are usually referred to as non-adiabatic couplings. They depend critically upon the velocity of the nuclei as can be seen in Eq. 8.

Having arrived at the point where a set of PES have formally been defined the problem now to be addressed is what choice of electronic basis is most applicable. The Adiabatic representation is the most frequently used choice of representation and uses as basis states the eigenfunctions of $\mathcal{H}_0$,

$$\left|\mathcal{H}_0 - V_k\right| \phi_k = 0 \ . \tag{9}$$

Then from Eq. 7 it is seen that the eigenvalues V_k themselves are the required PES. The off diagonal elements $U_{kk'}$ are identically zero leaving only terms $\mathcal{T}'_{kk'}$ to couple motion between different electronic states. The states $\phi_k(\mathbf{r};\mathbf{R})$ then satisfy the variational principle and may be calculated using, for example, density functional methods. In what follows, and indeed for the overwhelming majority of hyperthermal gas-surface reaction dynamical studies, the terms $\mathcal{T}'_{kk'}$ will be set to zero and nuclear motion will evolve on the ground-state adiabatic PES.

Having established the origin of the black line in **Fig. 2**, it remains to determine an expression for the desorption rate. The 'reactants' signify the chemisorbed species and 'products' the bare surface plus desorbed molecule. The development follows that of transition state theory which takes as its starting point an equilibrium gas-surface system where the probability of finding momenta $\{\mathbf{V}\}$ and positions $\{\mathbf{R}\}$ for the particles is,

$$P(\mathbf{V},\mathbf{R}) \sim \exp\left[\frac{-\left(T_{\mathbf{R}}(\mathbf{V}) + V(\mathbf{R})\right)}{kT_s}\right] . \tag{10}$$

If one now draws a line in phase-space S at the top of the activation barrier, or more generally at the crossing seam, then at equilibrium the flux of particles crossing S in the forward and backward directions must be equal [6]. The dividing line may be parameterized by an arc length s, with $\mathbf{R}_s$ a point on the curve having a local normal $\mathbf{n}_s$ pointing toward the products. The classical thermal rate constant is then proportional to the flux of trajectories passing through S from reactants to products,

$$k = Q^{-1}\int \mathrm{d}\mathbf{V}\,\mathrm{d}s\ (\mathbf{V}\cdot\mathbf{n}_s)\,\chi_r(\mathbf{V},\mathbf{R}_s)\exp\left[\frac{-H(\mathbf{V},\mathbf{R}_s)}{kT_s}\right] , \tag{11}$$

where H (= $T_{\mathbf{R}}$(V) + V(R)) is the Hamiltonian of the nuclear subsystem, and Q is the partition function of the reactants. $\chi_r(\mathbf{V}, \mathbf{R}_S)$ is the characteristic function of reactive phase points and is unity if $(\mathbf{V}, \mathbf{R})$ lies on a reactive trajectory and zero otherwise. In transition state theory, the total rate constant k_{ts} is obtained by replacing $\chi_r(\mathbf{V},\mathbf{R}_S)$ with $\chi_+(\mathbf{V}, \mathbf{R}_S, \mathbf{n}_S)$ which is defined by,

$$\chi_+(\mathbf{V},\mathbf{R},\mathbf{n}) = \begin{cases} 1, & \text{if } \mathbf{V}\cdot\mathbf{n} > 0; \\ 0, & \text{otherwise.} \end{cases} \tag{12}$$

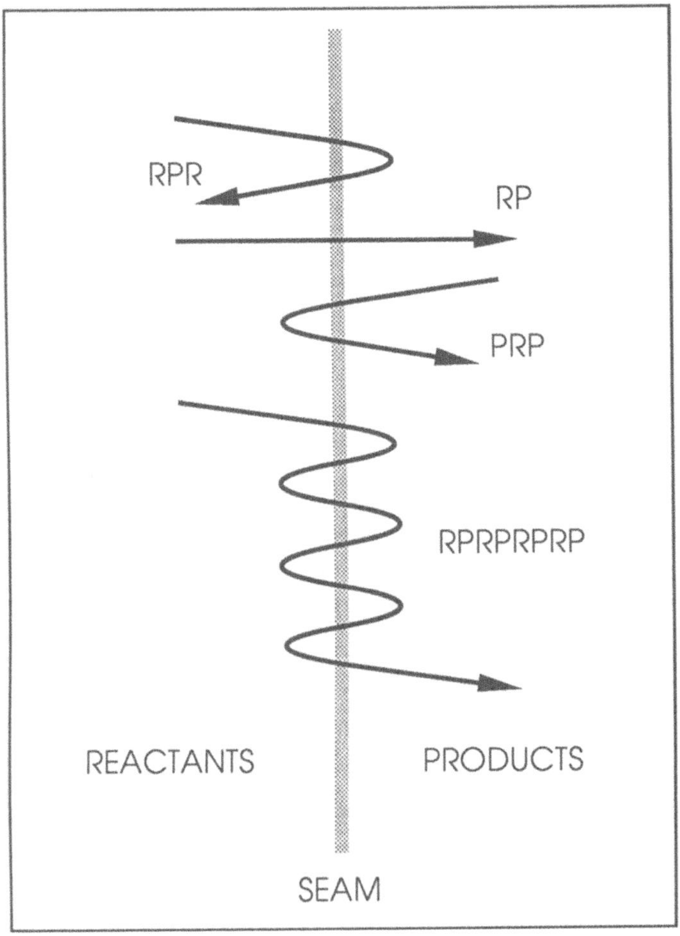

Figure 4: *A set of representative trajectories near the seam in the PES for an equilibrium gas-atom surface system..*

Clearly k_{ts} will be larger than k because it includes, *(i)* non-reactive trajectories which begin as reactants, cross S but then re-cross and end again as reactants (RPR), *(ii)* trajectories that start and end as products, but cross S (PRP) and *(iii)* true reactive trajectories with multiple crossings (either RP. . . . , RP or PR. . . . , PR). These trajectories are illustrated in **Fig. 4**. Grimmelmann *et al.* [7] addressed this issue by writing the true rate constant as the product

$$k = k_{ts} F, \tag{13}$$

where the factor $F(< 1)$, contains all of the re-crossing dynamics in the problem. By choosing a dividing surface in the asymptotic region of the products, they invoked microscopic reversibility and equated F with the sticking probability. It has been shown that by constraining S, one can obtain an upper bound to k_{ts}, a procedure often referred to as variational transition state theory [8]. If the transition state is located at a bottleneck which truly does not allow re-crossings, the variational procedure yields an exact result and $k_{ts} = k$.

To make further progress, an explicit form for $V(\mathbf{R})$ is required. If the surface is modelled by a system of near-neighbour potentials, and the desorbing species interacts with a single site upon the surface via $V(\mathbf{R}_0\text{-}\mathbf{R}_1)$, then it is possible to integrate Eq. 11 over the phonon states of the surface and derive a simple expression for the rate constant [9],

$$k_{ts}^{-1} = \sqrt{\frac{M}{2\pi k T_s}}\, \frac{1}{r_{w\text{-}s}^2} \int d(\mathbf{R}_0\text{-}\mathbf{R}_1) \left\{ \exp\left(\frac{-V(\mathbf{R}_0\text{-}\mathbf{R}_1)}{kT_s} \right) -1 \right\} . \tag{14}$$

Here M is the mass of the molecule and $r_{w\text{-}s}$ is a surface Wigner-Seitz radius. The desorption energy is approximately equal to the well depth and for a 'reasonable' parameter choice, this expression yields pre-exponents in the picosecond range.

3. Time dependent dynamical simulations

Within the last five years there has been a dramatic increase in the development of methods to study quantum dynamics in the time-domain [10]. While it is possible to extend classical simulations to ever increasing numbers of atoms [11], it must be remembered that there are real limitations for such studies. There is a quite general question regarding the validity of classical mechanics that is frequently pigeon-holed with the question of how important is tunneling in chemical reactions. This is particularly misleading since even in Eyring's paper on transition state theory, [2] he identified the key areas where quantum characteristics would manifest themselves in reactive encounters: *(i)* zero-point energy effects and quantized states in general, *(ii)* below barrier tunneling and above barrier reflection, and *(iii)* coherence effects.

Assume that a molecule prepared in some particular ro-vibrational state is approaching a surface with an activation barrier; it experiences the potential and reacts accordingly. In the classical sense, the barrier functions like an on-off switch [12]. If the initial kinetic

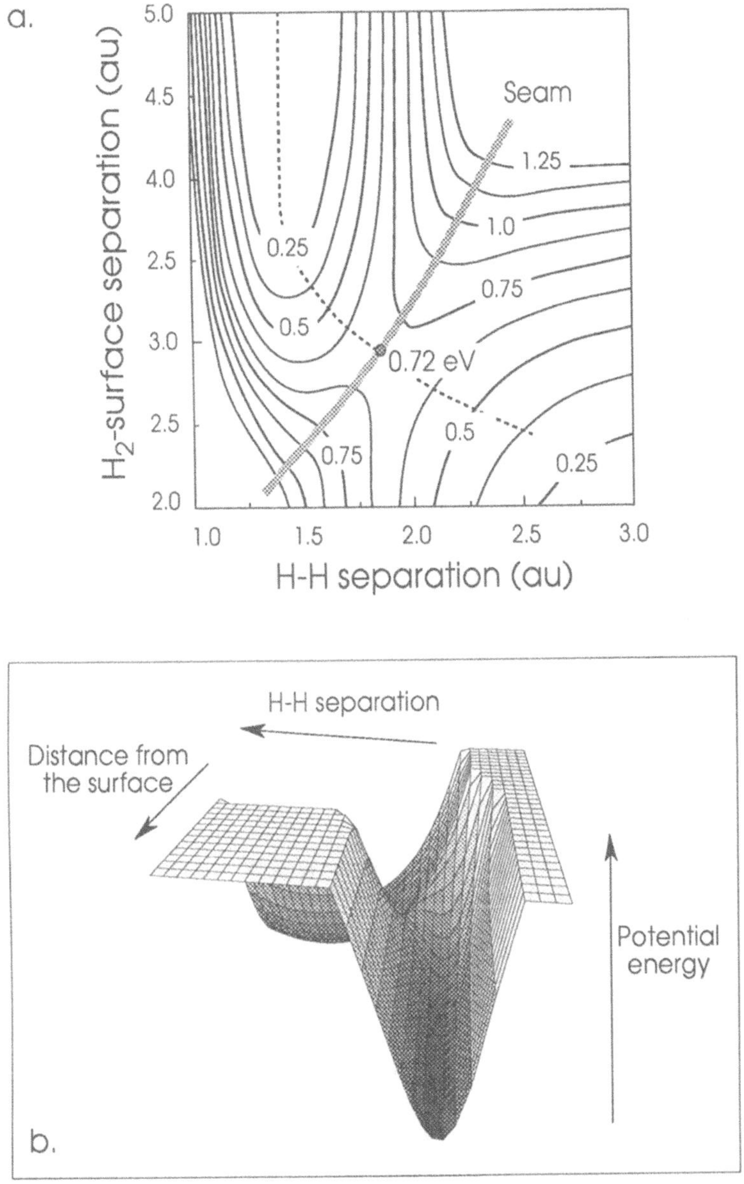

Figure 5: *The H2-Cu interaction potential used for studies mentioned in the text. a) A contour plot showing the main topological feature namely an activation barrier to dissociation located at a H-H separation in excess of the gas-phase equilibrium value. The broken line is the reaction path. b) a perspective view which corresponds to looking down the entrance channel.*

energy is greater than the barrier height then the trajectory will in general cross into the product state having no memory of its encounter [9]. Depending upon the position of the barrier along the reaction coordinate [13] there may be some slight rotation-to-translation (R-T) or vibration-to-translation (V-T) coupling in the region of approach to the surface which will result in a lower 'effective' barrier. Quantum mechanically this is quite different with substantial reflection occurring for even quite modest barrier heights [12]. In addition because of the lack of quantization in either the product or the reactant portions of the PES it is quite possible for a large number of initial trajectories to scatter either forward or backwards with vibrational energies below the zero-point threshold value. This is clearly nonsensical and for the case of H_2 can give rise to large differences in results obtained for either quantum or classical simulations on the *same* PES. Proponents of classical molecular dynamical methods claim that if quantum calculations could be extended to include many more degrees of freedom, then results from the two methods would be similar due to the increased degree of averaging occurring. There appear to be no concrete grounds to support this point of view.

Coupling quantum and classical degrees of freedom is a problem with a rich history. Thus far there is no consensus on how to treat this problem although there appear in the literature several attempts to do this. In particular, coupling quantum wavepackets with a classical solid has been investigated and produced interesting results [14, 15]. In parallel to this, the driven oscillator model, which combines a classical projectile with a bath of quantum oscillators, has also produced results of considerable importance. Philosophically it seems most sensible to combine an 'exact' treatment of the projectile with a semiclassical treatment of the surface [16], since for most problems, detailed information of the substrate degrees of freedom is totally unnecessary. At this moment in time, however, there exist no theoretical methods to accomplish this.

This can only be a brief introduction to time dependent quantum studies and for a more detailed account the interested reader is referred to my article in [10]. Let us assume that a PES, $V(\mathbf{R})$ has been 'obtained' from somewhere (there will be more said about this in what follows). This potential is then placed on a suitably sized grid. The problem is then to find a method to solve the time dependent Schrödinger equation

$$i \frac{\partial \Psi(\mathbf{R};t)}{\partial t} = \mathcal{H}\Psi(\mathbf{R};t) , \tag{15}$$

where $\mathcal{H}$ is the Hamiltonian for the problem,

$$\mathcal{H} = \mathcal{T} + V(\mathbf{R}) , \tag{16}$$

and $\mathcal{T}$ the kinetic energy operator.

A commonly used technique is the split operator method which involves dividing the time evolution operator

$$\Psi(\mathbf{R};t+\Delta t) = \mathcal{U}\,\Psi(\mathbf{R};t) = \exp\left[-i\,\Delta t\,\mathcal{H}\right]\Psi(\mathbf{R};t). \tag{17}$$

in the following way [17]

$$\mathcal{U}\,\psi(\mathbf{R};t) \approx \exp\left[-i\frac{\Delta t}{2}\mathcal{T}\right]\exp\left[-i\,\Delta t\,\mathcal{V}\right]\exp\left[-i\frac{\Delta t}{2}\mathcal{T}\right]\psi(\mathbf{R};t) \tag{18}$$

An efficient method to evaluate the second order spatial derivative in the kinetic energy employs the Fast Fourier Transform (FFT) algorithm [18]. A spatial derivative is mapped into reciprocal space as a multiplication by ik, where k is the wave vector. Similarly, the second derivative is mapped as a multiplication by $-k^2$. Therefore, by transforming the wavefunction from its real-space to its momentum-space representation with a FFT, the kinetic energy operator simply becomes multiplicative. In this way it is possible to apply each operator in the representation in which it is diagonal. The great efficiency of the FFT algorithm makes the method computationally tractable.

4. Dynamical simulations

4.1. VIBRATIONAL EFFECTS IN DISSOCIATIVE ADSORPTION

The observation that in H_2 scattering experiments only a certain fraction of the molecules that are incident 'stick,' is frequently taken to be evidence that there exists an activation barrier located between the entrance (H_2 + Surface) and exit channels (2H + Surface) of the PES [10]. An example of this form of topology is shown in **Fig. 5**. Assuming the surface to be a static entity, there are six degrees of freedom which are required to describe the interaction of a diatomic with that surface. Since it is only possible to present any two of these sensibly on a piece of paper, the particular slice of the PES shown in **Fig. 5** confines the molecular bond to be parallel to the surface plane. The two selected coordinates are the H-H separation (x) and the distance of the molecular centre-of-gravity from a point in the surface, (Z). This particular cut highlights the translational-to-vibrational (T $\rightarrow$ V) coupling experienced during a gas-surface encounter. The functional form of the PES is of little importance for present purposes and may be found elsewhere [19]. Suffice to say that it is comprised from two diabatic states that have been smoothly linked: a physisorption potential describing the interaction of the closed shell H_2 species with the surface and a chemisorption potential which accounts for the interaction of the dissociated fragments (H-atoms) with the surface [10]. The location of the activation barrier along the reaction path is crucial for an understanding of the dynamics [13]. The PES shown in **Fig. 5** is an example of a barrier located toward the exit channel and is based upon a calculation for H_2 interacting with a small Cu cluster [20].

The asymptotic initial state wave function was taken to be a product of a Morse oscillator in vibrational state n and a minimum uncertainty Gaussian function centred at Z_i with a mean initial momentum p_Z,

$$\psi(x,Z;t=0) = \chi_n(x)\, g(Z\text{-}Z_i,p_Z). \tag{19}$$

This state is propagated in time using Eq. 18 until an asymptotic final state is attained. The two possible fates for the molecule are reflection from the surface in the vibrational state n' (not necessarily $= n$) or dissociative adsorption. Calculational details have been

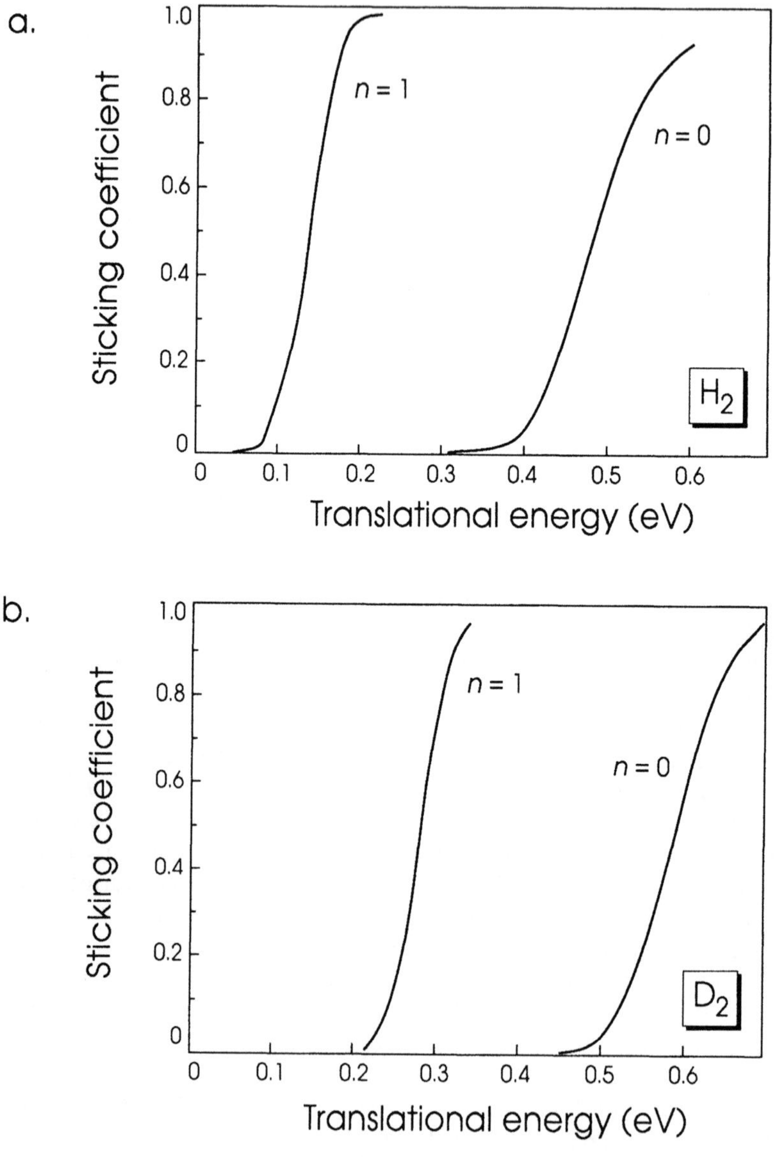

Figure 6: *Dissociation probability as a function of initial translational energy for a)* H_2 *and b)* D_2 *scattering from a Cu surface. The curves refer to incidence with the molecule in either its vibrational ground state or first excited state. It is quite clearly seen that for the PES shown in the previous figure that vibrational energy significantly enhances dissociative adsorption.*

presented elsewhere [19, 21, 22]. **Fig. 6** shows the results for the dissociation probabilities as a function of translational energy for H_2 and D_2. The following features are immediately obvious: *(i)* increasing the translational energy always enables molecules to dissociate, ultimately with unit probability; *(ii)* placing energy in vibrational degrees of freedom enhances dissociation; and *(iii)* for a given translational energy and vibrational state, more H_2 dissociates than D_2. These results can be understood quite simply by considering the change in the zero point energy as the reaction path is traversed (see **Fig. 5**). At each point along the reaction path, the motion will evolve vibrationally adiabatically. Since the frequency of vibration of the H_2 'entity' will be reduced along this path, there will be significant V $\rightarrow$ T exchange [13]. For a barrier located in the entrance channel of the reaction, this will result in a negligible effect; for a barrier located further around the reaction path this will have a considerable effect, as evidenced by the data shown in **Fig. 6**. This data may be comprehensively understood working within this framework: translational energy will always be effective simply because of the basic topology of the surface; vibrational energy will act to reduce the height of the effective barrier because of the increased V $\rightarrow$ T occurring for the states with higher n; the vibrational states of the heavier isotopes are lower and, as a consequence, the perceived activation barrier will appear higher [23].

Using this basic model it is possible to interpret a wide range of recent molecular beam data for the dissociative adsorption of H_2 on Cu surfaces [24-26]. It has been experimentally verified that by putting excess energy in the vibrational coordinate, it is possible to increase the probability of dissociative adsorption and, as a consequence, decrease the survival probability for $n = 1$ molecules [27].

Fig. 7 shows the consequences of scattering a $n = 0$ beam of hydrogen from a similar PES [28]. It may be seen that as soon as the energetic threshold permitting $0 \rightarrow 1$ excitation is exceeded ($\varepsilon_{trans} = 0.514$ eV), the scattered molecules contain a relatively large $n = 1$ fraction. This is a graphic manifestation of the non-adiabatic coupling discussed above. As the molecules reflect from the barrier, the PES strongly mixes translational with vibrational motion and the molecules undergo a $0 \rightarrow 1$ transition [29, 30]. Again, were the barrier located early in the reaction zone, this would not be observed [31].

4.2. SURFACE TEMPERATURE EFFECTS IN METHANE DISSOCIATION

There have been a vast number of experimental investigations of the dissociative adsorption of CH_4 on a variety of metal surfaces [32-35]. In these studies almost every conceivable experimental parameter has been varied; gas temperature (kinetic energy), angle of incidence, surface orientation, surface temperature, etc., etc., in an attempt to clarify the primary dissociation mechanism. Up until recently, the discussion in the literature was particularly vitriolic, with different experimental groups presenting their viewpoint in the strongest of terms. There were several areas of conflict but in particular there was strong disagreement as to the extent to which a tunnelling mechanism was required to explain the strong isotope dependence. In addition the role played by the surface temperature in the formation of the methyl radical was also unclear.

Just over a year ago there appeared a dynamical model for the H-atom abstraction from CH_4 catalysed by a Pt(111) surface [36]. There appeared to be overwhelming

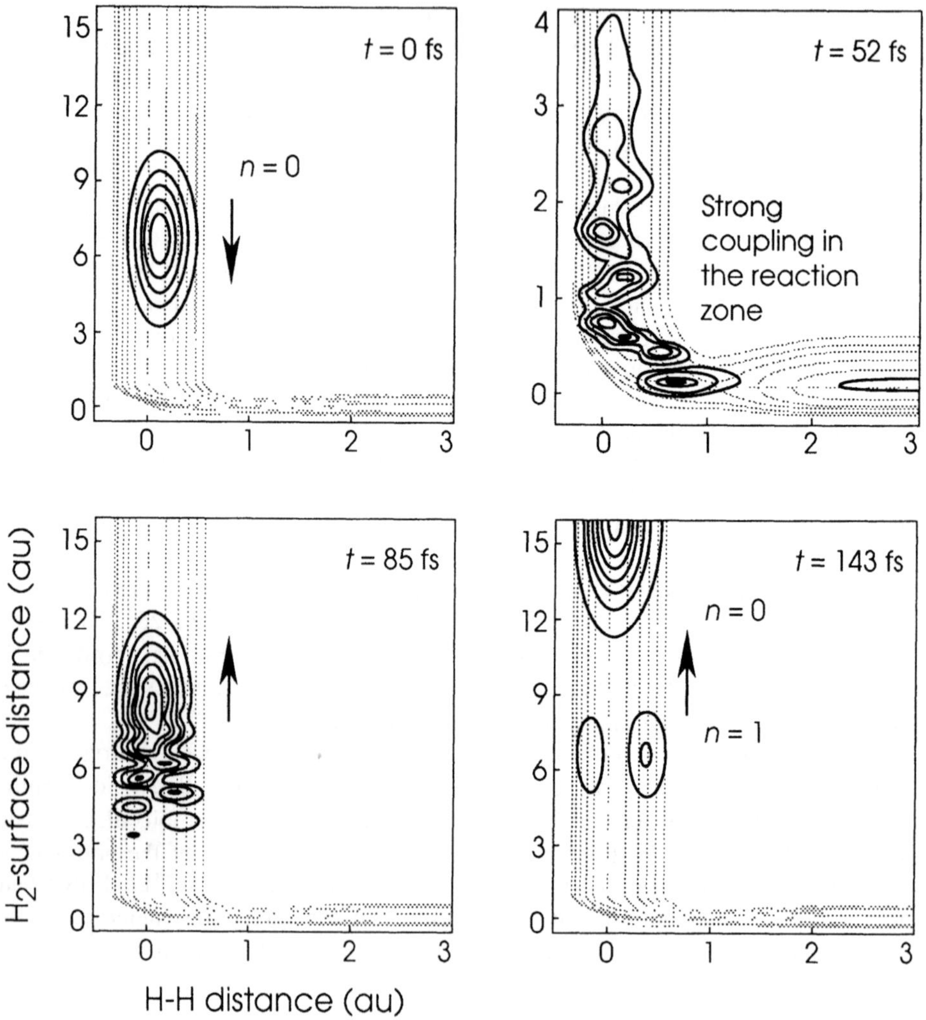

Figure 7: *Four frames showing the time development of a wavepacket representing a H_2 molecule in its ground vibrational state incident with 0.7 eV translational energy on a PES similar to the late-barrier one shown in Fig. 5. As the wavepacket 'turns the corner' it undergoes simultaneous dissociation and vibrational excitation. The scattered fraction emerges with a fast component still in the vibrational ground state and a slow part that is vibrationally excited.*

experimental evidence that the molecule, upon impact with the surface, undergoes a reaction $CH_4 \rightarrow CH_3 + H$. This was modelled using a PES very similar to that shown in **Fig. 5**, reducing the molecular dimensionality to that of a heteronuclear diatomic. What this amounts to is describing the methyl radical as an inert mass 15 pseudo-atom having no internal degrees of freedom. This 'diatomic' was then coupled to the surface which was represented by an Einstein oscillator, able to take up vibrational energy via an harmonic coupling to the molecule. This is shown schematically in **Fig. 8** Reasonable choices for the PES parameters were found and a quantum dynamical simulation was performed in a way discussed in Sec. 4.1 The only new ingredient required was that the initial wavefunction (Eq. 19) needed to be modified so as to include the initial eigenstate of the surface oscillator. Surface temperature could then be easily included by performing a Boltzmann weighting of the final state-resolved results.

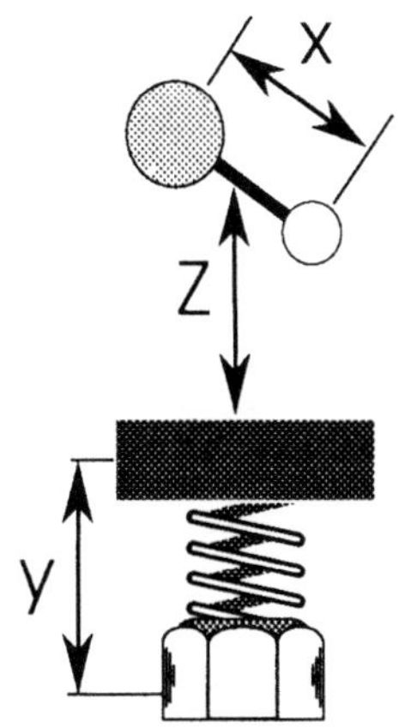

Figure 8: *Schematic representation of the degrees of freedom involved in the calculation for the dissociation of methane on a Pt(111) surface. The molecule is represented by a heteronuclear diatomic comprised of a methyl radical and an H-atom. A surface atom is taken to be a soft quantum cube with a harmonic interaction with the bulk. Energy may be freely exchanged between translation, vibration and surface motion.*

Fig. 9 shows the results for the incidence energy dependence of the dissociation probability for CH_4 and CD_4. In **(a)** it can be seen that there is a strong dependence on T_s with results given for 0 and 800 K. Experimental results confirming this behaviour [37] are shown in **(b)**. The value of the activation barrier in the static PES is 0.76 eV and so the results indicate that there is a considerable tunnelling which accounts for the isotopic behaviour. The dependence upon the surface temperature is more clearly seen in **Figs. 9(c) and (d)**, where the dissociation probability for a variety of incidence energies is plotted as a function of the inverse surface temperature [37]. The key effect to arise from these low-dimensional dynamical calculations is the modulation of the 'effective' barrier height which arises from the surface motion. Because of the large mass disparity in the heteronuclear CH_3-H 'diatomic,' in the appropriate mass-scaled coordinate system, the barrier appears in the entrance channel. The motion of the surface approximates that due to a Rayleigh mode which cycles the activation barrier in the *z*-direction, parallel to the reaction path. Thus an incoming molecule will, in the two limits, perceive a barrier which is either moving toward or away from it; resulting in a lower or higher effective barrier. In comparison with a 0 K surface, the latter case is irrelevant but the reduction in the barrier gives rise to a *thermally assisted tunelling* capable of rationalizing the diverse experimental data.

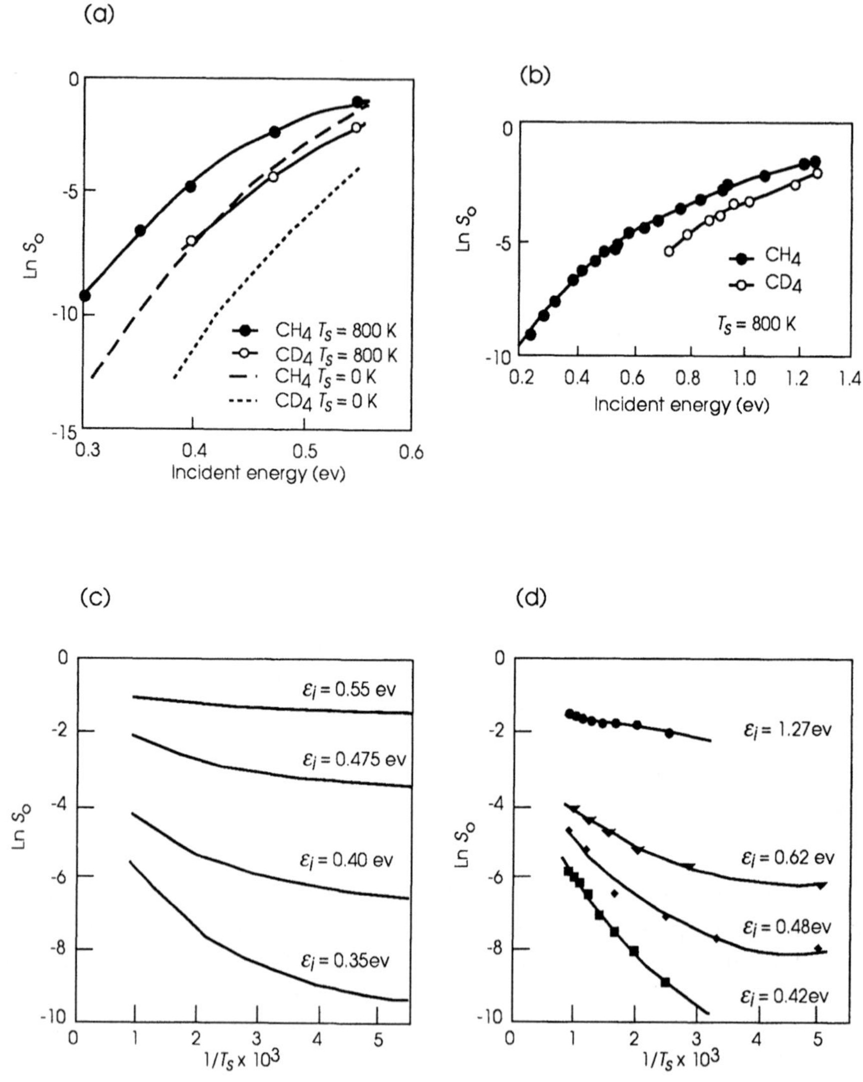

Figure 9: *a) Dissociation probability for CH_4 and CD_4 as a function of incidence translational energy at a surface temperature of 800 K. The dashed lines are fits through the 0 K calculations. b) Experimental results for $CH_4/Pt(111)$. c) Dissociation probabilities as a function of inverse surface temperature for a range of incidence energies. d) Experimental results for $CH_4/Pt(111)$.*

5. Summary

In this short article an attempt has been made to describe some of the most recent theoretical studies in gas-surface dynamics, particularly those that might have an impact on the field of heterogeneous catalysis. The most obvious example has been the dissociative adsorption of molecules on metal surfaces. This is a fast-growing area of research both experimentally and theoretically. It would seem clear that over the next five to ten years our understanding of process dynamics will improve markedly on account of this type of research.

References

[1] P. W. Atkins, *Physical Chemistry* (Oxford University Press, Oxford, 1982).
[2] H. Eyring, J. Chem. Phys. **3**, 107 (1935).
[3] J. A. Barker and D. J. Auerbach, Surf. Sci. Rep. **4**, 1 (1979).
[4] H. Zacharias, Appl. Phys. **A47**, 37 (1988).
[5] J. E. Lennard-Jones, Trans. Faraday Soc. **28**, 333 (1932).
[6] E. E. Nikitin, *Theory of Elementary Atomic and Molecular Processes in Gases* (Clarendon, Oxford, 1974).
[7] E. K. Grimmelmann, J. C. Tully and E. Helfand, J. Chem. Phys. **74**, 5300 (1981).
[8] R. D. Levine and R. B. Bernstein, *Molecular Reaction Dynamics and Chemical Reactivity* (Oxford Univ. Press, Oxford, 1987).
[9] J. Harris, S. Holloway, T. Raman and K. Yang, J. Chem. Phys. **89**, 4427–4439 (1988).
[10] S. Holloway, in *Rettner review* edited by M. R. Ashfold and C. T. Rettner (Royal Society of Chemistry, London, 1991), p. 88.
[11] A. E. DePristo, in *Rettner review* edited by M. R. Ashfold and C. T. Rettner (Royal Society of Chemistry, London, 1991), p. 47.
[12] B. Jackson and H. Metiu, J. Chem. Phys. **86**, 1026 (1986).
[13] D. Halstead and S. Holloway, J. Chem. Phys. **93**, 2859 (1990).
[14] C. Cerjan and R. Kosloff, Phys. Rev. B **34**, 3832 (1986).
[15] B. Jackson, J. Chem. Phys. **88**, 1383 (1987).
[16] G. Drolshagen and E. J. Heller, Chem. Phys. Lett. **104**, 129 (1984).
[17] J. A. Fleck, J. R. Morris and M. D. Feit, Appl. Phys. **10**, 129 (1976).
[18] D. Kosloff and R. Kosloff, Lett. Comput. Phys. **52**, 35 (1983).
[19] M. R. Hand and S. Holloway, J. Chem. Phys. **91**, 7209 (1989).
[20] J. Harris and S. Andersson, Phys. Rev. Lett. **55**, 1583 (1985).
[21] M. R. Hand and S. Holloway, Surf. Sci. **211/12**, 940 (1989).
[22] M. R. Hand and J. Harris, J. Chem. Phys. **92**, 7610 (1990).
[23] S. Holloway, A. Hodgson and D. Halstead, Chem. Phys. Lett. **147**, 425 (1988).
[24] B. E. Hayden and C. L. A. Lamont, Phys. Rev. Lett. **63**, 1823 (1989).
[25] H. F. Berger, M. Leisch, A. Winkler and K. D. Rendulic, Chem. Phys. Lett. **175**, 425 (1991).
[26] C. T. Rettner, D. J. Auerbach and H. Michelsen, Phys. Rev. Lett. **68**, 1164 (1992).
[27] A. Hodgson, J. Moryl and H. B. Zhao, Chem. Phys. Lett. **182**, 152 (1991).

[28] S. Holloway and G. R. Darling, Comments in At. Mol. Phys. **27**, 335-348 (1992).
[29] A. Hodgson, J. Moryl, P. Traversaro and H. Zhao, Nature **356**, 501 (1992).
[30] C. T. Rettner, D. J. Auerbach and H. Michelsen, Phys. Rev. Lett **68**, 2547 (1992).
[31] G. R. Darling and S. Holloway, J. Chem. Phys. **97**, 734 (1992).
[32] C. N. Stewart and G. Ehrlich, J. Chem. Phys. **62**, 4672 (1975).
[33] H. F. Winters, J. Chem. Phys. **62**, 2454 (1975).
[34] C. T. Rettner, H. E. Pfnür and D. J. Auerbach, Phys. Rev. Lett. **54**, 2716 (1985).
[35] M. B. Lee, Q. Y. Yang and S. T. Ceyer, J. Chem. Phys **87**, 2724 (1987).
[36] J. Harris, J. Simon, A. C. Luntz, C. B. Mullins and C. T. Rettner, Phys. Rev. Lett. **67**, 652 (1991).
[37] A. C. Lunz and D. S. Bethune, J. Chem. Phys. **90**, 1247 (1989).

CHEMICAL REACTIVITY AS A FUNCTION OF COVERAGE

V.P. ZHDANOV
Boreskov Institute of Catalysis
Pr. Akademika Lavrentieva, 5
Novosibirsk 630090
Russia

ABSTRACT. The kinetics of rate processes on solid surfaces, as a rule, is substantially nonideal owing to the heterogeneity of the substrate, lateral interactions between adsorbed particles or adsorbate-induced changes in the surface. To describe these effects, one can use various approaches developed in statistical physics, including the quasi-chemical and Bethe-Peierls approximations, transfer-matrix technique, renormalization-group method, and Monte Carlo simulations. The review presented outlines the main ideas and mathematical formalism of the above mentioned approaches and their application in the theory of kinetic processes in adsorbed overlayers.

1. INTRODUCTION

Study of the kinetics of elementary surface rate processes such as adsorption, desorption, and the simplest chemical reactions is of great practical interest because at present heterogeneous catalysis is the cornerstone of the chemical industry, and optimization of catalytic processes in this field is impossible without detailed data on their kinetics. Moreover, study of the kinetics of rate processes on solid surfaces is also of interest academically because there are many interesting problems here.

The simplest model for describing the kinetics of rate processes on a surface is that of an ideal adsorbed layer (lattice-gas without lateral interaction). This model seems to be the first suggested by Langmuir in his classic study of surface phenomena. On the basis of the Langmuir model, the kinetics of elementary processes is described by simple power-law equations: the monomolecular desorption rate is proportional to the coverage of the surface by adsorbed particles, the rate of bimolecular reactions between adsorbed particles is proportional to the product of their coverages, etc. The model of an ideal adsorbed layer is attractive because of its simplicity. For this reason, it is widely used for a qualitative analysis of surface phenomena [1]. A real adsorbed layer, as a rule, is substantially nonideal, owing to the heterogeneity of the surface, lateral interactions between adsorbed particles, or adsorbate-

R. W. Joyner and R. A. van Santen (eds.), Elementary Reaction Steps in Heterogeneous Catalysis, 359–376.

induced changes in the surface. It is important to note that these factors, resulting in strong coverage dependence of rate constants of various kinetic processes, are often mutually interconnected. For example, lateral interactions between adsorbed particles and substrate atoms may result in surface reconstruction [2]. In turn, the latter process may result in formation of a set of new adsorption sites. A typical example here is spontaneous or adsorbate-induced reconstruction of the (110) surface of fcc metals. In this case, formation of the missing-row structure on the surface is facilitated by strongly anisotropic lateral interaction between metal atoms in the topmost layer of the substrate [3], and this structure contains a lot of different adsorption sites.

The simplest method for describing a heterogeneous surface is based on the idea that the surface consists of a set of ideal regions. The nonideality of the kinetics in this case is obtained by averaging the ideal kinetics over some parameters, for example, the surface binding energy of adsorbed particles. This method is a formal one. The construction of more detailed theory is precluded by the fact that the structure of a nonuniform surface is often too complicated to be adequately described theoretically. Some new ideas in this field are connected with application of fractal geometry to simulating the rate processes on heterogeneous surfaces [4]. However, the latter approach is also not universal.

In the case of a uniform surface, i.e., a surface consisting of equivalent unit cells, the nonideality of the kinetics is determined by lateral interactions between adsorbed particles or by adsorbate-induced changes in the surface. In this review, we present the basic principles for describing the kinetics of elementary processes on a uniform surface taking into account lateral interactions. The main physical ideas employed in this field have recently been discussed in detail [5]. In the present review, we focus our attention on the mathematical formalism. In particular, compared to ref. [5], we discuss in more detail the transfer-matrix technique and renormalization-group method.

2. LATTICE-GAS MODEL

The detailed description of particles adsorbed on a uniform surface depends on whether the lateral interaction energy between them is greater or lower than the activation barrier for diffusion along the surface. In the former case, the adsorbed particles can occupy sites which are not correlated with the relief of the substrate, and their arrangement can be incommensurate with the arrangement of the surface atoms. The theory of rate processes in such systems is in the seminal state [6]. In the latter case, the arrangement of adsorbed particles is commensurate with the relief of the substrate. We shall study primarily the commensurate structures, since such structures are encountered most often in chemisorbed overlayers and the rate theory here is well developed. The potential barrier for surface diffusion of chemisorbed particles is usually high in comparison with temperature (we set $k_B = 1$). In this case, the state of elementary surface sites can be described by the occupation number: $n_i = 1$ if site i is occupied by a particle and $n_i = 0$ if not. Thus, the Hamiltonian of adsorbed particles has the simple form

$$H = \sum_{ij} \epsilon_{ij} n_i n_j \ , \tag{2.1}$$

where ϵ_{ij} is the lateral interaction energy between particles occupying sites i and j.

Hamiltonian (2.1) corresponds to the two-dimensional lattice-gas model - one of the classical models of statistical physics [7]. This model can be employed to describe many phenomena occuring on solid surfaces, including phase diagrams of adsorbed particles and kinetics of various processes [5].

The general properties of lateral interactions between particles adsorbed on metals (including Van der Waals interaction, electrostatic dipole-dipole interaction, as well as indirect interaction via the substrate) have now been thoroughly studied [8]. On the whole, however, reliable calculations of the values of lateral interactions are the task for the future. At present, the lateral interactions appearing in Hamiltonian (2.1) must be regarded as empirical parameters.

3. GENERAL KINETIC EQUATIONS

Calculation of the rate of an elementary surface process involves description of both the dynamics and the statistics of adsorbed particles. In the first approximation, the effect of the particles surrounding a particle or a pair of particles entering directly into the elementary transformation reduces to a shift of the bottom of the potential well and the top of the potential barrier owing to lateral interactions. In this approximation, the dynamic and statistical parts of the problem are decoupled, and the dependence of the rate of the process on the surface coverages by adsorbed particles is determined wholly by the statistical distribution of these particles on the surface. Analysis of the dynamics of the nucleus motion is necessary only to determine the absolute magnitude of the preexponential factor of the rate constant of the process. In this review, we are interested only in the statistical part of the problem. The dynamics of adsorbed particles is considered in [5].

The assumption that the dynamic and statistical parts of the problem can be separated is correct provided the time that characterizes the motion of the reacting particles near the top of the potential barrier is shorter than the time that characterizes the rearrangement of the nearest-neighbour particles. In the case of chemisorption, this condition is usually fulfilled.

The other main assumptions, employed in the framework of the lattice-gas model for describing the effect of lateral interactions on the kinetics of surface rate processes, are usually the following. A single type of sites is used. A given site is vacant or occupied by a single adsorbed particle. Particles can be in the ground or in the activated state. Lateral interactions between neighbouring adsorbed particles are taken into consideration. It is also assumed that activated complexes interact with neighbouring adsorbed particles. Repulsive interactions are assigned positive values. At any temperatures and coverages, surface migration is customarily believed to be rapid compared to other processes, so that thermodynamic equilibrium is attained in the adsorbed overlayer. The number of activated complexes is assumed to be small, so that the probabilities of distribution of particles on the surface may be calculated without taking into consideration activated complexes.

The general formulas for describing the rates of elementary processes in the lattice-gas model have a simple form [5]. The rate of an elementary process, as usual, is expresses in terms of the energy difference between the bottom of the potential well and the top of the potential barrier for the motion of particles. For example, the kinetics of the bimolecular reaction between adsorbed particles A and B (the Langmuir-Hinshelwood mechanism) is described by

$$d\theta_A/dt = d\theta_B/dt = -W_{LH}(\theta_A,\theta_B) \tag{3.1}$$

with

$$W_{LH} = \nu \sum_i \mathcal{P}_{AB,i} \exp[-(E_{LH} + \Delta\epsilon_i)/T)] , \tag{3.2}$$

where θ_A and θ_B are the adsorbate coverages; E_{LH} the energy difference between the activated complex A^*B^* and the pair AB provided the neighbouring sites are empty; $\mathcal{P}_{AB,i}$ the probability that a pair of nearest-neighbouring sites are occupied by particles A and B, with the environment denoted by index i; $\Delta\epsilon_i = \epsilon_i^* - \epsilon_i$; ϵ_i the lateral interaction of the pair AB with the environment; and ϵ_i^* the lateral interaction of the activated complex $A^*A^*$with the same environment.

Formulas of type (3.1) and (3.2) are not so much a solution as a formulation of the problem of calculating the rate of elementary processes, since the main difficulty lies in calculating the probabilities appearing in these formulas. Only in some particular cases, one can carry out the sum over all configurations and derive compact analytical expressions for the rate constant of various processes. For example, if the activated complex does not interact with the environment ($\epsilon_i^* = 0$), the desorption rate constant can be represented as (see eqs. (4.2.12) or (4.5.8) in [5])

$$k_d = k_d^o \exp(\mu/T)(1 - \theta)/\theta , \tag{3.3}$$

where k_d^o is the desorption rate constant at low coverages, and μ is the chemical potential of adsorbed particles. Under the same assumption, the chemical diffusion coefficient can be expressed as (see eq. (7.3.16a) in [5])

$$D = D^o \mathcal{P}_{00} \exp(\mu/T) \frac{1}{T} \frac{\partial\mu}{\partial\theta} , \tag{3.4}$$

where D^o is the diffusion coefficient at low coverages, and $\mathcal{P}_{00}$ is the probability that two nearest-neighbour sites are empty.

To calculate probabilities of various arrangements of adsorbed particles and other parameters (e.g., the chemical potential), one can use various approaches developed in statistical physics.

4. MEAN-FIELD APPROXIMATION

The simplest statistical approximation is the mean-field one. Within the framework of this approximation, the arrangement of particles is assumed to be random, the energy of interaction between a particle and other particles is replaced by the average interaction, and the coverage dependence of rate constants is described by simple exponential functions [5]. The mean-field approximation was very popular due to its simplicity. This approximation is however rather rough because it does not take into account correlations in the arrangement of adsorbed particles. For this reason, we will not discuss the mean-field approximation in more detail.

5. QUASI-CHEMICAL APPROXIMATION

5.1. Formalism

The simplest statistical approach that takes into consideration correlations in the arrangement of particles is the quasi-chemical approximation. The most direct way to the quasi-chemical equations is based on application of the grand canonical distribution to the two-site cluster. Taking into account the interaction of the particles located in the cluster and replacing the interaction between these particles and the environment by the average self-consistent interaction, one easily obtains

$$\mathcal{P}_{AA}\,\mathcal{P}_{OO}/\mathcal{P}_{AO}^2 = 0.25\,\exp(-\epsilon_{AA}/T)\ ,$$

where $\mathcal{P}_{AA}$, $\mathcal{P}_{AO}$ and $\mathcal{P}_{OO}$ are the probabilities that two adjacent sites are occupied by two, one or zero particles, and ϵ_{AA} is the nearest-neighbour lateral interaction. In addition, we have two balance equations

$$\mathcal{P}_{AA} + \mathcal{P}_{AO} + \mathcal{P}_{OO} = 1\ ,$$

$$2\,\mathcal{P}_{AA} + \mathcal{P}_{AO} = 2\theta_A\ .$$

The analytical solution to these equations is

$$\mathcal{P}_{AA} = \theta_A - \alpha\ ,\quad \mathcal{P}_{AO} = 2\alpha\ ,\quad \mathcal{P}_{OO} = 1 - \theta_A - \alpha\ ,$$

where

$$\alpha = \{1 - [1 - 2\beta\theta_A(1 - \theta_A)]^{1/2}\}/\beta$$

and

$$\beta = 2[1 - \exp(-\epsilon_{AA}/T)] .$$

By analogy, for coadsorption of particles A and B, we have [5]

$$\mathcal{P}_{AA} + \mathcal{P}_{AB} + \mathcal{P}_{AO} + \mathcal{P}_{BB} + \mathcal{P}_{BO} + \mathcal{P}_{OO} = 1 ,$$

$$2\,\mathcal{P}_{AA} + \mathcal{P}_{AB} + \mathcal{P}_{AO} = 2\theta_A ,$$

$$2\,\mathcal{P}_{BB} + \mathcal{P}_{AB} + \mathcal{P}_{BO} = 2\theta_B ,$$

$$\mathcal{P}_{AA}\,\mathcal{P}_{OO}/\mathcal{P}^2_{AO} = 0.25\exp(-\epsilon_{AA}/T) ,$$

$$\mathcal{P}_{BB}\,\mathcal{P}_{OO}/\mathcal{P}^2_{BO} = 0.25\exp(-\epsilon_{BB}/T) ,$$

$$\mathcal{P}_{AA}\,\mathcal{P}_{BB}/\mathcal{P}^2_{AB} = 0.25\exp[-(\epsilon_{AA} + \epsilon_{BB} - 2\,\epsilon_{AB})/T] ,$$

where ϵ_{AA}, ϵ_{BB} and ϵ_{AB} are the A-A, B-B and A-B lateral interactions.

The latter system of equations has no analytical solution. However, it can be easily solved numerically employing the following iteration method. If $\mathcal{P}^{(n)}_{AA}$, $\mathcal{P}^{(n)}_{BB}$ and $\mathcal{P}^{(n)}_{OO}$ are the probabilities on step n, step (n+1) is given by

$$\mathcal{P}^{(n+1)}_{AO} = 2[\,\mathcal{P}^{(n)}_{AA}\,\mathcal{P}^{(n)}_{OO}\exp(\epsilon_{AA}/T)]^{1/2} ,$$

$$\mathcal{P}^{(n+1)}_{BO} = 2[\,\mathcal{P}^{(n)}_{BB}\,\mathcal{P}^{(n)}_{OO}\exp(\epsilon_{BB}/T)]^{1/2} ,$$

$$\mathcal{P}^{(n+1)}_{AB} = 2\{\,\mathcal{P}^{(n)}_{AA}\,\mathcal{P}^{(n)}_{BB}\exp[(\epsilon_{AA} + \epsilon_{BB} - 2\,\epsilon_{AB})/T]\}^{1/2} .$$

Step (n+2) is defined as

$$\mathcal{P}^{(n+2)}_{AA} = \mathcal{P}^{(n)}_{AA}2\theta_A/[2\,\mathcal{P}^{(n)}_{AA} + \mathcal{P}^{(n+1)}_{AB} + \mathcal{P}^{(n+1)}_{AO}] ,$$

$$\mathcal{P}^{(n+2)}_{BB} = \mathcal{P}^{(n)}_{BB}2\theta_B/[2\,\mathcal{P}^{(n)}_{BB} + \mathcal{P}^{(n+1)}_{AB} + \mathcal{P}^{(n+1)}_{BO}] ,$$

$$\mathcal{P}^{(n+2)}_{OO} = \mathcal{P}^{(n)}_{OO}\,2(1 - \theta_A - \theta_B)/[2\,\mathcal{P}^{(n)}_{OO} + \mathcal{P}^{(n+1)}_{AO} + \mathcal{P}^{(n+1)}_{BO}] ,$$

and so on.

According to the quasi-chemical approximation, the probabilities that a given A particle has an A particle, a B particle, or an empty site near it (i.e., in the two-site cluster) are

$$P_{A,A} = \mathcal{P}_{AA}/(\mathcal{P}_{AA} + 0.5\,\mathcal{P}_{AB} + 0.5\,\mathcal{P}_{AO}),$$

$$P_{A,B} = 0.5\,\mathcal{P}_{AB}/(\mathcal{P}_{AA} + 0.5\,\mathcal{P}_{AB} + 0.5\,\mathcal{P}_{AO}),$$

$$P_{A,O} = 0.5\,\mathcal{P}_{AO}/(\mathcal{P}_{AA} + 0.5\,\mathcal{P}_{AB} + 0.5\,\mathcal{P}_{AO}).$$

Using these equations and taking into consideration that different pairs of sites are assumed to be independent in the quasi-chemical approximation, one can easily calculate the probabilities of various arrangements of adsorbed particles. For example, the probability that the A particle has in the nearest-neighbour sites n A particles and m B particles is

$$P_{A,nm} = \frac{z!}{n!\,m!\,(z-n-m)!}\,P^{n}_{A,A}\,P^{m}_{A,B}\,P^{z-n-m}_{A,O}.$$

Employing formulas of this kind, it is possible to carry out the sum in equations of type (3.2) and to obtain analytical expressions for the rate constants of various processes. In particular, such expressions for monomolecular adsorption and desorption, dissociative adsorption and associative desorption, and for the simplest chemical reactions (the Eley-Rideal and Langmuir-Hinshelwood mechanisms) are presented in [5].

Finally, it should be noted that the quasi-chemical approximation gives an accurate solution for the one-dimensional lattice-gas model with nearest-neighbour lateral interactions. To improve the quasi-chemical approximation in the two-dimensional case, we can consider the cluster that contains a central site and other adjacent sites. This leads to the Bethe-Peierls approximation. The latter approximation is known to be equivalent to the quasi-chemical approximation in the case of a square lattice. For more complex lattices, the Bethe-Peierls approximation is more accurate than the quasi-chemical one.

5.2. Applications

The quasi-chemical approximation was widely used to simulate the coverage dependence of the sticking coefficients, kinetics of desorption in isothermal conditions, thermal desorption spectra, and kinetics of various chemical reactions, including, e.g., CO oxidation on Ir, hydrogen-oxygen reaction on Pt, NO decomposition on different metals, and HCOOH decomposition on Ni. The detailed review of studies in this field is given in [5].

6. TRANSFER-MATRIX TECHNIQUE

6.1. Formalism

The transfer-matrix technique yields an accurate solution for the one-dimensional lattice-gas model. This technique can also be very effectively used to solve the two-dimensional problems [5,9]. Let us consider, for example, adsorption on a rectangular lattice with anisotropic nearest-neighbour lateral interactions ϵ_x and ϵ_y. The Hamiltonian corresponding to this case can be represented as (we include the chemical potential into the Hamiltonian)

$$H = \sum_{i,j} (\epsilon_x n_{n,j} n_{i+1,j} + \epsilon_y n_{i,j} n_{i,j+1} - \mu n_{i,j}) , \qquad (6.1)$$

where $n_{i,j}$ is the occupation number of site (i,j). The grand partition function for the model under consideration can be calculated as follows. The lattice is replaced by the strip of finite width n and N along the X and Y directions. The periodic boundary conditions are used along the Y direction. The vector $Z_k^{(n)}$ is so introduced that its components are the grand partition functions for the strip with fixed occupation numbers of sites located on the right end of the strip (the arrangement of particles in these sites is marked by index k). Adding an additional row of sites to the right end of the strip and examining all possible occupation numbers of sites in rows n and n+1, we have the recursive relationship

$$Z_k^{(n+1)} = \sum_l T_{kl} Z_l^{(n)}, \qquad (6.2)$$

where T_{kl} is the transfer matrix. At $n \to \infty$, the probability to find a given arrangement of particles on the right end of the strip is not dependent on n. Thus, at $n \to \infty$, the vector $Z_k^{(n+1)}$ should be proportional to $Z_k^{(n)}$, i.e.

$$Z_k^{(n+1)} = \lambda_o Z_k^{(n)} . \qquad (6.3)$$

Substituting this relationship into eq. (6.2), we can conclude that at $n \to \infty$ $Z_k^{(n)}$ is actually an eigenvector and λ_o is an eigenvalue of the matrix T_{kl}. The latter matrix can be represented as

$$T = SDS^{-1}, \qquad (6.4)$$

where S is a nonsingular matrix and **D** is a diagonal matrix with elements that are equal to eigenvalues of the matrix T. Substituting eq. (6.4) into eq. (6.2), one can easily derive that at $n \to \infty$ the grand partition function, Z, is expressed as

$$Z \sim \lambda_o^n \quad \text{or} \quad \ln Z = n \ln \lambda_o , \qquad (6.5)$$

where λ_o is now the largest eigenvalue of T. Comparing eqs. (6.3) and

(6.5), we can infer that in eq. (6.3) λ_o is also the largest eigenvalue of T and accordingly $Z_k^{(n)}$ is actually the eigenvector corresponding to this eigenvalue. Thus, the calculation of the grand partition function is reduced to the calculation of the largest eigenvalue of the transfer matrix. The latter problem can be solved by using standard numerical methods (e.g., by the iteration procedure [5]).

The grand canonical distribution yields the following relationship between coverage and the chemical potential

$$\theta = \frac{T}{nN}\frac{\partial \ln Z}{\partial \mu} = \frac{T}{N}\frac{\partial \ln \lambda_o}{\partial \mu} .$$

Using this relationship, one can calculate the coverage dependence of the chemical potential.

Employing Hamiltonian (6.1) as an example, let us discuss now a way of constructing the transfer matrix. If we use this Hamiltonian directly the transfer matrix will be nonsymmetric. In practice, however, it is more convenient to employ symmetric matrices. To construct a symmetric transfer matrix, we can rewrite Hamiltonian (6.1) in the following form

$$H = \sum_{ikl} (0.5\, H_{i,k} + 0.5\, H_{i+1,l} + U_{i,k,l}) , \qquad (6.6)$$

where

$$H_{i,k} = \sum_j (\epsilon_y n_{i,j}^{(k)} n_{i,j+1}^{(k)} - \mu n_{i,j}^{(k)}) ,$$

and

$$U_{i,k,l} = \sum_j \epsilon_x n_{i,j}^{(k)} n_{i+1,j}^{(l)} .$$

Physically, the terms $H_{i,k}$ and $H_{i+1,l}$ describe rows i and i+1 in states k and l, respectively, and $U_{i,k.l}$ describes the interaction between rows i and i+1. Representation (6.6) is symmetric with respect to indexes k and l. Accordingly, the transfer matrix constructed for this representation is symmetric as well. Note also, that at given indexes k and l the terms $H_{i,k}$, $H_{i+1,l}$ and $U_{i,k,l}$ are the same for different rows. For this reason, calculating these terms, we may omit index i. Then, the transfer matrix is given by

$$T_{kl} = \exp[-(0.5\, H_k + 0.5\, H_l + U_{k,l}) /T] .$$

For example, let us construct the transfer matrix for the simplest two-dimensional strip (N = 2) described by Hamiltonian (6.1). In this case the two sites located on the right end of the strip, may be in four states described by occupation numbers (0,0), (1,1), (0,1) and (1,0). The corresponding transfer matrix is given by

$$T_{11} = 1, \quad T_{22} = \exp[(2\mu - 2\epsilon_x - 2\epsilon_y)/T] ,$$

$$T_{33} = T_{44} = \exp[(\mu - \epsilon_x)/T],$$

$$T_{12} = \exp[(\mu - \epsilon_y)/T], \quad T_{13} = T_{14} = \exp(\mu/2T),$$

$$T_{23} = T_{24} = \exp[(1.5\mu - \epsilon_y - \epsilon_x)/T], \quad T_{34} = \exp(\mu/T).$$

Employing the transfer-matrix technique, one can also calculate the probabilities of various arrangements of particles. For example, the probability to find an arbitrary row in state k can be calculated as follows. Let us first put this row on the right end of the strip containing n rows. Taking into account relationship (6.3), we can represent the grand partition function for this arrangement as $\lambda_o^n V_k$, where V_k is the appropriate component of the eigenvector corresponding to the largest eigenvalue of the transfer matrix. If we fix the arrangement of particles in a given row and then add n additional rows to the right end of the strip, the grand partition function for this arrangement will be just $\lambda_o^{2n} V_k^2$. Thus, the probability to find an arbitrary row in state k can be represented as

$$P_k = V_k^2 / \sum_k V_k^2 .$$

Using the same arguments and the properties of the transfer matrix, it is also possible to calculate the probabilities of more complex arrangements of particles. Let us, for example, calculate the probability to find two arbitrary nearest-neighbour rows in states k and l. If at first the row in state k is located on the right end of the strip containing n rows, the grand partition function for this arrangement will be $\lambda_o^n V_k$ Adding an additional row in state l on the right end of the strip, we can represent the grand partition function for the latter arrangement as $\lambda_o^n T_{lk} V_k$. If we fix the arrangement of particles in these two rows and then add n additional rows to the right end of the strip, the grand partition function for the strip will be $\lambda_o^{2n} V_l T_{lk} V_k$. Thus, the probability to find two arbitrary nearest-neighbour rows in states k and l is given by

$$P_{kl} = V_k T_{kl} V_l / (\sum_{kl} V_k T_{kl} V_l) = V_k T_{kl} V_l / (\lambda_o \sum_k V_k^2).$$

For three rows, we accordingly have

$$P_{klm} = V_k T_{kl} T_{lm} V_m / (\lambda_o^2 \sum_k V_k^2) .$$

The probability to find n nearest-neighbour rows in given states can be calculated by analogy.

Employing the probabilities defined above, one can calculate the rate constant for various processes. In addition, it is possible to derive that at large distances the correlation function of the density fluctuations is given by

$$\langle \Delta n(x_i)\, \Delta n(x_j) \rangle \sim (\lambda_1/\lambda_0)^{|x_i - x_j|/a},$$

where a is the lattice spacing, and λ_1 the second largest (by its absolute value) eigenvalue. Accordingly, the correlation length is defined by

$$\xi = a/\ln(\lambda_0/|\lambda_1|).$$

The latter expression and the scaling hypothesis yield the possibility to construct a phase diagram of the system [9,10].

6.2. Applications

The transfer-matrix technique has been effectively employed to calculate phase diagrams of adsorbed overlayers (see, e.g., [10]). The application of this method to simulating rate processes with participation of adsorbed particles is just beginning. In particular, the transfer-matrix technique has recently been used to describe thermal desorption spectra [11], the effect of surface reconstruction on the apparent Arrhenius parameters for desorption [12], and the coverage dependence of the chemical diffusion coefficient [13]. Experience shows that the transfer-matrix technique yields very good results already for small N like N = 4.

7. RENORMALIZATION-GROUP METHOD

The renormalization-group method represents the most significant advance in statistical mechanics in the last two decades [14,15]. The main idea of this method lies in transformation between the system of interest and another whose partition function can be more easily obtained [15]. The transformation renormalizes parameters in the grand partition function without changing the form of the Hamiltonian. The latter is achieved by removing some of the degrees of freedom of the system. Then, the iteration procedure yields the possibility to calculate the grand partition function and other properties of the system under consideration. The method has two main branches: the momentum-space renormalization group and the real-space renormalization group. We will discuss the latter approach because it is of greater practical importance. At present, there are two main versions of the real-space renormalization-group method: the site-cell renormalization and the bond-moving renormalization.

7.1. Site-cell renormalization

To describe the main ideas of the site-cell renormalization, let us consider, for example, adsorption of particles with nearest-neighbour interactions on the honeycomb lattice. The grand partition function in this case can be represented as usual.

$$Z = \sum \exp[-H(\epsilon, \mu, N)/T] , \qquad (7.1)$$

where H is the Hamiltonian, ϵ the nearest-neighbour interaction, μ the chemical potential, N the number of sites. The summation in (7.1) is carried out over all the possible configurations. To calculate Z, the original lattice is divided into cells containing three sites (Fig. 1). Every cell is described by the effective occupation number, introduced so that n=1, if the sum of occupation numbers of sites belonging to a given cell is equal to 2 or 3, and n=0 in other cases. The honeycomb lattice of cells is characterized also by the effective nearest-neighbour interaction ϵ_1 and by the effective chemical potential μ_1. Then, the summation over occupation numbers in (7.1) is replaced by summation over cells (i.e., over effective occupation numbers). This procedure is assumed to be represented as

$$Z = \sum \exp([-H(\epsilon_1, \mu_1, N/3) + NG_1]/T) , \qquad (7.2)$$

where G_1 is the renormalization constant. Thus, the partition function of the initial system containing N sites is replaced by the partition function of the system containing N/3 sites and the effective interaction and chemical potential. Repeating the renormalization procedure many times, we have

$$Z = \exp[N\sum_{n=1}^{\infty} G_n/(3^{n-1}T)] . \qquad (7.3)$$

The analytical properties of Z are seen to be dependent on the properties of the renormalization procedure. Studying the behavior of the renormalization constants, it is possible to construct a phase diagram of the system. First of all, however, it is necessary to obtain the equations which connect the original and renormalized parameters of the Hamiltonian. These equations can be derived employing the cluster approach. For example, let us choose the simplest cluster containing two nearest-neighbour cells with periodic boundary conditions. After renormalization, this cluster may be in four states with occupation numbers (0,0), (1,1), (1,0) and (0,1). The corresponding partition functions are

$$\Sigma_1 = \exp(6G_1/T) , \quad \Sigma_2 = \exp[(2\mu_1 - 3\epsilon_1 + 6G_1)/T] ,$$

$$\Sigma_3 = \Sigma_4 = \exp[(\mu_1 + 6G_1)/T] .$$

Before renormalization, the same states of the cluster are described by the following partition functions

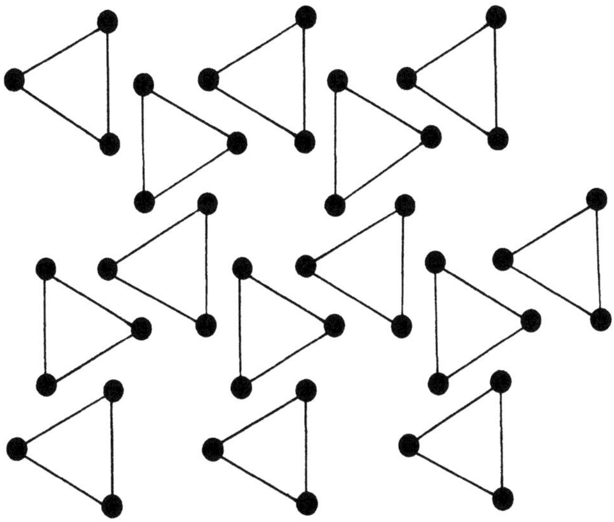

Figure 1. Three-site cells on a honeycomb lattice.

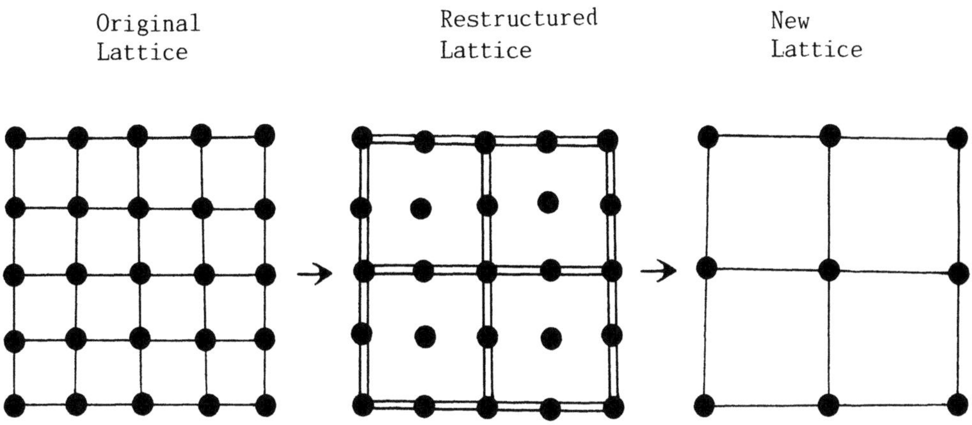

Figure 2. Bond-maving steps for a square lattice.

$$\Sigma_1 = 1 + 6\exp(\mu/T) + 9\exp[(2\mu - \epsilon)/T] ,$$
$$\Sigma_2 = \exp[(6\mu - 9\epsilon)/T] + 6\exp[(5\mu - 6\epsilon)/T] + 9\exp[(4\mu - 4\epsilon)/T],$$
$$\Sigma_3 = \Sigma_4 = \exp(3\mu/T) + 3\exp(2\mu/T) + 3\exp[(4\mu - 3\epsilon)/T] + 9\exp[(3\mu - 2\epsilon)/T].$$

Assuming that the partition functions before and after renormalization are the same, one can easily derive the equations connected ϵ_1, μ_1 and G_1 with ϵ and μ.

7.2. Bond-moving renormalization

The bond-moving (or Migdal-Kadanoff) renormalization is based on analyzing the contribution of different bonds to the grand partition function. To describe the main ideas of this approach, let us consider, for example, adsorption of particles A with nearest-neighbour interaction on the square lattice. The partition function in this case can be represented as

$$Z = \sum \exp[(\mu N_A - \epsilon N_{AA})/T] , \tag{7.4}$$

where ϵ is the A-A interaction, μ the chemical potential, N_A the number of particles, and N_{AA} the number of the A-A pairs. As usual, the summation in (7.4) is carried out over all the possible configurations. To represent expression (7.4) in a more symmetric form with respect to A-A, A-O and O-O bonds, we can use the balance relationship

$$zN_A = 2N_{AA} + N_{AO} , \tag{7.5}$$

where z is the number of nearest-neighbour sites. Substituting eq. (7.5) into expression (7.4) and taking into account that z=4 for a square lattice, we have

$$Z = \sum \exp[(K_{AA}N_{AA} + K_{AO}N_{AO})/T] , \tag{7.6}$$

where $K_{AA} = \mu/2 - \epsilon$ and $K_{AO} = \mu/4$. To calculate Z, the original lattice, containing N sites, is replaced by the lattice containing N/4 sites (Fig. 2). A new representation of Z is assumed to be

$$Z = \sum \exp[(K^1_{AA}N_{AA} + K^1_{AO}N_{AO} + \mathcal{N}G_1)/T] , \tag{7.7}$$

where K^1_{AA} and K^1_{AO} are the effective interactions, G_1 is the renormalization constant, and $\mathcal{N} = 2N$ is the total number of bonds in the original lattice. Repeating the renormalization procedure, we obtain (cf. eq.

(7.4))

$$Z = \exp[\mathcal{N}\sum_{n=1}^{\infty} G_n/(4^{n-1}T)] . \tag{7.8}$$

The transition from the original lattice containing N sites to the new lattice containing N/4 sites can be carried out via the restructured lattice obtained by moving the bonds of the original lattice (Fig.2). Assuming that the partition functions of the original and restructured lattices are the same, neglecting the contribution of the decoupled sites of the restructured lattice to Z, and comparing the contributions of different bonds of the restructured and new lattices to Z, it is possible to obtain the equations connecting the original and renormalized parameters of Z. For example, the contribution of the A-A bond of the new lattice to Z is

$$\exp[(K^1_{AA} + 4\,G_1)/T] .$$

This bond is formed from the A=A=A and A=O=A bonds of the restructured lattice. The contribution of the latter bonds to Z is

$$\exp(4K_{AA}/T) + \exp(4K_{AO}/T) .$$

Thus, we get

$$\exp[(K^1_{AA} + 4G_1)/T] = \exp(4K_{AA}/T) + \exp(4K_{AO}/T) .$$

Analyzing by analogy the formation of A-O and O-O bonds of the new lattice, one can easily derive two additional equations defining the renormalization procedure

$$\exp[(K^1_{AO} + 4G_1)/T] = \exp[(2K_{AA} + 2K_{AO})/T] + \exp[(2K_{AO}/T)] ,$$

and

$$\exp(4G_1/T) = 1 + \exp(4K_{AO}/T) .$$

7.3. Applications

Above, we have described the simplest renormalization procedures. In practical applications, these procedures can be modified to improve the quantitative results while at the same time retaining the underlying simplicity [15].

During the last two decades, the renormalization group method has been widely employed in various branches of condensed matter physics [15]. Application of this approach in the field of surface science is just beginning. In particular, the site-cell and bond-moving renormalizations have been used [16,17] to calculate phase diagrams of adsorbed overlayers. The site-cell renormalization have recently been employed to study the effect of the order-disorder phase transition on the coverage dependence of the chemical diffusion coefficient [18].

The renormalization-group method is attractive because it describes accurately phase transitions in the lattice models. On the other hand, this approach is not relevant for calculating the probabilities of various arrangements of particles. For this reason, the renormalization-group method can in fact be used to simulate the kinetic processes on the surface only in situations when the rate of a given process can be expressed via the chemical potential of adsorbed particles (see, e.g., eqs. (3.3) and (3.4)).

8. MONTE CARLO SIMULATIONS

Monte Carlo simulations are increasingly popular for treating complex many-body systems. This method is particularly effective for describing adsorbed overlayers owing to their low space dimensionality. The first step in simulation is to replace an infinite lattice by a finite cluster (typically, beginning with 40 x 40 to 100 x 100). The next step is to introduce the probabilities of transitions between various configurations. As a rule, the transition from configuration i to configuration j is assumed to be allowed if these configurations differ in a position of one particle. The transition probabilities W_{ij} should satisfy the detailed-balance principle

$$W_{ij} = W_{ji}\exp[-(E_j - E_i)/T] ,$$

where E_i and E_j are the energies of configurations. Moreover, the probabilities should correspond to the rate process under consideration if we study the kinetics of a process. In other words, the probabilities should be calculated taking into consideration detailed information on various transitions. On the other hand, if we study the equilibrium properties of a system, the probabilities may be arbitrary and a continuum of Monte Carlo schemes is possible. In the latter case, the following simple algorithm is frequently employed. If the system is initially in state i, a change in the system is generated randomly within some space of allowed states. The change is always allowed if the energy is lowered, and is allowed with probability $\exp[-(E_j - E_i)/T]$ if the energy is increased by the change. The new configuration is used as the initial state for the next step. Then, a desirable observable is

$$\langle A \rangle = \sum_i A_i P_i ,$$

where P_i is the probability of configuration i averaged over time.

An excellent review of the Monte Carlo algorithms has been presented by Binder [14]. During the last decade, these algorithms have been very widely employed to simulate the kinetics of various kinetic processes in adsorbed overlayers, including adsorption, desorption, surface diffusion, domain growth, molecular beam epitaxy, and chemical reactions. The results obtained in these fields are surveyed in refs. [5,20].

9. CONCLUSION

Finally, we should again emphasize that kinetic processes in real adsorbed overlayers, as a rule, are substantially nonideal, owing to the heterogeneity of the surface, lateral interactions between adsorbed particles and adsorbate-induced changes in the surface. All these factors result in the strong dependence of the rate constants of kinetic processes on coverage. Our goal was to describe the main statistical approaches which may be used to simulate such nonideal kinetic processes. We hope that in the nearest future these approaches (e.g., the transfer-matrix technique) will be more widely used for simulations of rate processes in adsorbed overlayers.

ACKNOWLEDGMENTS

The author thanks Prof. R.A. Van Santen for the proposal to prepare this review. The discussions with Dr. A.V. Myshlyavtsev are also acknowledged.

REFERENCES

1. Boudart, M. and Djega-Mariadassou, G. (1984). Kinetics of Heterogeneous Catalytic Reactions, Princeton University Press, Princeton, New Jersey.
2. Somorjai, G.A. and Van Hove, M.A. (1980) 'Adsorbate-induced restructuring of surfaces', Progr. Surf. Sci. 30, 201-231.
3. Roelofs, L.D., Foiles, S.M., Daw, M.S., and Baskes, M.I. (1990) 'The (1x2) missing-row phase of Au(110): Energetics determined from an extended embedded atom method', Surf. Sci. 234, 63-71.
4. Rothschild, W.G. (1991) 'Fractals in heterogeneous catalysis', Catal. Rev.-Sci. Eng. 33, 71-107.
5. Zhdanov, V.P. (1991) Elementary Physicochemical Processes on Solid Surfaces, Plenum, New York.
6. Zhdanov, V.P. (1991) 'Rate processes on moderately smooth surfaces', J. Chem. Phys. 95, 2162-2168.
7. Baxter, R.J. (1982) Exactly Solved Models in Statistical Mechanics, Academic Press, London.
8. Einstein, T.L. (1991) 'Multisite lateral interactions and their consequences', Langmuir 7, 2520-2527.
9. Derrida, B., de Seze, L., and Vannimenus, J. (1981) in Lecture Notes in Physics, Vol. 149, Springer, Berlin, p. 46.
10. Kinzel, W., Selke, W., and Binder, K. (1982) 'Phase transitions on centered rectangular lattice gases', Surf. Sci. 121, 13-31.
11. Myshlyavtsev, A.V. and Zhdanov, V.P. (1989) 'The effect of nearest-neighbour and next-nearest-neighbour lateral interactions on thermal desorption spectra', Chem. Phys. Lett. 162, 43-46;

Myshlyavtsev, A.V., Sales, J.L., Zgrablich, G., and Zhdanov, V.P. (1990) 'The effect of lateral interactions on thermal desorption', J. Stat. Phys. 58, 1029-1039;

Payne, S.H., Kreuzer, H.J., and Roelofs, L.D. (1991) 'Isosteric heat of adsorption for repulsive interactions', Surf. Sci. 259, L781-786;

Kreuzer, H.J. (1992) 'Thermal desorption kinetics', Langmuir 8, 774-781.

12. Myshlyavtsev, A.V. and Zhdanov, V.P. (1990) 'The effect of adsorbate-induced surface reconstruction on the apparent Arrhenius parameters for desorption', J. Chem. Phys. 92, 3909-3916;

Zhdanov, V.P. (1991) 'Arrhenius parameters for rate processes on solid surfaces', Surf. Sci. Rep. 12, 183-242.

13. Myshlyavtsev, A.V. and Yablonskii, G.S. (1990) 'Application of the transfer-matrix method to calculating the diffusion coefficient of interacting particles', Poverkhnost No 12, 36-43 (in Russian).

14. Wilson, K.G. (1983) 'Renormalization group and critical phenomena', Rev. Mod. Phys. 55, 583-600.

15. McMahon, P.D., Gland, E.D., and Walker, J.S. (1988) 'Renormalization group theory in solution thermodynamics', Chem. Eng. Sci. 43, 2561-2586.

16. Claro, F. and Kumar, V. (1982). 'Phase diagrams for a square lattice with two- and three-body interactions', Surf. Sci. 119, L371-377.

17. Nagai, K. (1984) 'Phase diagram of hydrogen adsorbed on Ni(111)', Surf. Sci. 136, L14-18.

18. Tarasenko, A.A. and Chumak, A.A. (1989) 'Position-space renormalization-group study of diffusion in a 2D lattice gas', Poverkhnost No 11, 98-104; (1991) 'Diffusion of adsorbed atoms on a triangular lattice', Poverkhnost No 3, 37-44 (in Russian).

19. Binder, K. (1979), in Monte Carlo Methods in Statistical Physics, ed. Binder, K. Springer, Berlin.

20. Lombardo, S.J., and Bell, A.T. (1991) 'A review of theoretical models of adsorption, diffusion, desorption, and reaction of gases on metal surfaces', Surf. Sci. Rep. 13, 1-72.

SOLVENT EFFECTS ON THE DIELS-ALDER REACTION OF METHYL VINYL KETONE AND CYCLOPENTADIENE FROM COMPUTER SIMULATIONS

WILLIAM L. JORGENSEN, DONGCHUL LIM, and JAMES F. BLAKE
Department of Chemistry
Yale University
New Haven, Connecticut 06511-8118
USA

ABSTRACT. Free energy of solvation profiles for the Diels-Alder reaction of cyclopentadiene and methyl vinyl ketone have been obtained in liquid propane, methanol, and water at 25 °C. Four transition states were located for the gas-phase reaction from ab initio calculations with the 3-21G and 6-31G(d) basis sets. Calculations at the MP3/6-31G(d)//6-31G(d) level confirmed that the endo-cis transition state is lowest in energy. The gas-phase minimum energy reaction path was followed in periodic cells containing 260 - 500 solvent molecules in Monte Carlo simulations, which yielded the changes in free energies of solvation. Essentially no solvent effect was found in propane; however, hydration is computed to lower the free energy of activation by 4.2 ± 0.4 kcal/mol, similar to experimental observations. Intermediate effects are computed and observed for methanol. The rate accceleration in water is traced to both a hydrophobic effect and enhanced hydrogen bonding to the carbonyl group in the transition state.

1. Introduction

The traditional notion that the rates of Diels-Alder reactions are insensitive to solvent effects[1] has undergone revision during the last decade, particularly for reactions in aqueous media.[2-6] For example, rate accelerations of 741 and 6805 have been obtained in water relative to alkane solvents for the reactions of cyclopentadiene (CP) with methyl vinyl ketone (MVK)[2a] and 2-methoxy-1,6-naphthoquinone.[6] The origin of the acceleration in water has received much scrutiny.[2-7] The leading possibilities are hydrophobic association,[2,4,6] micellar catalysis,[3a,3b] high internal solvent pressure,[3c] solvent polarity,[5] and hydrogen bonding.[7] In order to probe this phenomenon further, we have carried out Monte Carlo simulations to compute the changes in free energy of solvation (ΔG_{sol}) during the reaction of CP and MVK in liquid propane, methanol and water. The approach is an updated version of our efforts on S_N2, addition and association reactions.[8] First, ab initio molecular

R. W. Joyner and R. A. van Santen (eds.), Elementary Reaction Steps in Heterogeneous Catalysis, 377–387.

orbital calculations are used to locate the transition state(s) for the reaction and to obtain a minimum energy reaction path (MERP) in the gas phase. The ab initio calculations are also used to provide partial charges for the reactants along the reaction path, which are needed for the potential functions that describe the intermolecular interactions between the reacting system and solvent molecules. The reacting system is then placed in a periodic cell with hundreds of solvent molecules and the changes in free energies of solvation along the reaction path are computed from Monte Carlo simulations and statistical perturbation theory.

2. Results of the Ab Initio Calculations

Houk and co-workers previously found only minor variations in transition state (TS) structure for the reaction of 1,3-butadiene and acrolein when optimized at the 3-21G, 6-31G(d), and MP2/6-31G(d) levels.[9] We located the four transition states for CP plus MVK corresponding to MVK being s-cis or s-trans and the approach being exo or endo with the 3-21G and 6-31G(d) basis sets, and calculations to levels as high as MP3/6-31G(d)//6-31G(d) were subsequently performed to provide a better estimate of the energetics.[10] The four transition states are illustrated in Figure 1. Only slight differences are again found between the 3-21G and 6-31G(d) optimized structures. For example, going around the forming 6-membered ring, the 6-31G(d) optimized bond lengths of the endo-cis transition state (Figure 1) are 2.05, 1.39, 2.38, 1.38, 1.39 and 1.40 Å, while the 3-21G optimizations yielded 2.09, 1.38, 2.34, 1.38, 1.40, and 1.39 Å. There is little asymmetry in the computed transition state structures; the two forming C-C bonds have lengths that differ by 0.3 Å with the shorter one occurring at the terminal carbon of the dienophile, as expected from resonance arguments. Consistent with the acrolein precedent,[9] the endo-cis TS was found to be lowest in energy. The exo-cis, endo-trans, and exo-trans alternatives have higher energies by 0.32, 2.60, and 5.04 kcal/mol, respectively, at the 6-31G(d)//6-31G(d) level and by 0.24, 2.46, and 4.82 kcal/mol at the MP3/6-31G(d)//6-31G(d) level. As discussed in full elsewhere,[11] the computed activation energy is sensitive to the choices of basis set and correlation energy treatment. For the endo-cis transition state, the results are 24.8 kcal/mol (3-21G//3-21G), 35.5 kcal/mol (6-31G(d)//6-31G(d)), and 15.9 kcal/mol (MP3/6-31G(d)//6-31G(d)). The last value is within a few kcal/mol of experimental estimates.[11] Another point is the small calculated energy difference between s-cis and s-trans methyl vinyl ketone. The s-cis form is favored at the 6-31G(d)//6-31G(d) level by 0.25 kcal/mol. The preference diminishes to 0.02 kcal/mol at the MP3/MP2/6-31G(d)//6-31G(d) level; however, experimental data indicate that the s-trans form has the lower energy by 0.8 kcal/mol.[12]

The endo-cis TS then provided a starting point for the reaction path following procedure in GAUSSIAN 90 that traces the MERP.[13]

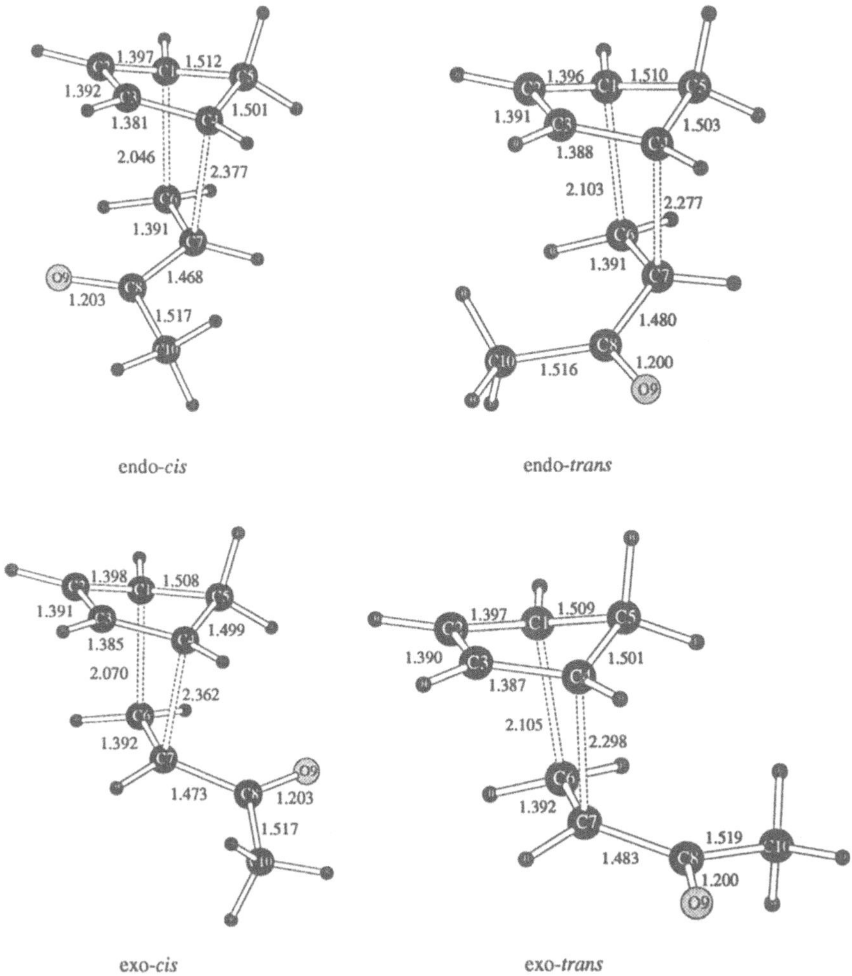

Figure 1. Computed structures for the four transition states from 6-31G(d) optimizations. Selected distances are given in Å.

Essentially, a movie containing 65 frames was obtained covering reaction coordinate values from 1.5 to 8.2 Å. The reaction coordinate, r_C, has been defined here as the average of the lengths of the two forming C–C bonds. Four frames from the "movie" are condensed in Figure 2. The leftmost frame is at the largest separation considered (r_C = 8.2 Å), the third frame is the transition state, and the fourth frame illustrates the product. The geometry optimizations for the reaction path were carried out with the 3-21G basis set. 6-31G(d)//3-21G single-point calculations were then performed for the 65 frames in order to obtain better wave functions for deriving the partial charges.

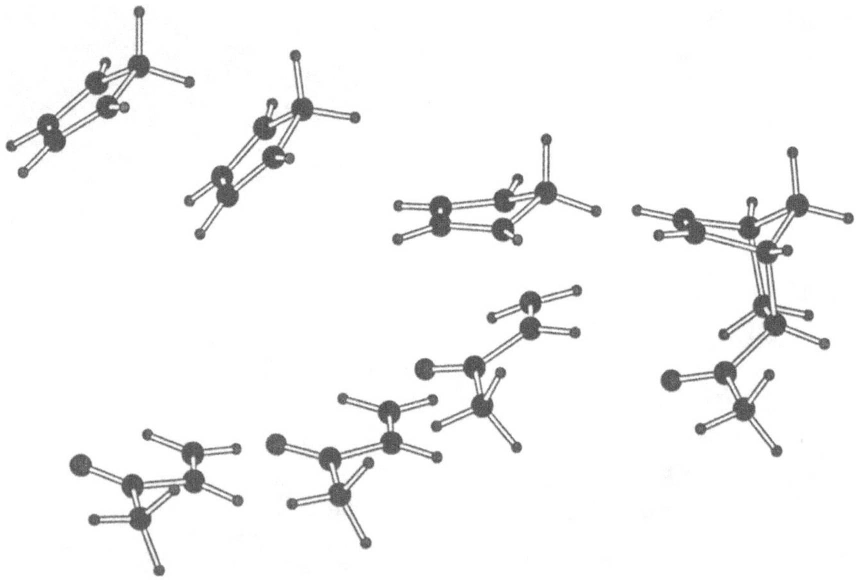

Figure 2. Structures along the reaction path at reaction coordinate values of 8.2, 6.8, 2.2, and 1.6 Å.

3. Intermolecular Potential Functions

Besides water, methanol was chosen as a solvent for the calculations in view of Breslow's observation that the rate acceleration in this alternative hydrogen-bonded medium is only 13 versus 741 in water.[2a] Furthermore, the reference rate for these two figures was obtained in isooctane, which has been simplified here to propane. Well-proven potentials for the solvents are available; the TIP4P model was adopted for water along with the OPLS potentials for propane and methanol.[14] The latter employ united-atom CH_2 and CH_3 groups; however, an all-atom representation was used for the solutes. The intermolecular potential energy consists of Coulombic and Lennard-Jones interactions, as in eq 1.[14] The interaction energy between molecules a and b is represented as a double sum over the interactions between all of the sites i in molecule a and the sites j in molecule b. The A and C parameters may

$$\Delta E_{ab} = \sum_i \sum_j (q_i q_j e^2/r_{ij} + A_{ij}/r_{ij}^{12} - C_{ij}/r_{ij}^6) \qquad (1)$$

be expressed in terms of the Lennard-Jones parameters, σ and ϵ, such that $A_{ii} = 4\epsilon_i\sigma_i^{12}$ and $C_{ii} = 4\epsilon_i\sigma_i^6$, and the combining rules are $A_{ij} = (A_{ii}A_{jj})^{1/2}$ and $C_{ij} = (C_{ii}C_{jj})^{1/2}$. The Lennard-Jones parameters are relatively transferable[15] and standard values for hydrocarbon and carbonyl groups were used;[16] they were scaled linearly between reactants and product. The more serious issue is the partial charges, q_i. For MVK and cyclopentadiene, there is a partial charge on each atom, which can vary along the reaction path. Previously, we fit the atomic charges to reproduce ab initio results on complexes of the reacting system with a water molecule.[8] In view of the size of the present system and the modest anticipated charge shifts, simpler alternatives were considered, namely, Mulliken charges or charges fit to electrostatic potential surfaces (EPS).[17,18] A somewhat better correlation (r = 0.99) exists between 6-31G(d) Mulliken charges and the OPLS charges for neutral molecules with first-row atoms than for 6-31G(d) EPS charges (r = 0.93).[19] Smaller basis sets give poorer fits. Though 6-31G(d) Mulliken charges were used primarily for the reacting system in this study, the effect of switching to the EPS charges was also evaluated. The charges were obtained from 6-31G(d)//3-21G calculations on the 65 frames. In both cases, the charge shifts are small, as shown in Figure 3 for the reactants and transition state. For example, the Mulliken charges for the carbonyl O and C progress from -0.56 and 0.53 for the reactants to -0.61 and 0.56 for the TS and to -0.54 and 0.55 for the product. It may be noted that the 6-31G(d) dipole moments for MVK, cyclopentadiene, the TS, and product are 3.06, 0.29, 3.44, and 2.89 D. By this criterion, the transition state is somewhat more polarized than MVK, which is consistent with the small charge shifts for the carbonyl group.

4. Monte Carlo Simulations

Finally, the Monte Carlo simulations were executed in the NPT ensemble at 25 °C and 1 atm with Metropolis sampling and periodic boundary conditions. The cubic cells contained 500 water, 260 propane or 260 methanol molecules plus the reacting system. The edges of the simulation cells were then ca. 25, 27, and 33 Å long, respectively. 43 of the 65 frames were used, spaced roughly 0.15 Å apart. The BOSS program[20] was used to perturb the system between adjacent frames and to compute the change in free energy of solvation, ΔG_{sol}, via statistical perturbation theory.[8b,21] Double-wide sampling was used meaning that a simulation was run for frame i with perturbations to frames i + 1 and i - 1.[22] Then, the next simulation was at frame i + 2 with perturbations to frames i + 1 and i + 3. Some free energy increments were calculated in both directions (i → i + 1 and i + 1 → i) and showed excellent accord. Each simulation involved 10^6

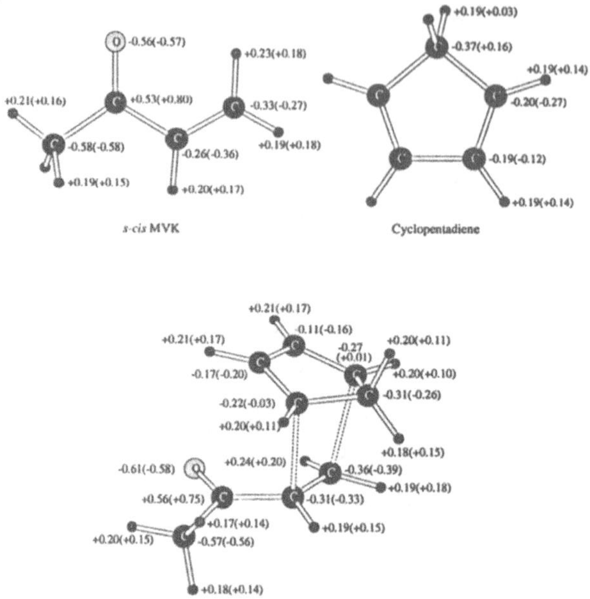

Figure 3. Computed partial charges for methyl vinyl ketone, cyclopentadiene, and the transition state. Mulliken charges and charges fit to the electrostatic potentials (in parentheses) are provided from the 6-31G(d) wavefunctions.

configurations of equilibration and 4 x 10^6 configurations of averaging. A solute - solvent molecule interaction was included if any solute atom was within 9, 10, and 12 Å of the central atom in water, methanol, and propane.

The key results are in Figure 4, which shows the variation in ΔG_{sol} along the reaction path. Such curves are often called potentials of mean force (pmf's). Solvent effects are negligible in propane, as witnessed by the essentially flat pmf. However, in water there is strong stabilization (-4.2 ± 0.4 kcal/mol) of the TS, which occurs at the minimum at 2.22 Å. The minimum is followed by a 3.1 kcal/mol uptick to bring the product to -1.1 ± 0.4 kcal/mol. The gas-phase energy profile for the reaction is flat to 3.3 Å, so the minimum at 6.8 Å in Figure 4 predicts a solvent-separated intermediate. A water molecule is found between the reactants with a hydrogen often pointing towards the center of the cyclopentadiene ring. An analogous orientation has been observed in the hydration of benzene rings.[16b,23] The predicted stabilization of the TS in water relative to propane of 4.2 kcal/mol compares well with the observed rate data, which give a 3.8 kcal/mol

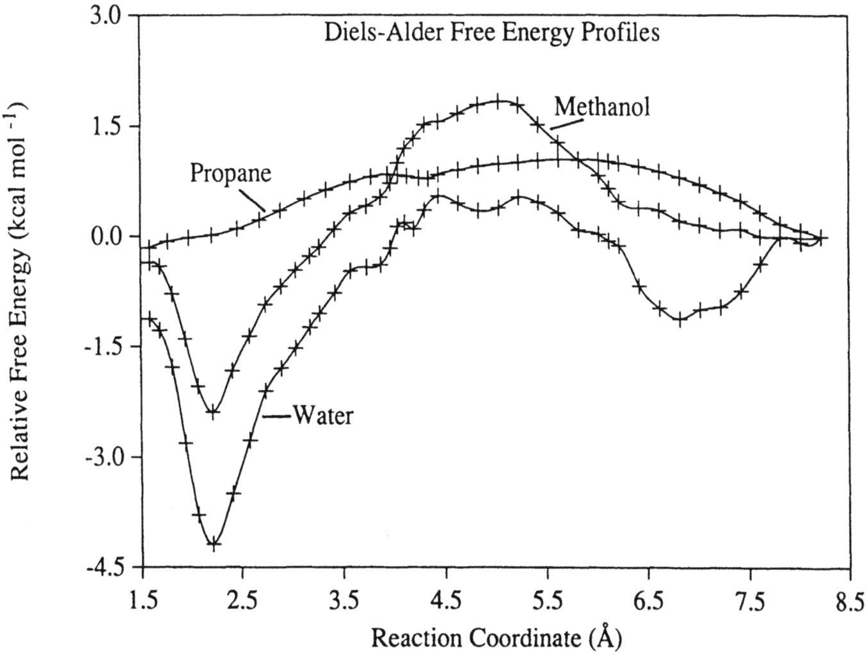

Figure 4. Computed changes in free energies of solvation (kcal/mol) along the minimum energy reaction path. These results are from the series of Monte Carlo simulations in the three solvents.

lowering of the free energy of activation in water relative to isooctane.[2a] The situation in methanol is intermediate; the predicted stabilization of the TS is 2.4 ± 0.3 kcal/mol, while the rate data indicate a 1.5 kcal/mol shift.[2a] These comparisons assume a lack of solvent dynamical effects on the barrier crossing.[24]

The effect of switching to the EPS charges was evaluated by perturbing between the Mulliken and EPS charges for the reactants and transition state in water. The EPS charges were found to raise the free energies of hydration of MVK, cyclopentadiene, and the TS by 2.9, 4.0, and 4.8 kcal/mol. The net effect is that the transition state is predicted to be stabilized by an additional 2.2 ± 0.2 kcal/mol relative to the reactants with the EPS charges. Thus, the EPS charges lead to poorer accord with the experimental data. Though these results illustrate the sensitivity of the computed results to the choice of charges, they also show that both charge models indeed find pronounced acceleration for the reaction in water.

Efforts were made to unravel the origin of the acceleration in water. Clearly, micelles are not required. To try to gauge electrostatic effects, the charges for the TS were perturbed to the reactant and product values while maintaining the TS geometry. In going to the product charges, ΔG_{hyd} increases by 2.7 ± 0.1 kcal/mol, which accounts for most of the 3.1 kcal/mol up-tick. The increase is smaller (0.8 ± 0.1 kcal/mol) in going to the reactant charges, though the significant geometrical change is a complication. Further analysis found that although the number of hydrogen bonds to the carbonyl remains at 2 - 2.5 during the reaction, the strength of each bond is 1-2 kcal/mol greater at the TS. This reflects the sensitivity of hydrogen bonding to small charge variations.[22] The effect is illustrated in Figure 5, which shows the distributions of interaction energies for individual water molecules with the reacting system. The bands below ca. -4 kcal/mol represent the water molecules that are hydrogen-bonded to the carbonyl oxygen, while the spikes centered at 0 kcal/mol reflect the many weak interactions between the reacting system and distant water molecules. The distributions for the transition state and product are similar in form, consistent with the structural similarity, and the low-energy bands integrate to reveal the 2 hydrogen bonds. The low-energy band is not as well resolved for the reactants. The band for the hydrogen bonds to MVK is likely merging with a band for particularly favorable water - cyclopentadiene interactions, near -4 to -5 kcal/mol, in the geometry with a water hydrogen directed towards the ring center. The striking difference is that the hydrogen bonds to the carbonyl oxygen in water are clearly stronger at the transition state than for either the product or reactants. In methanol, the same patterns are found. There are still two hydrogen bonds to the carbonyl oxygen; however, the shift in

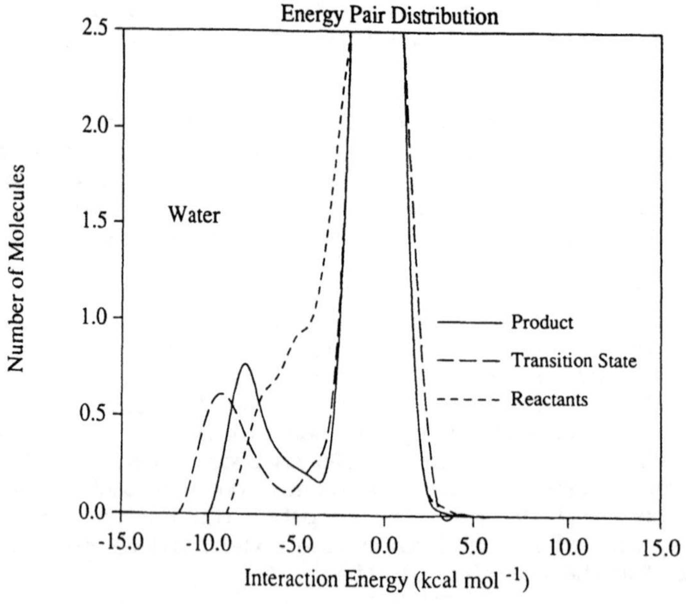

Figure 5. Distributions of interaction energies between water molecules and the reactants, transition state and product of the Diels-Alder reaction. The number of water molecules is given on the y-axis that interacts with the reacting system with the interaction energy shown on the x-axis. Y-axis units are molecules per kcal/mol.

strengths is not as pronounced as in water.

The contribution of hydrophobic effects was approached by computing a free energy profile with all the solute partial charges set to zero. ΔG_{hyd} declines smoothly from reactants to product by 4.4 ± 0.4 kcal/mol. This could account for the rate acceleration; however, the chargeless model undoubtedly overestimates the hydrophobic component. A more appropriate measurement of the hydrophobic component can come from solubility data. The changes in ΔG_{hyd} for the reactions of 1,3-butadiene + ethylene → cyclohexene and isoprene + ethylene → 1-methylcyclohexene are -1.5 and -1.3 kcal/mol.[25] These values agree well with the overall change of -1.1 kcal/mol in Figure 4. The TS and product are close in geometry; a significant difference in hydrophobicity seems unlikely.

Consequently, the present results support the idea that the aqueous acceleration for the CP + MVK reaction contains a significant non-hydrophobic component stemming from enhanced polarization of the TS that leads to stronger hydrogen bonds at the carbonyl oxygen. The variations for the accelerations of aqueous Diels-Alder reactions with cyclopentadiene[2,6] can then be ascribed to variations in the polarization of the dienophile at the TS with the hydrophobic component contributing about a factor of 10 to the rates. If this is the case, then Diels-Alder reactions such as cyclopentadiene dimerization or cylopentadiene with isoprene should show modest acceleration in water since only the hydrophobic component would be available. This notion is being tested computationally.

Acknowledgment

Gratitude is expressed to the National Science Foundation for support of this research.

References

(1) Sauer, J. and Sustmann, R. (1980) Angew. Chem., Int. Ed. Engl. 19, 779.

(2) (a) Rideout, D. C. and Breslow, R. (1980) J. Am. Chem. Soc. 102, 7816. (b) Breslow, R., Maitra, U. and Rideout, D. (1983) Tetrahedron Lett. 24, 1901. (c) Breslow, R. and Maitra, U. (1984) Tetrahedron Lett. 25, 1239. (d) Breslow, R. and Guo, T. (1988) J. Am. Chem. Soc. 110, 5613. (e) Breslow, R. and Rizzo, C. J. (1991) J. Am. Chem. Soc. 113, 4340. (f) Breslow, R. (1991) Acc. Chem. Res. 24, 159.

(3) (a) Grieco, P. A., Garner, P. and He, Z. (1983) Tetrahedron Lett. 24, 1897. (b) Grieco, P. A., Yoshida, K. and Garner, P. (1983) J. Org. Chem. 48, 3137. (c) Grieco, P. A., Nunes, J. J. and Gaul, M. D. (1990) J. Am. Chem. Soc. 112, 4595.

(4) (a) Schneider, H.-J. and Sangwan, N. K. (1986) J. Chem. Soc., Chem. Commun., 1787. (b) Sangwan, N. K. and Schneider, H.-J. (1989) J. Chem. Soc., Perkin Trans. II, 1223.

(5) Cativiela, C., Garcia, J. I., Mayoral, J. A., Avenoza, A., Peregrina, J. M. and Roy, M. A. (1991) J. Phys. Org. Chem. 4, 48.
(6) Blokzijl, W., Blandamer, M. J. and Engberts, J. B. F. N. (1991) J. Am. Chem. Soc. 113, 4241.
(7) Rodgman, A. and Wright, G. F. (1953) J. Org. Chem. 18, 465. Kelly, T. R., Meghani, P. and Ekkundi, V. S. (1990) Tetrahedron Lett. 31, 3381.
(8) For reviews, see: (a) Jorgensen, W. L. (1988) Adv. Chem. Phys., Part II 70, 469. (b) Jorgensen, W. L. (1989) Acc. Chem. Res. 22, 184.
(9) Birney, D. M. and Houk, K. N. (1990) J. Am. Chem. Soc. 112, 4127. Storer, J. W., Raimondi, L. and Houk, K. N., to be published.
(10) The ab initio calculations were performed with GAUSSIAN 90, Revision F: Frisch, M. J., Head-Gordon, M., Trucks, G. W., Foresman, J. B., Schlegel, H. B., Raghavachari, K., Robb, M., Binkley, J. S., Gonzalez, C., Defrees, D. J., Fox, D. J., Whiteside, R. A., Seeger, R., Melius, C. F., Baker, J., Martin, R. L., Kahn, L. R., Stewart, J. J. P., Topiol, S. and Pople, J. A., Gaussian, Inc., Pittsburgh PA, 1990.
(11) Jorgensen, W. L., Lim, D. C. and Blake, J. F. (1993) J. Am. Chem. Soc. 115, 0000.
(12) (a) Durig, J. R. and Little, T. S. (1981) J. Chem. Phys. 75, 3660. (b) de Smedt, J., Vanhouteghem, F., van Alsenoy, C., Giese, H. J., van der Veken, B. and Coppens, P. (1989) J. Mol. Struct. 195, 227.
(13) Gonzalez, C. and Schlegel, H. B. (1990) J. Phys. Chem. 94, 5523.
(14) (a) Jorgensen, W. L., Chandrasekhar, J., Madura, J. D., Impey, R. W. and Klein, M. L. (1983) J. Chem. Phys. 79, 926. (b) Jorgensen, W. L., Madura, J. D. and Swenson, C. S. (1984) J. Am. Chem. Soc. 106, 6638. (c) Jorgensen, W. L. (1986) J. Phys. Chem. 90, 1276.
(15) Jorgensen, W. L. and Tirado-Rives, J. (1988) J. Am. Chem. Soc. 110, 1657.
(16) (a) Jorgensen, W. L., Briggs, J. M. and Contreras, M. L. (1990) J. Phys. Chem. 94, 1683. (b) Jorgensen, W. L. and Severance, D. L. (1990) J. Am. Chem. Soc. 112, 4768.
(17) Cox, S. R. and Williams, D. E. (1981) J. Comput. Chem. 2, 304. Williams, D. E. and Yan, J. M. (1988) Adv. Atomic. Mol. Phys. 23, 87.
(18) Besler, B. H., Merz, K. M., Jr. and Kollman, P. A. (1990) J. Comput. Chem. 11, 431.
(19) Nguyen, T. B. and Jorgensen, W. L., unpublished results.
(20) Jorgensen, W. L. (1990) BOSS, Version 2.9, Yale University, New Haven, CT.
(21) Zwanzig, R. W. (1954) J. Chem. Phys. 22, 1420.
(22) Jorgensen, W. L. and Ravimohan, C. (1985) J. Chem. Phys. 83, 3050.
(23) Atwood, J. L., Hamada, F., Robinson, K. D., Orr, G. W. and Vincent, R. L. (1991) Nature 349, 683. Suzuki, S., Green, P. G., Bumgarner, R. E., Dasgupta, S., Goddard, W. A. and Blake, G. A. (1992) Science 257, 942.
(24) (a) Truhlar, D. G., Hase, W. L. and Hynes, J. T. (1983) J. Phys.

Chem. 87, 2664. (b) Gertner, B. J., Wilson, K. R. and Hynes, J. T. (1989) J. Chem. Phys. 90, 3537.

(25) Hine, J. and Mookerjee, P. K. (1975) J. Org. Chem. 40, 292.

MOLECULAR MODELING AND MOLECULAR GRAPHICS OF SORBATES IN MOLECULAR SIEVES

D. P. VERCAUTEREN, L. LEHERTE[a], D. J. VANDERVEKEN[a],
J. A. HORSLEY[b], C. M. FREEMAN[c], and E. G. DEROUANE[a]
[a]*Institute for Studies in Interface Science (ISIS), Chemistry Department,*
Facultés Universitaires Notre-Dame de la Paix,
rue de Bruxelles, 61, B-5000 Namur, Belgium
[b]*Catalytica, Inc.,*
430 Ferguson Drive, Mountain View CA 94043, USA
[c]*BIOSYM Technologies, Inc.,*
9685 Scranton Road, San Diego CA 92121, USA

ABSTRACT. This contribution demonstrates the usefulness of molecular graphics, molecular mechanics, and molecular dynamics in the study of the structural, thermodynamic, and transport properties of different sorbates within various molecular sieves. Five significant applications to zeolite chemistry are reviewed.

1. Introduction

For the last decade, the usefulness of computer-aided molecular modeling has been evident largely in the design of drugs, agrochemicals, and other important biological systems. It is likely that the understanding and modeling of catalysts in general and of the catalytic behaviour at the molecular level in particular can benefit from real-time interactive computer graphics, as well as from accurate quantum mechanical calculations and large scale molecular statistics simulations. Although theory has so far had a limited impact on the understanding and prediction of catalytic processes, the advent of increasingly powerful computing facilities allowing more realistic representation and simulation of the systems of interest has made this area of research one of great potential and rapid growth. More particularly, zeolites are interesting systems for such studies as their structures are defined and well known in most cases. "Fundamental" zeolite chemistry and physical properties have been well assessed by the full arsenal of spectroscopic and other physico-chemical techniques.

In the following paragraphs, five examples of possible applications of molecular modeling applied to zeolitic systems will be briefly reviewed, i.e.,:

1. the interactive construction and manipulation of ZSM-5 starting from crystallographic data,
2. the modeling of structural recognition and pre-organization of n-hexane within a zeolite L-based catalyst,
3. the simulation of the shape-selective isopropylation of naphtalene over mordenite,

R. W. Joyner and R. A. van Santen (eds.), Elementary Reaction Steps in Heterogeneous Catalysis, 389–401.

4. the interactive docking of water within a ferrierite channel with real-time monitoring of the energetics, and
5. the visualization of the self-diffusion character and trajectories of water molecules in ferrierite pores computed by microcanonical molecular dynamics simulation.

2. Construction of 3D Zeolite Frameworks

In order to understand the structural chemistry of zeolites it is essential to be able to build three-dimensional (3D) models which allow to emphasize the framework topology, i.e., the shape and size of cages and channels [1]. Computer graphics is an ideal tool for such a task as it allows easy construction of the 3D net from a complete mathematical description and it also allows manipulation and modification of the built model for consideration in further molecular mechanical, quantum mechanical, or molecular statistical investigations. 3D frameworks can easily be built and modified starting from monocrystal x-ray diffraction data: the cell dimensions, the space group symmetry, and the fractional coordinates of each atom in the asymmetric unit. A complete description of the cell structure (one unit cell) also needs the definition of the connectivity (the interatomic bonds) which is, most often, given as a list of pairs of connected atoms or computed by comparing the interatomic distance to the sum of the covalent radii of each atom. In Figure 1, the 3D topochemistry of ZSM-5 is exemplified including the cell description represented as a perspective 3D net model. The atoms and half of the interatomic bonds formed with the nearest neighbour are coloured according to the atomic type (in Figure 1, oxygens are in grey, silicons in white).

The user interacts with the graphic model via various positional menus. For example, in Figure 1, the left menu allows the user to display or hide the cell limits (Tog box), to find the symmetry characteristics (Sym query) associated with a particular atom (chosen with the mouse), and to add, remove, and edit (Sym add, remove, edit) the symmetry cards. The right menu allows the selective addition (+ cell) or removal (- cell) of cells along any of the three crystallographic axes (a, b, c).

3. Modeling of Structural Recognition and Preorganization in Zeolite Catalysis

Using a simple van der Waals (VDW) model for the interaction energy of a molecule sorbed into a pore of molecular dimension, Derouane *et al* [2] have indicated that: (i) a molecule may acquire supermobility when its dimension matches closely that of the surrounding pore, a case referred to as the "floating molecule" and that (ii) "creeping motion" of chain molecules along the channel walls should be expected in other situations. This confinement model separates the effect of intrinsic molecular properties from that of the extrinsic geometrical curvature of the substrate. It thus applies to all microporous solids, zeolites or other molecular sieves and insists on the role played by structural complementarity for determining the conformation of the host molecule which eventually could form a supermolecule together with its environment.

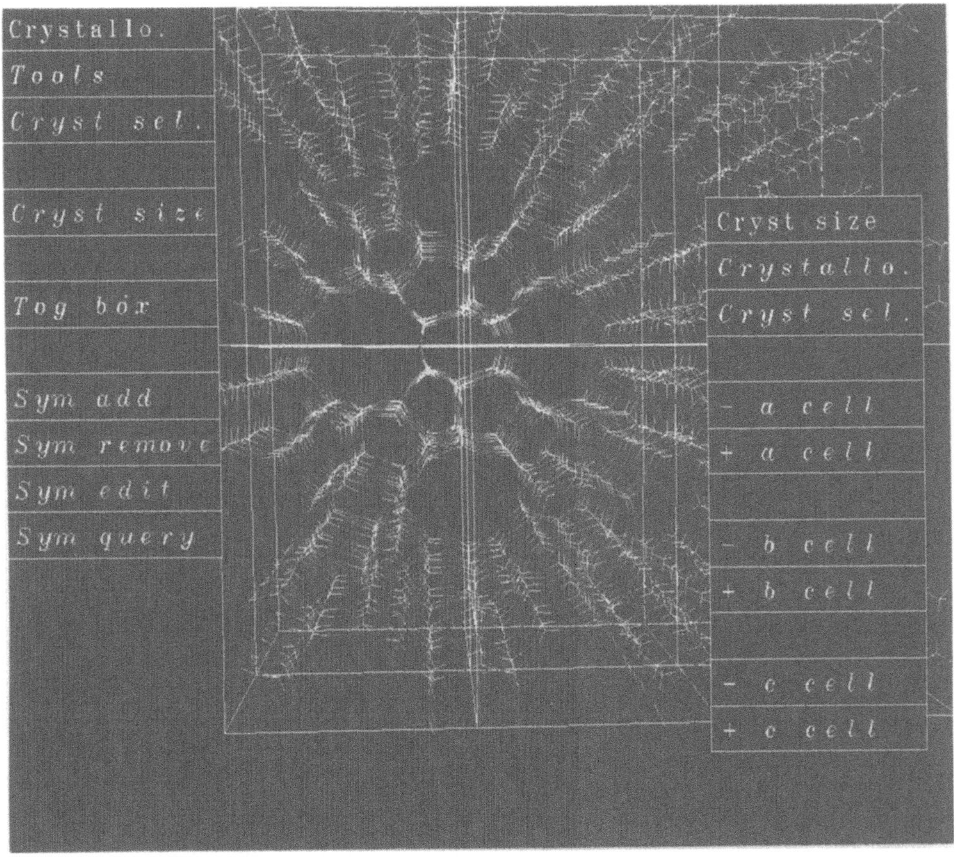

Figure 1: Perspective view of the 3D crystalline structure including the cell of ZSM-5 represented as a wire-frame model

In particular, the unique channel properties of (Ba,K)-exchanged zeolite L (LTL) may play a role in its unmatched selectivity (>90%) for the dehydrocyclization of n-hexane into benzene. Indeed, non-binding interactions of n-hexane with the active environment leads to the recognition of the particular LTL cavity shape and the preorganization of n-hexane as a pseudo-cycle. The supermolecule is transformed into a metallocycle which can yield cyclic compounds and ultimately benzene.

These stages have been conveniently described and schematized by molecular graphics simulations [3]. The orthographic representation in Figure 2 allows the visualization LTL (colored wire-frame model + grey VDW dots surface) along three perpendicular axes simultaneously: the upper right square corresponds to a front view (projection into the xy plane), the lower right to a bottom view (projection into xz), and the upper left to a side view (projection into zy).

Figure 2: Orthographic views of the unidimensional channel structure of LTL (colored wire-frame model + grey VDW dots surface) and interaction of n-hexane (grey wire-frame model + white VDW dots surface) with a single platinum atom of a Pt_{15} cluster (clipped white VDW dots surface) entrapped in the adjacent cage of the tubular building

By moving the front and rear clipping planes, i.e., the two yellow parallel horizontal and vertical lines in the bottom and side views, respectively, one interactively defines a "slab" in order to focus on the main unidimensional 12T channel of LTL; by doing so one clearly portrays the tubular assembly as shown in the front view which only displays the contents of the slab. The menu shows the various volume options, VDW or Connolly dots surface envelopes, stick or stick-and-ball backbones, with or without transparency.

The aromatization pathway in which structural recognition and preorganization play an important role, was modeled as follows. First, the preferred conformations of n-hexane (grey wire-frame model + white VDW dots surface) are achieved by minimizing the VDW contacts between the occluded species and the LTL environment. The next step consists of the diffusion of n-hexane through the open LTL channels, essentially free of bulky cations and acidic sites. n-Hexane ultimately meets head-on with a Pt atom most probably from a Pt_{15} cluster (clipped white VDW dots surface) protruding through a window from the adjacent cage, and interacts with its terminal CH_3. Then, n-hexane or its already partially dehydrogenated product (guest) bound to platinum curves itself in the groove of the LTL cage (host) in order to optimize its interaction with the surrounding framework. The CH_3 in position 6 is thereby forced back to the vicinity of the Pt active site. Structural recognition and preorganization thus appear as central determinants for the 1-6 ring closure.

The higher the guest-host binding, and the more the supermolecule resembles the transition state configuration, the lower the energy of the transition state relative to the reactants and products should be. Consequently, chain lengths longer than C_6 or branched chains should congest and destabilize to various degrees the molecule-zeolite (guest-host) complex and thus decrease catalytic activity. This is indeed observed experimentally. Once the 1-6 metallocycle is formed on the tip of the Pt_{15} cluster, the reaction proceeds via the chemistry that has been extensively demonstrated in the literature [4].

To the best of our knowledge such a simulation is the first clear example of confinement (conformation) zeolite catalysis, not to be mistaken with a molecular shape-selective, zeolite-catalyzed organic conversion.

4. Simulation of Shape-Selective Isopropylation of Naphthalene Over Mordenite

Linear dialkylated polynuclear aromatics, such as 2,6 diisopropyl naphthalene (2,6-DIPN) are important intermediates in the manufacture of high strength specialyzed polyesters and liquid crystal polymers. 2,6-DIPN is presently manufactured by the isopropylation of naphthalene over amorphous silica-alumina catalysts, which leads to a mixture of isomers. The principal components of the mixture are 2,6 and 2,7-DIPN which are produced in equal amounts. Separation of the 2,6 and 2,7 isomers is difficult and costly. One of the factors that limits the growth of markets for the high strength polymers is the cost of the monomers, so there is a strong incentive for the development of a catalyst that would selectively produce the 2,6 isomer.

Zeolites are obvious candidates as catalysts for the selective production of 2,6-DIPN. Steric differences between the 2,6- and 2,7-DIPN could lead to large differences in diffusion rates within the pores of certain zeolites which will be reflected in the product yields. Zeolites with a wide range of pore sizes are available and thus a screening process is necessary in order to select the most promising zeolite for testing in a reactor.

The screening process was initially carried out using molecular graphics. Visual inspection of the fit of 2,6- and 2,7-DIPN in the pores of various zeolites revealed that mordenite is the best candidate for the selective production of the 2,6 isomer. Figure 3 shows views of 2,6- and 2,7-DIPN in mordenite. It can be seen that the 2,6 isomer fits almost perfectly in the elliptical mordenite pore. The excellent fit suggests that this isomer may be an example of a "floating molecule", as defined by Derouane *et al* [2], with very little resistance to movement along the pore. The fit for the 2,7 isomer is less good, with some overlap of the van der Waals (VDW) spheres of the molecule and zeolite at several points. This suggests that there could be significant barriers to diffusion for this isomer.

In Figure 3, the menu shows various advanced visualization tools to improve the three-dimensional appearance, i.e., lights, hidden line and hidden surface removal (HLHSR), transparency, and depth cueing (decrease of intensity with depth). The VDW spheres are set transparent to allow the superimposition of the molecular skeleton of both interacting species.

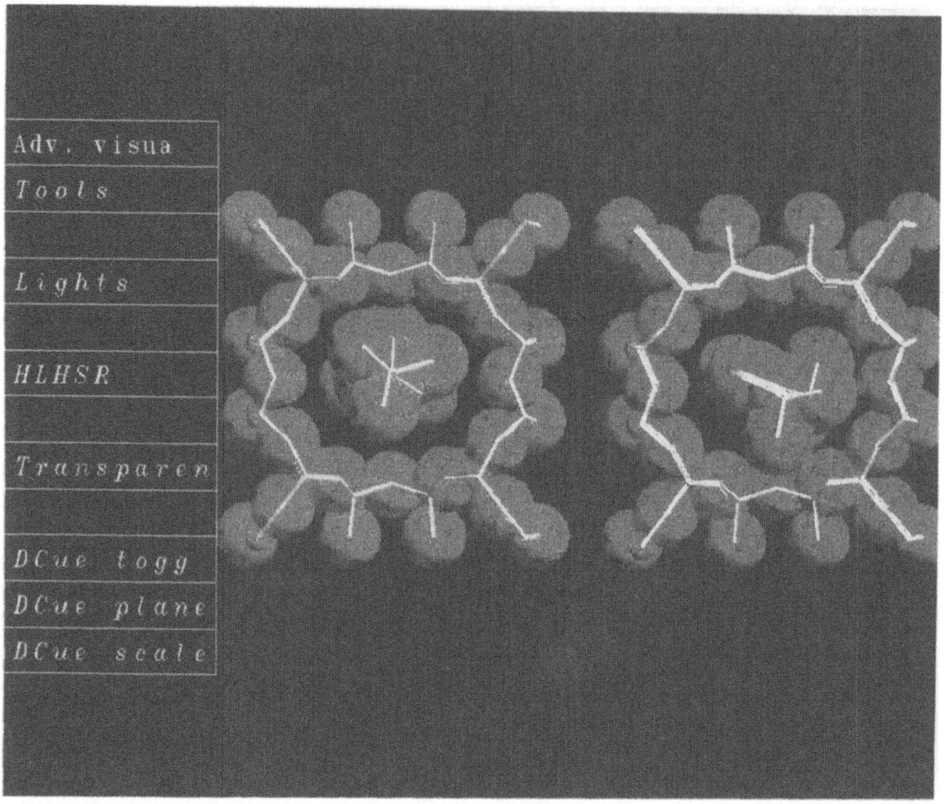

Figure 3: Parallel view of two volumic representations (VDW solid envelope) depicting 2,6-DIPN (left) and 2,7-DIPN (right) in mordenite. The VDW envelopes are set transparent to allow the superimposition of the molecular skeleton of both interacting species

In order to examine the potential shape selectivity of mordenite in a more quantitative way, a computational study of the diffusion of 2,6- and 2,7-DIPN in the mordenite pore has been carried out. Therefore, the approach that has been adopted assumes a general diffusion direction along the pore and uses empirical parameters to calculate the minimum energy pathway along this direction [5]. The diffusion path is defined by a pair of points on the channel axis at opposite ends of the section of the channel under investigation. In a given step of the calculation, the sorbate is constrained by a strong harmonic potential to lie at a fixed distance from each of these extreme points. The sorbate molecule is shifted along the channel axis in the direction of the diffusion in steps of 0.2 Å and the energy is minimized at each step. The path that the sorbate follows is therefore the minimum energy path in the specified direction.

The zeolite lattice was held rigid at its crystallographically determined geometry during the minimization, while the sorbate was allowed complete freedom to optimize the interaction with the host, subject to the above distance constraint. The potential parameters used in the calculation were taken from the consistent valence force field of BIOSYM Technologies, Inc [6]. The calculated minimum energy profiles for the diffusion of 2,6- and 2,7-DIPN in mordenite are shown in Figure 4. It is evident that the diffusion of 2,6-DIPN, with an energy barrier of 4 $kcal.mol^{-1}$, is significantly less hindered than the diffusion of 2,7-DIPN with an energy barrier of 18 $kcal.mol^{-1}$. Thus, the minimum energy path calculations support the prediction based on molecular graphics that mordenite should be selective for the production of 2,6-DIPN in the isopropylation of naphthalene.

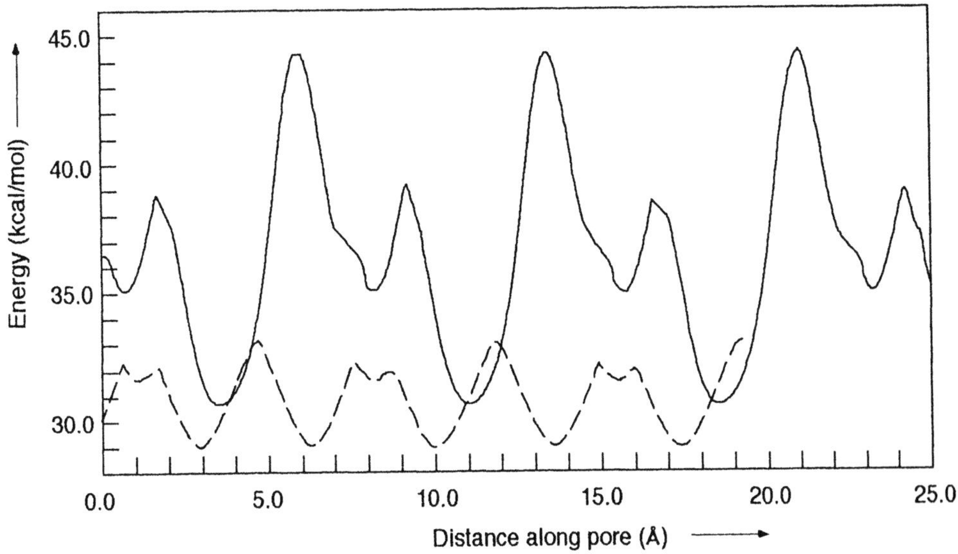

Figure 4: Minimum energy profiles for diffusion of 2,6-DIPN (dotted) and 2,7-DIPN (solid) in mordenite

Experimental tests have borne out this prediction. In a batch autoclave reactor the reaction of naphthalene with propylene over mordenite at 275 C gave a 2,6/2,7 ratio of 2.9 for the two DIPN isomers. This value is significantly higher than the ratio obtained with any other zeolite tested, i.e., 0.8 for zeolite L. The combination of molecular graphics and molecular mechanics appears thus to be a powerful and reliable method for screening and selection of zeolite catalysts in shape selective reaction.

5. Interactive Docking of Small Molecules within Zeolites

It should be possible to move small guest molecules interactively through the cages and channels of a host zeolite. Simultaneously, one should be able to monitor in real-time the interaction energy associated with the location of the guest versus the host, and display the formation of inter-molecular bonds. To evaluate the interaction energy in real-time, one requires a potential function that is reasonably accurate (the interaction energy has to be correlated with experimental values such as sorption heats) and simple enough for real-time docking.

Let us illustrate here the interaction between a water molecule and a ferrierite framework with a Si/Al ratio equal to 8. The role of water in zeolites is important in two ways: to complete the coordination of the cations present in the pores and to minimize the electrostatic repulsion between the framework oxygens. The amount of physisorbed water depends essentially on the Si/Al ratio. In our model the total interaction energy U is limited to two-body terms:

$$U = \sum_{i}^{3N} \sum_{j}^{M} U_{ij}$$

where i and j denote water and zeolite atoms, respectively, and 3N and M are the number of water and framework atoms, respectively. Each contribution U_{ij} is composed of a dispersion (D), a repulsion (R), and an electrostatic (E) interaction energy term:

$$U_{ij} = -U_{ij\,D} + U_{ij\,R} + U_{ij\,E}$$

In considering the dispersion energy, U_D, we adopted the Kirkwood-Müller formula which involves experimentally measured or computable parameters:

$$U_D = \sum_{i\,j}^{3N,M} (6mc^2\alpha_i\alpha_j) / ((\alpha_i/\chi_i)+(\alpha_j/\chi_j))\, r_{ij}^6 = \sum_{i}^{3N,M} A_{ij}/r_{ij}^6$$

where m is the electron mass, c the velocity of light in vacuum, r_{ij} the interatomic distance between atoms i and j, α_i and α_j their polarizabilities, and χ_i and χ_j their diamagnetic susceptibilities. Such a formula has been widely used for the particular calculation of interaction energies of several molecules within a variety of zeolites.

The electrostatic energy term U_E is also computed within the pair-wise approximation between all the water molecule atoms (3N) and all zeolite atoms (M):

$$U_E = \sum_{i\,j}^{3N,M} C_{ij}\, q_i\, q_j / r_{ij}$$

where q_i and q_j are the Mulliken *ab initio* STO-3G atomic charges determined for significant pentameric models of the zeolite. The application of such partial charges is preferred due to the fact that the assumption of a complete ionicity of the framework, i.e., Na^+, O^{2-}, Si^{4+}, Al^{3+}, has been shown to overestimate the adsorption heats, leading to the conclusion that the framework is rather covalent. The repulsion term U_R is evaluated as:

$$U_R = \sum_{i\,j}^{3N,M} (A_{ij}/2)\,(r_i + r_j)^6/r_{ij}^{12} = \sum_{i\,j}^{3N,M} B_{ij} / r_{ij}^{12}$$

where r_i and r_j are the van der Waals and ionic radii of the water and ferrierite atoms, respectively. The values used in our model for the constants A_{ij}, B_{ij}, and C_{ij}, for the polarizabilities, α, for the diamagnetic susceptibilities, χ, and for the radii, r, of the water and zeolite atoms are available in reference 7.

In Figure 5, the interactive docking between water and a framework composed of 522 atoms is illustrated. In this simple example, only one water molecule (N=1) is considered and one assumes that the framework is not disturbed by the introduction of the guest molecule, i.e., our simple potential does not allow the simulation of intramolecular and intraframework degrees of freedom. The oxygen atoms of ferrierite and the one of water are presented in grey whereas the silicon (or aluminum) and hydrogen atoms are in white.

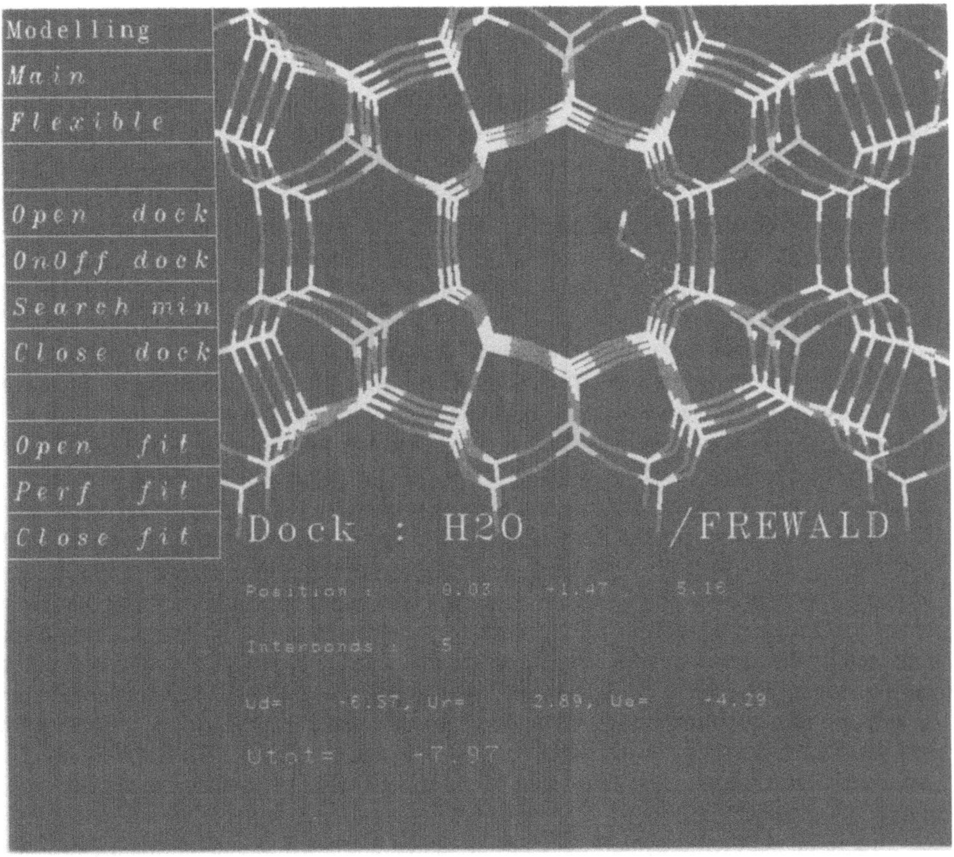

Figure 5: Perspective view of the 3D crystalline structure of ferrierite and visualization of the interactive docking of water inside the zeolite pore (oxygen atoms are in grey; silicon, aluminum, and hydrogen atoms are in white)

As water is guided into the channel, the lower menu monitors the position of the water molecule, the dispersive (Ud), the repulsive (Ur), the electrostatic (Ue), the total energy (Utot) in kcal.mol^{-1}, and the number of close contacts (dotted lines) between the guest molecule and the framework. From such simple real-time interaction simulation, one observes that the stabilization sites (lowest potential energies) are localized close to the $H^+AlO_4^-$ areas and that the center of the main 10T channel does not correspond to a stable position which is in agreement with first principles studies [2].

The menu displays the various options for the least-square fitting and docking applications. Particular options allow to toggle on/off the real-time computation part of the energy (OnOff dock) and to reach local minimas by minimization (Search min).

6. Analysis of Molecular Dynamics Simulations of Small Molecules within Zeolites

Molecular computer graphics is also essential as post-processing tool, especially to analyze the huge number of conformations and/or configurations generated by large-scale Monte Carlo (MC) simulations or to analyze the self-diffusion character and trajectories of small molecules moving in zeolite pores computed by molecular dynamics (MD). In this domain, our laboratory has applied extensive MC and microcanonical MD simulations to various water coverages inside ferrierite at room temperature in order to detail the structural, energetic, and dynamical behaviour of physisorbed water and compare it to liquid water [7].

The water - ferrierite interaction potential considered is the one described previously in the docking application: a Lennard - Jones contribution plus a Coulomb term. However, the necessity to take into account periodicity and long-range effects to obtain total configuration energy and self-diffusion coefficient values in agreement with the experimental values has been shown. This was achieved by the Ewald summation technique. Water - water interactions were described by the largely tested Matsuoka - Clementi - Yoshimine potential [8] adapted from non-empirical configuration interaction calculations on water dimers. Neither intra-molecular nor intra-framework vibrations were allowed. As here only physisorption processes are studied, one assumes that there is no coupling between the sorbent and the sorbate. Starting from equilibrated configurations obtained by the Metropolis MC technique, the phase space trajectories were computed for various water coverages (N=23, 33, and 41) interacting within a 522 ferrierite atoms simulation box by numerically solving the classical Newton and Euler equations for the translational and rotational motions acting on each water molecule. The production stage was carried out with a time step of 0.75 fs during 35,000 steps with the results stored every 3 fs for analysis.

Structural and thermodynamic properties can be presented as conventional one-, two-, or three-dimensional graphs. Structural properties are generally analyzed as 1D and 2D occurrence probability distributions of the guest atoms along the host channel axes and in selected windows, respectively. An alternative is to compute radial distribution functions between the guest atoms and the framework atoms or particular points such as the center of a cage or a channel. Thermodynamic properties can, for example, be decomposed into energy probability distributions of particular guest molecules per various channels.

For dynamical properties, one needs to find other ways to analyze for example time sequences, collective motions, and trajectories and phase space of selected moving atoms or molecules. Nowadays graphics workstations are appropriate tools for the interpretation of results coming from large-scale simulations as they allow real-time animation. Therefore, special functionalities to visualize the diffusion processes as calculated by our MD programs has been developed.

Figure 6 depicts part of the successive positions of some (17 out of 33) of the water molecules (water oxygens are in grey, hydrogens in white) within the 522 ferrierite atoms framework computed from the 26.25 ps MD simulation at 298K.

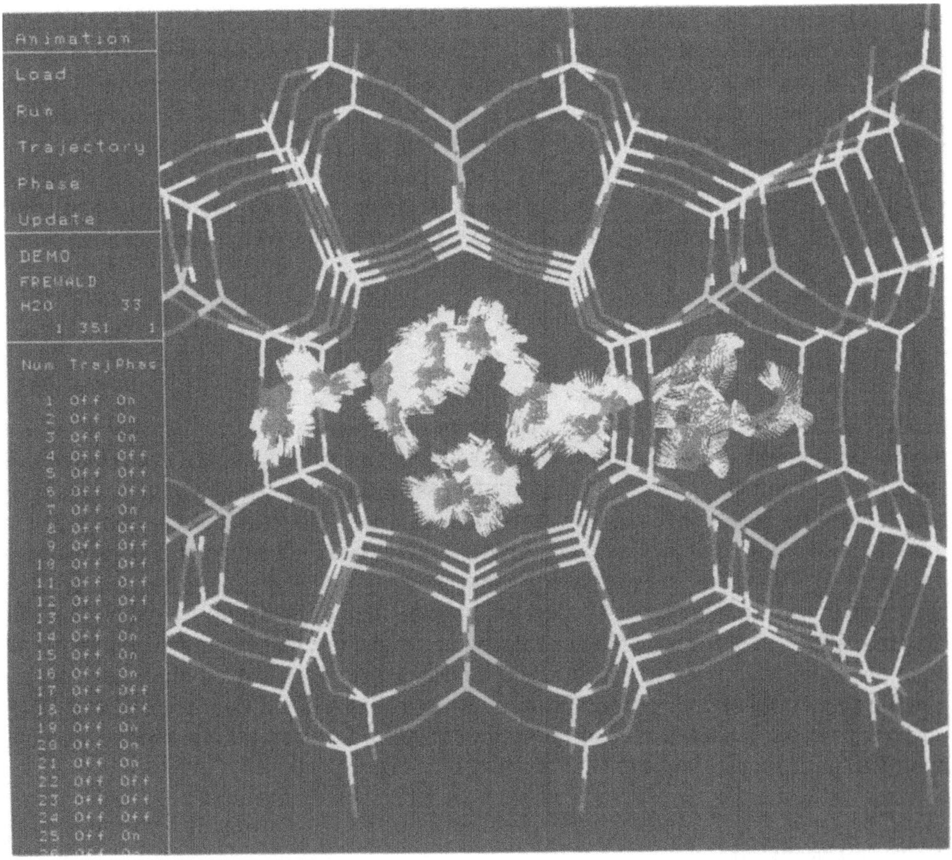

Figure 6: Perspective view of the 3D crystalline structure of ferrierite and visualization of the successive positions of some (17 out of 33) of the water molecules (water oxygens are in grey, hydrogens in white) within the 522 ferrierite atoms framework computed from the 26.25 ps MD simulation at 298K

From such view, one can clearly observe the different behaviours of the various water molecules. Some are trapped into the zeolite stabilization sites, while other may diffuse into the channels. Also as mentioned earlier, the center of the main 10T channel does not correspond to a stable position; the molecules in the center are linked to each other by hydrogen bonds and form a well defined cage. Looking to the phase space also allows the translational and rotational motions of the water molecules inside the 10T and 8T channels to be differentiated. As the water molecular geometry is kept rigid during the MD simulation, the only vibrational movement corresponds to the center-of-mass (COM) vibrations. Compared to liquid water, sorbed molecules present a more "solid-like" behavior characterized by a restricted translational motion and a higher COM vibration frequency.

The top part of the menu contains the possible animation options. The "Run" option applies translational and rotational matrices to the molecules in order to see them moving in real-time as a movie; "Trajectories" builds the COM trajectories by drawing a colored marker for each COM successive position; "Phase" builds the complete molecular structure (colored wire-frame model) for all steps of the MD simulation as presented in Figure 6. The second part displays the three integers defining the length of the animation to visualize, i.e., the starting point, end point, and step (here from 1 to 351 with a step of 1, each step corresponding to a time step of 3 fs). The bottom menu is a scrollable datalist of visibility flags to control the display of the trajectories (traj) and phases (phas) of some selected molecules; these flags are turned on/off by picking with the mouse (here, one presents the phases of 17 out of the 33 water molecules).

7. Conclusions

From this short review, it is clear that molecular graphics, molecular statistics, and molecular dynamics have become unique and indispensable tools for teaching zeolite fundamental chemistry, for pre- and post-processing data (generating input, interpreting results), and for prospective purpose (anticipate, examine, test the chemical good sense of both the theoreticians and experimentalists). As both modeling software and particularly the hardware technology of the graphics workstation will continue to improve rapidly, it is now time to start to think about cooperative processing, i.e., analyzing visual output with the possibility of steering a calculation running in parallel on another computing client. One example of such an interface in this field would be the visualization of the trajectories resulting from the MD simulation in real-time and to be able to steer the calculation by changing various parameters of the potential function, the temperature, the translational and rotation steps, the time step and total length of the simulation,...

Acknowledgments

DPV, LL, DJV, and EGD acknowledge the Belgian National Foundation for Scientific Research (FNRS), IBM-Belgium, and the FUNDP for the use of the Namur Scientific Computing Facility. Part of this work was funded by the Belgian Program on Interuniversity Attraction Poles initiated by the Belgian State Prime Minister Office (Science Policy Programming).

References

[1] a) Meier, W. M. and Olson, D. H. (1978) 'Atlas of Zeolite Structure Types', Structure Commission of the International Zeolite Association, Juris Press, Zurich; b) Mortier, W. M. (1982) 'Compilation of Extra Framework Sites in Zeolites', Butterworths, Guilford; c) Smith, J. V. (1988) 'Topochemistry of Zeolites and Related Materials 1. Topology and Geometry', Chem. Rev. 88, 149-182.

[2] Derouane, E. G., André, J. M., and Lucas, A. A. (1988) 'Surface Curvature Effects in Physisorption and Catalysis by Microporous Solids and Molecular Sieves', J. Catal. 110, 58-73.

[3] Derouane, E. G. and Vanderveken, D. J. (1988) 'Structural Recognition and Preorganization in Zeolite Catalysis: Direct Aromatization of n-Hexane on Zeolite L-Based Catalysts', Appl. Catal. 45, L15-L22.

[4] Gault, F. G. (1981) 'Mechanisms of Skeletal isomerization of Hydrocarbons on Metals', Advan. Catal. 30, 1-96 and references therein.

[5] a) Pickett, S. D., Nowak, A. K., Thomas, J. M., and Cheetham, A. K. (1989) 'Computer Simulation of the Adsorption and Diffusion of Benzene in Silicalite, Theta-1, and a New Zeolite, EU-1', Zeolites 9, 123-128; b) Schroeder, K. P. and Sauer, J. (1990) 'Computer Simulation of Benzene in Silicalite. 1. Low-Coverage Sorbate Structures and Diffusion Barriers', Z. Phys. Chemie, Leipzig 271, 289-296 c) Nakazaki, Y., Goto, N., and Inui, T. (1992) 'Simulation of Dynamic Behaviors of Simple Aromatic Hydrocarbons Inside the Pores of a Pentasil Zeolite', J. Catal. 136, 141-148.

[6] Hagler, A. T., Lifson, S., and Dauber, P. (1979) 'Consistent Force Field Studies of Intermolecular Forces in Hydrogen-Bonded Crystals. 2. A Benchmark for the Objective Comparison of Alternative Force Fields', J. Am. Chem. Soc. 101, 5122-5130.

[7] a) Leherte, L., André, J. M., Derouane, E. G., and Vercauteren, D. P. (1991) 'Study of the Water Behavior into a Ferrierite Zeolite by Molecular Dynamics Simulations', Comp. Chem. 15, 273-285; b) Leherte, L., André, J. M., Derouane, E. G., and Vercauteren, D. P. (1991) 'Self-Diffusion of Water into a Ferrierite-Type Zeolite by Molecular Dynamics Simulations', Molecular Transport in Confined Regions and Membranes, The Royal Society of Chemistry, Faraday Division Symposium 26, J. Chem. Soc. Faraday Trans. 87, 1959- 1970; c) Leherte, L., André, J. M., Derouane, E. G., and Vercauteren, D. P. (1992) 'What Does Zeolitic Water Look Like?: Modelization by Molecular Dynamics Simulations', Internat. J. Quantum Chem. 42, 1291-1326.

[8] Matsuoka, O., Clementi, E., and Yoshimine, M. (1976) 'CI Study of the Water Dimer Potential Surface', J. Chem. Phys. 64, 1351-1361.

Session Four, *Theory in Heterogeneous Catalysis*

Rapporteur E. SHUSTOROVICH
Corporate Research Laboratories
Eastman Kodak Company
Rochester
New York 14650-2001
USA

Theory is the language of science. Without experiment there are no facts to talk about, but without theory we do not know what we are talking about. Heterogeneous catalysis has always been an interdisciplinary field, but in recent years, practitioners feel overwhelmed by the number and diversity of the techniques and approaches used. As a result, the ability to communicate across the field is deteriorating, and this makes the role of theory increasingly important, despite (and maybe because) of the palpable scepticism of many practitioners of heterogeneous catalysis about the relevance of model theoretical studies.

Activity and selectivity of catalytic reactions are determined by reaction rates, where the values of the activation barriers are usually of critical importance. This makes calculations of reaction energies (ultimately, of the reaction energy profile) the primary theoretical task. The session fairly represented the state of affairs in this field covering advances in the most important areas such as cluster-type *ab initio* quantum chemical calculations of metals, metal oxides and zeolites, molecular dynamics and Monte Carlo simulations, lattice-gas models of chemical reactivity, and reaction dynamics on metal surfaces.

The perennial issues in cluster-type simulations of metal surfaces of the type presented by Pacchioni is what size and geometry of a cluster are proper to model a given surface phenomenon. Because there are no definitive criteria, there is quite a freedom of choice. The presented results for Cu clusters were very reasonable. For the most part, the questions were concerned with one aspect, how the computational findings for a cluster of one metal can be extrapolated for surfaces of another metal. Specifically, the question was how the calculated binding energy shifts (BES) of Cu atoms due to adsorption of K, which were found to conform the conventional picture of distinct change transfer, will square with the recent iconoclastic projections by other authors (eg. Citrin) that there is no charge transfer for potassium or cesium adsorbed on tungsten. Difficulties in generalising the above computations of BES stem from the fact that there are several contributions of opposite sign to BES (in particular, adsorption of an alkali atom changes not only the metal electronic charge but also the metal coordination number), so that they may or may not cancel out, depending on their magnitude. No consensus has been reached.

This raised the general question of how to correlate various observables, which was discussed in detail for the propensity of NO to dissociate, NO_S--->N_S+O_S, on transition metal surfaces. It was stressed that numerous attempts to correlate empirically the dissociation activation barrier $\Delta E^*_{NO,s}$ with various energetic parameters such as the gas phase bond energy D_{NO}, the heat of molecular chemisorption Q_{NO}, or the heats of atomic chemisorption Q_N and Q_O, have failed to produce a meaningful relation. This failure clearly shows that $\Delta E^*_{NO,s}$ depends on more than one parameter, ie. the potential energy surface (PES) is more than one-dimensional, and only theory can suggest which correlations (which combinations of parameters) may be informative. In particular, this

R. W. Joyner and R. A. van Santen (eds.), Elementary Reaction Steps in Heterogeneous Catalysis, 403–405.

Rapporteur's work on the bond-order conservation Morse-potential (BOC-MP) model provides the following complex interrelation:

$$\Delta E^*_{NO,s}=1/2[D_{AB}+Q_{AB}-Q_A-Q_B+Q_AQ_B/(Q_A+Q_B)]$$

in good agreement with experiment. This relationship can shed light on why open surfaces facilitate the dissociation of NO. In the experimental literature it has been suggested that open surfaces are preferable because they exhibit a larger heat of NO chemisorption. This argument alone is not correct from the BOC-MP viewpoint, since a larger values of Q_{NO} makes the enthalpy for NO dissociation larger, and likewise, the activation barrier $\Delta E^*_{NO,s}$. What should be recognised is that the larger values of Q_{NO} seen on open surfaces relative to a close-packed surface are a manifestation of an even larger increase in the value of Q_N. It is the increase in Q_N and Q_O on open surfaces, and not the increase in Q_{NO} per se, that results in lowering of $\Delta E^*_{NO,s}$. For example, on the kinked Pt(321) surface the low-coverage value of Q_O is 94 Kcal/mol compared to Q_O = 85 Kcal/mol on Pt(111).

The reaction dynamics calculations of Holloway on dissociation of H_2 on Cu surfaces stirred up a lively discussion. The dimensionality of PES appears to be critical for the mechanism of direct dissociation from the gas phase. The conventional one-dimensional (Lennard-Jones) diagram relating the potential energy to some reaction coordinate (usually the surface-adsorbate distance) can serve for pedagogical purposes only. The minimally meaningful diagram is two-dimensional, including the M-H_2 and H-H distances. Experimental data on dynamics of H_2 dissociation can be understood only within the comprehensive PES, revealing translational-vibrational and even rotational coupling. Moreover, for dissociation of H_2 the tunnelling appears to play an important role as well. This complex and essentially quantum mechanical picture, however, may be relevant only to direct dissociation from the gas phase. As was stressed during discussion, the precursor-mediated dissociation, which includes the most important (chemically and catalytically) cases of thermodissociation from the chemisorbed state, may have a very different PES, with a different activation barrier. For H_2/Cu, experiment shows no evidence of the molecular precursor, but for other systems, say N_2/W or O_2/Cu, the precursors have been identified. Moreover, in the case of N_2/W(100), both dissociation mechanisms may operate depending on the molecular kinetic energy. Molecular adsorption followed by thermodissociation appears to prevail at low energies, but direct dissociation from the gas phase becomes predominant at high energies. Reaction dynamics calculations of molecular trapping followed by competing desorption and dissociation are a tough challenge for the future.

For lattices of largely covalent bonds such as are found in many metal oxides and zeolites, cluster-type calculations of bond geometry and charge distribution usually do a good job, and the papers presented further confirmed this expectation. The result which attracted most attention was the redistribution of bond-lengths in the zeolite Al-O-Si-O rings under protonation of oxygen, namely the oscillation of weaker (longer) and stronger (shorter) O-X bonds, revealing bond-order conservation. Particularly accurate were projections of various vibrational frequencies and force constants [1].

The status and the promise of large-scale computer simulations using approaches such as molecular dynamics (MD) and Monte Carlo (MC), presented by Vercauteren, have also been discussed. The common feature is the interplay of computations at three different levels: (1) state-of-the-art *ab-initio* calculations of small molecules (clusters, fragments, etc); (2) use of the scaled semiempirical two-centre potentials (Coulomb, van der Waals, Morse, Lennard-Jones, etc) for intermolecular interactions; (3) MD and/or MC simulations of various properties of the ensemble of molecules. Combined with the irresistible beauty of colour molecular graphics, these simulations are being more and more widely applied. Given the increasingly stringent budget limitations, a general question was which part of these expensive simulations is most worthy of the money spent. A specific question was how the accuracy of the scaled intermolecular potentials (level 2) may affect the results. Although Vercauteren suggested that the thousands of averaging events during MC simulations make the accuracy of the potentials less critical, the prevailing view was that without accurate intermolecular potentials simulations even with massive sampling may be a wasteful exercise. Also, although the current MD/MC simulations are impressively efficient in treating some reactions in solutions, as indicated in the paper presented by Blake, and in studying structural, thermodynamic and transport properties of different adsorbates in zeolites, nothing comparable has been developed yet for reactions and processes on surfaces of metals and metal oxides.

The overall tone of the discussions reflected the desire of theorists and experimentalists for further mutual understanding and useful interaction. The need to make extra efforts toward this goal, particularly by the realistic modelling of phenomena of catalytic importance and by designing experiments to test theoretical projections, was emphatically stressed.

1. See paper by van Santen presented in session 1.

REACTION MODELLING ON THE BASIS OF THE PUBLISHED POTENTIAL ENERGY DIAGRAM

K. C. Waugh
ICI C&P Ltd
P O Box 1
BILLINGHAM
Cleveland TS23 1LB

ABSTRACT. This paper describes in detail the methodology within which the rates of a catalytic reaction of an industrial process operating at elevated temperatures and pressures are predicted on the basis of its fundamental elementary kinetics obtained under ultra-high vacuum conditions using the techniques of surface science. The methodology is exemplified using the elementary kinetics contained in the published potential energy diagram of ammonia synthesis produced by Ertl and co-workers. The predicted rates fall within an order of magnitude of those found experimentally. Furthermore, the overall activation energy predicted by the model (62kJ mol^{-1}) is also in good agreement with that found experimentally (81kJ mol^{-1}) on iron single crystals at 20bar.

1. THE ENERGETICS

The potential energy diagram [1] shown in Figure 1 is the compilation of a series of experiments conducted by Ertl et al for the determination of the detailed energetics of adsorption, desorption and surface reaction in ammonia synthesis. The energetics have been determined under ultra high vacuum by the techniques of temperature programmed desorption and by temperature dependent Auger spectroscopy on iron(111), (110) and (100) single crystals at, or near, the limit of zero coverage. It is reasonable therefore to question their relevance to the industrial process which is carried out at 100 bar and 720K on a doubly promoted (K_2O/Al_2O_3) polycrystalline iron catalyst; the extrapolation in pressure alone is roughly fourteen to fifteen orders of magnitude.

This extrapolation has been carried out, by calculation, in a computer program [2]. The program solves the sets of non-linear equations implied in the statement of the Arrhenius parameterised elementary reactions of which the potential energy diagram is comprised. Table 1 lists the eight reversible elementary reactions which constitute the mechanism of ammonia synthesis. The activation energies of each of the forward and reverse reactions listed in Table 1 are a re-expression of the potential energy diagram of Figure 1 in

R. W. Joyner and R. A. van Santen (eds.), Elementary Reaction Steps in Heterogeneous Catalysis, 407–422.

TABLE 1. Energetics of the Elementary Reactions Derived from the Potential Energy Diagram (Figure 1) of Ammonia Synthesis

E_f/kJmol^{-1}		E_r/kJmol^{-1}	
0	$N_2 + Fe \rightleftarrows N_{2(a)}$	46	(1)
46	$N_{2(a)} + Fe \rightleftarrows N_{(a)} + N_{(a)}$	226	(2)
0	$H_2 + Fe \rightleftarrows H_{2(a)}$	0	(3)
0	$H_{2(a)} + Fe \rightleftarrows H_{(a)} + H_{(a)}$	109	(4)
126	$H_{(a)} + N_{(a)} \rightleftarrows NH_{(a)} + Fe$	20.9	(5)
54	$H_{(a)} + NH_{(a)} \rightleftarrows NH_{2(a)} + Fe$	20.9	(6)
62.7	$H_{(a)} + NH_{2(a)} \rightleftarrows NH_{3(a)} + Fe$	20.9	(7)
50.2	$NH_{3(a)} \rightleftarrows NH_3 + Fe$	0	(8)

The subscript (a) refers to an adsorbed component.

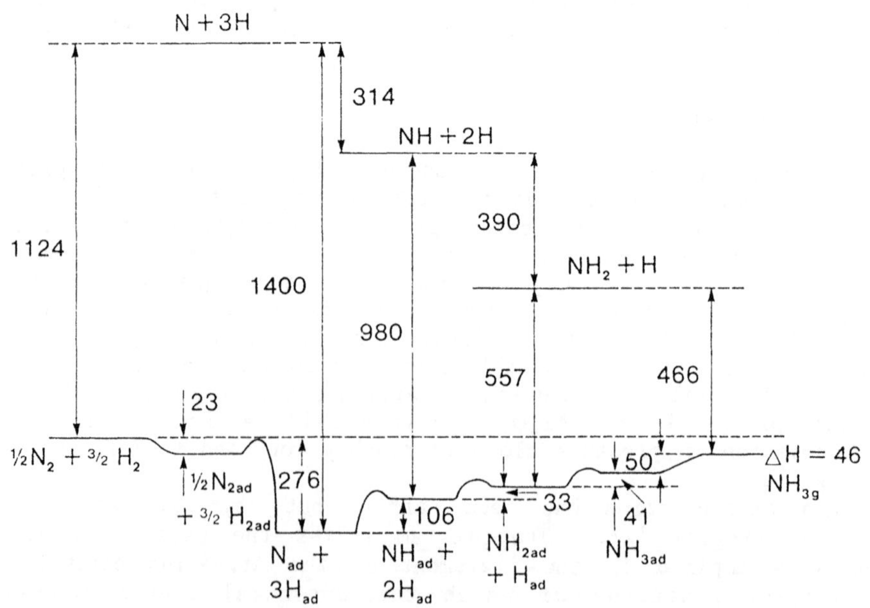

Figure 1 Potential Energy Diagram for Ammonia Synthesis on a Promoted Iron Catalyst

elementary reaction terms for the synthesis of ammonia on a doubly promoted (K_2O/Al_2O_3) iron catalyst. This will be verified below.

The activation energy (E_{-1}) for the desorption of the molecularly held nitrogen precursor state is set at 46kJ mol^{-1}, the value determined by Ertl et al for the desorptin of molecular nitrogen from potassium promoted Fe(111) [3] and potassium promoted Fe (100) [4]. The activation energy for the dissociation of the molecularly held nitrogen precursor state (E_2) is set at 46kJ mol, producing a net activation energy of zero for dissociative adsorption of nitrogen - the value found for nitrogen adsorption for the three low index faces of iron promoted by potassium [5]. The desorption activation energy of the atomically held nitrogen has been determined by Ertl et al [6] by temperature programmed desorption to be 226 kJ mol and is therefore the value given to E_{-2}.

Reactions (3) and (4) describe the dissociative adsorption of hydrogen. The desorption activation hydrogen (E_{-4}) is 109kJ mol^{-1} [7], having zero activation energy to dissociative adsorption (E_3, E_4).

From the energetics of reactions 1-4 listed in Table 1 the well depth for one adsorbed nitrogen atom (113 kJ mol^{-1}) and three hydrogen atoms (163 kJ mol^{-1}) is 276kJ mol^{-1}. This is the energy minimum shown in Figure 1, it is also the value reported by Ertl [1, 5].

The subsequent activation energies of the surface hydrogenation steps (reactions (5), (6) and (7)) are predetermined by the shape of the reactant and product parts of the potential. At the product end of the potential the overall reaction is exothermic by 46kJ mol^{-1} [8], and the desorption activation energy of ammonia from Fe(111) is 50 $kJmol^{-1}$ [9] which is the activation energy for reaction (8) (E_8) of Table 1.

Adsorbed ammonia is therefore 96kJ mol^{-1} below zero potential energy and this state has to be joined to that of the adsorbed nitrogen and three hydrogen atoms at 276kJ mol^{-1} below zero. The reactions (6) and (7) are endothermic by 33 and 41kJ mol^{-1} respectively (E_6-E_{-6} and E_7-E_{-7} of Table 1), values which are identical to those published by Ertl in his potential energy diagram [1]. The reverse reactions - the dehydrogenation steps, E_{-6}, E_{-7} (and E_{-5}) - are all given nominal activation energies of 21kJ mol^{-1}, a value which has not been specified by Ertl even though they are indicated in the published potential energy diagram [1]. Since the dehydrogenation reaction of the adsorbed NH species has been given a value of 21$kJmol^{-1}$ (E_{-5}), to complete the potential energy diagram, the activation energy for the addition of an adsorbed hydrogen atom to an adsorbed nitrogen atom (E_5) must be 126kJ mol^{-1}.

Therefore the energetics listed in Table 1 are the exact re-expression, in elementary reaction terms, of the potential energy diagram published by Ertl [1] for ammonia synthesis.

2 THE VALUES OF THE PRE-EXPONENTIAL TERMS

Although it appears to be a statement of the obvious, it is important to note that the rates and net activation energies predicted by the energetics listed in Table 1, depend critically on the values assigned to the pre-exponential terms of the forward and reverse rate constants of the eight reversible reactions. Transition state theory provides an extremely useful formalism for the evaluation of pre-exponential terms, allowing explicit statement to be made of the assumptions involved in the calculation. This will now be outlined.

2.1 The Desorption Reaction Pre-exponential

The reactions listed in Table 1, adsorption, surface reaction and desorption are bimolecular (the former two) or unimolecular (desorption). In transition state theory the rate of a gas phase unimolecular reaction

$$AS \rightarrow A + S \tag{9}$$

is given by

$$\text{Rate} = \frac{kT}{h} \frac{q^{\circ\#}_{AS}}{q^{\circ}_{AS}} C_{AS} e^{-E^{\#}/RT} \tag{10}$$

(The units of rate in equation 10 are number per unit volume per second since the partition functions and the concentrations are in units of number per unit volume; h (Planck's constant has units of erg seconds, k (Boltzmann's constant) has units of ergs per degree and T is the absolute temperature).

The same equation applies for a unimolecular surface reaction (a desorption) except that now the units of rate are number per cm^2 per second since both the partition functions and the surface concentrations are in units of number per cm^2. Changing to molar units the C's and g's are divided by Avogadro's number, N, so that the rate of desorption is now given by

$$\text{Rate of desorption/mol cm}^{-2}\text{ s}^{-1} = \frac{kT}{h} \frac{q^{\circ\#}_{AS}}{q^{\circ}_{AS}} \rho_{AB}\, e^{-E^{\#}/RT} \tag{11}$$

Where ρ_{AS} is the surface density in units of mol cm^{-2}. Since the partition functions $q^{\circ\#}_{AS}$ and q°_{AS} differ only in respect of one vibrational degree of freedom which, in the activated complex $AS^{\#}$ is assumed to be a translation in the reaction coordinate, they cancel, so that the pre-exponential term for a desorption is given by $\frac{kT}{h}$ ($\sim 10^{13}$ s^{-1}) except in the circumstances where it has been determined to be otherwise (vide infra).

2.2 The Bimolecular Surface Reaction Pre-exponential

Calculation of the pre-exponential term of a bimolecular surface reaction is also carried out by anology with the gas phase reaction. The rate of the bimolecular gas phase reaction

$$A + B \rightleftarrows AB \quad (12)$$

is

$$\text{Rate/number per unit vol per sec} = \frac{kT}{h} \frac{q^{\circ\#}_{AB}}{q^{\circ}_{A} q^{\circ}_{B}} C_A C_B e^{-E^{\#}/RT} \quad (13)$$

(The units of the partition functions and concentrations are number per unit volume in equation 13.) For a surface bimolecular reactions these units became number per cm^2 so that changing to molar units equation 13 becomes.

$$\text{Rate/mol } cm^{-2}\ s^{-1} = N \frac{kT}{h} \frac{q^{\circ\#}_{AB}}{q^{\circ}_{A} q^{\circ}_{B}} \rho_A\, \rho_B\, e^{-E^{\#}/RT} \quad (14)$$

the ρ's again being surface densities i.e mol cm^{-2}. The surface partition functions can be factored into their translational, vibrational and rotational components such that

$$q^{\circ}_{i} = q^{\circ}_{i(trans)}\, q^{\circ}_{i(vib)}\, q^{\circ}_{i(rot)}$$

Since $q^{\circ}_{i(trans)} = \Sigma_i g_i \exp(-E_{i(trans)}/kT)$ (g_i is the degeneracy of the ith state).

2.2.1 Bimolecular Surface Reaction between Fixed Adsorbates producing a Fixed Transition State

For fixed adsorbates and intermediate $E_{i(trans)}$ is zero and $q^{\circ}_{i(trans)}$ reduces to its degeneracy

i.e. $q_{i(trans)} = \Sigma_i\, g_i$ (the degeneracy of the ith state)

$= N_s$ (the number of surface sites cm^{-2} ($\sim 10^{15}$ sites cm^{-2})

The equation for the rate of a bimolecular surface reaction now becomes

$$\text{Rate} = \frac{N}{Ns} \frac{kT}{h} \left\{ \frac{q^{\circ\#}_{AB(vib)}\, q^{\circ\#}_{AB(rot)}}{q^{\circ}_{A(vib)}\, q^{\circ}_{A(rot)}\, q^{\circ}_{B(vib)}\, q^{\circ}_{B(rot)}} \right\} \rho_A \rho_B e^{-E^{\#}/RT} \quad (15)$$

Since the vibrational and rotational partition functions of the transition state complex and of the reactants roughly cancel, the bimolecular surface A-factor (A_{BS}) now becomes

$$A_{BS} = \frac{N}{Ns} \; \frac{kT}{h}$$

$$\cong 10^{21}$$

The reactions in Table 1 which probably involve a fixed adsorbate and intermediate are

(i) the recombination of the nitrogen atoms (reaction (2)) and
(ii) the dehydrogenation of the NH species (reaction (5)), the NH species probably being immobile due to its high heat of adsorption.

2.2.2 Bimolecular Surface Reaction between Mobile Adsorbates producing a Mobile Transition State

For reactions involving mobile adsorbates and intermediates (reactions such as the hydrogen atom recombination (reaction (4)) and the hydrogenation of the NH_2 species (reaction (7)) the rate of the bimolecular surface reaction is given by

$$\text{Rate} = \frac{NkT}{h} \left\{ \frac{q^{\circ\#2}_{AB(trans)} \; q^{\circ\#}_{AB(vib)} \; q^{\circ\#}_{AB(rot)}}{q^{\circ 2}_{A(trans)} \; q^{\circ 2}_{B(trans)} \; q^{\circ}_{A(vib)} \; q^{\circ}_{A(rot)} \; q^{\circ}_{B(vib)} \; q^{\circ}_{B(rot)}} \right\}$$

$$\rho_A \rho_B e^{-E^{\#}/RT} \qquad (16)$$

Where $q^2{}_{trans} = \dfrac{2\pi mkT}{h^2}$

$$\cong 10^{16}$$

Again, cancelling the contributions of the vibrational and rotational partition functions, the bimolecular surface reaction A-factor A_{BS} now becomes

$$A_{BS} = N \frac{kT}{h(2\pi mkT/h^2)}$$

$$\cong 10^{21}$$

2.2.3 Bimolecular Surface Reaction between one Fixed and one Mobile Adsorbate producing either a Mobile or Fixed Transition State

Two other types of bimolecular surface reaction can exist in the elementary reactions listed in Table 1. These are: (i) reaction between a mobile adsorbate and a fixed adsorbate leading to a fixed intermediate, e.g. reaction (2) ($N_{2(a)}$ + Fe → Na + Na) and reaction (5) (Na + N(a) → NH(a) + Fe) and (ii) reaction between a mobile adsorbate and a fixed adsorbate leading to a mobile intermediate e.g. reaction (4) ($H_{2(a)}$ + Fe → Ha + Ha). For the former the bimolecular surface A-factor is

$$A_{BS} = N \frac{kT}{h\, q^{o2}_{A(trans)}} \tag{17}$$

$$= N \frac{kT}{h(2\pi mkT/h^2)} \tag{18}$$

$$\cong 10^{21}$$

and for the latter it is

$$A_{BS} = N \frac{kT}{h} \frac{q^{o\#2}_{AB(trans)}}{Ns\, q^{o2}_{A(trans)}} \tag{19}$$

$$= \frac{N}{Ns} \frac{kT}{h} \tag{20}$$

$$\cong 10^{21}$$

2.3 The Values of the Pre-exponential Terms of the Eight Elementary Reactions Involved in Ammonia Synthesis (Table 2)

Therefore the values of all the bimolecular surface reaction A-factors listed in Table 2 are 10^{21} with the exception of cases where they have been experimentally determined to be otherwise. The A-factor values listed in Table 2 are those used in the calculation of ammonia synthesis rates at 100 and 107 bar.

The A-factors which differ significantly from the predicted values are those for dissociative nitrogen adsorption (reactions (1) and (2)) and for ammonia decomposition. The initial sticking probability into the molecularly held precursor state was determined by Ertl et al to be 10^{-2} [3] (a value derived from the adsorption isotherms recorded via the work function change) and so A_1 has a value of 10^{11} and not 10^{13}. The desorption A-factor from this molecularly held state is given the value of 10^{10} s^{-1}, a value assigned to it by Ertl et al [10] on the assumption that the value of the pre-exponential term for desorption from the promoted state would be the same as that determined for the unpromoted state [3].

The pre-exponential for the dissociation of the molecularly held nitrogen precursor state (A_2) is 10^{-6} below the "theoretical" value of 10^{21}. This is in accord with the measured extremely low sticking probability of atomic nitrogen [3, 10].

The initial sticking probability of ammonia on an iron(111) surface has been measured by Ertl et al [9] to be 0.1 and so the ammonia adsorption A-factor (A_{-8}) has been given a value of 10^{12}.

The pre-exponential terms A_5, A_6, A_6 have all been given values which differ slightly from 10^{21} so that the model accurately reflects the entropy change of the reaction.

Taken in combination, Tables 1 and 2 are the exact expression of the kinetics and mechanism of ammonia synthesis on a potassium promoted iron(111) surface derived in a considerable body of work by Ertl et al [1, 3, 4, 5, 6, 7, 9, 10, 11 ,12]. This system has been claimed to be a model for ammonia synthesis on a doubly promoted (K_2O/Al_2O_3) iron

TABLE 2. Values of the A-factors for the Elementary Reactions Involved in Ammonia Synthesis

$\log_{10}$(A-factor (forward))	Reaction	$\log_{10}$(A-factor (reverse))	
11.0	$N_2 + Fe \rightleftarrows N_{2(a)}$	10.0	(1)
15.0	$N_{2(a)} + Fe \rightleftarrows N_{(a)} + Na$	21.0	(2)
13.0	$H_2 + Fe \rightleftarrows H_{2(a)}$	13.0	(3)
21.0	$H_{2(a)} + Fe \rightleftarrows H_{(a)} + H_{(a)}$	21.0	(4)
20.8	$H_{(a)} + N_{(a)} \rightleftarrows NH_{(a)} + Fe$	21.0	(5)
20.7	$H_{(a)} + NH_{(a)} \rightleftarrows NH_{2(a)} + Fe$	21.0	(6)
21.6	$H_{(a)} + NH_{2(a)} \rightleftarrows NH_{3(a)} + Fe$	21.0	(7)
13.0	$NH_{3(a)} \rightleftarrows NH_3 + Fe$	12.0	(8)

catalyst. The kinetics, therefore, ought to be able to predict rates of ammonia synthesis under conditions of temperature (~720K) and pressure (~100 bar) at which ammonia is synthesised industrially. As outlined previously this prediction is carried out in a computer program which solves sets of non-linear equations implied in the elementary reactions of Tables 1 and 2. The program accepts, as input, statements of these elementary reactions together with their Arrhenius parameters from which it sets up mass balance equations for each of the components (gas phase and surface). It then solves these equations by Newton-Raphson iteration, a convergent solution being deemed to have occurred when the sum of the squares of the residuals is less than a pre-set very small value (usually taken to be 10^{-20}).

Additional input to the program consists of the reaction pressure, the inlet gas composition, flow rates at given reaction temperatures, the free gas volume, and the weight of catalyst and its surface area. This latter was taken to be $10 m^2 g^{-1}$ which is a value which is typical of doubly promoted iron catalysts. The total area of the catalyst is parameterised in $10 Å^2$ sites of which 15% are taken to be clean iron on which the hydrogen and nitrogen are designated to adsorb competitively.

The rationale for the value of 15% of the total number of surface sites being clean iron and for the concept of competitive adsorption of hydrogen and nitrogen is again to be found in the work of Ertl et al [11, 12]. Using X-ray photoelectron spectroscopy Ertl et al [11] showed that on a reduced BASF catalyst 15 atom % of the surface was free iron on which the presence of atomic hydrogen inhibited the rate of dissociative nitrogen chemisorption. The ultimate predicted steady state coverage of the surface by hydrogen and nitrogen atoms by the program, however, depends solely on the kinetics of their chemisorption and of their removal by reaction.

3 THE PREDICTION

3.1 Predicted Reaction Rates

At 720K, 107 bar and a gas hourly space velocity GHSV of 16000 hr^{-1} the exit ammonia concentration predicted by the kinetics of the potential energy diagram of Figure 1 is 1.7%. The experimentally determined value is 13.2% [13]. The agreement is remarkably good considering all the parameters involved and that a minor variation of one of the pre-exponential or activation energies could well result in complete agreement.

3.2 Predicted Overall Activation Energy

A further test of the model is the value of the net activation energy it predicts. This is done simply by increasing the GHSV by 10^2 and carrying out the calculation at 720 and 680K. The predicted exit ammonia concentrations are now 0.145% and 0.079% from which an overall activation energy of 62kJ mol^{-1} is obtained. This value is intermediate between that (45kJ mol^{-1}) determined by Nielsen [14] by analysis of his rate data obtained near equilibrium using the

Temkin-Pyzhev equation [15] and 81kJ mol^{-1} obtained by Somorjai et al [16] on an iron single crystal (Fe(111)) at 20 bar. Again the agreement is remarkable.

3.3 Predicted Surface Coverages

Since the program calculates the steady state concentrations of all the species involved (gas phase and surface) it therefore predicts the surface composition of the operating catalyst. It is:

Surface Composition

Hydrogen ($H_{(a)}$)/atom cm^{-2}	Nitrogen ($N_{2(a)}$)/molecule cm^{-2}
1.8×10^{13}	2.3×10^{12}
Nitrogen ($N_{(a)}$)/atom cm^{-2}	Nitrogen Hydride ($NH_{(a)}$)/species cm^{-2}
1.3×10^{14}	2.7×10^{12}

The predicted nitrogen atom coverages (1.3×10^{14} atoms cm^{-2}) are, again, in remarkably good agreement with those determined experimentally [12]. At 580K, under 150 torr N_2, 450 torr H_2 and 6 Torr NH_3 (equilibrium conditions) the nitrogen atom coverage of an iron single crystal (Fe(111)) was 7×10^{13} atoms cm^{-2}.

4 ALTERNATIVE MODELS

4.1 Lowered Nitrogen Atom Recombination Activation Energy and Pre-exponential

In a series of papers [17-19] Stoltze and Norskov have propounded an alternative model to the one described by the combination of Tables 1 and 2 above, differing in the values of only one of the parameters in each of the Tables. The parameters which are changed are the pre-exponential term and the activation energy for nitrogen atom recombinative desorption. The rationale for changing these parameters is that since the sticking coefficient for the adsorption of nitrogen atoms is very low ($\sim 10^{-6}$), the pre-exponential for the desorption reaction must be equally low [17]. Since the nitrogen atom desorption activation energy had been determined by Ertl et al [6] in a temperature programmed desorption experiment by solution of the Redhead equation [20/21] at the peak maximum temperature for an assumed value of $10^{13} s^{-1}$ for the desorption pre-exponential, lowering the value of the desorption pre-exponential by Stoltze et al [17], required a concomitant lowering of the desorption activation energy to maintain the value of desorption rate constant at the peak maximum temperature (880K).

The value of the desorption activation energy used by Stoltze and Norskov was 161kJ mol^{-1} (cf 226kJ mol^{-1} of Table 2) from which a value of 7.9×10^{9} for the desorption pre-exponential can be derived. This is a lowering of the normal desorption A-factor by 4.9×10^{-4}, considerably less than the value of 10^{-6} found for the sticking probability. Nevertheless, using this combination of values Stoltze and Norskov predict ammonia synthesis rates which are in exact agreement with those found experimentally.

Several points can be made about this. Firstly one is no longer dealing with the published potential energy diagram. The energy minimum of a nitrogen atom plus three hydrogen atoms is now raised by 32kJ mol^{-1}.

Secondly, there is no need to lower the desorption A-factor commensurate with the lowered adsorption A-factor. It is at this point that the transition state theory formalism is particularly helpful. The dissociation of the molecularly held mobile precusor state through a fixed intermediate which exists on only a small fraction (10^{-3} say) of the total number of surface sites is

$$A_{diss} = N \frac{kT}{h} \frac{Ns \times 10^{-3}}{q^{o2}_{N_2(trans)}\ q^{o}_{N_2(rot)}\ q^{o}_{N_2(vi}} \tag{21}$$

The recombination of the fixed adsorbed nitrogen atoms to a fixed intermediate again the latter being adsorbed on a small fraction of the surface results in the following A-factor.

$$A_{recom} = N \frac{kT}{h} \frac{Ns \times 10^{-3}}{(Ns \times 10^{-3})} \tag{22}$$

The ratio of the adsorptive and recombinative A-factors is

$$\frac{A_{diss}}{A_{recom}} = \frac{Ns \times 10^{-6}}{q^{o2}_{N_2(trans)}\ q^{o}_{N_2(vib)}\ q^{o}_{N_2(rot)}} \tag{23}$$

$\cong 10^{-6}$ to 10^{-7} depending on the degree of rotational freedom accorded to the mobile precursor. Therefore having A-factors for a forward and reverse reaction which differ by a factor of 10^{6} is not a violation of the principle of microscopic reversibility as has been suggested (18).

Finally, the overall activation energy predicted by this potential energy diagram is 17kJ mol^{-1} which is considerably lower than the experimental values of 45kJ mol^{-1} of Nielsen [14] or the 81kJ mol^{-1} of Somorjai et al [16].

4.2 Activated Dissociation of the Molecular Nitrogen Precursor State

Several authors have reported that the dissociative adsorption of nitrogen on iron is activated. Ertl et al [6] have reported that on clean Fe(100) the adsorption of nitrogen atoms has an activation energy which increases from 20kJ mol^{-1} at a coverage of 0.05ML to 45kJ mol^{-1} at 0.2ML. This result is in good accord with the coverage dependent activation energy reported by Scholten et al [22] for nitrogen atom adsorption on an alumina promoted iron catalyst. The activation energy was found to increase linearly with coverage from 40kJ mol^{-1} at a coverage of 0.05Ml to 96kJ mol^{-1} at a coverage of 0.25ML, remaining constant at that value up to a coverage of 0.7ML. Baiker et al [23] from a kinetic analysis of the rates of ammonia synthesis over amorphous Fe_{91} Zr_9 (temperature range 623 to 723K at 4 bar) calculate an activation energy of nitrogen atom adsorption of 85kJ mol^{-1}. Finally, Emmett and Brunauer [24] using a volumetric technique to measure amounts adsorbed, obtained nitrogen atom adsorption activation energies of 58.6, 60.6, 64.4, 71.5, 75.3 and 66.9kJ mol^{-1} for the adsorption of 1.7, 2.0, 2.3, 3.2 and 3.6ml of nitrogen on 16.46g of a doubly promoted (K_2O/Al_2O_3) iron catalyst in the temperature range 546 to 723K. (Taking a nominal value of $10m^2g^{-1}$ for the surface area of the catalyst, those volumes adsorbed correspond to a coverage range of 0.026ML to 0.055ML and so the activation energies to nitrogen adsorption accord reasonably well with those found by Scholten et al [22].) Indeed, in the same paper [24], Emmett and Brunauer found that the rate of ammonia synthesis and nitrogen atom uptake on the catalyst were the same (ammonia synthesis rate 0.21 ml min^{-1}, nitrogen atom uptake in first minute 0.26ml). They tentatively concluded from this that nitrogen atom adsorption might be the rate determining step in ammonia synthesis. Furthermore it is an important tenet of the Temkin Pyzhev equation (an equation which accurately reflects the rate of ammonia synthesis and decomposition on doubly promoted iron catalysts) that nitrogen adsorption (as atoms) has a coverage dependent activation energy.

It is important therefore that the consequences (in terms of predicted rates) of the activated adsorption of atomic nitrogen be examined. The potential energy diagram of Figure 2 has been constructed using the high coverage (80% nitrogen atom coverage adsorption/desorption kinetics of Scholten et al [22]. The activation energy for dissociation of the molecularly held precursor state has been given the value of 96kJ mol^{-1} while the reverse reaction (nitrogen atom recombination) has been given the value of 134kJ mol^{-1}. The hydrogen desorption activation energy (E_{-4}) has been lowered from 109kJ mol^{-1} (the low coverage value [7]) to 62.7kJ mol^{-1} (the high coverage model [7]) so that the potential accurately reflects a high coverage model. A complete listing of the energies involved is given in Table 3.

The striking feature of the potential energy diagram so produced (Figure 2) is the shallowness of the well (136kJ mol^{-1}) for one nitrogen atom (42kJ mol^{-1}) and three hydrogen atoms (94kJ mol^{-1}) adsorbed on the surface. From this point now only 90kJ mol^{-1} is required to produce gas phase ammonia.

4.3 The Prediction

Using standard values for the pre-exponential terms (10^{13} for adsorption and unimolecular desorption reactions and 10^{21} for bimolecular surface reactions) at 107 bar and 723K this model predicts an exit ammonia composition of 14.2% (cf 13.2% experimental [13]) for a GHSV of 16000 hr^{-1}.

TABLE 3. Energetics of the Elementary Reactions of a High Coverage, Activated Nitrogen Adsorption Model of Ammonia Synthesis used to Construct the Potential Energy Diagram of Figure 2.

E_f /kJ mol^{-1}	Reaction	E_r /kJ mol^{-1}	
0	$N_2 + Fe \rightleftarrows N_{2(a)}$	46	(1)
96	$N_{2(a)} + Fe \rightleftarrows N_{(a)} + N_{(a)}$	134	(2)
0	$H_2 + Fe \rightleftarrows H_{2(a)}$	0	(3)
0	$H_{2(a)} + Fe \rightleftarrows H_{(a)} + H_{(a)}$	62.7	(4)
40.9	$H_{(a)} + N_{(a)} \rightleftarrows NH_{(a)} + Fe$	20.9	(5)
30.9	$H_{(a)} + NH_{(a)} \rightleftarrows NH_{2(a)} + Fe$	20.9	(6)
30.9	$H_{(a)} + NH_{(a)} \rightleftarrows NH_{3(a)} + Fe$	20.9	(7)
50.2	$NH_{3(a)} \rightleftarrows NH_3 + Fe$	0	(8)

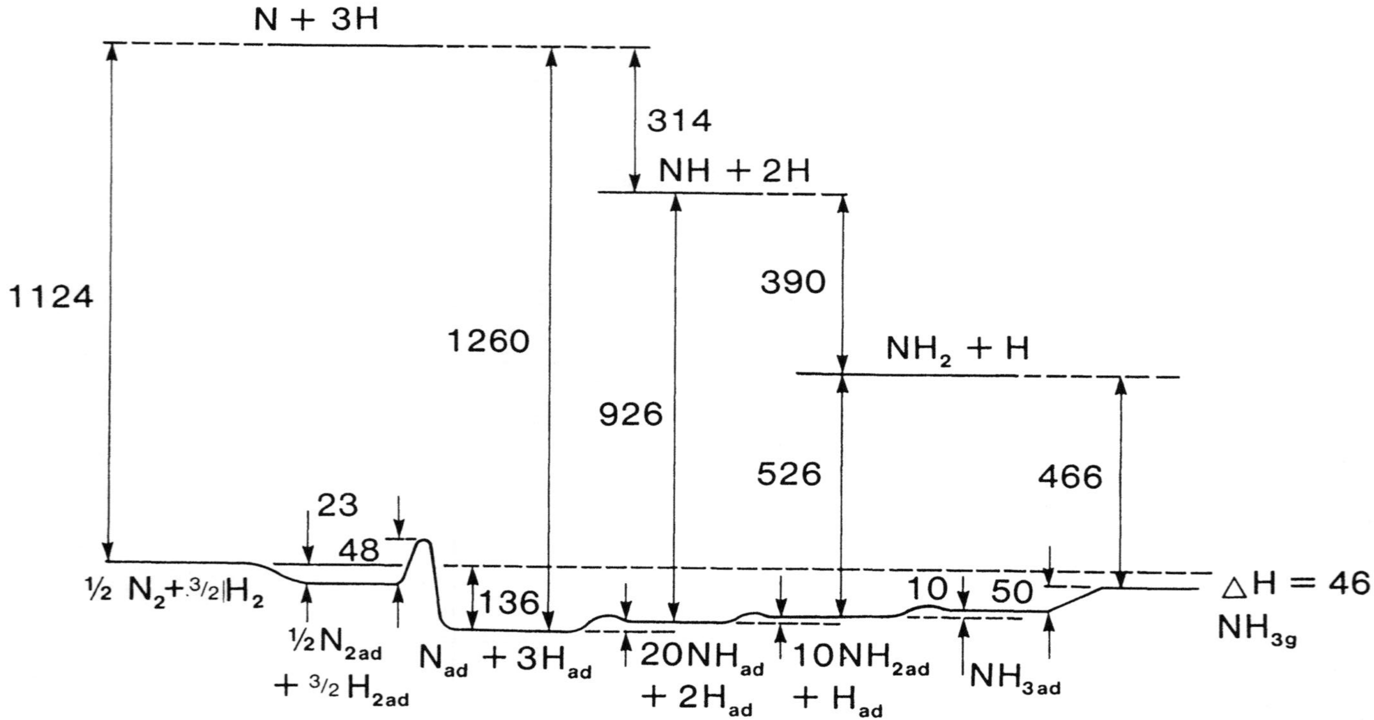

Figure 2 **Potential Energy Diagram for Ammonia Synthesis on a Promoted Iron Catalyst with Activated Adsorption**

The surface coverage corresponding to that rate is: $N_{2(a)}$ = 4.4 x 10^{12} molecule cm^{-2}, $N_{(a)}$ = 8.1 x 10^{13} atom cm^{-2}, $NH_{(a)}$ = 1.2 x 10^{13} species cm^{-2}, $NH_{2(a)}$ = 8.4 x 10^{12} species cm^{-2}, $NH_{3(a)}$ = 5.7 x 10^{12} molecule cm^{-2}, $H_{(a)}$ = 3.3 x 10^{13} atom cm^{-2}, Fe = 5.3 x 10^{12} sites cm^{-2}. The nitrogen atom coverage of this model accords very well with that found by Ertl et al (7 x 10^{13} atom cm^{-2}) [12].

Although the overall activation energy predicted by the model is 117kJ mol^{-1} is higher than the (81kJ mol^{-1}) found by Somorjai et al [16], the model is remarkably good.

5. CONCLUSIONS

1 The potential energy diagram (Figure 1) derived from the detailed kinetics of ammonia synthesis obtained from a considerable body of work by Ertl and co-workers [1, 3, 4, 5, 6, 7, 9, 10, 11, 12] in combination with experimentally derived A-factors, predicts rates of ammonia synthesis within a factor of 10 of those found experimentally when the kinetics are extrapolated by calculation to pressures (107 bar) and temperatures (723K) used in the industrial manufacture of ammonia. Furthermore the overall activation energy predicted by it (62kJ mol^{-1}) is also in good agreement with that (81kJ mol^{-1}) found by Somorjai et al [16] at 20 bar.

2 The potential energy diagram constructed on the premise that the adsorption of nitrogen as atoms on a doubly promoted (K_2O/Al_2O_3) iron catalyst is activated also predicts rates of ammonia synthesis at 107 bar and 723K which are in good agreement with those found experimentally but with a too high overall activation energy.

3 This lack of uniqueness of the energetics which can predict rates of ammonia synthesis and the debate as to whether the adsorption of nitrogen on a doubly promoted iron catalyst is activated, indicates that the detailed kinetics of nitrogen adsorption on a doubly promoted catalyst should be determined using modern surface science techniques.

6. REFERENCES

1 Ertl, G. (1983) Catalysis, Science and Technology, 4, 273.

2 Bowker, M., Parker, I. B., and Waugh, K. C. (1985) Appl. Catal., 14, 101.

3 Ertl, G., Lee, S. B., and Weiss, M. (1982) Surf. Sci., 114, 527.

4 Ertl, G., Weiss, M., and Lee, S. B. (1979) Chem. Phys. Lett., 60, 391.

5 Ertl, G. (1980) Catal. Rev-Sci. Eng., 21(2), 201.

6 Bozso, F., Ertle, G., Grunze, M., and Weiss, M. J. Catal., (1977) 49, 18.

7 Bozso, F., Ertl, G., Grunze, M., and Weiss, M. (1977) Appl. Surf. Sci., 1, 103.

8 Nielsen, A. (1968) An Investigation on Promoted Iron Catalysts for the Synthesis of Ammonia, Third Edition p21 Jul., Gjellerups Forlag

9 Grunze, M., Bozso, F., Ertl G., and Weiss, M. (1978) Appl. Surf. Sci., 1, 241
10 Ertl, G., Lee, S. B., and Weiss, M., (1982) Surf. Sci. 114, 515.
11 Ertl, G., and Thiele, N., (1979) Appl. Surf. Sci., 3, 99.
12 Ertl, G., Huber, M., Lee, S.B., Paal, Z., and Weiss, M. (1981) Appl. Surf. Sci., 8, 373.
13 Reference 8 p 51
14 Reference 8 p 115
15 Temkin, M.I., and Pyzhev, V., (1939) J. Phys. Chem. (USSR), 13, 851.
16 Spencer, N. D., Schoonmaker, R. C., and Somorjai, G. A., (1982) J. Catal., 74, 129.
17 Stoltze, P., and Norskov, J. K. (1985) Phys. Rev. Lett., 55, 2502.
18 Stoltze, P. (1987) Physica Scripta, 36, 824.
19 Stoltze, P., and Norskov, J. K. (1988) J. Catal., 110, 1.
20 Redhead, P. A. (1961) Trans. Faraday Soc., 57, 641.
21 Redhead, P. A. (1962) Vacuum 12, 203.
22 Scholten, J. J. F., Zweitering, P., Konvalinka, J. A., and de Boer, J. H. (1959) Trans. Faraday Soc., 55,2166.
23 Baiker, A., Baris, H., and Schlogl, R. (1987) J. Catal., 108, 467.
24 Emmett, P.H. and Brunauer, S. (1934) J. Amer. Chem. Soc., 56, 35.

TRANSIENT KINETICS IN CATALYSIS BY METALS

A. FRENNET *
Catalyse Hétérogène.- Université libre de Bruxelles
CP243 Campus Plaine ULB
B 1050 Bussels
Belgium

ABSTRACT. The various kinetic methods and concepts that were successively developped since Langmuir are summarized. Transient kinetics concerning some systems, namely CO-H_2 conducted on metal catalysts are very briefly reviewed.
The conditions of applicability of the transient method is analyzed in terms of TOF and coverage in active intermediates. Some results concerning informations on the alloying effect of Rh with Ib metals (Cu, Ag) provided by the transient method applied to the kinetic studies of the CH_4-D_2 exchange reaction are presented. In the case of ethane hydrogenolysis, one of the most studied reaction as model for highly demanding reaction, it is analyzed how the independent use of each of the kinetic methods may be misleading in the analysis of the reaction mechanism. On the contrary, the combined informations provided by steady state kinetic studies and transient kinetic studies, both conducted with labelled molecules, makes it possible, in this particular case, not only to determine the reaction scheme, but to derive the rate, back and forth, of the succesive elementary steps of that reaction. It is concluded that, studied under such conditions, the ethane hydrogenolysis reaction can be considered as a chemical probe of the catalytic properties of metal surfaces in their catalyticaly working conditions.

1. - Introduction

From the early work of Ostwald at the end of last century, it is well known that catalysis may not modify the thermodynamics of a chemical reaction [1].
The most important properties of a good catalyst are certainly related to its selectivity. Selectivity concerns the activity of the catalyst for a given reaction path, which of course is related to the reaction mechanism. The understanding of the reaction mechanism is thus the final key in catalysis.
Within the various probes used to try to understand this reaction mechanism, kinetic studies are certainly the most important ones and have been the most widely used [2].
During about half a century, these studies were based on the concepts of chemisorption as introduced succesively in two papers by Langmuir [3, 4]. They consisted of classical kinetic studies, in which rates are measured under steady conditions. The procedure is then to study the influence of the partial pressure of the reactants, and to render account for the measured pressure dependences, using an adequate combination of reasonable elementary steps, as done in homogeneous kinetics. The mechanisms considered most generally for these elementary steps are of the Langmuir-Hinshelwood type or sometimes of the Eley-Rideal type, assuming that the elementary steps preceding the rate determining surface reaction are in equilibrium, and that the elementary steps following that rate determining step are not significant for the rate equation [5]. Unfortunately, very

* Senior Research associate, National Fund for Scientific Research, BELGIUM.

R. W. Joyner and R. A. van Santen (eds.), Elementary Reaction Steps in Heterogeneous Catalysis, 423–439.

often, more than one mechanism may produce roughly similar rate equations or equations able to render account for the same partial pressure dependences. Of course, in homogeneous reactions each elementary step is characterized by real rate constants in both directions and the mechanism of the overall reaction may be elucidated on the basis of an adequate combination of elementary steps. In heterogeneous reactions, which are essentially composed of gazeous reactants in presence of a solid catalyst, the succesive elementary steps are taking place at the surface of the catalyst and/or between that surface and the fluid phase of the reactants. The important differences between homogeneous and heterogeneous kinetics comes from the fact that, in general, the surface of the catalyst is not uniform or behaves as such.

From the first chemisorption model of Langmuir, where the surface is treated as uniform, various concepts of non uniformity have been introduced based on a priori or on induced heterogeneity. If all the sites are not equivalent, it becomes possible to introduce the idea that only a fraction of these sites are active (H. S. Taylor [6]).

If the surface behaves as non uniform, whatever the reason of non uniformity (a priori or induced), a very important difference between homogeneous and heterogeneous kinetics is that in homogeneous kinetics, the rate constants of the elementary steps are only dependent on the temperature, whereas in heterogeous kinetics, they are also dependent on θ, that depends not only on the temperature but also on the partial pressures and thus also on the conversion! Treatments of non uniform surfaces were made by Temkin succesively in a series of papers from 1957 [7] till 1979 [8]. More recently it was further introduced that through the simultaneous effects of competion for adsorption on the same sites and of composition of the active site by an"ensemble" of a priori chemisorption sites, partial pressure dependences of some elementary steps can be of the same type as those characteristic of the overal catalytic reaction [9].

Thus, although the kinetic behaviour of the overal reaction reflects the mechanism of the reaction, one may conclude that classical kinetic studies are not sufficient to elucidate the reaction mechanism.

Much hope has been put in the study of the surface. The development, during these last decades, of the UHV technology and of electronics allowed to investigate more and more deeply the nature and the properties of model and real surfaces. Nevertheless, these studies involve in most cases to work under pressure conditions extremely different from those under which catalysis takes place.As the amount, and even the nature, of the various adsorbed species is dependent on the experimental conditions, the informations obtained by these surface science techniques, if they are important, must be used with much care.

It is important to remember the remark made by Tamaru [10] that "***the most important properties of a catalyst are not those of the bare surface, but those of the surface in its catalyticaly working state***".

As catalysis concerns essentially the reaction path, kinetic studies are anyway of prime importance, and interesting alternatives to classical kinetic studies are the different possibilities of transient kinetic studies.

2. - Transient Kinetic Studies.

In spite of the large number of years from which the first research works using transient techniques have been published [11], the number of papers devoted to transient kinetic studies is still small as compared to the number of papers dealing with classical kinetic studies. Anyway, a series of excellent reviews of transient effects in heterogeneous catalysis have been published. Let us mention the works by Kobayashi [12,13], Wagner [14], Bennet [15], Happel [16] and recently Tamaru[17].

CO hydrogenation into hydrocarbons, mainly into methane, is probably one of the most extensively studied real reaction these last twelve years using transient effects by several research teams: Happel [18, 19, 20], Tamaru [21, 22, 23], Bell [24,25], Biloen [26, 27], O Bennet [28, 29], van Santen [30, 31].

Metal catalysts are widely used in hydrocarbon reactions. Nevertheless, transient kinetics were not often used for these reactions. Most of the work on "real" hydocarbons, that is C_5 to C_8 hydrocarbons, was conducted by the group of Margitfalvi [32 to 37]. It is to be noticed that all these studies were made with unlabeled molecules.
The hydrogenolysis of ethane is, on the other hand, one of the model reactions the most widely used because of its extremely important demanding character and sensitivity to structure and alloying effect. In most of the works where that reaction was used as a probe reaction to study these structure and alloying effects, to simple ideas and reaction schemes have been applied [9].
Very different ideas concerning the reaction mechanism that came out in our laboratory when studying that reaction with labeled molecules under transient conditions have been published [9, 38]. In chapter 3 of this paper, we analyze what the various kinetic methods are able to provide as informations concerning the elementary steps of that reaction.

2.1. THE PARAMETERS TO WHICH A STEP FUNCTION MAY BE APPLIED

Transient response of a catalytic system implies of course to work under dynamic conditions, that is in a flow reactor. The most important aspect of transient kinetic measurements lies in the study of the simultaneous variations of both activity and selectivity together with variations of the surface composition during its adaptation to the gazeous system. Correlations between activity and selectivity on one hand, and surface composition on the other hand can only be made if the surface composition is uniform enough along the catalyst bed. This implies to work with a well mixed reactor.
When working under low pressure conditions (molecular flow conditions, $P < 10^{-2}$torr), the mixing may be achieved by self diffusion. Such conditions have been used with model catalysts such as evaporated metal films [39 to 44].
When working arround atmospheric pressure, much care need to be taken on that point. A good solution is to use a recirculation type reactor.
The transient method consists in the study of the adaptation of a catalytic system to a perturbation of one of the parameters controling the kinetics of the heterogeneous catalytic reaction.
A step of the following parameters may be imposed :
- 1)- the temperature. Reaction following a step of that parameter is not very much used. Continuous variations of temperature are much more used, but in that case one refers to "temperature programmed variation reactions", and generally not to transient studies.
- 2)- the partial pressure of one of the reactants. In the case, most widely studied, where one of the reactant is hydrogen, the surface of the catalyst is equilibrated with that reactant under flowing conditions, and a step of the partial pressure of the second reactant (CO, N_2 or a hydrocarbon) is made. The reaction is then measured until steady activity is observed. After having reached steady activity, a complementary transient effect may be studied by stopping the inlet of the second reactant. The relaxation of the system may then be studied.

An alternative to that procedure is to make a pulse of the second reactant, which corresponds to a very similar procedure as the preceding one, except that the inlet of the second reactant is stopped before steady conditions are attained.
Both of these methods may be applied using labelled molecules for one of the reactants. For example one may replace normal hydrogen by deuterium. This allows to differentiate the transient kinetics concerning the adsorption - desorption of the reactant from the transient kinetics of the surface and/or desorption steps of the products of the catalytic reaction.
- 3)- the isotopic composition. For example, in the hydrogenation of CO, one may substitude, at steady activity of the catalyst, normal CO by CO molecules labelled with ^{13}C. This method offers the advantage to maintain the chemical composition of the gas phase constant. This may be of great importance, because in this case the chemical composition of the surface will also stay constant. One is so sure that the local properties of the various sites will stay constant during the study of the transient effect.
This is to be opposed to the preceding case, where the chemical composition of the surface radicals may be modified in an important way.
Partial pressure steps offer simultaneously advantages and desadvantages. The desadvantages lie in that, due to non uniformity of the surface, the energetic term characteristic of some constants (rate and/or equilibrium constants) of some elementary steps is dependent on the coverage, making the value of these "constants" to vary during a transient experiment. The advantages consist mainly in the possibility to relate the variations of activity and/or selectivity to the surface composition during its variations.

2.2. PARAMETERS CONTROLING THE POSSIBILITY TO APPLY THE TRANSIENT METHOD.; CORRELATIONS WITH THE **TOF** UNDER STEADY CONDITIONS.

In order to be able to decompose the reaction mechanism in its elemetary steps, it is important to work under experimental conditions such that, when steady state is attained, the conversion is sufficiently small to assume initial conditions for the reaction kinetics to be achieved.
Two important different situations must be distinguiched concerning the intrinsic properties of the adsorbed species, depending on their variations or not, during the transients measured, variations associated with that of the range of coverage θ during the transient.

2.2.1.*No variations of the specific properties of the chemisorbed radicals with coverage.* One may expect the specific properties of the chemisorbed active surface intermediates to remain constant during the transient following a step of the partial pressure of one of the reactants if, when steady state is attained, their surface coverages is small or vary very little.
Let us consider experimental conditions similar to that used in the transient effects applied in the study of the hydrogenolysis of ethane with D_2 [41 to 44]. An important advantage of that system lies in the very fast adsorption and desorption rates of hydrogen as compared to those of hydrocarbons, so that adsorption desorption equilibrium for D_2 may be considered as attained at any moment

- *Build-up in surface active intermediates.* In such experiments the surface of the metal catalyst is equilibrated with flowing D_2 at the temperature of reaction. The transient kinetic study starts with a step of the partial pressure of C_2H_6. The initial coverage in hydrocarbon radicals is thus equal to zero. The rate of formation of the various deuterated isomers of ethane (associated with the adsorption desorption of that reactant) and of methane (associated with the formation and/or desorption of that product of the catalytic reaction) are followed as a function of time until steady values are measured.

If R_∞ is the rate of formation of any of these products at steady activity and R_t the rate at any time t, the variations of R_t with time obey the following equation

$$R_\infty - R_t = R_\infty . \exp(-t/\tau) \quad (1)$$

where τ is the time constant needed for the surface intermediate, implied in the formation of the considered product, to reach steady coverage.

- *Desorption transient.* For the same systems this is achieved by measuring the decrease of the rate of formation of the various products with time after having turned of the inlet of hydrocarbon. This is normaly made after steady state had been attained. If steady state is not attained, one is then dealing with the pulse method. The rate of formation of the various products obey then the following equation

$$R_t = R_\infty . \exp(-t/\tau) \quad (2)$$

If such a law is observed, we may also write that the coverage θ_i in the most abundant surface intermediate (**masi**) or the most abundant reactive intermediate (**mari**) responsable for the formation of product **i** will decrease according to

$$d\theta_i/dt = \nu_i . \theta_i \quad (3)$$

with $$\nu_i = 1/\tau_i \quad (4)$$

There is a range of values of the ratio of flux of reactant to the extent of active surface area of the catalyst in which the total coverage in hydrocarbon residues may be measured from in-out mass balance. This total coverage in carbonaceous residues may then be compared with the coverage in surface intermediates responsable for the formation of product i that may be calculated using the equation

$$R_{i\infty} = \nu_i . \theta_i \quad (5)$$

In this equation, $R_{i\infty}$ is the rate of formation of product i at steady activity, ν_i is the inverse of τ_i as measured in a transient experiment of the desorption type. Then θ_i is the coverage in the masi or in the mari responsable for the formation of product i at steady value of the activity.

If now $R_{i\infty}$ is expressed in terms of TOF, the coverage will be expressed in terms of $(na)_i/(n_m)_H$, where $(n_m)_H$ represents.the number of hydrogen chemisorption sites.

Windows of characteristics of the system allowing to measure transient kinetics.

Let us remember that this analysis applies if, at steady state, the conversion is sufficiently small to assume initial conditions for the reaction kinetics to be achieved.

We have to take into account some characteristics of the whole experimental device:

- the volume of the reactor and the value of the volumic flow rate used define a time constant τ_g to fill in that volume. The value of τ_g is always larger than 1s., and may be as large as 100s. when working on model catalysts handeled in an UHV type apparatus (large dead volume). To be able to measure transient effects characteristic of the adaptation of the surface of the catalyst to the gas phase, the time constant characteristic of the surface τ_s need to be larger or at least of the same order as that of the gas phase τ_g.
- on the other hand, it seems reasonable to consider the transient measurements to be achieved within one day, thus to last no longer than a few hours. The value of τ_s needs not to exceed to much 10^3 s.

It comes that the range of usefull values of v, that is equal to $1/\tau_s$, lies between $1 s^{-1}$ and 10^{-4} s^{-1}.

Values of TOF under steady conditions, coverage θ in active intermediates and probability of reaction v are interelated through eq.5.

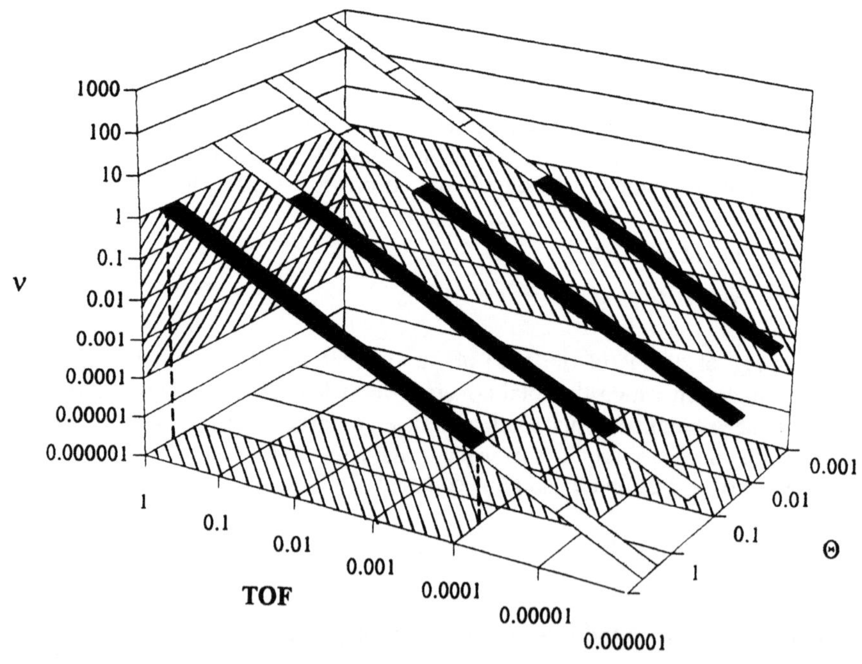

Fig.1 Range of coupled values of TOF and θ at steady conditions allowing to apply the transient method.

One sees that if the coverage θ is large, larger than 0.1, the transient method may be applied to systems where the TOF has a value ranging from 10^{-4} up to 1. On the other hand, if the coverage is smaller than 1% ($0.01>\theta>0.001$), the range of values of TOF of

10^{-7} and 10^{-3} molec. s^{-1}.
The range of TOF, as well as that of coverage, are generaly dependent on the working temperature and pressure ranges.
Anyway, it may be concluded that the transient method is applicable to a rather wide range of experimental conditions that are overlapping most of those that are used practically for laboratory kinetic studies.

2.2.2. *Variations of the specific properties of the chemisorbed radicals with coverage.*
This is often the case when the coverage variations during a transient effect are important. The results obtained under such conditions are not simple to analyse. It is anyway possible to get some ideas about the fraction of adsorbed species that are and that are not intermediates in the reaction scheme [30].

3. Analysis of the mechanism of ethane hydrogenolysis by various kinetic methods.

In order to learn as much as possible about the effects:
- of the surface structure of the metal
- of the association of two or more metals
- of the nature of the support

on the catalytic properties, one generally uses a demanding reaction [5]. Ethane hydrogenolysis is probably one of the most demanding reaction that has been widely used in this frame [9]. We therefore have chosen that reaction to analyze what the kinetic methods can bring us as usefull informations concerning the understanding of the effects mentioned above.
An analysis of the mechanisms proposed in the literature on the basis of classical kinetic studies conducted in laboratory type flow microreactors has recently been published [9]. We will anyway briefly review the various interpretations that have been proposed for the reaction mechanism of that model reaction.
We will try to evidence the insufficiencies of the informations provided by every kinetic method used individually.

3. 1. THE CLASSICAL STEADY STATE KINETIC METHOD.

In order to elucidate the mechanism, the only information that this method can provide is the value, and its variations, of the dependence of the overal reaction rate on the partial pressure of the reactants.
In most cases, the applied procedure is the following [5]:
- one of the possible elementary steps is assumed to be the rate determining step (rds)
- the steps preceding the rds are treated as being at quasi equilibrium
- the steps following the rds are not significant for the rate equation

The rds for ethane hydrogenolysis is assumed to be the C-C bond breaking.
The measured inhibiting term in hydrogen pressure is then interpretated in terms of degree of dehydrogenation of the mari [45 to 51]. The differences between these proposed mechanisms lie in:
- the role of molecular hydrogen, of dissociatively chemisorbed hydrogen, or without any role of hydrogen on the rds, the breaking of the C-C bond [45 to 48]
- the degree of dehydrogenation of the mari following the experimental conditions [51]
- the existence of two types of chemisorption sites, one for hydrogen, the other one for the hydrocarbon [50]

When the hydrogen pressure dependence is measured in a sufficiently large range, it is observed that the order of the reaction rate with respect to the H_2 pressure do vary with that pressure, but this variation do correspond to an increase of the inhibiting effect of H_2 with the H_2 pressure [9]. In a parallel way, this inhibiting term is the most important the lowest the temperature. In the frame of the proposed mechanism, this leads to the very curious conclusion that the mari is the most dehydrogenated the largest the H_2 pressure is and the lowest the temperature! [9].

3. 2. THE TRANSIENT KINETIC METHOD.

Recently the ethane hydrogenolysis has been studied over Ru/Al_2O_3 using both the steady state and the transient method [52]. In this work, classical assumptions are madė.

- the rate determining step is the breaking of the C-C bond
- the coverage in C_1 radicals is negligible, since the steps following the rds are fast
- there is competition between hydrogen and ethane for adsorption on the same sites

The equations considered are thus still:

$$H_{2(g)} + 2S_v \Leftrightarrow 2H_{(s)} \tag{6}$$

$$C_2H_{6(g)} + 2S_v \Leftrightarrow C_2H_{x(2s)} + aH_{2(g)} \tag{7}$$

$$C_2H_{x(2s)} + H_{2(g)} \Rightarrow 2CH_{y(s)} \tag{8}$$

It comes that the following rate equation, that is considered to apply by the authors

$$R = \frac{k.K_E.P_E.P_H}{P_H^a.(K_H.P_H)^{0.5} + K_E.P_E + P_H^a} \tag{9}$$

where: -k is the rate constant of the rds
-K_E is the adsorption constant of ethane
-K_H is the adsorption constant of hydrogen
-P_E is the ethane pressure
-P_H is the hydrogen pressure

is sufficiently complex to be able to render account for the measured kinetic features.

3. 3. STEADY STATE KINETICS WITH LABELLED MOLECULES

We have studied the C_2H_6 hydrogenolysis on Rh catalysts with the hydrogen in the form of its stable isotope deuterium under steady conditions [41]. From hereon the symbol H will be used when no isotopic distinction is made, whereas when isotopic distinction need to be made, D is used for deuterium and P is used for light hydrogen or protium.

The interest of using deuterium is to get informations concerning the interaction of the ethane molecule with the metal surface. This is facilitated by the fact that the H_2-D_2 heterogeneous exchange reaction is very fast as compared to the D_2 exchange rate with the hydrocarbon or to the adsorption rate of the hydrocarbon.As an exemple, results concerning the C_2H_6-D_2 reaction on Rh.presented on fig.2 show that three regions, depending on the value of the hydrogen pressure, have to be distinguished, in each of which the ratio of the value of the rate of some elementary steps do vary in an important way. When going from right to left, it is observed that:

- the adsorption rate of ethane increases progressively, but that the rate of increase is progressivelly decreasing

- the single exchange rate (R_{SE}), corresponding thus to the formation of the mono-deuterated isomer of ethane (C_2P_5D), is varying in a parallel way to the adsorption rate
- on the right hand side, most of the adsorbed ethane molecules desorb in the form of highly deuterated ethanes (R_{ME}), and that only a small fraction of these adsorbed ethane molecules see the C-C bond broken
- around a value of about 10^{-2}Torr of the hydrogen pressure , the rates of multiple exchange and of hydrogenolysis (R_H) are equivalent. The value of the hydrogen pressure at which the ratio R_{ME} / R_H is equal to 1 is increasing with the temperature
- on the left hand side, hydrogenolysis prevails and a progressively small fraction of the adsorbed ethane molecules comes back in the gas phase

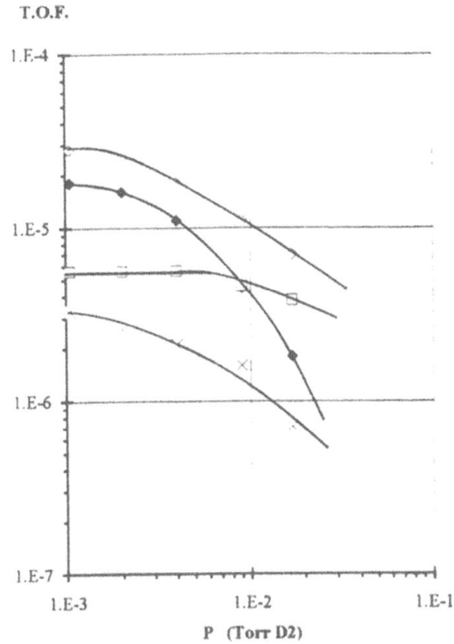

Fig.2. C_2H_6-D_2 reaction on Rh. Variations of R_A (◇), R_{SE} (x), R_{ME} (□), R_H (♦), with p_{D_2},TOF in molec/s.Hsite

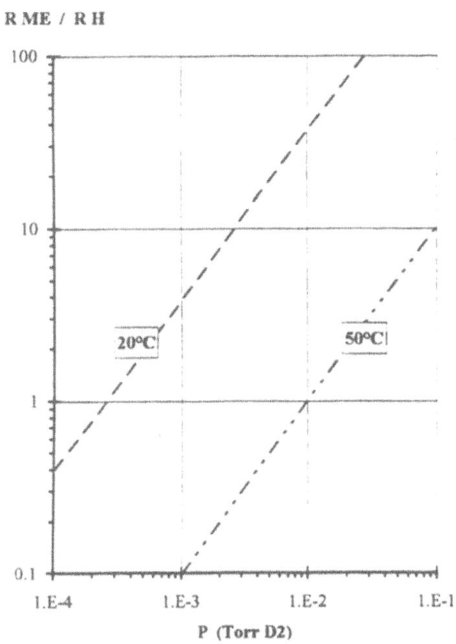

Fig. 3.C_2H_6-D_2 reaction on Rh.

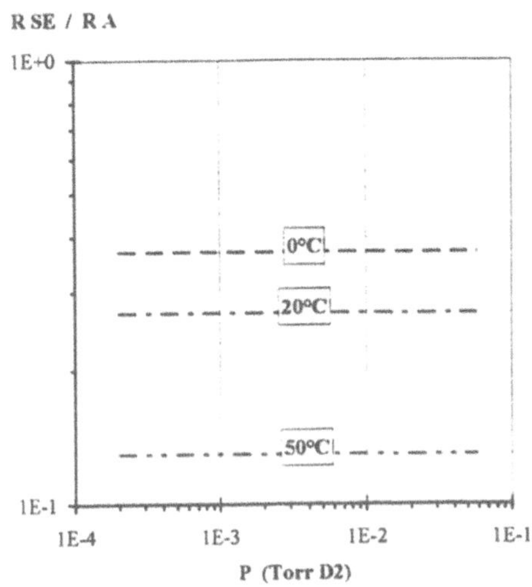

Fig 4. C_2H_6-D_2 reaction on Rh.

The same results are plotted, together with results at 20°C, a different way in fig.3 and 4, showing that:

- the ratio R_{SE}/R_A (fig.4) is independent of the hydrogen pressure not only at 50°C, supporting the idea that the single exchange proceeds through a mechanism that do not involve the chemisorption of the hydrocarbon [38]
- again not only at 50°C, the probability of the C_2 adsorbed radicals to desorb in the form of highly deuterated isomers of ethane instead of undergoing the breaking of the carbon-carbon bond (R_{ME} / R_H) is stronly dependent on the hydrogen pressure (Fig.3).

If one wants to learn the maximum about the reaction mechanism, the experimental conditions under which the reaction should be studied under transient conditions are those for which the rates of most of the steps are of the same order of magnitude. On rhodium, at 50°C, this corresponds to an H_2 pressure between 5 10^{-3} Torr and 10^{-2} Torr.

3. 4. TRANSIENT KINETICS WITH LABELLED MOLECULES.

Transient kinetics of the ethane hydrogenolysis with deuterium have been measured on Rh films [42]. We hereafter summerize, as an exemple, the main type of informations that such measurements are able to provide.

The particular exemple treated here was experimentaly conducted under conditions chosen according to the criterium given above, that conditions where the rate of exchange with deuterium is of the same order as that of hydrogenolysis.

In practice, the experimental conditions were: T = 50°C, D_2 pressure = $4.3.10^{-3}$Torr, ethane pressure = 5.10^{-5}Torr. Transient studies were conducted as follows: the catalyst is equilibrated with flowing deuterium at reaction temperature. A first transient is measured from the time where the inlet of ethane starts until steady activity is measured. A second transient is then measured after stopping the inlet of ethane..

At any moment, the value of the different rates (R_A, R_{SE}, R_{ME}, R_H) is calculated using the following equation

$$R_i=(\phi_{out})_i - (\phi_{in})_i + V\frac{dp_i N_a}{dt\ RT} \qquad (10)$$

where ϕ represents the flow out and in of molecules type i in molec./s.

In the particular case presented here, the experimental conditions are such that the value of the TOF at steady activity and the hydrocarbon coverage satisfy the conditions for making it possible to measure transient kinetics as given on fig.1 of this paper.

Furthermore, it is possible to calculate the coverage in hydrocarbon radicals by integration on time of the in-out carbon mass balance

$$(\theta_c)_t=\int_0^t(\Sigma R_i) \qquad (11)$$

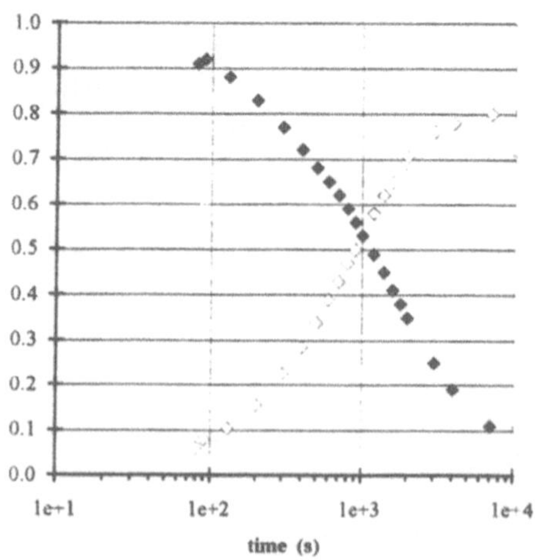

Fig. 5. Build up in Hydrocarbon coverage Variations of $\theta_c.2.10^{-2}$ (◇) and of $d\theta_c/dt$ $.2.10^{-5}$(◆) with time.

It is interesting to notice here that the hydrocarbon coverage at steady state is only in the range of 2%. The low value of hydrocarbon residues in reactions involving alkanes (on the contrary to unsaturated hydrocarbons) has been discussed elsewhere [53].

The informations given by the build up experiment presented on fig. 6, are the following:

- after a period of about 100s., corresponding roughly to the delay needed to fill the volume of the reactor, the adsorption rate of ethane reaches a steady value
- the single exchange rate varies proportionally to the adsorption rate
- the hydrogenolysis rate increases proportionally to the total hydrocarbon content of the surface
- the multiple exchange rate presents more complex variations; R_{ME} reaches very quickly a value that remains constant for about 200s.. During that period of time, the coverage in hydrocarbon radicals remains very small. When that coverage increases significantly, the value of R_{ME} starts to increase in a parallel way to the hydrogenolysis rate and to the total carbon coverage.

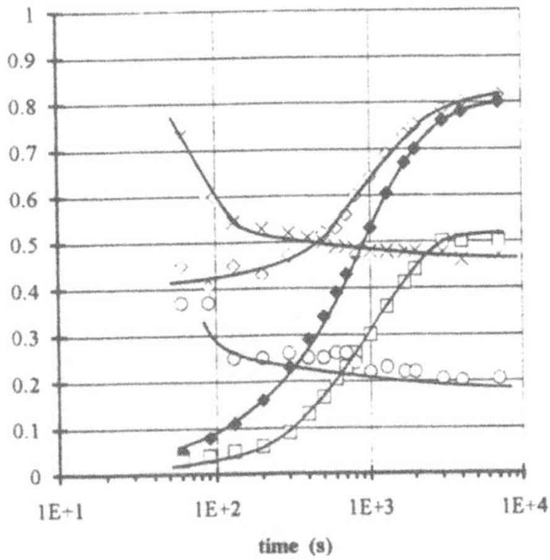

Fig.6. C_2H_6-D_2 reaction on Rh, Transient kinetics. Variations with time of $\theta_C.2.10^{-2}$ (♦), $R_A.2.10^{-5}$ (×) $R_{SE}.2.10^{-6}$ (○), $R_{ME}.4.10^{-6}$ (◇) $R_H.1.10^{-5}$ (□)

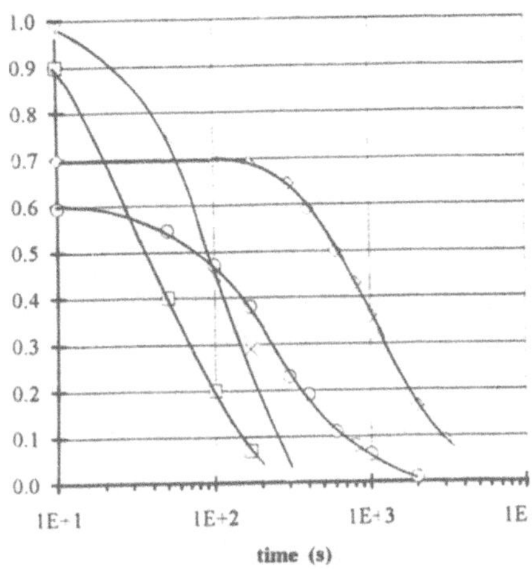

Fig. 7 C_2H_6-D_2 reaction on Rh, Desorption transient. Variations of p in Torr $C_2P_6.5.10^{-5}$(□), $C_2P_5D.2.10^{-6}$ (×), $C_2D_6.1.10^{-5}$ (○), $CD_4.2.10^{-5}$(◇) with time

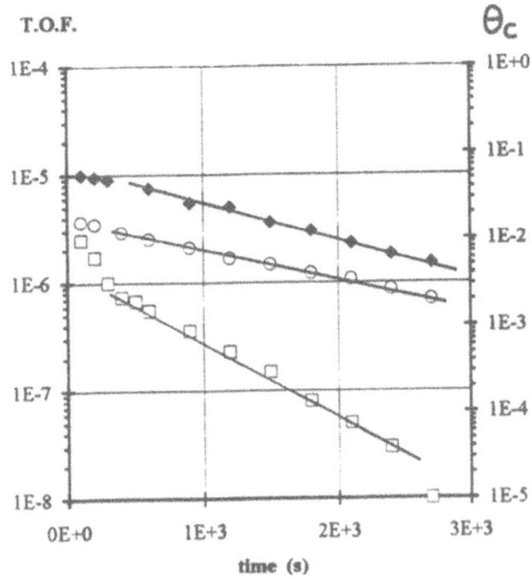

Fig. 8 C_2H_6-D_2 reaction on Rh, Desorption transient. Variations of θ_C (○), R_{ME} (□) and R_H (♦) with time

After steady state is attained, the inlet of ethane is turned of, and the relaxation of the system is measured.One sees on fig.7 that the pressure in the different molecules decrease differently. The decrease in C_2P_6 is controlled by the sum of evacuation and of adsorption rates. The decrease in C_2P_5D is controlled by the same rates minus its rate of formation that is proportional to the adsorption rate of C_2P_6. The decrease in C_2D_6 is much slower and the decrease in the methane pressure even much more slower.

These results, transformed in terms of rates (R_{ME} and R_H) using the equation 10, are plotted together with the carbon coverage θ_C calculated from in-out mass balance, in a log. scale as a function of time on fig.8. The time constant characteristic of the decrease of R_H, is the same as that of the decrease of θ_C.

It is remarkable that the total coverage in carbon atoms derived from in-out mass balance during the build up transient as well as during the desorption transient have the same value as the one calculated by applying equation 5 to the value of the time constant measured by the desorption transient and of tha TOF measured at steady state [42, 43].

The decrease of R_{ME} decomposes in two exponentials, one is characterized by a time constant about half of the one of R_H, the first one is much smaller and of the same order as that of the volume of the reactor [43]. It is anyway characteristic of the desorption from the surface as the evacuation of the volume has already been taken into account by the equation mentionned above, used to calculate the value of the various desorption rates. When plotting as a function of , both in log. scale (fig.9), one observes a second order dependence. These results are thus showing that the recombination of C_1 radicals, leading to the formation of ethane as final product, is not completely negligeable. It also comes that, under the conditions used here, the masi is a C_1 radical.

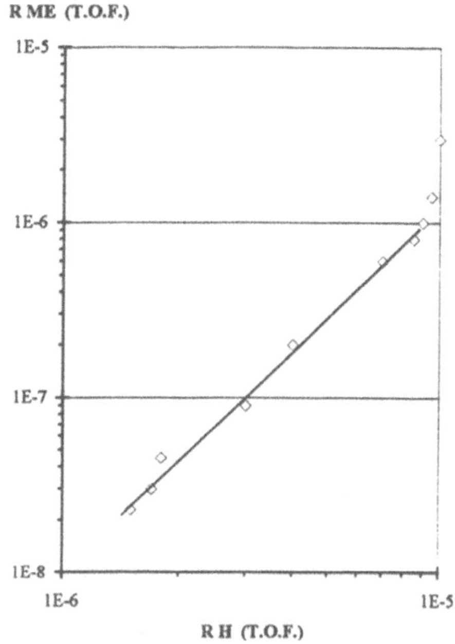

Fig. 9 C_2H_6-D_2 reaction on Rh, R_{ME} as a function of R_H during a transient desorption

3.5. COMBINED STEADY STATE AND TRANSIENT KINETICS WITH LABELLED MOLECULES.

Let us first consider the informations provided by the steady state kinetic measurements. The use of deuterium allows, besides the determination of the hydrogenolysis rate, to obtain informations on the adsorption-desorption step of ethane. The rate of adsorption can be measured; it varies monotoneously with the hydrogen pressure, whatever the reaction that follows the adsorption step, exchange or hydrogenolysis. Furthermore, the hydrogen pressure has an inhibiting effect of the same type as that characterizing the hydrogenolysis rate, but that can not been interpretated in terms of dehydrogenation of

surface intermediates. The proposed inerpretation is based on competition for adsorption of hydrogen and ethane and the need for the hydrocarbon to adsorb to find an ensemble of z hydrogen adsorption sites simultaneously free.
If one remembers that the rate of single exchange is independent of both the hydrogen pressure and of the hydrocarbon coverage during transients, one comes to the mechanism already proposed [9, 38].

$$\overset{\text{step 1}}{(C_2H_6)_g + (D)_a + zS \Leftrightarrow (C_2H_6D)_a} \overset{\text{step 2}}{\Leftrightarrow (C_2H_5)_a + (HD)_g + wS} \qquad (12)$$

where the SE do not involve any chemisorbed hydrocarbon, and is the reaction backword of step1. The chemisorption step is then step2.
The chemisorbed C_2H_5 radical may then dehydrogenate more or less or may simply exchange with surface hydrogen, that under the conditions used is deuterated. The deuterated radical may then desorb, giving raise to ME, or may decompose, leading to hydrogenolysis.
Thus, from steady state kinetic measurements conducted with D_2, we may derive:

- the direct rate of step 1 $\quad \mathbf{R_1} = R_a \quad$ (13)
- the backword rate of step1 $\quad \mathbf{R'_1} = R_{SE} + R_{ME} \quad$ (14)
- the direct rate of step2 $\quad \mathbf{R_2} = R_a - R_{SE} \quad$ (15)
- the backword rate of step2 $\quad \mathbf{R'_2} = R_{ME} \quad$ (16)

Let us now consider the <u>informations given by the transient experiments:</u>

the complete analysis lead us to propose the following reaction scheme [42]

$$(C_2H_6)_g \overset{\text{step1}}{\Leftrightarrow} (C_2H_6D)_a \overset{\text{step2}}{\Leftrightarrow} (C_2H_x)_a \overset{\text{step3}}{\Leftrightarrow} (C_2H_y)_a \overset{\text{step4}}{\Leftrightarrow} 2(C_1)_a \overset{\text{step5}}{\Rightarrow} 2(CH_4)_g \qquad (17)$$

where the detailed mechanism of each step is not considered.
During the first part of the transient in build up, the coverage in species other than $(C_2H_x)_a$ is negligeable, and the ME has reached a first steady value. Thus R'3 is negligeable and we may write

$$d\theta\, C_2H_x/dt = R_2 - R'_2 - R_3 = 0 \qquad (18)$$

it comes that $\quad (R_3 = R_2 - R'_2)_0 \quad$ (19)

during that first period. As R_3 is expected to be proportional to the coverage in C_2H_x as well as R_{ME}, it is possible to calculate the value of $\mathbf{R_3}$ at steady activity by multiplying the value of R_3 as calculated by eq.19 by the ratio of R_{ME} in the final steady activity to the value of R_{ME} at first steady activity.
By extrapolating to t=0 the slow part of the decrease in ME during a transient experiment in desorption, one obtains the value of R'_4.
Of course $\mathbf{R_5}$ is equal to the rate of methane formation in the gaz phase. Under the conditions used, readsorption of methane is completely negligeable.

As, at final steady state:

$$d\theta_{C_1}/dt = R_4 - R'_4 - R_5 = 0 \qquad (20)$$

It finally comes that:

$$\mathbf{R_4} = R_5 + R'_4 \qquad (21)$$

and, as:

$$d\theta\ C_2H_x\ /dt = R_2 + R'_3 - R'_2 - R_3 = 0 \qquad (22)$$

and

$$\mathbf{R'_3} = R_3 + R'_2 - R_2 \qquad (23)$$

When the infomations obtained by steady state and transient kinetics, both conducted with D_2, are simultaneously treated it is possible to determine the rates in both directions, and the net rate, of a series of steps of the catalytic reaction, as shown on fig.10.

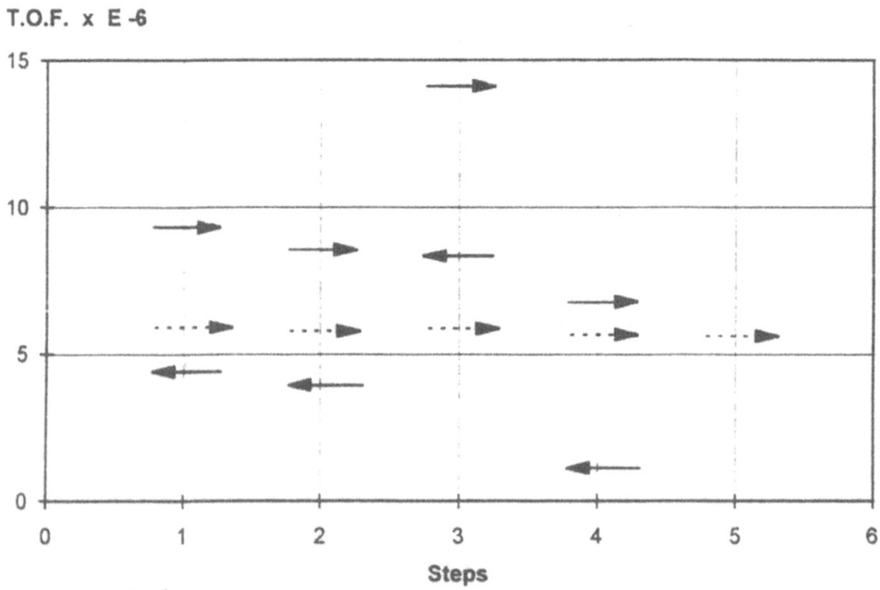

step1 step2 step3 step4 step5

$(C_2H_6)_g \Leftrightarrow (C_2H_6D)_a \Leftrightarrow (C_2H_x)_a \Leftrightarrow (C_2H_y)_a \Leftrightarrow 2(C_1)_a \Rightarrow 2(CH_4)_g$

Fig.10. Values of the rates of the successive elementary steps of C_2H_6 hydrogenolysis on Rhodium, as derived from the association of steady state and transient kinetics with D_2.

4. Transient Kinetics of the CH_4 - D_2 Exchange Reaction

Besides the hydrogenolysis of ethane with D_2, we also have studied the exchange reaction between methane and D_2 [39, 54, 55].

The most important features are:

- for both hydrocarbons, the single exchange rate is independent of the hydrocarbon coverage during its build-up as well as during its desorption. This constitues one of the arguments in favour of the mechanism that we have proposed for the adsorption step of alkanes on metal surfaces [9, 38].

- for both hydrocarbons, multiple exchange presents a dependence on the hydrocarbon coverage. In the case of methane the multiple exchange rate is proportional to the hydrocarbon coverage. The situation is more complex in the case of ethane, [9, 41 to 44]
- from transient measurements of the desorption type, it appears that the dependence of the desorption rate to the hydrogen pressure is important and may not be related to the variations of hydrogen coverage, which is of the Temkin type [9]. The measurement of the hydrogen pressure effect and of temperature on the desorption rate of the chemisorbed hydrocarbon radicals represents detailed informations non accisible by classical kinetic studies
- another type of informations only available from transient experiments, but of the "build-up type", with such systems are the variations of the adsorption rate of the alkane with the hydrocarbon coverage during its build-up [40].

As an exemple, we present on fig.11 results of desorption transients of C_1 radicals (that is the step 5 in equation 17), when measured on two different catalysts after CH_4-D_2 steady state exchange (Rh and Rh-Ag alloy) and after ethane hydrogenolysis on Rh. The value of v characteristic of step5 exhibit an important positive dependence on the hydrogen pressure in all cases.

On both catalysts, the effect of the temperature on v characteristic of C_1 created from methane is negligeable between 150°C and 175°C.. This is to be compared with the similar value of v characteristic of the desorption of C_1 created on the surface by ethane hydrogenolysis that, at 50°C, is only a factor two smaller than in the preceding case. On the other hand, at the same hydrogen pressure and the same temperature, the probability v to desorb the C_1 radical from the Rh-Ag catalyst is about 2.5 times more important than on the Rh catalyst.

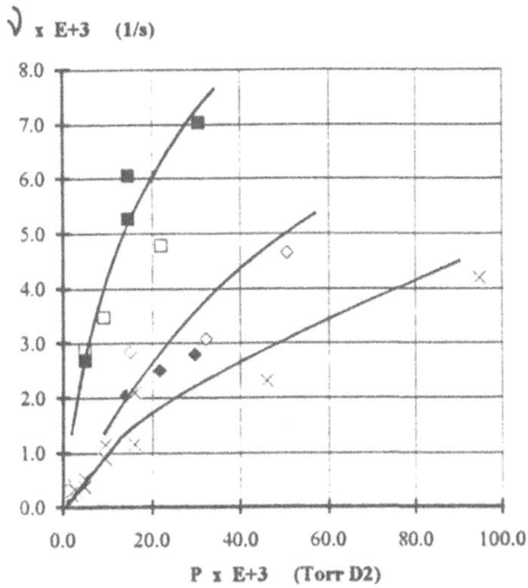

Fig.11. C_1 desorption transients from CH_4 on Rh- Ag at 150°C (□), at 175°C (■) from CH_4 on Rh at 150°C (♦), at 175°C (◇) from C_2H_6 on Rh at 50°C (✕)

5. - Conclusion

The reactions of methane and of ethane with deuterium when studied by both steady state and transient kinetics may be used as a powerfull probe to test the catalytic behaviour of metal surfaces. It is possible to determine the effects of partial pressure, of coverages and of temperature not simply on the overall reaction rate of the catalytic reaction, but on at least some of the elementary steps of that reaction. This may thus constitute a very interesting kinetic probe to test a series of effects like structure effects, alloying effects,

support effects,....Furthermore, **the informations obtained by this method are characteristic of the surface in its catalyticaly working state.** This chemical probe is able to determine how and where in the reaction scheme these effects are modifying the properties of the catalyst.
Till now it has been scarcely used. Besides methane and ethane, it should be necessary to apply it to propane. There are indications that the modifications of the carbon skeleton of higher hydrocarbons proceed through a chemisorbed hydrocarbon radical that is bond to the surface by two carbon atoms that are in 1-3 position [56, 57, 58].
Rather few work has been made on more "real" hydrocarbons (C_5 to C_8). This is certainly due to the high complexity of the analytical problems. No reactions of such hydrocarbons have been studied under transient conditions with deuterium.

The help of J.M. Bastin, A. Löfberg, M. Nouinou and L. Seyfried in the preparation of this document is here acknowledged

References

1 F.G. Donnan, (1933) Ostwald Memorial Lecture, J. Chem. Soc., 316.
(1951) Chemical Society Memorial Lectures,
The Chemical Society, London, 4, 1.
2 R. L. Burwell, (1991) Determination of Mechanism in Heterogeneous Catalysis, in Catalysis, Science and Technology, ed. J.R. Anderson and M. Boudart, 9, 1.
3 I. Langmuir, (1915) J. Amer. Chem. Soc., 37, 1139.
4 I. Langmuir, (1922) Trans. Faraday Soc., 17, 621.
5 M. Boudart and G. Djega-Mariadassou, (1984) "Kinetics of Heterogeneous Catalytic Reactions", Princeton Univ. Press, Princeton, NJ.
6 H.S. Taylor, (1925) Proc. Roy. Soc. (London), A108, 105.
7 M.I. Temkin, (1957), Zhur. Fiz. Khim., 31, 1.
8 M.I. Temkin, (1979), Adv. Catal. Relat. Subj., 28, 173.
9 A. Frennet, (1988) General Kinetics of Hydrogen Effects: Hydrocarbon Transformations over Metals as Model Reactions, in "Hydrogen Effects in Catalysis", ed.Z. Paal and P.G. Menon, 399, Marcel Dekker, NY
10 K. Tamaru, (1964) Adsorption Measurements during Surface Catalysis, in Adv. in Catalysis, 15, 65
11 C. Wagner and K. Hauffe, (1939) Z. Elektrochem., 45, 409.
12 H. Kobayashi and M. Kobayashi, (1974) Catal. Rev-Sci. Eng., 10, 139.
13 M. Kobayashi, (1982) Chem. Eng. Sci., 37, 393.
14 C. Wagner, (1970) Adv. in Catalysis, 21, 323.
15 C. O Bennet, (1976) Catal. Rev., 13, 121.
16 J. Happel, E. Walter and Y. Lecouturier, (1986) Ind. Eng. Chem. Fundam., 25, 704.
17 K. Tamaru, (1991) Dynamic Relaxation Methods in Heterogeneous Catalysis, in Catalysis, Science and Technology, ed. J.R. Anderson and M. Boudart, 9, 87.
18 J. Happel, I. Suzuki, P. Kokayeff and V. Fthenatis, (1980) J. Catal., 65, 59.
19 J. Happel, H. Y. Chek, M.Otarod, S. Ogawa, A. J. Severdia, T. Yoshida and V. Fthenatis, (1982) J. Catal., 75, 314.
20 J. Happel, V. Fthenatis, T. Yoshida and S. Ogawa, (1981) Proc. 7th Intern. Cong. Catal., 542, Tokyo.
21 H. Yamasaki, Y. Kobori, S. Naito, T. Onishi and K. Tamaru, (1981) J. Chem. Soc. Faraday Trans.I, 77, 2913.
22 Y. Kobori, H. Yamasaki, S. Naito, T. Onishi and K. Tamaru, (1982) J. Chem. Soc. Faraday Trans.I, 78, 1473.

23 S. Naito, Y. Kobori, H. Yamasaki, T. Onishi and K. Tamaru, (1982) Chem. Soc. Japan, 170.
24 N. W. Cant andA.T. Bell, (1982) J. Catal., 73, 257.
25 M. de Pontes, G.H. Yokomizo and A.T. Bell, (1987) J. Catal., 104, 147.
26 P. Biloen, J. N. Helle, G. A. van den Berge and W. M H. Sachtler, (1985) J. Catal., 81, 450.
27 P. Biloen, (1986) J. Catal., 97, 330.
28 A. M. Efstathiou and C. O Bennet, (1989) J. Catal., 120, 137.
29 A. M. Efstathiou and C. O Bennet, (1989) J. Catal., 120, 118.
30 T. Koerts, W. J. J. Welters and R and A. van Santen, (1992) J. Catal., 134, 1.
31 T. Koerts and R. A. van Santen, (1992) J. Catal., 134, 13.
32 J.L. Margitfalvi, P. Szedacsek, M. Hegedüs and F. Nagy. (1985). Applied Catal., 15, 69.
33 J.L. Margitfalvi, P. Szedacsek, E. Kern-Talas and F. Nagy. (1985). Acta Chim. Sci. Hung., 119, 213.
34 J.L. Margitfalvi, P. Szedacsek, E. Kern-Talas and F. Nagy. (1985). 9th. North American Meeting of the Catalysis Soc. Houston, Texas, 24.
35 P. Szedacsek, E. Talas, M. Hegedüs and J.L. Margitfalvi, (1987). Proc. 6th. Int. Symp. Heter. Catal., Sofia, part1, 82.
36 J.L. Margitfalvi, P. Szedacsek, M. Hegedüs, E. Talas and F. Nagy. (1988). Proc. 9th. Int. Cong. on Catalysis, Calgary, Alberta (Canada), 3, 1283.
37 J.L. Margitfalvi and M. Hegedüs, (1991). Symp. on Alkylation, Aromat., .Division of Petroleum Chemistry, Inc. Am. Chem. Soc., New-York City Meeting, 887.
38 A. Frennet, (1992) Catal. Today, 12, 131.
39 A. Frennet, G. Liénard, A. Crucq and L. Degols, (1981) Proc. 7th Intern. Congres Catal., Tokyo 1980, Part B, 1482.
40 A. Frennet, G. Liénard, A. Crucq and L. Degols, (1980) Proc. ECOSS III, ed. D. A. Degras and M. Costa, 1, 419.
41 A. Frennet, G. Liénard, A. Crucq and L. Degols, (1982) Acta Chim. Acad. Sci. Hung., 111, 499.
42 A. Frennet, G. Liénard, A. Crucq and L. Degols, (1984) Proc. 9th Ibero-American Symposium on Catal., Lisboa, 493.
43 G. Liénard and L. Degols, (1990) Bull. Soc. Chim. Belg., 99 (10), 803.
44 G. Liénard and L. Degols, (1990) Bull. Soc. Chim. Belg., 99 (10), 817.
45 A. Cimino, M. Boudart and H. Taylor, (1954), J. Phys. Chem., 58, 796.
46 J.H. Sinfelt and W.F. Taylor, (1968), Trans. Faraday Soc., 64, 3086.
47 J.H. Sinfelt, (1977), J. Catal., 27, 468.
48 G. Leclercq, L. Leclercq and R. Maurel, (1974), Bull. Soc. Chim. Fr. , 11, 2329.
49 G. Leclercq, L. Leclercq and R. Maurel, (1976), J. Catal., 44, 68.
50 M. Boudart, (1972), AIChE J., 18, 465.
51 B.S. Gudkov, L. Guczi and P. Tétényi, (1982), J. Catal., 74, 207.
52 S.B. Shang and C.N. Kenney, (1992), J. Catal., 134, 134.
53 F. Garin, G. Maire, S. Zyade, MN. Zauwen, A. Frennet and P. Zielinski, (1990), J. Molec. Catal., 58, 185.
54 A. Frennet, (1974), Cata. Rev-Sci. Eng., 10, 37.
55 A. Frennet, G. Liénard, A. Crucq and L. Degols, (1978), J. Catal., 53, 150.
56 J. R. Anderson, (1973), Adv. Catal., 23, 1.
57 F. G. Gault, (1981), Adv. Catal., 30, 1.
58 E.H. van Broekhoven and V Ponec, (1985), Pogr. Surf. Sci.. 19, 351.

REACTION KINETICS AND INDUSTRIAL CATALYSIS

Dr. Jens Rostrup-Nielsen
Haldor Topsoe A/S
2800 Lyngby
Denmark

ABSTRACT. Reaction kinetics is an important tool for industrial catalysis. An empirical approach based on simplified kinetics has been very useful. However, more detailed knowledge is required to predict selectivity patterns of complex reactions and to describe conversion profiles in environmental processes, where one or more reactants are removed to the ppm level. Micro-kinetic analysis based on input from surface science and "in-situ" kinetic studies has provided a better basis for formulating steady state kinetics with more predictive power. Advanced mathematical reaction engineering models have provided a strong instrument for using kinetic data for reactor design. However, pilot tests at industrial mass velocity are required to determine constants for the models and to identify key design parameters related to catalyst deactivation.

The problems are illustrated by examples from various reactions including ammonia synthesis, selective catalytic reduction of NO, methanation and steam reforming of natural gas.

1. Empirical Approach and Microkinetics

1.1. The Empirical Approach

1.1.1. *Conversion Plots*: Industrial catalysis needs reaction kinetics to predict catalyst volumes for industrial processes. Many industrial reactions are very complex with no detailed kinetics available and in many cases empirical methods can be effectively used with great savings in time.

Even complex catalytic systems may approximate a pseudo first order of relationship (Hougen et al., 1959) when only space velocity or the feed rate is varied keeping temperature pressure and feedstock composition constant. If so,

$$\frac{W}{F} = \frac{X_{eq}}{k} \ln \frac{1}{1-X/X_{eq}} \qquad (1)$$

and linear relationships should result from plotting W/F against $\ln(1/(1-X/X_{eq}))$ for low conversions and from plotting their reciprocals for high conversions. By intrapolation, the W/F values for given conversions are determined. It is evident that this empirical method requires a lot of testing and has limited predictive power.

1.1.2. *Transport Restrictions*: Many early kinetic studies were influenced by uncontrolled mass and heat transfer restrictions with little use for scale-up to industrial conditions as well as for studying the mechanism of the reaction. Today, most studies are carried out on catalyst powder in reactor systems resulting in gradientless conditions. If experiments are to be carried out on catalyst pellets, it is essential to use full size pellets and that the conditions reflect industrial mass velocities and heat

R. W. Joyner and R. A. van Santen (eds.), Elementary Reaction Steps in Heterogeneous Catalysis, 441–460.

transfer rates. At very high degrees of diffusion restrictions (effectiveness factor $\eta < 0.1$), it can be shown (Satterfield, 1970 and Froment et al., 1979) that the intrinsic rate expression was easily related to the rate equation obtained on the catalyst pellet:

$$r_{eff} = \eta \cdot r_i = \left(\frac{1}{\Phi_L}\right)(k_i \cdot C^m)$$

$$\Phi_L = d_{p,s}\sqrt{\frac{k_i \cdot C^{m-1}}{D_{eff}}}, \quad \text{i.e} \tag{2}$$

$$r_{eff} = \frac{1}{d_{p,s}} \cdot (C_s^{(m+1)/2}) \cdot \sqrt{D_{eff} \cdot k_i}$$

Hence, it may be possible to derive intrinsic rate expressions from measurements on catalyst pellets at these conditions (Toettrup, 1988).

Diffusion restrictions may have a severe impact on the selectivity of catalysts. For a series type of reactions:

$$A \rightarrow B \rightarrow C \tag{3}$$

being typical for partial oxidation, pore diffusion will result in a lower yield of B. As an example, the selectivity of B will drop to 50% of the maximum obtainable for an η = 0.3. (Satterfield, 1970). Hence, it is necessary to evaluate catalyst selectivity on an intrinsic and full pellet size basis as well.

The pore diffusion should have no influence on selectivity for parallel reactions:

$$A \begin{cases} \nearrow B \\ \searrow C \end{cases} \tag{4}$$

However, deactivation from reactants (parallel poisoning, coking) may be overlooked in many bench-scale reactors, because film diffusion results in much lower reactant concentration on the catalyst surface than that experienced in the industrial plant.

In conclusion, evaluation of catalyst selectivity should involve intrinsic kinetic tests as well as tests on industrial catalyst pellets at industrial mass velocities.

1.1.3. *Parallel Gas Phase Reactions*: The situation may become even more complex when the catalytic reaction is accompanied by homogeneous reactions in the void spaces in the catalyst bed (Holton et al., 1976) and in the preheating and cooling section of the reactor. An example is **oxidative coupling** of methane to ethylene.

$$2CH_4 + ½O_2 = C_2H_4 + H_2O \tag{5}$$

It is generally accepted (Lunsford, 1992) that methyl radicals formed on the catalyst surface are reacting in the gas phase to ethane which is converted into the ethylene:

$$\begin{array}{lll} CH_4 + O_2 + * & = CH_3^{\cdot} + HO_2^{\cdot} + * & a \\ 2CH_3 + M & = C_2H_6 + M & b \\ C_2H_6 & = C_2H_4 + H_2 & c \end{array} \tag{6}$$

in which M indicates a wall effect.

Although the kinetics of the conversion of ethane into ethylene is well-described (Albright et al., 1983), the ethylene yield is related in a complex way to the so-called kinetic severity function, KSF (de Blieck et al., 1971 and Zdonik et al., 1967):

$$KSF = \int f(T)dt \tag{7}$$

which describes the residence time/temperature history of ethane in the reactor. It is evident that this is difficult to control the KSF in laboratory reactors and that scale-up could easily be misleading.

1.2. Rate Equations

1.2.1. *The Paradox of Heterogeneous Catalysis*: With the development of the chemical reactor engineering, rate expressions were established from experimental work, varying feed composition and temperature and using the Langmuir/Hinshelwood approach (Hougen et al., 1959 and Froment et al., 1979) for formulating possible mechanisms and their equivalent rate equations:

$$r = \frac{(\text{kinetic factor}) \cdot (\text{driving force})}{(\text{adsorption group})} \tag{8}$$

The rate equation showing the best fit or the minimum residual squares of conversion was chosen and often this was taking as proof for the mechanism behind the expression. This could easily be misleading because even if it made the mechanism plausible, the fit by itself would be no proof of its assumptions. It is the classical philosophical problem of judging a theory by its causes rather than by its effects: "No doubt, it is more common to wish to arrive at an agreeable result than to wish to arrive at a true result" (Russell, 1961).

Also, it was shown that many of the rate data could be expressed by simple power rate expressions (Weller, 1956, Bohlbro, 1966):

$$r = k \cdot p_A^{\alpha} \cdot p_B^{\beta} \cdot p_c^{\gamma} \tag{9}$$

which could reflect the simpler Freundlich isotherm. This isotherm is related to the Langmuir isotherms by the approximation:

$$\frac{Ap}{1+Ap} \sim B \cdot p^n, \quad 0<n<1 \tag{10}$$

Therefore care should be taken in drawing conclusions about reaction sequences from the form of the kinetic expression only.

As an example (Kwan, 1956), it was shown that for instance the kinetics for the **ammonia synthesis** derived on basis of the Temkin isotherm could be derived on basis of the Langmuir or the Freundlich isotherm as well.

In principle, the different isotherms represent different distributions of adsorption sites, reflected by different variation of the heat of adsorption with coverage. Nevertheless, Boudart showed that even when assuming a wide distribution of heat of adsorption, the resulting kinetics would be close to that obtained when assuming that the reaction took place only on the most active sites (Boudart, 1968). This is the background for what he termed the "paradox of heterogeneous catalysis" (Boudart, 1956) stating that irrespective of the complexity of the catalytic surface, very often kinetic data can be represented by the Langmuir isotherm with assumptions far from reality. Hence, eq. (8) and eq. (9) should be considered merely as appropriate mathematical forms accounting for the kinetic data.

Boudart showed that many reactions can be represented by simplified reaction sequences when it is possible to assume or identify a most abundant surface intermediate (masi) and either a rate determining step (rds) or two irreversible steps (Boudart, 1972). On this basis, simplified kinetic expressions can easily be derived at.

In many situations, it is possible to assess the relevance of the kinetic constants. The active surface area may be determined by chemisorption and the activity may then be referred to the number of active sites in term of the turnover frequency. Also, the physical relevance of the equilibrium constants can be assessed by estimating the enthalpy and entropy of adsorption (Boudart et al. 1967). This simplified approach has proven powerful for many industrial reactions. Nevertheless, irrespective of high sophistication of the mathematical modelling of chemical reactors, most work was still based on semi-empirical Langmuir Hinshelwood kinetics (Froment et al. 1979).

However the rds as well as the masi may change, through the reactor as function of conversion or temperature. If it changes with conversion, it will also change inside the catalyst pellet, if the effectiveness factor is low. Also, one may speculate about the meaning of $*$ in the kinetic sequence (Roberts, 1992). The size of ensemble required may change with conversion and temperature. Also, the surface may be reconstructed under reaction conditions (Somorjai, 1991) depending on the composition of the gas at various positions in the reactor. The supported liquid phase catalyst for SO_2-oxidation for sulphuric acid may be considered as an extreme. The composition of the vanadium potassium salt melt changes not only through the reactor, but also inside the catalyst pellet (Villadsen et al., 1978). Also, the reaction may show oscillations on the surface as observed recently for CO-oxidation on palladium (Ehsasi et al., 1992). In conclusion,there are several reasons that the implied Langmuir Hinshelwood approach may break down.

1.2.3. *Steady State Kinetics and Surface Science. Microkinetic Analysis*: The classical way overcoming the situation described above is to use steady state kinetics with no assumption of a rate determining step. This approach easily leads to complicated equations with a number of constants difficult to determine in catalytic reactions (Christiansen, 1953). However, the progress in surface science has provided strong tools for estimating the kinetic constants of the various reaction steps.

For the **ammonia synthesis** (Stoltze et al., 1985), it was possible to deduct industrial rates from measurements of the sticking coefficient of nitrogen on iron, a potassium-promoted surface (Bozco et al. 1977) in ultra-high vacuum equipment. The adsorption of nitrogen was assumed as the rds and the equilibrium constants for the equilibrated steps were estimated from other measurements. The resulting correlation is shown in Fig. 1. Still, this analysis was based on the ideal Langmuir assumptions and the rate data were taken from initial sticking coefficients. Later, work has indicated that more complex correlations of rates with surface coverages of nitrogen and potassium may have to be considered (Tornqvist, 1991 and Fastrup, 1992).

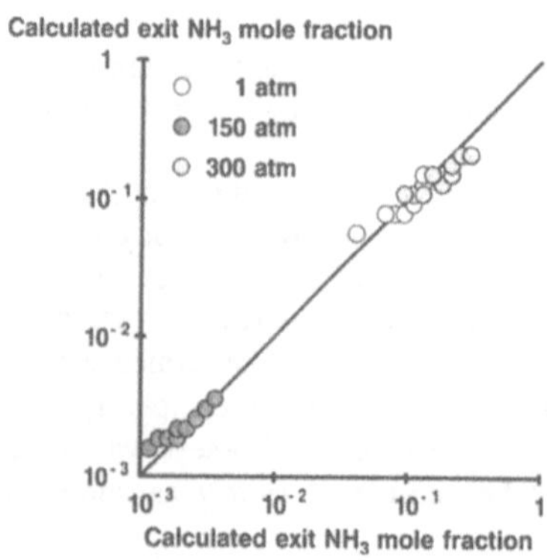

Fig. 1. Ammonia Synthesis. Calculated and Measured Conversions (Stoltze et al. 1985).

This example of bridging the gap between surface science at ultra-high vacuum and industrial conditions is based on a well-known process with a well-established rate determining step. For reactions with a less established basis, it may be helpful to use in-situ techniques where the catalyst structure or the surface intermediates are studied under reaction conditions.

This is illustrated by the **selective catalytic reduction** over a vanadium/titania catalyst.

$$NO + NH_3 + \frac{1}{4}O_2 = N_2 + 1\frac{1}{2} H_2O \quad (11)$$

This reaction was studied in a FTIR cell (Topsoe et al., 1991) as shown in Fig. 2. The unit consists of a feed system, the FTIR cell working as a catalytic reactor connected to the FTIR spectrometer, and a mass spectrometer for analysis of products.

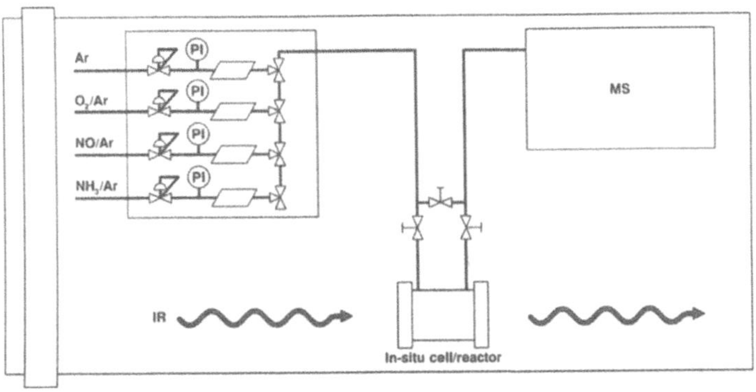

Fig. 2. In-situ FTIR for SCR-reaction (Topsoe et al., 1992).

Fig. 3 shows FTIR-spectra of the vanadium-titanic catalyst at 350 and 300°C, respectively (Topsoe et al, 1991). The ammonia absorption bands from Brondsted and Lewis acidic sites are present at both temperatures, but the band intensity is lower at 350°C than that at 300°C. This indicates that the same adsorbed ammonia species are present on the surface at the two temperatures, but that the steady state coverage is lower of the higher reaction temperature.

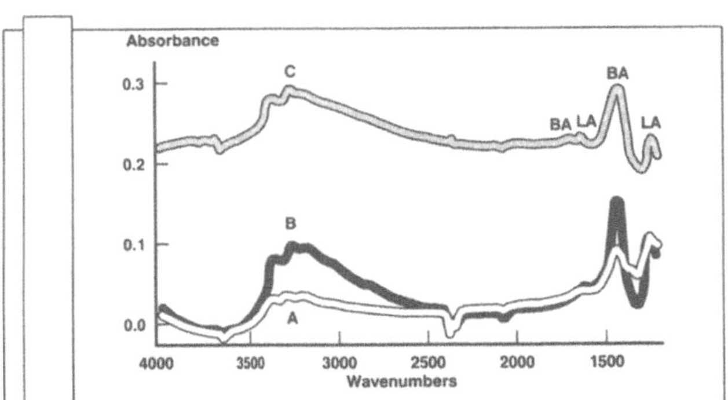

Fig. 3. FTIR Spectra of SCR-catalyst (Topsoe et al., 1992).

These observations were used to formulate a micro-kinetic model for the SCR-reaction involving a rate-determining step of gaseous NO reacting with adsorbed ammonia being in equilibrium with the gas phases (Dumesic et al., 1992).

$$\begin{array}{llll} NH_3 + * \rightleftharpoons NH_3* & \text{(fast)} & a & \\ NO + NH_3* \rightarrow \text{product} + * & \text{(slow)} & b & \end{array} \quad (12)$$

The model resulting in the rate expression:

$$r = k_2 \cdot p_{NO} \cdot \frac{K_1 \cdot p_{NH_3}}{1 + K_1 \cdot p_{NH_3}} \quad (13)$$

could predict the NO-conversions derived from the on-line MS-measurements

The input from surface science to micro-kinetics has made it meaningful to solve the steady state reaction equations, which is possible with modern computer techniques (Dumesic et al., 1987). Fig. 4 shows results from a micro-kinetic analysis of **cracking of isobutane** over a zeolite catalyst (Dumesic et al. in press). The pre-exponential factors in the mechanism were estimated using transition state and collision theory and these values were not adjusted in the subsequent analysis of the experimental data. Thermodynamic consistency was achieved by using statistical thermodynamics for estimating equilibrium constants (i.e. partition functions and transition state theory). The reaction proceeds via a complex network with 3 different pathways of 21 steps with 7 being essentially irreversible and kinetically significant. The relative rates of three pathways are responsible for the olefin/paraffin ratio in the product. Fig. 4 shows how the rates of the initial steps of the three pathways vary through the reactor. This trend and hence product selectivity could be influenced by changing the acidity of the catalyst.

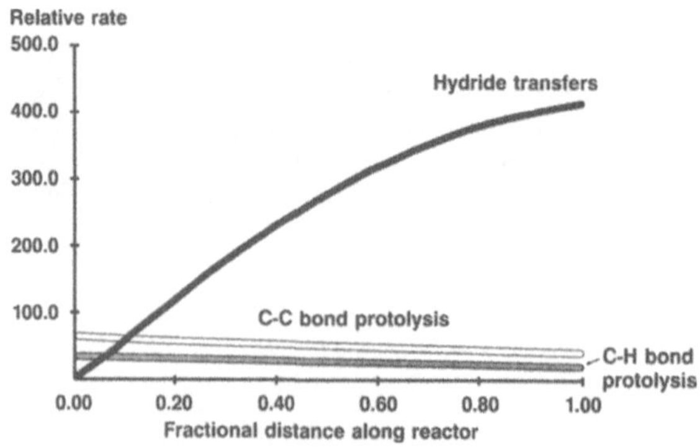

Fig. 4. Cracking of Isobutane over Zeolite. Rates of Three Pathways as Function of Distance from Reactor Inlet (Dumesic et al., 1992).

1.3. Kinetics and Environmental Catalysis

1.3.1. *Break-down of Simplified Approach*: The break-down of the simplified approach is evident in particular for processes in which one component should be completely removed down to ppm-level. This type of reactions are becoming more important in environmental processes such as selective catalytic reduction for removal of NO_x, catalytic combustion of carbon monoxide, hydrocarbons and other organic compounds in waste gas streams.

For **oxidation of carbon monoxide** on noble metals, the reaction order changes from minus one at high CO-coverages to first order at very low paretial pressures of carbon monoxide, which reflects a change in reaction mechanism (Boudart et al. 1984). Boudart's analysis showed that the availability of oxygen was critical at high CO-coverages meansing that the inhibited adsorption of oxygen was the rds. At low CO-coverages, the rate took place between adsorbed carbon monoxide and oxygen at the periphery of islands of adsorbed oxygen atoms with the surface reaction becoming the rds.

The simplified kinetics may well assist in establishing rate equations for the extreme conditions, but may hardly lead to a description of the wide transition zone. Also, Langmuir kinetics cannot explain the oscillations observed at very low pressures (Ehsasi et al., 1992).

The problem can be illustrated further by the well-known **methanation process** in ammonia plants, removing traces of carbon monoxide in the synthesis gas to below 5 ppm. The reaction which is carried out over a supported nickel catalyst may be represented by the following sequence:

$$\begin{array}{llr} H_2 + 2* = 2H* & a & \\ CO + * = CO* & b & (14) \\ CO* + 2* = C* + O* & c & \\ C* + H* = CH* = & d & \end{array}$$

It is generally accepted (Vannice, 1982) that the surface reaction of adsorbed carbon atoms and hydrogen atoms (step d) is rate determining which results in a kinetic expression of the form (15a). If steps c and d are assumed to be irreversible and C* to be the masi simple two-step kinetics results in the form of 15b.

$$r = \frac{r \cdot k \cdot p_{CO} \cdot p_{H_2}}{(1+K_1 \cdot p_{H_2} + K_2 \cdot p_{CO})} \qquad a \qquad (15)$$

$$r = \frac{k \cdot p_{CO}}{(1+K_3\, p_{CO}/p_{H_2})^2} \qquad b$$

However, at industrial conditions, the coverage of adsorbed carbon atoms is so large that the availability of adsorbed hydrogen atoms becomes rate limiting (Schoubye, 1969). As shown in Fig. 5, the reaction remains zero order with respect to carbon monoxide over a large range to become -0.5 at high CO-contents in the feed. Schoubye found that the data could be represented by the expression:

$$r = \frac{k(1+0.2 \cdot p_{H_2}^{0.2})}{(1+K\, p_{CO}/p_{H_2})^{0.5}} \qquad (16)$$

The simplified kinetics (15a & b) do not account for the slight positive reaction order 0.1-0.2 with respect to hydrogen increasing with the hydrogen pressure which was found over a wide pressure range (Schoubye, 1970a).

At very low concentrations of carbon monoxide, the adsorption of carbon monoxide becomes rate determining and the reaction should approach first order kinetics as also observed for the oxidation of carbon monoxide described above. Again, there is a change of rds throughout the reactor.

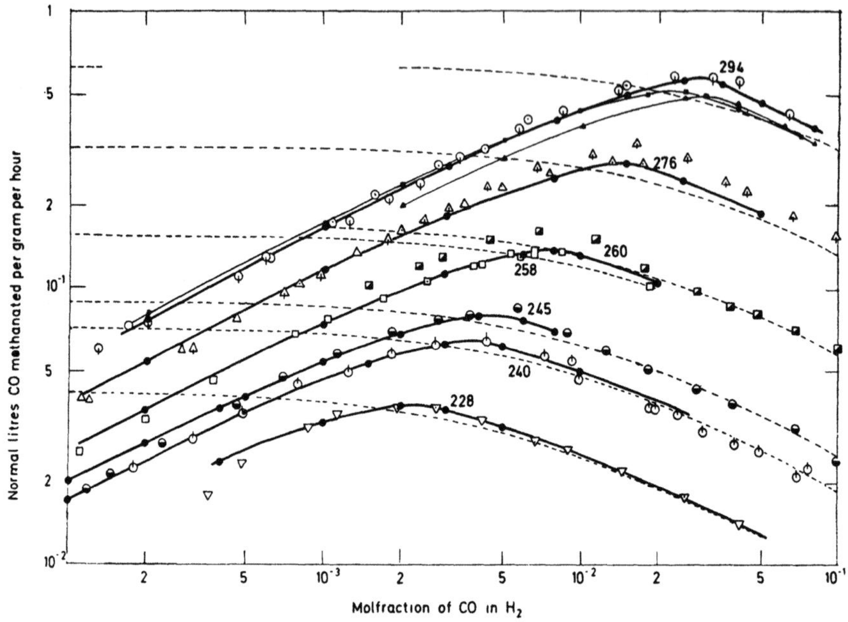

Fig. 5. Methanation on Nickel Catalyst (Schoubye, 1970b).
Dotted lines: data corrected for diffusion restrictions.

Fig. 5 indicates that the rate approaches 1st order with respect to carbon monoxide at very low concentrations of carbon monoxide. However, Schoubye showed that this observation was due to diffusion restrictions and when correcting for this as shown in Fig. 5, the reaction order remained zero down to CO contents of ca. 10 ppm which was the lower detection limit in his experiments (Schoubye, 1970b). Hence, it may not be possible to verify that the intrinsic rate approaches first order kinetics.

This discussion illustrates the complexity of dealing with the kinetics of reactions over a high range of conversion. It involves change of reaction mechanism throughout the reactor and a complex interaction with diffusion restrictions.

1.3.2. *Reaction Order and Effective Rates at Low Concentrations*: The interaction between diffusion restrictions and intrinsic kinetics becomes complex as the concentration of the reactants approaches zero. Table 1 shows how the impact on the effectiveness factor depends on the reaction order of the intrinsic reaction. A zero order reaction becomes diffusion limited, whereas a second order reaction of stoichiometric mixtures is determined by intrinsic kinetics. The relative rates have been estimated in Table 1 for a high conversion process assuming identical intrinsic rates at the reactor inlet. The actual rates were estimated assuming as an example $\eta = 0.3$ for the first order reaction. It is demonstrated that wrong kinetics can lead to estimated rates at high conversion being more than one size of order wrong.

TABLE 1: Relative Rates in High Conversion Processes.
Same Intrinsic Rate of Reactor Inlet. η = 0.3 for first order reaction.

Intrinsic Reaction Order	-1	0	1	2
Thiele modulus Φ_L 1) Effectiveness factor for C 0	$a \cdot C\sqrt{k_{-1}}$ 0	$a \cdot C\sqrt{k_0 C}$ 0	$a\sqrt{k}$ const.	$a\sqrt{k/C}$ 1.
Reactor Inlet X=0% - Intrinsic rate - Actual rate	 1.0 1.5	 1.0 1	 1.0 0.3	 1.0 0.1
Reactor Exit X=99% - Intrinsic rate - Actual rate	 100 0.02	 1 0.01	 0.01 0.003	 10^{-4} 10^{-4}

1) $\Phi_L = d_{p,s}\sqrt{k_i/D_{eff}} \cdot c^{(m-1)/2)} = a \cdot \sqrt{k_i/C^{1-n}}$

The **selective catalytic reduction** of NO by means of ammonia, eq. (11) may illustrate the complexity:

With surplus ammonia, the intrinsic rate should approach first order kinetics as also predicted by eq. (13). The reaction remains dominated by diffusion restrictions. However, since ammonia cannot be accepted in many flue gas cleaning processes, the reaction is carried out with stochiometric amounts of the reactants resulting in second order kinetics, when concentrations of the reactants approach zero (re. eq. (13)) and with the intrinsic kinetics being dominating (re. Table 1).

The consequences of assuming a wrong reaction order can have substantial impact on the estimated catalyst volume as illustrated in Fig. 6. The tentative conversion profiles are based on identical intrinsic rates at the inlet. The first order reaction remains first order irrespective of diffusion restrictions, whereas the effective reaction order of the second order reaction will start being less than two (ca. 1.5) and increasing to two as intrinsic kinetics takes over. The SCR-reaction with stoichiometric concentrations of reactants starts with a reaction order close to one (re. eq. (13)) and approaches second order.

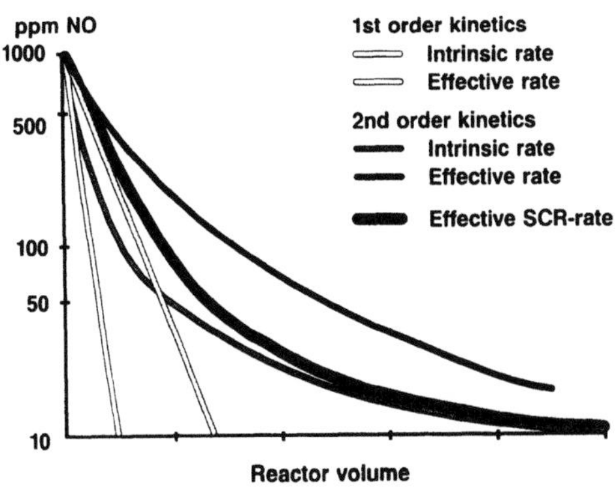

Fig. 6. Tentative Concentration Profiles assuming Different Reaction Orders (dotted lines: no diffusion restrictions)

The situation illustrated in Fig. 6 may be found in many environmental processes. It emphasizes the need to supplement differential kinetic studies with integral tests.

In general, reactor sizing becomes very critical also because at the very low concentrations of reactants, by-pass phenomena due to minor deviation from plug flow at reactor wall may play an important role.

For industrial methanation, the situation is simplified by a well-defined gas composition with surplus of hydrogen, but for instance in a reactor for selective catalytic reduction, where both reactants are present in ppm-amounts in the final part of the reactor, other components in the gas phase may dictate the concentration of surface species. This emphasizes the need for reaction expressions covering a much wider range of gas compositions in particular with only traces of reactants being present. Steady state kinetics may be required to describe the complex transitions described in Fig. 6.

1.4. The Need for Pilot Tests

For industrial application, the kinetics should be checked at integral conditions at pilot scale (i.e. full catalyst pellet size and industrial mass velocity) also to identify the impact of transport phenomena by-pass phenomena etc. as discussed above. Further, it is often at the pilot scale that the importance of secondary phenomena (poisoning, sintering) are recognized.

An example is the **high temperature methanation** for the manufacture of SNG. An economic process will allow a high temperature increase and minimum recycle resulting in a temperature profile as shown in Fig. 7. It would be difficult to assume that the reaction would proceed with the same rds and masi throughout the temperature range of reaction zone.

Much work reported in literature (Vannice, 1982) has dealt with studying the initial activity of the reaction and how to improve the sulphur resistance. One study (Della Batta, 1976) based on single stream tests at low conversion and at gradientless conditions claimed that sulphur resistance could be gained at conditions forming nickel carbide. Water was found to destroy this advantage at industrial conditions. Hence, the results are of little relevance to industrial operation, because water would be formed immediately even inside the first catalyst pellet. Also, sulphur can easily be removed to less than 5 ppb by conventional desulphurization technique (Rostrup-Nielsen et al., 1985) and little would be gained by improving the sulphur resistance to f.inst. 1 ppm of sulphur.

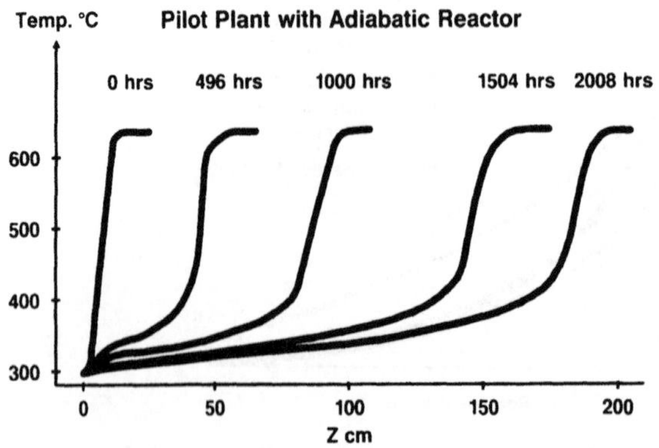

Fig. 7. High Temperature Methanation. Typical Temperature Profiles for Catalyst Exposed to Deactivation by beta-carbon. Pilot Tests at Mass Velocity 14000 kg/m^2/h (Gierlich et al., 1980).

The relevant question for the catalyst design was to study how to maintain low temperature activity on catalysts which had been exposed to long-time exposure to the high temperature (650°C), and secondly to assess the importance of formation of non-reactive carbon species (Rostrup-Nielsen et al., 1985), (beta-carbon (McCarthy et al., 1979)) at the inlet due to the high partial pressure carbon monoxide (re. Fig. 7).

It is evident that the size of reactor was not determined from kinetics, but from the impact of sintering and deactivation from beta-carbon and sulphur poisoning. Hence, the many initial rate measurements on the methanation reaction were of minor relevance to industrial operation.

In general, many industrial reactors are sized on basis of secondary phenomena such as heat transfer, flow distribution and deactivation.

1.5. New Approach in Industrial Research on Catalysis

The progress in surface science has surely resulted in a much better understanding of the catalytic phenomena, and computer methods have resulted in a much stronger modelling of complex reactions. This has formed the basis for a new approach in industrial catalysis (Rostrup-Nielsen, 1992).

Bench-scale → Fundamental Studies (in-situ)

Bench-scale → Large scale pilot operation

The traditional bench-scale testing simulates the process, but it gives neither results of fundamental character and seldom data useful for scale-up to industrial conditions. By applying fundamental methods, and a rapid scale-up to semi-industrial pilot scale instead of focussing on bench-scale, the R&D becomes more expensive, but far more effective.

The traditional scheme has been to start doing basic research for creating ideas for applied research and to complete the applied research before start of development and engineering. A more effective approach is to work with overlapping phases meaning that the applied research is supplemented by fundamental studies.

In the following, this will be illustrated by an example showing the interaction of the empirical approach and the deductive micro-kinetics.

2. STEAM REFORMING AT CARBON-FREE CONDITIONS

2.1. State-of-the-art

Steam reforming is a well-known process for converting natural gas into synthesis gas (Rostrup-Nielsen et al., 1992):

$$\begin{array}{llll} CH_4 + H_2O = CO + 3H_2 & -206\ \text{kJ/mol} & \text{a} & \\ CO + H_2O = CO_2 + H_2 & +41\ \text{kJ/mol} & \text{b} & (17) \\ CH_4 + H_2O = CO_2 + 4H_2 & -165\ \text{kJ/mol} & \text{c (a+b)} & \\ CO_2 + CH_4 = 2CO + 2H_2 & -247\ \text{kJ/mol} & \text{d (a-b)} & \end{array}$$

In order to supply the heat for the overall endothermic reaction, the catalyst is loaded into a number of high alloy tubes inside the tubular reformer. The balance between heat input through the reformer tube walls and the heat consumption in the endothermic reaction is the central problem in steam reforming. Even a slight increase in the maximum tube wall temperature may result in a serious decline of the expected tube lifetime. Catalyst particle size and shape should be optimized to achieve maximum activity and maximum heat transfer while minimizing the pressure drop. The high mass velocities necessitate a relatively large catalyst particle size to obtain a low pressure drop across the catalyst bed. The steam reforming catalyst is normally based on nickel. The catalyst properties are dictated by the severe operating conditions, i.e. temperatures at the inlet of ca. 450-500°C, and at the exit 800-950°C and steam partial pressures up to 30 bar. The activity depends on the nickel surface area and the nickel crystals will sinter quickly above the Tammann temperature being 590°C for nickel (Rostrup-Nielsen 1984).

2.2. Mechanism and Kinetics

The chemisorption of methane on nickel involves the direct breakage of a C-H bond with no adsorbed molecule as precursor (Ceyer et al., 1987 and Alstrup et al., 1990a). This requires that the molecule has sufficient energy to overcome a barrier of about 52 kJ/mol and that it heats the free nickel site with free neighbor sites. The adsorbed CH_3* species is converted into adsorbed carbon atoms by a step-wise dehydrogenation (Alstrup et al., 1992).

$$CH_4 = CH_3* \rightarrow CH_2* \rightarrow CH* = C* \tag{18}$$

The rate of adsorption is in an order of magnitude higher on the open (110) nickel surface plane than on the dense (111) surface plane (Beebe et al., 1987). Adsorbed oxygen atoms were found to have no impact the chemisorption of methane except for decreasing the number of sites available (Alstrup et al., 1990b).

Alstrup et al. were able to establish a micro-kinetic model for formation of carbon on the nickel surface exposed to methane and hydrogen (Alstrup et al., 1992) Equilibrium constants for the various dehydrogenation steps were estimated from measurements of the vibration frequences of the intermediate surfaces on nickel as well as a value for the carbon nickel bond energy leaving the rate constants for irreversible steps as variable parameters. It was found that the best fit was obtained when the second and first dehydrogenation step was assumed to be irreversible as shown in eq. (18).

The kinetics of the steam reforming of methane was studied by the Temkin group on nickel foils. Early work (Bodrov et al., 1964) assumed the methane adsorption to be rate determining, which fits well to the general assumption of a first order with respect to methane. Later work avoided the discussion of a rds by using a steady state approach. The effective stoichiometric number for the reaction was determined to be 1 from measurements close to equilibrium (Khomenko et al., 1970) and on basis of Temkin's general kinetic identity (Boudart et al., 1984), the following rate expression was obtained for the temperature range, 470-700°C (Khomenko et al., 1971):

$$r = \frac{k p_{CH_4} \cdot p_{H_2O} \, (1 - Q_R/K_p)}{f(p_{H_2O}, p_{H_2})(1 + K_{H_2O} \cdot (P_{H_2O}/p_{H_2}))} \tag{19}$$

in which Q_R and K_p are the reaction and $f(p_{H_2}, p_{H_2O})$ is a polynominal in p_{H_2}, p_{H_2O}. Eq. (19) contains five temperature dependent constants.

Recently, the Froment group (Xu et al., 1989) established a complex Langmuir/Hinshelwood expression by the classical approach on basis of 280 measurements but for a narrow range of parameters with temperatures 500- 575°C, pressures 3-10 bar and H_2O/CH_4 of 3-5. A number of rate equations were established on basis of a mechanism involving 13 steps, when assuming one of the steps being the single rds.

Expressions of the form of eq. (8) were derived independently for reaction 14b and 14c being the sum of 17a & b and for the reverse step 17e. The rate of reaction 17a could be represented by:

$$r = \frac{\dfrac{k\, p_{CH_4} p_{H_2O}}{p_{H_2}^{2.5}}(1-Q_R/K_p)}{1+K_{CO}p_{CO}+K_{H_2}+K_{CH_4}p_{CH_4}+K_{H_2O}(p_{H_2O}/p_{H_2})} \qquad (20)$$

Eq. (20) contains four temperature dependent equilibrium constants. The model implies an unlikely negative heat of adsorption of steam. This may be a result of the narrow range of H_2O/CH_4 and temperatures studied. As a consequence, eq. (20) was not able to predict the decrease in rate when replacing steam by carbon dioxide (eq. 17d) at high temperatures (Bak Hansen et al., submitted).

Simplified two-step kinetics (Rostrup-Nielsen 1984) may involve the following steps:

$$\begin{aligned} &CH_4 + n* \rightarrow CH_x - *_n + 0.5\,(4\text{-}x)\,H_2 && a \\ &CH_x - *_n + O\text{-}* \rightarrow CO + 0.5\,X\,H_2 + (n+1)* && b \\ &H_2O + * = O - * + H_2 && c \\ &H_2 + 2* = 2H - * && d \end{aligned} \qquad (21)$$

assuming steps a and b to be irreversible and O* to be the masi. On this basis, the following rate expression is obtained:

$$r = \frac{k p_{CH_4}(1-Q_R/K_p)}{(1 + K_{H_2}\sqrt{p_{H_2}} + K_{H_2O}\dfrac{p_{H_2O}}{p_{H_2}})} \qquad (22)$$

This expression contains the same elements as eq.(20). It can explain the retarding effect of hydrogen observed in some studies and as well the retarding effect of steam in other studies. Eq. (22) may also explain that the negative reaction order with respect to steam may vary from catalyst to catalyst as reflected by the size of the equilibrium constant K_{H2O} for steam adsorption (step c). Thus, the addition of alkali or the use of magnesia as support results in enhanced steam adsorption which is useful in eliminating carbon formation (Rostrup-Nielsen 1984) This can be explained by replacing step 21c by:

$$\begin{aligned} &H_2O + \text{support} = H_2O\text{-support} && e \\ &H_2O\text{-support} + * = O* + H_2 && f \\ &H_2O + * = O* + H_2 && g \end{aligned} \qquad (23)$$

However, since steam is also adsorbed directly on the nickel surface (step 23g), the over-all equilibrium cannot be influenced by spill-over from the support without violating the principle of microscopic reversibility (Rostrup-Nielsen 1984). Therefore, it is doubtful to assume that steam adsorption (step. 21c or 23 e-g) is at equilibrium and a dynamic spill-over stiuation is more likely as illustrated in **Fig. 8.** The relative rates shown in Fig. 8 may change significantly with temperature. Therefore, the simplified assumptions of one rds or two irreversible steps behind eq. (20) and eq. (22) seems doubtful. Also, it is questionable whether the complicated expressions of eqs. (19) and (20) can be justified when the constants are not based on micro-kinetic measurements - in particular in view of the complexity of the steam reformer with its strong concentration and temperature gradients.

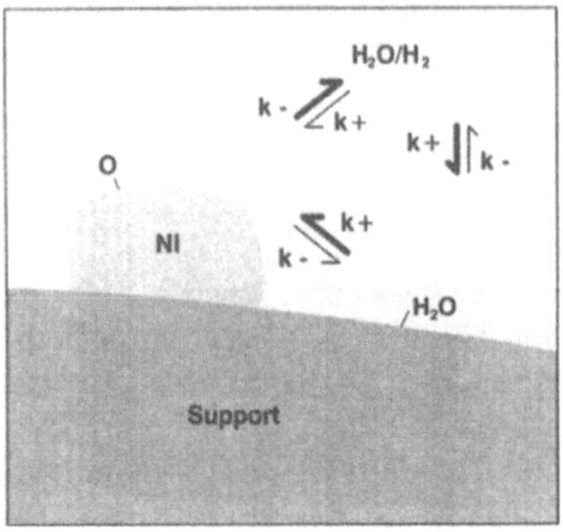

Fig. 8. Interaction of Steam with Steam Reforming Catalyst (Rostrup-Nielsen, 1984).

2.3. Empirical Approach

Industrial steam reforming is characterized by significant temperature and concentration gradients because of the relatively large catalyst pellets applied and because of the high heat of reaction. Fig. 9 shows measured temperature gradients and calculated concentration gradients in a Berty-type reactor (Rostrup-Nielsen et al., 1988).

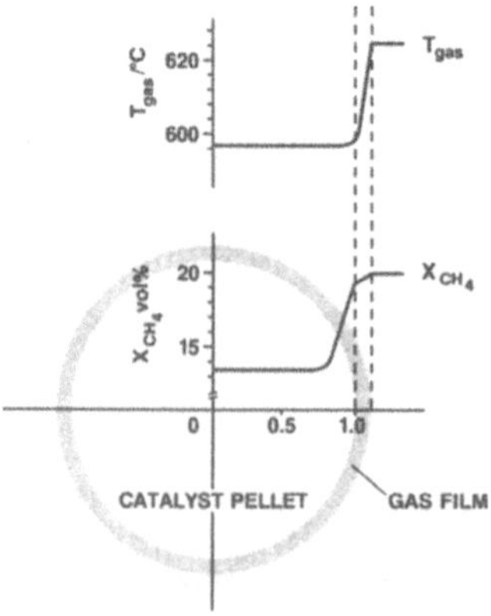

Fig. 9. Steam Reforming of Methane in Berty-type Reactor. Concentration and Temperature Profiles (Rostrup-Nielsen, et al. 1988).

The temperature drop over the gas film around the catalyst is more than 10°C, whereas the concentration gradient is significant in the outer shell of the catalyst pellet. The temperature gradient makes laboratory measurements on catalyst pellets questionable if the catalyst pellet temperature is not measured directly for instance by installing a thermocouple in the pellet.

Industrial reformer conditions were simulated (Rostrup-Nielsen, 1984) by a computer program, REACTOR, based on a one-dimensional heterogeneous model for the calculation of mass and heat transfer restrictions inside the catalyst particles and in the gas film surrounding the particles. The effectiveness factor, η, decreases from the reactor inlet with typical values below 0.1 as illustrated in Fig. 10. Mass transport restrictions are related mainly to pore diffusion, whereas the heat transfer restrictions are located in the gas film as observed in the Berty reactor (Fig. 9) (Rostrup-Nielsen et al., 1988). The strong endothermic reaction results in a temperature drop of 5-10°C over the gas film in spite of a high mass velocity, corresponding to Reynolds numbers ca. 10.000. As shown in Fig. 10, the bulk gas arrives rapidly at a composition close to the equilibrium when passing down the reformer tube which results in a reaction rate being proportional to the axial temperature gradient, dT/dZ, as a driving force (Rostrup-Nielsen, 1984).

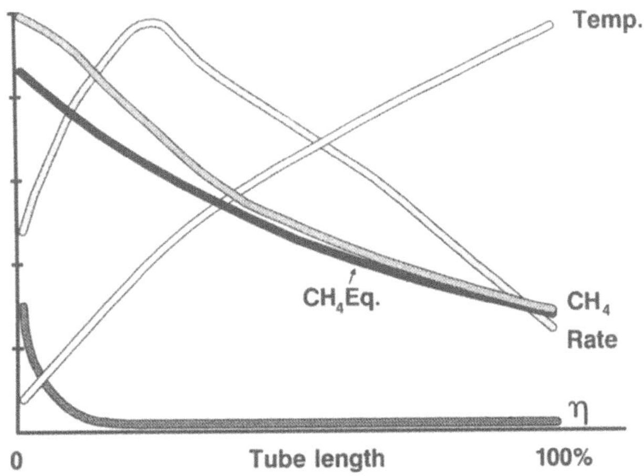

Fig. 10. Tubular Steam Reformer. Conversion and Rate Profiles (Rostrup-Nielsen, 1984)

As shown in Fig. 10, this results in a decreasing reaction rate in the lower part of the tube in spite of the increasing temperature as actual temperature gradient decreases.

The low effectiveness factor in a tubular reformer means that the effective rate can be expressed by a modified rate equation as eq. (2). This type of equation was based in a two-dimensional reactor model (Froment et al., 1979, Rostrup-Nielsen et al., 1978) which is able to simulate the actual as well as radial temperature and conversion profiles in the reformer. The radial temperature gradients are significant with high values of dT/dR at the tube wall, whereas radial concentration in gradients are insignificant. This means that close to tube wall, there would be significant deviation from equilibrium as illustrated in Fig. 11. The radial dispersion and reaction rates are too small to compensate for the steep gradient.

The advanced computer model is of little value without knowledge of the involved transport constants i.e. the radial Peclet number for mass and heat transfer (Froment et al., 1979) and the heat transfer coefficient at the tube wall. These parameters were determined in experiments carried out in a full-size monotube pilot plant equipped with thermocouples determining axial and radial temperature profiles as well as a number of samples points for gas analyses (Rostrup-Nielsen et al., 1988) . With the adjusted parameters, it was possible to predict gas compositions and temperatures at any position in the reformer tube. This empirical approach provided the basis for safe engineering of the coupling of heat transfer and chemical reaction in tubular reformers.

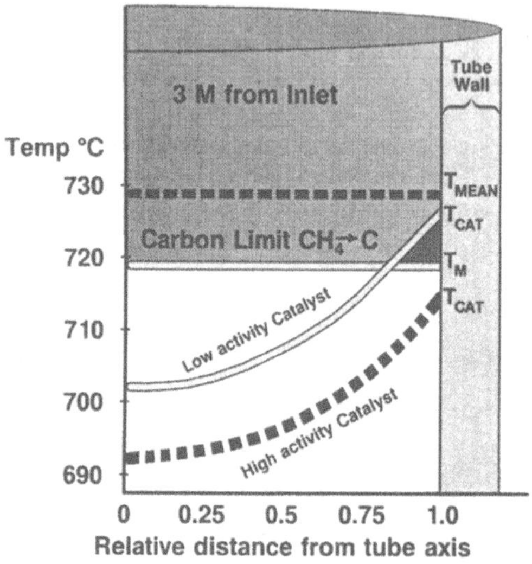

Fig. 11. Tubular Steam Reformer. Radial Gradients and Potential for Carbon Formation. Potential for Carbon for $T_{Cat} > T_M$. T_M is temperature for which $p^2_{H2}/p_{CH4} = K_p$ (24a) (Rostrup-Nielsen, 1984).

2.4. Selectivity for Carbon-free Operation

Steam reforming involves the risk of carbon formation by decomposition of methane or by the Boudouard reaction:

$$\begin{aligned} CH_4 &= C + 2H_2 \quad &a \\ 2CO &= C + CO_2 \quad &b \end{aligned} \tag{24}$$

The carbon grows as a fiber (whisker) with nickel whisker at the top (Rostrup-Nielsen 1984). Adsorbed carbon atoms which are not reacting to gaseous molecules are dissolved in the nickel crystal and solid carbon nucleates on the non-exposed side of the crystal, preferably from the dense (111) surface planes (Alstrup, 1988).

Carbon formation must be eliminated because it may lead to break-down of the catalyst and build-up of carbon deposits and broken catalyst may cause partial or total blockage of the reformer tube. For traditional applications for steam reforming, this problem is solved by using surplus steam, whereas the increased interest in C_1-chemistry needs synthesis gases with low H_2/CO ratios being manufactured by using low steam carbon ratios or by using CO_2 (Reaction (17d)) instead of steam (Bak Hansen et al., submitted and Rostrup-Nielsen, 1988). This increases the risk of carbon formation.

The selectivity problem in the reformer can be expressed as follows:

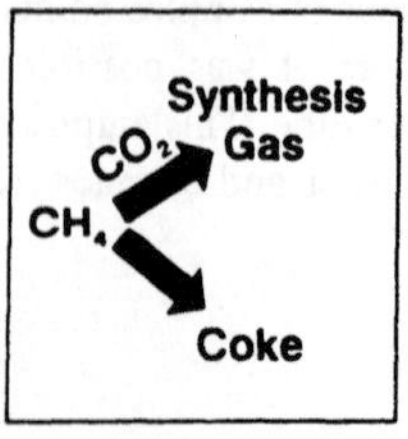

(25)

Methane may decompose to carbon instead of reacting with steam or CO_2, if the gas above the catalyst shows affinity for carbon formation according to the reaction (24). This can be expressed by the carbon activity a_c^{eq} expressed by:

$$a_c^{eq} = K_c \cdot \frac{p_{CH_4}}{p^2_{H_2}} \tag{26}$$

in which K_c is equilibrium constant for the methane decomposition (24a).

A safe, conservative design would be to require $a_c^{eq} < 1$ at any position in the reformer tube (Rostrup-Nielsen 1984). This is equivalent to require $T_{Cat} < T_M$ as shown in Fig. 11.

However, at the inlet of the reformer with almost no hydrogen present, a_c^{eq} is much larger than one ($T_{Cat} > T_M$), meaning that there is a potential for carbon formation, but in practice no carbon is formed. Therefore, an important question for the industrial operation is: what is the maximum inlet temperature allowed before carbon will be formed. A situation may exist at other positions in the reformer close to the tube wall as illustrated in Fig. 11. When is operation possible if $a_c^{eq} > 1$?

Whether a carbon is formed, is a question whether the **steady state** activity of carbon $a_c^s > 1$ (Rostrup-Nielsen, 1984). a_c^s can be expressed by balancing the rate of carbon formation without presence of steam with the rate of gasification of the adsorbed carbon atoms.

$$\begin{aligned} CH_4 &= C* + 2H_2 && \text{a} \\ C* + H_2O &= CO + H_2 + * && \text{b} \end{aligned} \tag{27}$$

$$\begin{aligned} r_c &= k_c \cdot f_1(p_{CH_4},p_{H_2}) - \frac{k_c}{K_c}\, f_2(p_{H_2}) \cdot a_c \\ &= \frac{k_c}{K_c} \cdot f_3(p_{CH_4},\, p_{H_2}) \cdot (a_c^{eq} - a_c) \end{aligned} \tag{28}$$

$$r_g = k_q \cdot g\,(p_{H_2O},p_{H_2}) \cdot a_c$$

a_c is equal to a_c^s for $r_c = r_g$, i.e.

$$a_c^s = \frac{a_c^{eq}}{1 + \dfrac{k_g \cdot g(p_{H_2O}p_{H_2})}{k_c/K_c \cdot f_3(p_{CH_4},p_{H_2})}} \tag{29}$$

This means that $a_c^s \leq a_c^{eq}$. Carbon is not formed for $a_c^s < 1$. The rate of reaction (27a) can be expressed by the rate expression found by by microkinetic analysis (Alstrup et al., 1992). The rate equation for reaction (27b) has not been published on a microkinetic basis. The functionf(p_{H_2O}, p_{H_2}) depends on the type of catalyst in question, reflected by the "effective" adsorption constant K_{H_2O} for steam as discussed above.

It is evident that an expression like eq. (29) should be derived on basis of a microkinetic analysis involving a detailed study of the individual reaction steps. This expression might then be used directly to analyze the situation at each position in the reformer tube when the gas composition is known from the two-dimensional homogeneous model described above.

3. CONCLUSIONS

The steam reforming example illustrates how reaction engineering and the empirical approach as well as micro-kinetics play a role in industrial catalysis. In general, there is a trend from bench-scale testing to on the one-hand side, fundamental studies giving more precise information of the individual reaction steps and on the other hand to pilot experiments at industrial mass velocity and full catalyst particle to establish the constants in the advanced chemical reactor engineering models.

In some situations, as for instance for ammonia synthesis, it is possible to bridge the gap from micro-kinetics to industrial design. In other situations like steam reforming, the solutions have to be approached from both ends with the micro-kinetic being in particular important to describe selectivity problems.

Microkinetic analysis makes the use of steady state kinetics meaningful. This may become of great industrial importance in environmental catalysis where one or two components are to the removed to the ppm level and where it is likely that simple kinetic assumptions may easily break down.

Acknowledgement
The author thanks Mr. P. Schoubye, Dr. He. Topsoe for useful discussions.

References

Albright, L.F., Crynes, B.L., Corcoran, W.H. Eds. (1983) "Pyrolysis, Theory and Practice", Academic Press, New York.
Alstrup, I. (1988), J.Catal. 109, 241.
Alstrup, I., Chorkendorff, I., and Ullmann, S. (1990a), Surface Sci. 227, 191.
Alstrup, I., and Chorkendorff, I., and Ullmann, S. (1990b), Surf.Sci. 234, 79.
Alstrup, I., Tavares, M.T. (1992), J.Catal. 135, 147.
Bak Hansen, J.-H. and Rostrup-Nielsen, J.R., J.Catal. (submitted).
Beebe, Jr., T.P., Goodman, D.W., Kay, B.D. and Yates, T.J. Jr., (1987) J.Chem.Phys. 87, 2305.
Bodrov, I.M., Apel'baum, L.O., Temkin, M.I. (1964), Kinet.Katal. 5, 696.
Bohlbro, H. (1966) "An Investigation on the Kinetics of the Conversion of Carbon Monoxide with Water Vapor over Iron Oxide Based Catalysts", Gjellerup, Copenhagen.
Boudart, M. (1956) AIChE-Journal, 2, 62.
Boudart, M. (1966) "Kinetics of Chemical Processes", Prentice-Hall, Englewood Cliffs.
Boudart, M., Mears, D.E., and Vannice, M.A. (1967), Ind.Chim. Belge 32 (1), 281.
Boudart, M. (1972) AIChE-Journal, 58, 465.
Boudart, M. and Djégo-Mariadasson, G. (1984) "Kinetics of Heterogeneous Catalytic Reactions", Princeton Univ. Press, Princeton, p. 105 & 79.
Bozso, F., Ertl, G., Grunze, M., and Weiss, M. (1977), J.Catal. 49, 18.
Ceyer, S.T, Yang, Q.Y., Lee, M.B., Beckerle, J.D. and Johnson, A.D. (1987), Stud.Surf.Sci.Catal. 36, 51.
Christiansen, J.A. (1953), Adv.Catal. 5, 311.
de Blieck, J.L., Grosens, A.G. (1971), Hydrocarbon Process. 50, (3) 76.
Della Betta, R.A., and Shelef, M. (1976) Preprint, ASC Div.Fuel.Chem. 21 (4), 43.
Dumesic, J.A., Milligan, B.A., Grepp, L.A. , Balse, V.R., Sarnowski, K.T., Beall, C.E., Kataoka, T., Rudd, D.F., Trevino, A.A. (1987), Ind.Eng.Chem.Res. 26, 1399.
Dumesic, J.A., Rudd, D.F., Aparicio, L.M. Rekoske, J.E., Trevino, A.A. (1992) "The Micro-kinetics

of Heterogeneous Catalysis", ACS-Series (in press)

Dumesic, J.A., Topsoe, N-Y., Slabiak, T., Morsing, P., Clausen, B.S., Tornqvist, E., and Topsoe, He. (1992), Proc. 10th Int. Congress on Catalysis, Budapest (preprint O91).

Ehsasi, M., Beedan, M., Karpowicz, A., Christmann, K., and Block, H.J. (1992), Proc. 10th Int. Congr. on Catalysis, Budapest 013.

Fastrup, B. (unpublished results)

Froment, G.F. and Bischoff, K.B. (1979), "Chemical Reactor Analysis and Design", John Wiley & Sons, New York.

Gierlich, H.H., Fremery, M., Skov, A. and Rostrup-Nielsen, J.R. (1980), Stud.Surf.Sci.Catal. 6, 459.

Holton, R.D., and Trimm, D.L. (1977), Proc. 6th Int. Congr. on Catalysis, London 1976. The Chemical Society, London , Vol. 2, p. 727.

Hougen, O.A., and Watson, K.M. (1959) "Chemical Process Principles", Vol. 3, J.Wiley, New York, p. 963.

Khomenko, A.A., Apel'baum, L.O., Shub, F.S., Temkin, M.I. (1970), Kinet.Katal. 11, 1480.

Khomenko, A.A., Apel'baum, L.O., Shub, F.S., Snagovskii, Y.S., Temkin, M.I. (1971), Kinet.Katal. 12, 423.

Kwan, T. (1956), J.Phys.Chem. 60, 1033.

Lunsford, J. (1992), Proc. 10th Int. Congr. on Catalysis, Budapest, P.5

McCarthy, J.G., and Wise, H. (1979), 57, 406.

Roberts, M.W. Catal. Today (1992), 12, 501.

Rostrup-Nielsen, J.R. and Christiansen, L.J. (1978), Proc. 6th Symposio Ibero-Americano de Catalise, Rio de Janeiro, p. 1615.

Rostrup-Nielsen, J.R. (1984), "Catalytic Steam Reforming", Catalysis, Science and Technology (Anderson, J.R. and Boudart, M. eds.), Springer Verlag, Berlin, Vol. 5.

Rostrup-Nielsen, J.R. and Hojlund Nielsen, P.E. (1985) in "Deactivation and Poisoning of Catalysts", Marcel Dekker, New York, chapt. 6.

Rostrup-Nielsen, J.R., Christiansen, L.J., Bak Hansen, J.-H. (1988a), Appl.Catal. 43, 287.

Rostrup-Nielsen, J.R. (1988b), Stud.Surf.Sci.Catal. 36, 73.

Rostrup-Nielsen, J.R., Dybkjær, I., Christiansen, L.J. (1992) "Steam Reforming. Opportunities and Limits of the Technology" in: NATO ASI Chemical Reactor Technology for Environmentally Safe Reactors and Products (de Lasa, H. ed.), Kluwer Academic Press, Dortrecht (in press).

Rostrup-Nielsen, J.R. (1992) Catalysis Today, (in press).

Russell, B. (1961) "Our Knowledge of the External World", G. Allen & Urwin, London, p.242.

Satterfield, C.N. (1970) "Mass Transfer in Heterogeneous Catalysis", MIT Press, Cambridge Mass.

Satterfield, C.N. (1980) "Heterogeneous Catalysis in Practice", McGraw Hill, New York, Chapt. 11.

Schoubye, P. (1969), J.Catal. 14, 238.

Schoubye, P. (1970a), J.Catal. 18, 118.

Schoubye, P. (1970b), Chem.Eng.World 5, 40.

Somorjai, C.A. (1991). Catal.Lett. 9, 311.

Stoltze, P. and Norskov, J.K. (1985), Phys.Rev.Lett. 55, 2502.

Tornqvist, E. (1991). Catal.Lett. 8, 359.

Tottrup, P.B. (1982), Appl.Catal. 4, 377.

Topsoe, N-Y, and Topsoe, H. (1991), Catal.Today, 9, 77.

Vannice, M.A. (1982) in "Catalysis, Science and Technology" (Anderson, J.R. and Boudart, M. eds.), Vol. 3, Springer, Berlin, Chap. 3.

Villadsen, J., and Livbjerg, H. (1978), Catal.Rev.-Sci.Eng. 17 (2), 203.

Weller, S. (1956) AIChE-Journal, 2, 59.

Xu, J., and Froment, G.F. (1989), AIChE Journal 35, 88.

Zdonik, S.B., Green, E.J., Haller, L.P. (1967), Oil Gas J. 65, 96.

List of Symbols

a_c	Activity of carbon
a_c^{eq}	Activity of carbon relative to methane decomposition equilibrium: $a_c^{eq} = p_{CH_4}/p^2_{H_2} \cdot K_c$
a_c^s	Carbon activity at steady state
A,B	Adsorption coefficient
C	Concentration of key component
$d_{p,s}$	Equivalent sphere diameter 6(vol/surf)
D_{eff}	Effective diffusion coefficient (m^2/h)
f	Function
F	Flow of key component
g	Function
k	Rate constant
m,n	Exponent
r	Rate
r_i	Intrinsic rate
r_{eff}	Effective rate
K	Equilibrium constant
Q_R	Reaction quotient
R	Radial distance
t	Time
T	Temperature
W	Weight of catalyst
Z	Axial distance from reactor inlet
X	Conversion
X_{eq}	Conversion of equilibrium
$\alpha,\beta\gamma$	Exponents
η	Effectiveness factor
Φ_L	Thiele modulus, $\phi = \frac{1}{d_{p,s}}\sqrt{\frac{D_{eff}}{k_i} C^{m-1}}$

Session Five, *Chemical Kinetics and Chemical Engineering*

Rapporteur J.C. VEDRINE
Institut de Recherches sur la Catalyse, CNRS
2 Av. Albert Einstein
F-69626 Villeurbanne Cedex
France

1. *INTRODUCTION*

This session was mainly devoted to kinetic analysis based on surface science data (Waugh), transient kinetics in catalysis by metals (Frennet) and microkinetics in heterogeneous catalysis (Rostrup-Nielsen).

2. *ANALYSIS*

The main subject of this discussion was how to reconcile surface science and laboratory scale data, with the actual catalytic reaction under industrial conditions. Two of the participants, Waugh and Rostrup-Nielsen were from industry, with Frennet representing academia. Simple reactions as ammonia synthesis, water gas shift, carbon monoxide hydrogenation or oxidation, methane oxidation, H_2-D_2 and CH_4-D_2 exchange were chosen.

Waugh has really achieved a 'tour de force' by showing the methodolgy to follow to predict catalytic reaction rates of ammonia synthesis at elevated temperatures, conversions and pressures, starting from kinetic data obtained under high vacuum conditions and with surface science methodolgy. For instance it was shown that data from surfce science-type study, eg. on iron (111), (110) and (100) single crystals at very low coverage could be extrapolated to high pressure, high coverage, high temperature and K_2O/Al_2O_3 polycrystalline iron catalysts. The main message is that, provided kinetics measurements are done correctly, extrapolation to industrial conditions can be done satisfactorily. Waugh has explained the role of K as a promoter and has insisted on the influence of the preexponential factor. The difference in values as presented was so huge (five orders of magnitude) that everyone could be easily convinced.

Frennet has developed kinetic models for exchange reaction on metals by using in parallel three different approaches:

i) Use of labelled compound;
ii) Kinetic studies at stationary state;
iii) Kinetics under transient conditions.

Dilution of Rh metal by a second metal such as Ag or Cu, and the theory of ensembles in metals were described for characterising the mechanism of such reactions on metal as hydrogenation, hydrogenolysis and exchange.

Rostrup-Nielsen has presented microkinetic data and *in-situ* kinetics with special emphasis on the caution the experimentalist must exercise, namely even if experimental data fit very well with a proposed mechanism, this does not necessarily mean that the mechanism proposed is right. This is obvious but should be recalled.

R. W. Joyner and R. A. van Santen (eds.), Elementary Reaction Steps in Heterogeneous Catalysis, 461–463.

3. *DISCUSSION*

A question was asked concerning the meaning of the asterisk, *, which is often used to indicate an active site. This is obviously a simplification when the actual site is unknown, but it indicates that adsorption and activated intermediate are needed.

A question arose concerning environmental problems and the importance in the future of a better (100%!) selectivity in many catalytic processes. The panel insisted on the need for improved kinetics determination and extrapolation to industrial scale.

Another point was raised concerning new concepts necessary not only considering activity or site strength, but also broadening our view in the catalysis field. The key is no longer activity, since this could always be solved, but selectivity as said above and poisoning phenomenon (sintering of metals, ß carbon type deposits).

Another question concerned the relevance of elementary steps for further applications. Can they really be applied? For the three speakers, relevance is high since if you can trust the data you can determine the catalyst changes. Detailed kinetics are necessary! For instance, diluting metallic rhodium by silver enhances the desorption, but does not change the reaction mechanism, and decreases the C-C bond breaking probability. Van Santen asked how close to '*in-situ*' determination are kinetics data at the laboratory scale. Waugh clearly indicated that if good rates are obtained, extrapolation to the industrial scale can be done. Rostrup-Nielsen noted that in the last two decades characterisation of catalysts has dominated, while now we really need good kineticians. Basset remarked that he has been trained to kinetics for determining reaction mechanism, but he had the feeling that the chemical engineering approach concerns mass and heat transfers and hydrodynamic problems rather than chemistry and reaction mechanism. Rostrup-Nielsen answered that for him chemical engineering involves both approaches.

For Kuipers, exploratory research is necessary for new and more complex reactions since optimisation of simple reactions (ammonia, sulphuric acid, etc) has already been done a long time ago. For Haller, relative composition in $CO+O_2$ reaction is not important, while it is important for $CO+H_2$. Can we understand clearly the reasons in addition to that due to the wider range of products in the later case? For Rostrup-Nielsen microkinetic measurements, as described by Dumesic [1], may answer the question.

For Centi, surface science and industrial scale problems were clearly related in the three presentations. However, other important problems when kinetics play a role are deactivation and activation, and these were not discussed.

For Somorjai and Haller elementary reactions steps have to be characterised by molecular beam experiments (to extrapolate from low pressure to high coverage regime), time resolved transient, and angular resolved techniques.

Expertise exists now in other scientific communities. We have to move to and understand their concepts. The empirical approach of Froment is certainly to be followed at the first stage. Waugh responded to the widely differing questions by noting that really a bridge is necessary and we have to think how each reaction molecule can be activated when coverage increases.

Someone asked if models of chemical engineering reaction other than Monte Carlo calculations do exist. For Rostrup-Nielsen the answer is yes and one of his colleagues is using other approaches for methane activation. Van Santen analysed the situation now. It was thought twenty to thirty years ago that kinetics could tell us a lot about the active sites. However, scientists were disappointed by the results and so interest decreased. Now with new molecular approaches it is felt that one may go further.

This last remark could be taken as a main conclusion of the discussion since it brings some realistic hopes to a very difficult but certainly important problem of determining the usefulness of (micro) kinetic measurement for both reaction mechanism determination and industrial scale extrapolation.

Disappointments arose from the difficulty of correlating the nature of active sites and the kinetics of the reaction. This resulted in a declining interest in kinetic studies and in modelling at university level. However, now, the new molecular approach of the active sites and of its action with reactant molecules allow us to go further in the relationship between active sites and catalytic reaction. It offers a new prospective for the future. It was also emphasised that time resolved transient species techniques, angular resolved techniques and molecular beam chemical reaction do have their own scientific communities. The surface science and catalysis communities have to move towards the former ones to improve our approach of catalytic reaction and kinetics modelling.

4. *CONCLUSIONS*

Kinetic studies have been in strong decline for the last decade since their role in understanding the catalyst behaviour and in extrapolation to an industrial scale turn out to be rather low. The importance of the three presentations was in showing that detailed surface studies and determination of kinetic parameters are of prime importance as far as the measurements are good and mathematical expressions are relevant. It appears that for the last decade the characterisation of the catalyst, unfortunately rarely in *in-situ* conditions, outweighed the kinetic studies. It is also obvious that proposals for reaction mechanism taken from kinetic studies may well not be unambiguous, ie. other proposals may fit the experimental data and thus make the problem more difficult to solve. However, the presentations by the three speakers show that solutions are possible and even allows extrapolation to industrial conditions.

These kinetic approaches corresponded to a steady state regime (even in transient kinetics) and should thus be extended to activation and to deactivation, which are of prime importance in industrial processes.

Chemical reactions on an industrial scale not only involve kinetic laws, as described in the presentations, but also chemical engineering requirements, particularly mass and heat transfers (effectiveness, Thiele modulus, etc), size, shape and function of reactors. Their implication in the kinetics is important and obviously very complex. Exploratory studies, including chemical engineering requirements and extension to more complex reactions than those described by the three authors, should represent the future in catalysis.

[1] J.A. Dumesic, D.F. Rudd, L.M. Aparicio, J.E. Rekoske and A.M. Trevino, *The Microkinetics of Heterogeneous Catalysis,* ACS, Washington DC, 1993.

TOWARD NEXT CENTURY'S INDUSTRIAL CATALYSIS

J. J. LEROU and L. E. MANZER
Corporate Catalysis Center
Central Science & Engineering
Du Pont Company
Wilmington, Delaware 19880-0262 U.S.A.

Introduction

Environmental issues will be determining the future of the chemical industry around the globe. The signs are very clear: it is becoming increasingly difficult to obtain permits, eliminate waste, construct incinerators and receive and ship toxic materials. The enviro-economics will guide corporate decisions in technology and will aim to improve productivity. Prices will be set to reflect the cost of using environmental resources, and this will clearly affect the relative importance of environmental vs variable costs for current and future plants. Catalysis will play a key role in providing solutions to these environmental challenges by designing processes with 100% yield (by whatever process); by improved catalyst recovery, regeneration and recycle; by heterogenization of homogeneous catalysts; by increasing importance of chiral pharmaceutical and agrochemicals; by polymer recycle and by synthesizing environmentally safer products. Several of these aspects will be highlighted in this paper.

High Yield-Low Waste Processes

Elimination of byproducts and process waste is becoming a major issue and will clearly determine the viability of future chemical processes. Those processes which strive for zero emissions and very high process yields at the lowest cost will be winners in the 21st century. A first example involves the the synthesis of HCFC-141b, a result of Du Pont's work on the development of alternatives to the ozone depleting chlorofluorocarbons (CFCs). It is a replacement for several solvent blends and azeotropes, and for blowing foams:

$$\underset{VCl_2}{CH{=}CCl_2} + HF \longrightarrow \underset{141b}{CH_3CFCl_2} + \underset{142b}{CH_3CF_2Cl} + \underset{143b}{CH_3CF_3} \qquad (1)$$

Historically, low yields to HCFC-141b have been reported since the reaction is thermodynamically in favor of HFC-143a. A specially prepared AlF_3 catalyst was discovered [1] which is remarkably selective in producing HCFC-141b in nearly quantitative yield. The reactor was divided into two stages such that a high degree of conversion of VCl_2 was obtained in the first step. The second stage conditions are adjusted to provide a liquid film on the catalyst. As a result, the yield of HCFC-41b is 99.5%, with <500 ppm of VCl_2 remaining in the product. As a result, it is possible to go directly from the reactor into a

R. W. Joyner and R. A. van Santen (eds.), Elementary Reaction Steps in Heterogeneous Catalysis, 465–472.

product tank with no refining and recycle. The product directly from the reactor meets the purity specifications defined by an international toxicity panel.

The second process which is representative of this trend involves the manufacture of hydrogen peroxide. For many years Du Pont [2] has been developing catalytic technology which directly combines hydrogen and oxygen:

$$H_2 + O_2 \longrightarrow H_2O_2 \longrightarrow H_2O \qquad (2)$$

The advantage of this process, compared to the conventional alkyl anthraquinone process is that the only byproduct is water, so that, in essence there is zero waste. The remarkable effect of catalysis is very clear. Pd is a know peroxide decomposition catalyst but by using Pd and Pd alloys [3], in aqueous media, with selective promoters, it is possible to produce hydrogen peroxide concentrations of 10-45% with selectivities to H_2O_2 of >70%. However, not all processes produce these remarkable results so that effective utilization of waste is an attractive option.

Waste Minimization and Resourcing

Unfortunately, modern technology does not allow to economically achieve develop chemical processes that have 100% yield and operate under conditions that require minimal energy. The challenge is to effectively manage process byproducts and waste through minimization or conversion to salable products (resourcing). Unused products have traditionally be disposed of by incineration, atmospheric venting or underground injection. Historically there has been little incentive to recover and utilize these byproducts. Increasing environmental awareness and concern is now providing the initiative to reconsider these disposal options. A good example comes from Du Pont's adiponitrile process which is used in the manufacture of hexamethylenediamine, a precursor to nylon 6.6. The main process steps are shown in Figure 1.

<u>Main Chemistry</u>

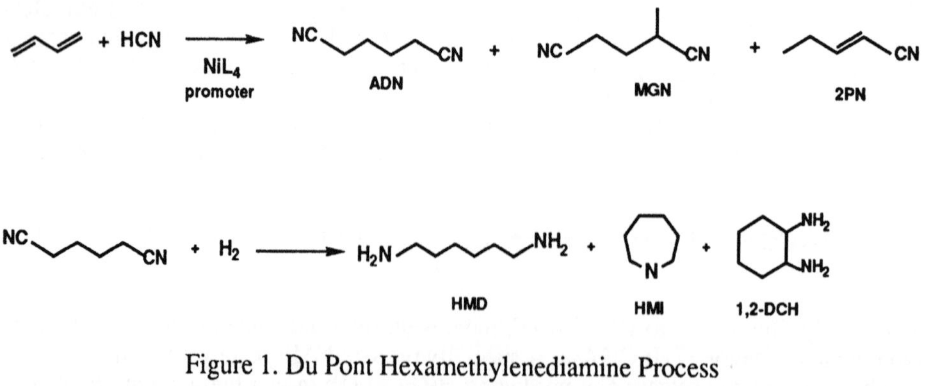

Figure 1. Du Pont Hexamethylenediamine Process

Two moles of hydrogen cyanide are added to butadiene, in a two-step process, using a zero valent nickel phosphite catalyst to produce the linear (ADN) and branched (MGN) C_6-dinitriles in very high yield. A small amount of the conjugated, undesirable, 2-pentenenitrile (2PN) has historically been separated and incinerated. The ADN is catalytically hydrogenated to hexamethylenediamine (HMD) is very high yield, although small amounts of other cyclic amines such as HMI, and 1,2-DCH are produced. Several years ago we began searching for opportunities to convert these minor byproducts into usable products instead of incinerating them.

The linear, conjugated, 2PN, was found to react readily with amines, (HX) to give 2-cyanobutylated amines as shown in equation 3.

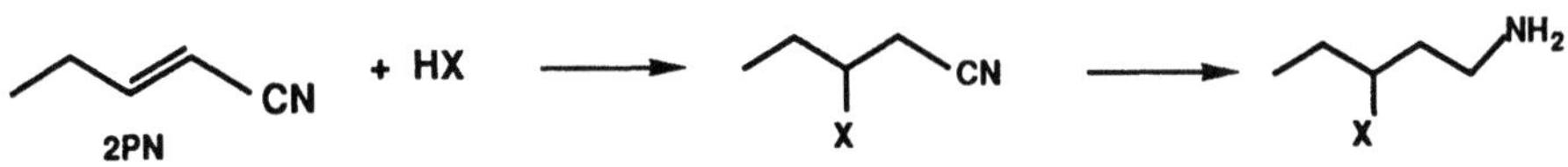

(3)

Hydrogenation of the aminonitrile with a Raney® catalyst leads to a family of branched diamines. Because of the branching, most of the aminonitriles and diamines are liquids at low temperature and have low freezing points. They have found markets as comonomers or curatives, since they lower polymer viscosity, crystallinity, and glass transition temperature.

Catalytic hydrogenation of MGN with a Raney catalyst gives the branched-amine methylpentamethylenediamine, MPMD, or 3-methylpiperidine (3MP) (4):

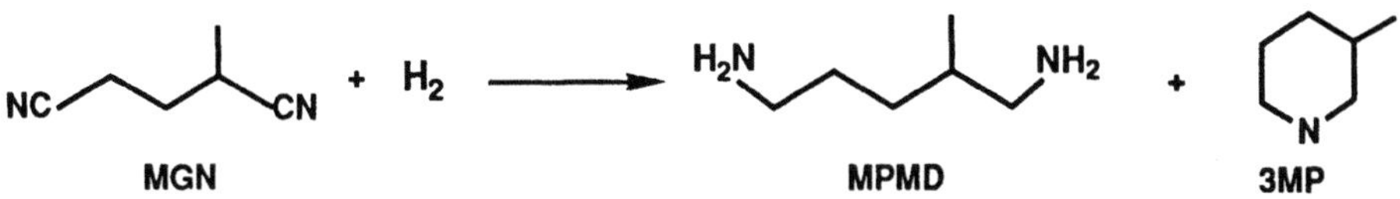

(4)

The product is dependent on conditions and choice of catalyst. The MPMD was initially isolated from plant streams to develop the market. Many applications were found as a polymer additive in urethanes and epoxies to reduce crystallinity and viscosity, as a water treatment chemical and as a monomer is polyamides. This material was once incinerated for its fuel value, but after a 10 year development effort it's use has grown so that a dedicated commercial facility is now used to produce this valuable coproduct, on purpose.

Two byproducts, 1,2-diaminocyclohexane (DCH) and hexamethyleneimine (HMI) are produced during ADN hydrogenation [4], resulting from a cyclic deamination reaction. These compounds were initially isolated from plant streams and slowly introduced into the marketplace. Through a long development program they have found growing applications as epoxy-curing agents and agricultural intermediates. A new dedicated commercial catalytic process has been started up to satisfy this new market.

Many other examples from Du Pont's polyester, Kevlar®, and adipic acid processes could be presented to show the clear value of catalytic technology to convert waste byproducts to valuable coproducts. Even a 1% yield loss, in a billion lb/year process can generate 10 million lbs/yr of material that can be burned as a byproduct or converted to a higher value in use coproduct. Waste resourcing makes sense if there is a very strong, long term corporate commitment to do so. This commitment by Du Pont for the HMD process, for example has taken over 15 years. These efforts across the entire coproduct business, currently generate several hundred million dollars in sales. Further efforts are expected to expand this new business to $ 1 billion/year in sales by 2000, an effort that relies very heavily on catalysis and process R&D.

Asymmetric Catalysis

Everyone is aware of the thalidamide tragedy in which one optically active isomer had good therapeutic effects and the other resulted in birth defects. One study [5] predicts that the trend in manufacture of optically active drugs over the next decade will increase, as shown below:

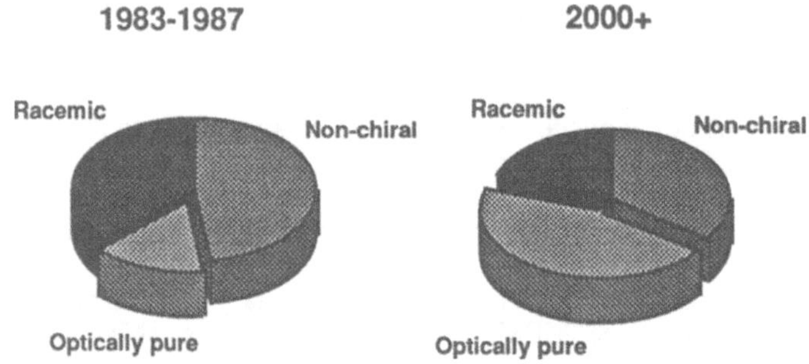

Figure 2: Trends in Optically Active Drugs

One example from Du Pont R&D [6] comes from our work on asymmetric hydrocyanation. The world's 5th largest prescription drug is Naproxen, which is popular anti-inflammatory compound. Asymmetric hydrocyanation technology offers the potential to produce a precursor to Naproxen (5):

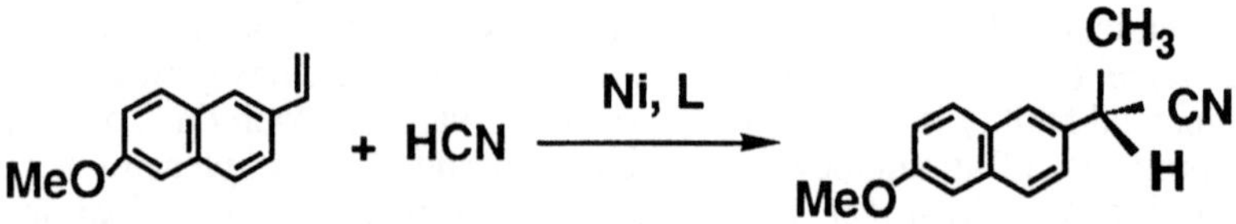

(5)

After considerable research, we settled on the use of chiral carbohydrate ligands (Figure 3) since their chemistry is well defined, they have several

Figure 3: Chiral Phosphinite Ligands for Hydrocyanation

asymmetric centers that can be adjusted to improve product selectivity, various substituents can be added to change catalytic selectivity and by converting them into phosphinite ligands, steric and electronic effects can be easily studied. A variety of substituents were studied and the effect was quite remarkable. Steric effects were minimal while electron withdrawing groups produced high ee's for the Naproxen precursor nitrile. The best reported results are >90% ee at 25°C. Over 4000 turnovers of catalyst have been shown, without deactivation at 100% conversion. After recrystallization, over 99% optical purity of the nitrile was observed. Although this is only one example, there are many other fine examples of asymmetric catalysis being developed that illustrate the power of catalysis in this area, particularly for drugs and longer term, chiral agrichemicals.

Hazardous and Toxic Materials Management

Hazardous and toxic materials such as HCN, HF, HCl, Cl_2, acrylonitrile, formaldehyde, ethylene oxide, sulfuric acid and phosgene, for example, are essential building reagents in the chemical industry since they often contain functionality or reactivity required for further chemical reactions. Future business practices must avoid or minimize the inventory and transportation of these materials.

Methylisocyanate (MIC) is familiar to us as a result of the tragic incident at Bhopal, India. It was produced by the phosgenation of methylamine:

$$CH_3NH_2 + COCl_2 \longrightarrow CH_3NCO + HCl \qquad (6)$$

As a consumer of MIC, Du Pont was concerned over the use and storage of this toxic material. Prior to Bhopal, research was initiated on a new process that would produce MIC from less hazardous materials and minimize its handling and storage. The proprietary [7] catalytic oxidative-dehydrogenation process, shown in equations 7-8 was discovered:

$$CH_3NH_2 + CO \longrightarrow CH_3NHCHO \qquad (7)$$

$$CH_3NHCHO + O_2 \longrightarrow CH_3NCO \qquad (8)$$

This innovative research has resulted in a commercial Du Pont process that makes MIC and converts it *in-situ* to an agrichemical product. Consequently, the potential for exposure is greatly reduced. This trend in *in-situ* manufacture and derivatization is clearly the way of the future for hazardous chemicals.

Another example involves HCN. Du Pont is a large producer of HCN for internal use and external sales. The nitrile functionality is extremely versatile in organic synthesis since it is easily converted to amides, acids, amines, esters, etc as shown in Figure 4.

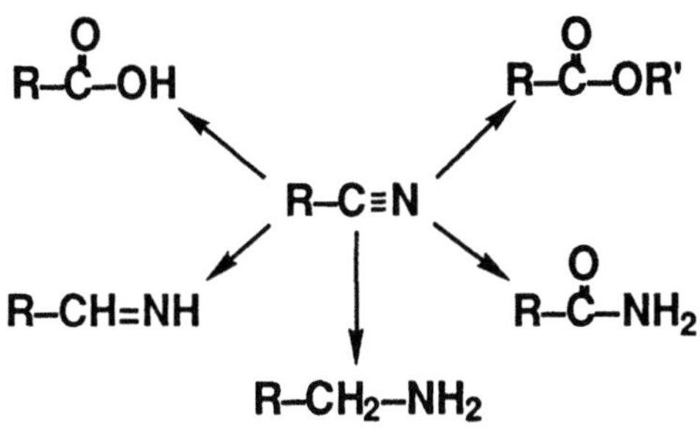

Figure 4. Versatility of Organic Nitriles

Conversion of the HCN, at our manufacturing sites would provide an environmentally safer option. We are using our proprietary hydrocyanation technology (similar to that shown in Figure 1) to prepare nitriles for transportation and customer functionalization. A vast number of nitriles have been produced by catalytic homogeneous hydrocyanation of mono, di, tri or polyolefins. Some representative examples are shown in Figure 5. The mono, di and tri nitriles are currently being evaluated

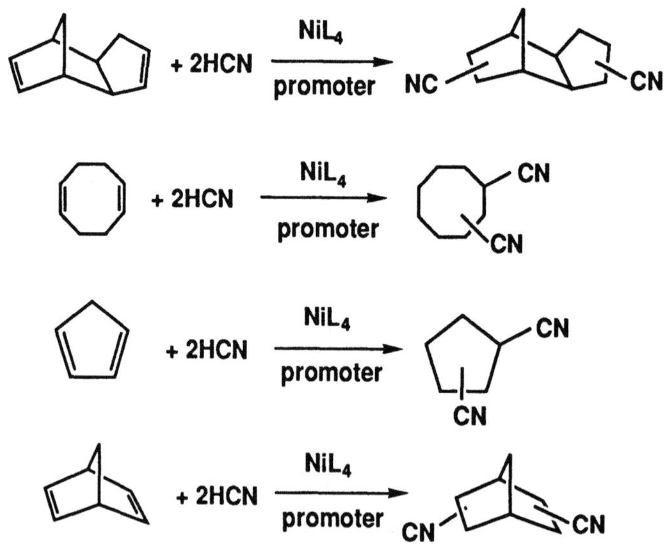

Figure 5: Examples of Diene Hydrocyanation

in the market place. Non-phosgene routes to isocyanates, and use of solid acids to avoid HF and H_2SO_4 as alkylation catalysts are other examples of research in progress to further minimize the use of hazardous materials.

Environmentally-Safer Products

Although products are introduced into the market to serve societal needs, their impact on the environment is not always predictable. The use of tetraethyl lead in gasoline provided a high octane gasoline for many years. However, lead has now been phased out in certain parts of the world, in favor of environmentally safer oxygenated organics such as methyl-t-butyl ether (MTBE). New catalytic technology is providing more isobutylene and MTBE for this large volume chemical. Another recent example [8-9] involves the recognition that ozone is being depleted by man-made chlorocarbons such as methylchloroform, carbon tetrachloride and chlorofluorocarbons (CFCs). Although these chemicals have served society very well it was not until atmospheric science developed during the late 1980's provided scientific evidence that they were causing significant ozone depletion (nearly 50 years after they were first introduced). As a result, industry has responded rapidly and is currently developing and commercializing safer products. Some of the products currently under development by the CFC industry to replace chlorofluorocarbons are shown in Table 1.

Table 1. Potential CFC Substitutes

Market	Current CFC	CFC-Alternative
Refrigerants	CFC-12 (CF_2Cl_2)	HFC-134a (CF_3CFH_2) HCFC-22 (CHF_2Cl) HFC-32 (CH_2F_2) HFC-125 (CF_3CF_2H) HCFC-124 (CF_3CHFCl) HFC-152a (CH_3CHF_2) Blends/Azeotropes
Blowing Agents	CFC-11 ($CFCl_3$)	HCFC-141b (CH_3CFCl_2) HCFC-123 (CF_3CHCl_2) HCFC-22 (CHF_2Cl) Blends/Azeotropes
Cleaning Agents	CFC-113 ($CF_2ClCFCl_2$)	Blends/Azeotropes New Compounds

These new products are much more complex than the CFCs they are replacing and require much more complicated catalytic technology. Most CFCs are produced in a single catalytic step while alternatives such as HFC-134a can require 2-5 complex catalytic steps as shown in Figure 6.

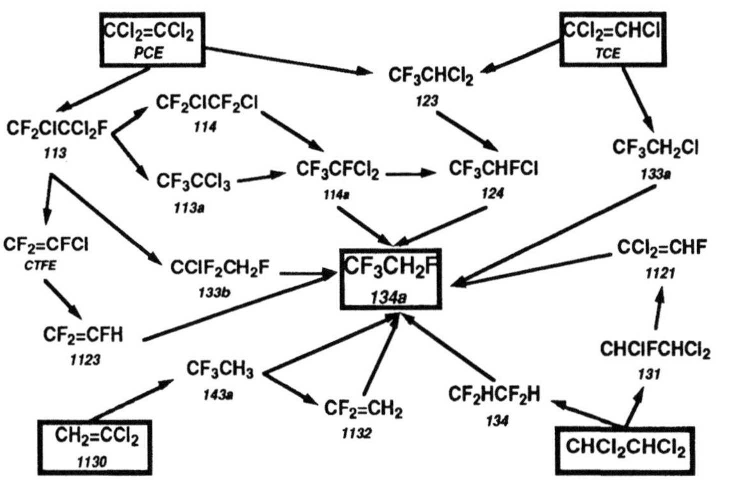

Figure 6: Potential Routes to HFC-134a

The key to their reduced ozone and global warming potentials is that they contain hydrogen atoms. As a result they have significantly shorter atmospheric lifetime because they are less stable. However, this also creates problems in their synthesis because they decompose readily, resulting in rapid catalyst deactivation. Longer lived catalysts have subsequently been developed. A detailed summary [10] of the catalytic chemistry reported for the preparation of the most significant CFC alternatives has been reported. Without rapid development of these new catalytic processes, CFCs would continue to be produced, resulting in further ozone depletion. As a result of Du Pont's effects to date, over $ 400 million have been spent and projected costs for full conversion of CFCs to alternatives is estimated to be about $ 1 billion dollars.

Similar process development is underway to find environmentally-safer substitutes for a variety of chemicals, polymers and products currently used by society.

Conclusions

The importance of catalysis and process research for developing environmentally safer processes and products for the future was demonstrated. This will become a condition for staying in business and those companies which commit resources will succeed. This provides an excellent opportunity for collaborations between government, academic and industrial laboratories. The long range research necessary to discover new chemistry and develop mechanistic understandings of current industrial processes can be exciting areas for joint collaboration. Universities and industry must work together to make this process easier.

References:

1. Ridley, M. G. et al. (1992) U.S. Patent (Du Pont): 5,105,033
2. Gosser, L. W.,(1989) U.S. Patent (Du Pont): 4,889,705
3. Gosser, L. W. and Schwartz, J. T. (1989) U.S. Patent (Du Pont): 4,832,938
4. Atadan, E. M. (1990) U.S. Patent (Du Pont): 4,937,336
5. Laird, T. (1989) Chemistry and Industry, 366
6. Casalnuova, A. and Rajanbabu T. V. (1992) U.S. Patent (Du Pont): 5,175,335
7. Heinsohn, G. E. and Rao, V. N. M. (1985) U.S. Patent (Du Pont): 4,537,726
8. Manzer, L. E., (1990) Science, 249, July, 31
9. Manzer, L. E., (1992) Catalysis Today, 13, 13
10. Manzer, L. E. and Rao, V. N. M., (1993) Advances in Catalysis (in press)

ELEMENTARY REACTION STEPS IN HETEROGENEOUS CATALYSIS: A WORKSHOP OVERVIEW

Gary L. Haller
Yale University
Department of Chemical Engineering
New Haven, CT 06520

George W. Coulston
E. I. duPont de Nemours and Co.
Experimental Station
Wilmington, DE 1988-0262

ABSTRACT. A brief synopsis of the workshop from an academic point of view is given. This includes expansion on the conventional definition of elementary step and mechanism. Some generalizations on the nature of heterogeneous catalytic sites are given with examples taken mostly from workshop presentations. This is followed by a discussion of approaches to catalyst characterization and future needs to identify elementary steps and associated mechanisms.

1. Introduction

1.1. STATEMENT OF OBJECTIVES

As outlined by Joyner at the beginning of the workshop*, our objective was to have a full discussion of how we understand and influence reaction mechanism. This question was to be addressed from the point of view of catalytic reactivity, surface science/catalysis, physico-chemical studies of reaction mechanism, chemical kinetics and theoretical studies. Our hope was to identify *key questions* which need to be addressed over the next five to ten year period and, if possible, to cast these in some appropriate form to allow the community of surface science and catalysis researchers to seek appropriate solutions. Joyner reminded us that all descriptions of catalytic events are approximations. Sometimes we use an extended-solid picture for our descriptions and, at other times, a molecular picture is used. We need to marry these approaches in such a way that they will allow us to achieve a more realistic description of elementary reactions steps understanding that all approximations or analogies, by definition, will be found wanting when compared against reality.

1.2. ABOUT ELEMENTARY REACTIONS STEPS AND MECHANISMS

We are all aware that the conventional definition of an elementary chemical reaction step is one that occurs as written. Usually, we are first introduced to this concept in a kinetics course and thus the implication is that if the reaction is an elementary one, we may write down the rate

* *see page 1.*

R. W. Joyner and R. A. van Santen (eds.), Elementary Reaction Steps in Heterogeneous Catalysis, 473–483.

expression by inspection by applying the law of mass action. (For heterogeneous catalysis, the definition is equivalent to assuming that one or more of the species in the reaction has sufficient surface mobility such that the law of mass action applies.) Indeed, sometimes one turns this around and says that if the rate expression can be written down based upon stoichiometry, the reaction is elementary. A corollary, however, would then be that one can prove the validity of a mechanism simply by showing that a suitable rate expression is derivable from it and this is known not to be the case based on the mathematical principle of uniqueness. The gas phase oxidation of NO is an example. The stoichiometric reaction:-

$$2NO + O_2 \rightarrow 2NO_2, \tag{1}$$

can be used to derive the correct rate expression, but the reaction step in (1) is not elementary. The rudimentary mechanism is actually the sequence of elementary steps shown in (2):-

$$NO + NO \rightarrow (NO)_2, \tag{2}$$

$$(NO)_2 + O_2 \rightarrow 2NO_2$$

Secondly, the inverse definition is wrong because it would not allow us to get beyond the *rudimentary mechanism*, a sequence of simple elementary steps that accomplishes the overall reaction as shown in (2) above (see Nieuwenhuys for a number of rudimentary mechanisms). A rudimentary mechanism provides a kinetic expression which may be useful for reactor design (see Rostrup-Nielsen for elaboration on this point). In turn, the kinetics constrain our imagination with regard to possible mechanisms, which must be consistent with the observed kinetics. However, understanding at the level of the rudimentary mechanism is unlikely to reveal the pressure points that would allow us to alter the activity or selectivity of a catalyst for a given reaction.

One of the authors has described one sort of mechanistic hierarchy in which each level of increased sophistication presumes a more complete knowledge of the mechanism [1]. While the levels of sophistication are arbitrary, for heterogeneous catalysis, a construction of six levels is quite natural. It is understood that all of the details included in lower levels are included in higher levels so that we arrive at a listing of incremental levels of knowledge added: i) rudimentary mechanism, ii) molecular structure of adsorbed species, iii) description of adsorption site structure, iv) description of the site environment, v) site-to-site interaction and vi) a full dynamic description of the reaction. The rudimentary mechanism has already been defined. We have reasonable confidence in our descriptions of several rudimentary mechanisms and, based on various vibrational spectroscopies (see Somorjai), some knowledge of the molecular structure of many adsorbed molecules which might be intermediates of reaction. The structure of chemisorbed CO comes immediately to mind here with the molecular picture of σ bonding to metal sites and back-bonding between the d atomic orbitals of the metal and the π^* orbitals of CO. Using the same example, we can claim some knowledge of the site structure for CO, at least with regard to atop or bridging between two (or three) metal atoms. By site environment, we have in mind the next nearest neighbor structure around the site, e.g., if CO is bridge bonded to two metal atoms, the reactivity of the CO is known to be different when these atoms are in a low index surface such as the (111) terrace of Pt in contrast to a step site [2]. By site interaction, we imply some information about how different entities on the surface interact, e.g., formation of islands of adsorbate, or how the reactants move together to complete a reaction step, e.g., surface diffusion. We have already noted that applying the law of mass action to surface species requires that one or more of the reactants have a diffusion rate high enough to sustain the reaction. Finally, we reach the level of sophistication where we may presume to say something about the dynamics of the reaction, which we have defined and discussed in another chapter of this book. Reaction dynamics at surfaces, from a theoretical point of view, is also the topic of Holloway's contribution.

To restate our objective somewhat more lyrically, an elementary reaction step in heterogeneous catalysis is the authentic dance of the atoms on the real potential energy stage. If we could see this dance clearly, we might then be able to choreograph the dance. This is our ultimate intent, but it is unlikely to be achieved in the time horizon we are considering. Our more modest goal is to identify some aspects of elementary steps in heterogeneous catalysis that can lead to mechanistic progress toward this goal.

2. Active Sites

From the comments made above on the levels of understanding of mechanism, it should be inferred that we can both write down the rudimentary mechanism and investigate some aspects of the molecular structure of adsorbed intermediates, while identifying the active site by the usual representation, an asterisk, *. We are in the habit of using this presentation for the "site" even when there are two (or more) species involved, e.g., adsorbed CO and O atoms in CO oxidation on metals, even when it is clear that the site for the two species cannot be the same. Clearly, the more sophisticated formulation of all elementary reaction steps in heterogeneous catalysis must involve a picture of the active site, but it is just at this point that we begin to stumble using the present techniques at our disposal.

2.1. SITE STRUCTURE AND DENSITY

Perhaps it is useful to begin by discussing what might be the most important aspects of sites to be characterized. Of course, we would like to know as much as we can about both the nature of the sites and their concentration. To understand the chemistry, the nature or structure would appear to be more important. However, most of the time we put a premium on knowing site density since that will allow us to compare catalysts between two laboratories. It is understood that almost always we have to be content with a measurement of site density which does allow a quantitative comparison between laboratories, but almost certainly has little to do with the actual site density involved in the reaction. That is, we usually measure the sites by a chemisorption probe which is assumed to have a 1:1 stoichiometry with the sites, e.g., H_2 dissociation on metals at room temperature. We are aware of the fiction involved here since there is ample evidence that the stoichiometry assumed, e.g., H/M = 1, is only rarely correct and even when it is on the virgin surface, it would not be accurate under the conditions of reactions [3]. There are one or two notable exceptions. The most obvious is the use of N_2 chemisorption on Fe at the reaction temperature to measure the site density for the NH_3 synthesis reaction [4]. Here it is known that the adsorption of N_2 is the rate limiting step in the reaction so that if one uses the adsorption isotherm at reaction temperature, one assumes that the observed site density must be a reasonable approximation to the correct one (although it will still be an over-estimate because under reaction conditions the steady-state coverage will be lower and because not all adsorption sites necessarily lead to reaction). Another case where the measured site density must be about right is that observed by ^{27}Al NMR for n-hexane over ZSM-5 [5] since the n-hexane cracking activity correlates with Al in the framework over several orders of magnitude and the acid sites are known to be associated with aluminum. Still, this kind of site density cannot be simply extended to other zeolites, e.g., to Y-zeolite since it is known that in this case the activity for cracking passes through a maximum when the aluminum site density it varied [6]. From the point of view of characterization, we might conclude that site density is not inherently very revealing except in so far as the order of magnitude will limit some possibilities, e.g., if it is as large as 10^{15} on a metal, the site cannot plausibly involve more than a few atoms.

2.2. LOCAL OR AN EXTENDED PICTURE OF A SITE

Depending on the surface, we might find a local or an extended picture of an adsorption site more useful. Consider H_2 adsorption on oxides and metals. On zinc oxide we have a rather local picture involving heterolytic dissociation on a cation/anion pair [7]:

$$H_2 + Zn^{2+}O^{2-} \rightarrow H\text{-}Zn^{+}OH^{-} \tag{3}$$

However, no simple local site picture of H_2 dissociative adsorption on Cu(111) is sufficient. It appears that this reaction may involve components of translation, vibration and rotation of the gas molecule and the interchange of these different kinds of energy during collision/dissociation on the surface [8]. It is likely that the pictures that are required to describe the sites for dissociative adsorption of H_2 reflect, in part, the techniques that have been used to investigate the different systems. In the case of zinc oxide, the site description is the result of infrared spectra of the hydride and hydroxyl which are the product of the dissociative chemisorption. These are normal chemical bonds and can therefore be described in the usual molecular (local) pictures of a bond. Such is surely the case for the homolytic dissociation of H_2 on copper also, but because of the dynamical studies we realize that the surmounting of the barrier to adsorption requires a rather extended picture of the solid and the interaction of the molecule with the solid surface. The lesson may be that dynamical descriptions of the activated complex formation for adsorption or reaction will require a rather extended picture while the stable chemisorbed species might be described in the usual molecular way. However, it is almost surely the case that sites on oxides are inherently more localized, because of the nature of bonding in oxides, than is the case for metals.

2.2.1 *Modification of the site by chemisorption.* One might state at the outset that there is a zero order effect of chemisorption in the Langmuirian sense that chemisorption of a molecule may be expected to block subsequent adsorption. This in not necessarily so. Consider CO adsorption on Pt or Pd. When CO is adsorbed the sites are modified in the Langmuirian sense for further adsorption of either CO or O_2. However, if O_2 is adsorbed first, there is little effect on subsequent adsorption of CO, so this is clearly non-Langmuir in behavior and the site modification by O_2 with respect to CO chemisorption is very subtle indeed [9]. Iwasawa gives several examples of the effect of chemisorption on subsequent reactivity of a site, e.g., co-adsorption of a second ethanol molecule on an isolated Nb site changes the selectivity from dehydration to dehydrogenation. Clearly, a local (molecular) picture is sufficient to picture this site modification. On the other hand, chemisorbed site modification may have long range effects on subsequent chemisorption. King* provided the example of pinning (stabilization toward restructuring) of the Pd (110) by adsorbing a little C. The Pd(110)(1x1) restructures to

* *Contribution not printed here.*

Pd(110)(1x2) when CO is chemisorbed because this increases CO binding, but then returns to Pd (110)(1x1) at full coverage. With pre-adsorbed C, there is no restructuring.

2.2.2 *Restructuring of surface (site) by reaction.* Many of the papers presented at this Workshop discussed explicitly, or implicitly, the restructuring of the surface or the site. We believe that one can find strong support for the hypothesis: *All catalytic reactions result in, indeed, require restructuring of the surface (site), although the restructuring may be local or long range.* Two examples have already been given in the previous section involving the restructuring (locally) of a Nb site by ethanol adsorption and (long-range) by CO adsorption on Pd (110). However, We believe it is axiomatic that there will always be some re-adjustment in the bond lengths in the vicinity of chemisorption and, if we take this as a kind of restructuring, we arrive at the generalization posed as a hypothesis. A theoretical description and explicit example is shown in Figure 2 of the chapter by van Santen et al. in the changes in the Si-O and Al-O bond lengths upon protonation of a silica-alumina cluster. Presumably a similar alteration of the bond lengths of Pd in the (110) surface occurs upon CO adsorption, but when the coverage is high enough that this propagates, it results in the restructuring to the Pd(110)(1x2) surface. Thus, the restructuring of a site is probably general; how local or long range it is will depend on the surface and is surely more likely to be long range in crystalline surfaces, e.g., zeolites or single crystal metals, rather than amorphous or dis-ordered surfaces because the ordered surface will provide a mechanism for propagation of the restructuring away from the adsorption (reaction) site.

3. Some Ways to Characterize Sites

3.1 CLASSICAL SITE AND KINETIC/REACTION PROBES

We are going to focus on physical methods of characterizing sites, but before we move into this, we need to remind ourselves of the catalytic chemical adage about site characterization, i.e., that a catalytic reaction is more sensitive and specific than any physical technique. One cannot dispute this, but it is worth adding that it is not always easy to deduce structural information from chemisorption or the reaction itself. Perhaps the most widely applied chemisorptive site probes are bases used to determine the strength/density of acid sites. The use of infrared spectroscopy with appropriate probe molecules is reviewed by Knözinger and will not be further discussed here. For this approach to be useful, one must have some chemical model for the site and, of course, the concept of Brønsted and Lewis acidity are such essential parts of fundamental chemistry that there is no need to describe these concepts. However, a specific structure for any site is still helpful, and perhaps the most widely known and accepted site picture that we have is that of the bridging hyoxyl for Brønsted acid sites in silica-alumina (10).

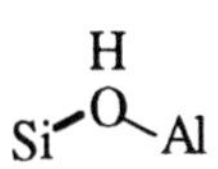

(4)

While there is general agreement that this is what we are talking about for both amorphous and crystalline (zeolite) silica-aluminas, there is no agreement on the explanation that allows the acidity of this site to be so different for amorphous and crystalline silica-aluminas (see Rabo and Gajda for a discussion of this point [11]).

A somewhat more powerful approach is to use a catalytic reaction to probe the sites. Often this means building some particular structure into the reactant molecule which will then reflect some aspect of the site structure by virtue of how reactive the molecule is. Organic chemistry offers a almost unlimited variety here from simple things like primary, secondary and tertiary hydrogen (or carbon) to the complexity of optical activity (Rooney provides some excellent examples of the use of molecular structure to probe sites). Even a relatively simple reaction can be diagnostic in the right hands. Haag et al. have developed the cracking of n-hexane on HZSM-5 such that it can now be used as a probe reaction to characterize other acid catalysts [12].

The general mechanism for acid catalyzed reaction envisions the addition of a proton to the molecule, the rearrangement of the protonated species, and the removal of the proton. For example, we can write for the isomerization of cyclopropane (c-C_3H_6) to propene (n-C_3H_6) on solid acid HA:

$$\begin{aligned} &\text{c-}C_3H_6 + HA \rightarrow \text{c-}C_3H_7^+ + A^- \\ &\text{c-}C_3H_7^+ + A^- \rightarrow \text{n-}C_3H_7^+ + A^- \\ &\text{n-}C_3H_7^+ + A^- \rightarrow \text{n-}C_3H_6 + HA \end{aligned} \tag{5}$$

This is a quite general mechanism and can be applied to many molecules. When applied to a paraffin, one arrives at a non-classical penta-coordinated carboniun ion. This leads to what is called the monomolecular cracking mechanism [12]. As noted by Haag *et al.*, one kind of evidence for this mechanism is that $C_6H_{15}^+$ formed from n-hexane on zeolite catalysts, e.g., HZSM-5, produces the same distribution of products as when this same ion is produced in the gas phase by chemical ionization by CH_5^+ in an ion cyclotron resonance mass spectrometer. The conditions which favor this mechanism are high temperature and low paraffin partial pressure and/or low conversion. The latter assures low olefin concentrations (as products). When significant olefin concentrations exist in the feed or are formed by the monomolecular cracking mechanism, these are more easily protonated than the paraffin and lead to the classical (bimolecular) mechanism involving a carbenium ion chain reaction which for n-hexane may be written:

$$\text{C-C-C-C-C-C} + \text{C-C}^+\text{-C} \rightarrow \text{C-C-C-C-C}^+\text{-C} + \text{C-C-C} \tag{6}$$

$$\text{C-C-C-C-C}^+\text{-C} \rightarrow\rightarrow \text{C-C(C)-C-C}^+\text{-C} \xrightarrow{\beta\text{-scission}} \text{C-C}^+\text{-C} + \text{C=C-C} \tag{7}$$

This is referred to as a bimolecular mechanism because the step (7) involving rearrangement of the carbenium ion and β-scission occurs very rapidly and the rate determining step is the hydrogen transfer reaction of step (6). The conditions where the bimolecular reaction path predominates are low temperature and high olefin concentration. In practice, both mechanisms may contribute to varying extents, depending on the composition of the feed and the reaction conditions. As discussed by Haag, *et al.*, the overall rate (which is a sum of the two mechanisms can be formulated such that in two limiting cases they produce apparent first order kinetics. However, the apparent first order rate constant represents the rate limiting formation of penta-coordinated carbonium ion (in the limit of high temperature and low olefin concentration) and formation of tri-coordinated carbenium ion (in the limit of low temperature and large olefin concentration).

3.2. SURFACE SCIENCE APPLICATIONS AND LIMITATIONS

3.2.1 *Oxide surfaces.* Surface science has made very significant contributions to our understanding of elementary steps on metals, but we need to see similar progress on oxides. Oxides are more difficult to study by surface science techniques for many reasons. Some oxides are difficult to grow as large single crystals; they are more prone to defects and can become non-stoichiometric at even modest temperatures in ultra-high vacuum. Of course, the fact that they are mostly insulators can compromise many of the electron spectroscopies by virtue of the static charge that results from electron emission. While they add complications to the experiments, all of these problems can be overcome and many laboratories have projects underway. Evans et al. (this book) have presented a study of Rh on single crystal $TiO_2(110)$ which combined many of the conventional surface science tools with reflection absorption infrared spectroscopy.

3.2.2. *High coverages.* While there are many examples of experiments performed under steady-state, high pressure conditions on single crystals, the vast majority of surface science experiments are performed in vacuum or very low partial pressures. The resulting low coverages can completely alter the selectivity and dominant reactions. While this can still be interesting surface chemistry, it would be even more interesting to use surface science to learn about practical reactions (at relatively high surface coverages).

In some recent work it has been shown that methyl radicals on Ni(111) readily recombine with sub-surface H atoms to form gas phase methane [13)] while the dominant reaction is decomposition of the methyl radicals with surface H atoms rather than recombination. One suspects this is the result of the rather low H surface coverages that can be accomplished in these experiments and that the decomposition would be inhibited at high surface coverages of H. Of course, one approach is to get the high coverages the same way as they are obtained in real catalysis, i.e., raise the pressure, but this means that in situ analysis has be be given up. We need to find innovative ways around this problem.

3.2.3. *Kinetic characterization on single crystal/model catalysts.* Of course, there are other reasons to go to high pressure other than just coverage considerations. This also makes possible kinetic characterization of the surfaces. While such studies might use kinetic probe reactions, the minimum we might expect to learn is whether the rate on the single crystal or model surface was comparable to that on high area conventional catalysts. This would then assure that whatever surface science characterization were possible, it might be relevant to the steady-state catalysis. A recent example of such an investigation was reported by Coulter and Goodman [14]. They investigated oxidative dimerization of methane on thin-film MgO and Li/MgO (on which electron spectroscopies could also be performed) and showed that kinetics on the model catalysts gave about the same rates as the polycrystalline Li/MgO under similar conditions.

3.3 TIME RESOLVED SURFACE SPECTROSCOPY

While many stable adsorbed species have been identified by various specproscopies, there is no guarantee that they play any role in the reaction even if they originate from one of the reactants. One approach to determining if a given species is kinetically significant, i.e., if it forms and reacts on a time scale sufficiently fast to sustain the steady state reaction, is to perform a transient experiment. Tamaru (15) has summarized the principles and provides some early examples and Frennet provides a more recent summary. The most effective approach is to have the transient occur effectively at steady-state by a switch in one or more of the reactants, isotopically labeled, and follow the evolution of the isotpically labeled intermediates on the surface, e.g., by infrared spectroscopy, or in the gas phase by mass spectrometry. Of course, a better experiment is to follow both the surface concentration and the gas phase simultaneously, but few of these experiments have been performed. A recent example using mass spectrometry to follow an isotope transient to estimate the rate coefficients for chain initiation, propagation and termination during Fischer-Tropsch synthesis over Ru/TiO_2 is given by Krishna and Bell [16].

There is a particularly elegant example of time resolved infrared spectroscopy applied to what is expected to be a true elementary step reported by Reutt-Robey et al [17]. As has already been mentioned, the law of mass action only applies to surface reactions involving two (or more) surface species if one of more of these species diffuse on the surface rapidly on the time scale of reaction. In the case of CO oxidation on Pt, the rate at hight temperature can be shown to be proportional to the product of $\theta_{CO}\theta_{O}$. Because oxygen atoms are known to diffuse slowly at reaction temperature, it is presumed that CO must diffuse rapidly. Just how rapidly this diffusion is on a microscopic dimension (20 atom length) has been shown by following the diffusion of CO initially on a Pt(111) terrace to a step. This can be followed by reflection infrared because the absorption frequency of CO on the terraces is different from CO on step sites. This migration was followed with a temporal resolution as fast as 5 msec and by varying temperatures, a diffusion barrier of only 4.4 kcal/mol was determined. While there are several examples of diffusion over macroscopic dimensions, it is microscopic experiments, such as this, which are probably of the most importance for catalytic kinetics. The macroscopic diffusion will necessarily average over defects where the diffusion rate is very low and because the fast diffusing species are likely to desorb before they can diffuse over a macroscopic distance (and therefore contribute little to reaction).

In the example given above of CO surface diffusion, the elementary step has been slowed down to the point where it could be followed by lowering the temperature. This is probably a reasonable thing to do for migration, but for a chemical reaction, the mechanism might change so we are in need of time resolved surface spectrocopies with much better time resolution. If it is possible to repeat the transient very fast, then it is possible to work out algorithms where data is collected at the same time into the transient and then to construct a spectrum at that "time". We have used this approach to look at the infrared emission of vibrational excited desorbed CO_2 from a Pd surface [18]. The transient was a pulse of reactant directed at a surface. The pulse was about 200 μs and spectra could be obtained at about 30 μs spacing through the transient using an efficient data collection algorithm that collects data continuously and then shuffled it into a matrix in which all the data at a given time into the transient could be transformed into a spectrum. In principle, this approach could be applied to other spectroscopies. Time-resolved measurement on a picosecond time scale have been demonstrated using infrared lasers to saturate and monitor the recovery of infrared absorptions on surfaces [19]. While this approach has not yet been applied to reactions on surfaces, this appears to be technically possible.

3.4 NEEDED SPECTROSCOPIC ATTRIBUTES

Most practical catalysts are either comprised of amorphous materials or small particles and, consequently, do not possess long-range order. Moreover, the existence of long range order does not in itself assist in the characterization of the relevant properties of the catalysts. For example, many industrially important oxidation catalysts contain particles of metal oxide that are large enough to exhibit diffraction patterns, but the bulk structure does not necessarily enlighten us about the surface structures which are relevant to catalysis. Thus, techniques which are element specific and provide local structural information without the requirement of long range order are very desirable. The element specificity is important so that we might tune to different components and interrogate them with respect to the surrounding order. Techniques which have these properties are NMR, X-ray absorption (EXAFS, XANES) and Mössbauer spectroscopies. As it happens, the wavelengths of the electromagnetic spectrum used by these spectroscopies are either very short (hard X-ray or γ-ray) or long (radio frequency) so that these spectroscopies do not require a vacuum and can be used at reactions pressures, and under some circumstances, at reaction temperatures. Unfortunately, this same property makes them bulk techniques so that special precautions must be taken to make them surface sensitive, e.g., have the element of interest entirely exposed to the reactants (as in zeolites) or in very small particles so that essentially all atoms may be exposed to the reactants in the fluid phase.

We have recently used solid state ^{27}Al, ^{31}P and ^{95}Mo NMR to investigate the interactions of the active phase (Mo) with the support (Al_2O_3) and PO_4^{-3} promoter in commercial $NiMoP/Al_2O_3$ HDS catalysts [20]. The object of this study was to understand the chemistry by which the active Mo interacts with the support (to form $Al_2(MoO_4)_3$), how this is moderated by the formation of $AlPO_4$ and the effect of hydration on these surface phases. While this is clearly not direct information on the active site, it is necessary structural information about the phase which defines the site. More direct information on active sites on $CoMo/Al_2O_3$ HDS catalysts has been obtained by Mössbauer spectroscopy [21]. Mössbauer spectroscopy of the cobalt, an experiment that involves the preparation of radioactive samples, shows that hydrodesulfurization activity parallels the formation of a particular cobalt species that does not resemble any of the known, bulk cobalt sulfides [21]. The same result appears in spectra of supported and unsupported mixtures of sulfided molybdenum and cobalt and has led the Topsøe group to propose a species they call "CoMoS" for the site of the hydrodesulfurization activity.

The attributes of synchrotron radiation, particularly as applied to EXAFS and XANES of catalysts, have been briefly reviewed by Joyner. He has emphasized the need for in situ measurements in parallel with catalyst activity and selectivity measurements. Recently, both EXAFS and XANES have been used to follow formation of bimetallic particles using in situ temperature programmed reduction (TPR). Bazin et al. [22] have followed the formation of Pt-Rh and Pt-Re on γ-Al_2O_3 using EXAFS and Jentys et al. [23] combined XANES with TPR. The EXAFS measurements have the advantage that they provide directly structural parameters (coordination numbers, internuclear distances), but the resolution is greatly degraded by Debye-Waller effects even at modest temperatures. XANES has the advantage that the spectra do not suffer from Debye-Waller degradation, but we do not have a good theory to interpret the electronic information that is, in principle, contained in the spectra. An approach which has much in common with XANES, but may have advantages when the whiteline intensity is not a linear function of oxidation state is the measurement of the shift in the X-ray absorption edge, from which the binding energy can be monitored in situ in a TPR experiment [24].

4. Theory

Given the interests and abilities of the Workshop organizers, theory can be said to have been well represented. This is appropriate because continuing reduction in cost and increase in speed of computing has brought "theory" into the laboratories of the novice as well as the expert. The word theory is placed in quotes here to imply that we take the catholic definition that includes everything from reaction modeling (Waugh) to molecular modeling/molecular graphics (Vercauteren et al.,) to quantum chemistry (Pacchioni and Bagus). The trend to make available calculational packages that can be used by anyone who can afford the modest cost of the hard and software will continue. This poses certain risks to those who use programs which they don't fully understand, but on balance it should be viewed as positive. From the point of view of characterizing heterogeneous catalyst sites, we believe it is essential that there be a close coupling between theory and experiment. One Workshop contribution which illustrated the power of this approach described the interpretation of scanning tunneling microscopy (STM) using theory as a guide (Sautet et al). This is essential until STM can be made element specific. Until STM is routinely used in such a way that composition as well as image (via either experimental improvements or coupled theory), it will be analogous to the early contributions of LEED where symmetry was easy and obvious, but true structure required the development of a firm theoretical underpinning. There is a recent and promising STM report in this regard where it was demonstrated that polyanion arrays (on a graphite surface) with different composition could be distingusihsed by a kind of density of states analysis [25].

5. Recommendations

The recommendations for the near future follow from the above discussion and are summarized succinctly. We seek surface science improvements that will make application to non-metals routine and allow electron spectroscopies to be applied to surfaces at high coverage. New techniques to analyze surface migration on a microscopic level under reaction conditions will be an essential part of sophisticated descriptions of elementary steps. We need new surface spectroscopic techniques which are element specific, provide local structure where no long range order exists, can be applied under reaction conditions and these would be more nearly ideal if they were also surface specific. Only STM has the latter property (among those discussed) but, as used today, lacks the element (or compositional) attribute. Whatever techniques we contemplate, making these techniques applicable under reaction conditions and developing time resolution will be important and essential challenge which should be given our immediate attention. While non-Faradaic electrochemical modification of catalytic activity (Vayenas) does not have many of the desired attributes, the ability to effectively have a tunable catalyst promoter is an exciting possibility suggesting that this approach should be combined with other surface structural analyses. Photocatalysis was not a Workshop discussion topic, but many would place it on a wish list where we need to make progress in the near future. And finally, the extension of even the present level of theoretical analysis to treat very large systems is essential for low symmetry problems represented by surfaces.

6. Acknowledgments

We wish to thank Robert Weber for a critical reading of a draft of this summary. We also acknowledge the sponsorship of the U.S. Department of Energy, "Research Needs Assessment for Advanced Heterogeneous Catalysts for Energy Applications", in which one of us is participating.

7. References

1. G. L. Haller, W. N. Delgass, in *Investigation of Rates and Mechanisms of Reactions* C. F. Bernasconi, Eds. (John Wiley, New York, 1986), vol. VI, pp. 951.
2. J. L. Gland, M. R. McClellan and F. R. McFeely, *J. Chem. Phys.* **79**, 6349 (1983).
3. B. J. Kip, F. B. M. Duivenvoorden and D. C. Koningsberger, R. Prins, *J. Catal.* **105**, 26 (1987).
4. H. Topsøe, N. Topsøe, N. Bohlbro and J. A. Dumesic, Eds., *Proc. 7th Intern. Congr. Catal.* (Kodansha, Tokyo, 1981).
5. W. O. Haag, R. M. Lago and P. B. Weisz, *Nature* **309**, 589 (1984).
6. R. Carvajal, P.-J. Chu and J. H. Lunsford, *J. Catal.* **125**, 123 (1990).
7. R. J. Kokes, A. L. Dent, C. C. Chang and L. T. Dixon, *J. Am. Chem. Soc.* **94**, 4429 (1972).
8. C. T. Rettner, D. J. Auerbach and H. A. Michelsen, *J. Vac. Sci. & Tech. A* **10**, 2282 (1992).
9. R. P. H. Gasser, *An Introduction to Chemisorption and Catalysis by Metals* (Clarendon Press, Oxford, 1985).
10. R. C. Hansford, *Ind. Eng. Chem.* **39**, 849 (1947).
11. J. Rabo and G. J. Gajda, *Catal. Rev. - Sci. and Eng.* **31**, 385 (1989-90).
12. W. O. Haag, R. M. Dessau and, R. M. Lago, *Studies in Surf. Sci. and Catal.* **60**, 255 (1991).
13. A. D. Johnson, S. P. Daley, A. L. Utz and S. T. Ceyer, *Science* **257**, 223 (1992).
14. K. Coulter and D. W. Goodman, *Catal. Lett.* **16**, 191 (1992).
15. K. Tamaru, *Dynamic Heterogeneous Catalysis* (Academic Press, New York, 1978).
16. K. R. Krishna and A. T. Bell, *J. Catal.* **139**, 104 (1993).
17. J. E. Reutt-Robey, D. J. Doren, Y. J. Chabal and, S. B. Christman, *Phys. Rev. Lett.* **61**, 2778 (1988).
18. D. A. Mantell, K. Kunimori, S. B. Ryali and G. L. Haller, *ACS Symposium Series* **288**, 465 (1985).
19. M. P. Casassa, E. J. Heilweil, J. C. Stephenson and R. R. Cavanagh, *M. Chem. Phys.* **84**, 2361 (1986).
20. O. H. Han, C. Y. Lin and G. L. Haller, *Catal. Lett.* **14**, 1 (1992).
21. R. Candia, B. S. Clausen and H. Topsøe, *J. Catal.* **77**, 564 (1982).
22. D. Bazin, H. Dexpert, J. P. Bournonville and J. Lynch, *J. Catal.* **123**, 86 (1990).
23. A. Jentys, B. J. McHugh, G. L. Haller and J. A. Lercher, *J. Phys. Chem.* **96**, 1324 (1992).
24. F. Hilbrig, C. Michel and G. L. Haller, *J. Phys. Chem.* **96**, 9893 (1992).
25. B. A. Watson, M. A. Barteau, L. Haggerty, A. M. Lenhoff and R. S. Weber, *Langmuir* **8**, 1145 (1992).

PRINCIPAL CONTRIBUTORS

Prof. J.M. Basset
Ecole Supérieure de Chimie Industrielle de Lyon
43 boulevard du 11 Novembre 1918
69100 Villeurbanne Cedex
France

Dr J. Blake
Dept of Chemistry
Yale University
New Haven CT 06511-8118
USA

Prof. Mme E. Bordes
Dept of Chemical Engineering
University of Technology
B.P. 649
60206 Compiègne Cedex
France

Dr G. Centi
Dept of Industrial Chemistry and Materials
Viale Resorgimento 4
40136 Bologna
Italy

Dr G.W. Coulston
Corporate Catalysis Center
Central Science & Engineering
Du Pont Company
Wilmington
Delaware 19880-0262
USA

Prof. T.M. Duncan
School of Chemical Engineering
Cornell University
Ithaca
New York 14853
USA

Prof. A. Frennet
Catalyse Hétérogène
Université libre de Bruxelles
CP243 Campus Plaine ULB
B 1050 Brussels
Belgium

Prof. G.L. Haller
Dept of Chemical Engineering
Yale University
New Haven CT 06520
USA

Dr B.E. Hayden
Dept of Chemistry
The University
Southampton SO9 5NH
United Kingdom

Dr S. Holloway
Surface Science Research Centre
University of Liverpool
PO Box 147
Liverpool L69 3BX
United Kingdom

Prof. Y. Iwasawa
Dept of Chemistry
The University of Tokyo
Hongo, Bunkyo-ku
Tokyo 113
Japan

Prof. W.C. Jorgensen
Dept of Chemistry
Yale University
New Haven CT 06511-8118
USA

Prof. R.W. Joyner
Leverhulme Centre for Innovative Catalysis
Dept of Chemistry
University of Liverpool
PO Box 147
Liverpool L69 3BX
United Kingdom

Prof. H. Knözinger
Institut für Physikalische Chemie
Universität München
Sophienstrasse 11
8000 München
Germany

Dr H.P.C.E. Kuipers
Koninklijke/Shell Laboratorium
P.O. Box 3003
1003 AA Amsterdam
The Netherlands

Dr J.J. Lerou
Corporate Catalysis Center
Central Science & Engineering
Du Pont Company
Wilmington
Delaware 19880-0262
USA

Dr B.E. Nieuwenhuys
Gorlaeus Laboratory
Leiden University
PO Box 9502
2300 RA Leiden
The Netherlands

Prof. G. Pacchioni
Dipartimento di Chimica Inorganica e
Metallorganica
Universita di Milano
Via Venezian 21
20133 Milano
Italy

Prof. V. Ponec
Gorlaeus Laboratory
Leiden University
PO Box 9502
2300 RA Leiden
The Netherlands

Prof. J.J. Rooney
School of Chemistry
The Queen's University
David Keir Building
Stranmillis Road
Belfast BT 5AG
Northern Ireland

Dr J. Rostrup-Nielsen
Haldor Topsoe A/S
Nymoellevej 55
2800 Lyngby
Denmark

Prof. R.A. van Santen
Schuit Institute of Catalysis
Eindhoven Technical University
P.O. Box 513
5600 MB Eindhoven
The Netherlands

Dr P. Sautet
Institut de Recherche sur la Catalyse
CNRS
2 Avenue Albert Einstein
F-69626 Villeurbanne Cedex
France

Dr E. Shustorovich
Corporate Research Laboratories
Eastman Kodak Company
Rochester
New York 14650-2001
USA

Prof. G.A. Somorjai
Dept of Chemistry, and Materials
Science Division
Lawrence Berkeley Laboratory
University of California
Berkeley CA 94720
USA

Prof. C.G. Vayenas
Institute of Chemical Engineering &
High Temperature Chemical Processes
University of Patras
Patras GR 26110
Greece

Prof. J.C. Vedrine
Institut de Recherche sur la Catalyse
CNRS
2 Avenue Albert Einstein
F-69626 Villeurbanne Cedex
France

Prof. D.P. Vercauteren
Institute for Studies in Interface Science
Dept of Chemistry
Facultés Universitaires Notre-Dame de
la Paix
rue de Bruxelles 61
B-5000 Namur
Belgium

Prof. K.C. Waugh
(present address)
Dept of Chemistry
UMIST
PO Box 88
Manchester M60 1QD
United Kingdom

Prof. P.B. Wells
Dept of Chemistry
University of Hull
Hull HU6 7RX
United Kingdom

Dr V.P. Zhdanov
Boreskov Institute of Catalysis
Pr. Akademika Lavrentieva 5
Novosibirsk 630090
Russia

INDEX

Bold entries refer to Chapters

GPSR Compliance

The European Union's (EU) General Product Safety Regulation (GPSR) is a set of rules that requires consumer products to be safe and our obligations to ensure this.

If you have any concerns about our products, you can contact us on ProductSafety@springernature.com

In case Publisher is established outside the EU, the EU authorized representative is:

Springer Nature Customer Service Center GmbH
Europaplatz 3
69115 Heidelberg, Germany

Batch number: 10607320

Printed by Printforce, the Netherlands